Lecture Notes in Computer Science 12301

Advanced Research in Computing and Software Science
Subline of Lecture Notes in Computer Science

More information about this series at http://www.springer.com/series/7407

Isolde Adler · Haiko Müller (Eds.)

Graph-Theoretic Concepts in Computer Science

46th International Workshop, WG 2020
Leeds, UK, June 24–26, 2020
Revised Selected Papers

 Springer

Editors
Isolde Adler (iD)
University of Leeds
Leeds, UK

Haiko Müller (iD)
University of Leeds
Leeds, UK

ISSN 0302-9743   ISSN 1611-3349 (electronic)
Lecture Notes in Computer Science
ISBN 978-3-030-60439-4   ISBN 978-3-030-60440-0 (eBook)
https://doi.org/10.1007/978-3-030-60440-0

LNCS Sublibrary: SL1 – Theoretical Computer Science and General Issues

This Springer imprint is published by the registered company Springer Nature Switzerland AG
The registered company address is: Gewerbestrasse 11, 6330 Cham, Switzerland

Preface

This volume contains the 32 papers presented at the 46th International Workshop on Graph-Theoretic Concepts in Computer Science (WG 2020). The workshop was held online during June 24–26, 2020, after the original plans to organise it in Leeds, UK, had to be abandoned due to COVID-19. Over 200 computer scientists and mathematicians from all over the world registered for the conference, although most talks were attended by 65–75 participants.

WG has a long-standing tradition. Since 1975, WG took place 24 times in Germany, 5 times in The Netherlands, 3 times in France, 2 times in Austria, 2 times in the Czech Republic, as well as in Greece, Israel, Italy, Norway, Slovakia, Spain, Switzerland, and Turkey one time each. This was the second time the workshop was organised in the UK.

WG aims to merge theory and practice by demonstrating how concepts from Graph Theory can be applied to various areas in Computer Science, or by extracting new graph theoretic problems from applications. The goal is to present emerging research results and to identify and explore directions of future research. The conference is well-balanced with respect to established researchers and junior scientists.

We received 96 submissions, 2 of which were withdrawn before entering the review process. The Program Committee (PC) provided 3 to 5 independent reviews for each submission. The PC accepted 32 papers – an acceptance ratio of nearly 1/3. As in previous years, due to strong competition and limited space there were papers that were not accepted although they deserved to be.

The prize for the Best Paper at WG 2020 was awarded to Jesper Nederlof, Michał Pilipczuk, Céline Swennenhuis, and Karol Węgrzycki for their paper "Hamiltonian cycle parameterized by treedepth in single exponential time and polynomial space." The prize for the Best Student Paper at WG 2020 was awarded to Sriram Bhyravarapu for his paper "Combinatorial bounds for conflict-free coloring on open neighborhoods," coauthored by Subrahmanyam Kalyanasundaram. The program included two inspiring invited talks, by Mihyun Kang (TU Graz, Austria) on "Topological aspects of random graphs" and by Jaroslav Nešetřil (Charles University, Prague) on "Three aspects of structural graph theory."

Moreover, many individuals contributed to the success of WG 2020. In particular our thanks go to:

- All authors who submitted their newest research results to WG
- The reviewers whose expertise supported the selection process
- The members of the PC, who graciously gave their time and energy
- The members of the Organizing Committee at the University of Leeds, UK
- The School of Computing at the University of Leeds, UK
- The EasyChair system for hosting the evaluation process

- Springer for supporting the Best Paper Awards
- The invited speakers, all speakers, the session chairs, and the participants for their contributions and support to make WG 2020 an inspiring event

July 2020   Isolde Adler
                                                                  Haiko Müller

Organization

Program Committee

Isolde Adler	University of Leeds, UK
Nikhil Bansal	Eindhoven University of Technology, CWI, The Netherlands
Cristina Bazgan	Université Paris Dauphine, France
Radu Curticapean	IT University of Copenhagen, Denmark
Thomas Erlebach	University of Leicester, UK
Celina de Figueredo	Federal University of Rio de Janeiro, Brazil
Pinar Heggernes	Bergen University, Norway
Bart Jansen	Eindhoven University of Technology, The Netherlands
Mamadou Kanté	Université Clermont Auvergne, France
Tereza Klimošová	Charles University Prague, Czech Republic
Łukasz Kowalik	University of Warsaw, Poland
Stephan Kreutzer	Technical University Berlin, Germany
Beppe Liotta	University of Perugia, Italy
Ross McConnell	Colorado State University, USA
Haiko Müller	University of Leeds, UK
Sang-il Oum	Institute for Basic Science, KAIST, South Korea
Viresh Patel	University of Amsterdam, The Netherlands
Alex Scott	University of Oxford, UK
Ryuhei Uehara	Japan Advanced Institute of Science and Technology, Japan
Sue Whitesides	McGill University, Canada

Organizing Committee

Isolde Adler	Haiko Müller
Martin Dyer	Alison Whiteley
Marc Heinrich	Samuel Wilson

Additional Reviewers

Mikkel Abrahamsen	Manu Basavaraju
Ernst Althaus	Julien Baste
André Nichterlein	Michael Bekos
Oliver Bachtler	Rémy Belmonte

Benjamin Bergougnoux
Therese Biedl
Thomas Bläsius
Hans L. Bodlaender
Stefan Boettcher
Nicolas Bonichon
Nicolas Bousquet
Andreas Brandstädt
Nick Brettell
Guido Brückner
Yixin Cao
Guilherme de Castro Mendes Gomes
Leroy Chew
Rafael Santos Coelho
Colin Cooper
Matthew Coulson
Bruno Courcelle
Christophe Crespelle
Konrad K. Dąbrowski
Clément Dallard
Sandip Das
Ewan Davies
Holger Dell
Alberto Espuny Díaz
Walter Didimo
Emilio Di Giacomo
François Dross
Vida Dujmović
Martin Dyer
Eduard Eiben
Jessica Enright
David Eppstein
Carl Feghali
Andreas Emil Feldmann
Asaf Ferber
Krzysztof Fleszar
Guilherme D. da Fonseca
Florent Foucaud
Viktor Fredslund-Hansen
Vincent Froese
Radoslav Fulek
Nicola Galesi
Archontia Giannopoulou
J. Pascal Gollin
Petr Golovach
Daniel Gonçalves

Niels Grüttemeier
Grzegorz Gutowski
Michel Habib
Michael Hamann
Tesshu Hanaka
Marc Heinrich
Duc A. Hoang
Jacob Holm
Andreas Holmsen
Md. Iqbal Hossain
Shenwei Huang
Davis Issac
Lars Jaffke
Pallavi Jain
Matthew Johnson
Mark Jones
Tomáš Kaiser
Mong-Jen Kao
Leon Kellerhals
Shuji Kijima
Eun Jung Kim
Ringi Kim
Philipp Kindermann
Evangelos Kipouridis
Masashi Kiyomi
Yasuaki Kobayashi
Yusuke Kobayashi
Christian Komusiewicz
Danny Krizanc
Arnaud Labourel
Michael Lampis
Van Bang Le
Orlando Lee
Erik Jan van Leeuwen
Paloma de Lima
Henri Lotze
Borut Lužar
Yingbin Ma
Gary Macgillivray
Raphael Machado
Ana Karolinna Maia
Thiago Marcilon
Andrea Marino
Nicolas Martins
Arnaud Mary
Tomáš Masařík

Tamara Mchedlidze
Kitty Meeks
George Mertzios
Martin Milanič
Pranabendu Misra
Kazuyuki Miura
Eiji Miyano
Hiroyuki Miyata
Hendrik Molter
Debajyoti Mondal
Fabrizio Montecchiani
Guilherme O. Mota
Marcin Mucha
Irene Muzi
Jesper Nederlof
Ofer Neiman
Alantha Newman
Nicolas Nisse
Jana Novotná
Eunjin Oh
Yoshio Okumoto
Mateus de Oliveira Oliveira
Aurélien Ooms
Sebastian Ordyniak
João Paixão
Fahad Panolan
Charis Papadopoulos
Pan Peng
Lucia Draque Penso
Thi Ha Duong Phan
Christophe Picouleau
Théo Pierron
Astrid Pieterse
Lionel Pournin
Arash Rafiey
Ashutosh Rai
Jean-Florent Raymond

Guus Regts
Jonathan Rollin
Ignaz Rutter
Paweł Rzążewski
Toshiki Saitoh
Gelasio Salazar
Maycon Sambinelli
Rudini Sampaio
Marcio Costa Santos
Rafael Schouery
Pascal Schweitzer
Aleksandar Shurbevski
Ana Silva
Murilo V. G. da Silva
Jefferson Elbert Simões
Blerina Sinaimeri
George Skretas
Uéverton Souza
Yann Strozecki
Asahi Takaoka
Alessandra Tappini
Anastasiia Tsvietkova
Tomáš Valla
Lluis Vena
Kevin Verbeek
Kristina Vušković
Kunihiro Wasa
Sebastian Wiederrecht
Samuel Wilson
Michał Włodarczyk
David R. Wood
Marcin Wrochna
Mingyu Xiao
Viktor Zamaraev
Meirav Zehavi
Anna Zych

The Long Tradition of WG

WG 1975	U. Pape – Berlin, Germany
WG 1976	H. Noltemeier – Göttingen, Germany
WG 1977	J. Mühlbacher – Linz, Austria
WG 1978	M. Nagl, H. J. Schneider – Burg Feuerstein near Erlangen, Germany

WG 1979	U. Pape – Berlin, Germany
WG 1980	H. Noltemeier – Bad Honnef, Germany
WG 1981	J. Mühlbacher – Linz, Austria
WG 1982	H. J. Schneider, H. Göttler – Neuenkirchen near Erlangen, Germany
WG 1983	M. Nagl, J. Perl – Haus Ohrbeck near Onasbrück, Germany
WG 1984	U. Pape – Berlin, Germany
WG 1985	H. Noltemeier – Schloß Schwanberg near Würzburg, Germany
WG 1986	G. Tinhofer, G. Schmidt – Stift Bernried near Munich, Germany
WG 1987	H. Göttler, H. J. Schneider – Kloster Banz near Bamberg, Germany
WG 1988	J. van Leeuwen – Amsterdam, The Netherlands
WG 1989	M. Nagl – Castle Rolduc, The Netherlands
WG 1990	R. H. Möhring – Johannesstift Berlin, Germany
WG 1991	G. Schmidt, R. Berghammer – Richterheim, Fischbachau near Munich, Germany
WG 1992	E. W. Mayr – Wilhelm-Kempf-Haus, Wiesbaden-Naurod, Germany
WG 1993	J. van Leeuwen – Sports Center, Papendal near Utrecht, The Netherlands
WG 1994	G. Tinhofer, E. W. Mayr, G. Schmidt – Herrsching near Munich, Germany
WG 1995	M. Nagl – Haus Eich, Aachen, Germany
WG 1996	G. Ausiello, A. Marchetti-Spaccamela – Cadenabbia near Como, Italy
WG 1997	R. H. Möhring – Bildungszentrum am Müggelsee, Berlin, Germany
WG 1998	J. Hromkovič, O. Sýkora – Smolenice Castle, Slovakia
WG 1999	P. Widmayer – Centro Stefano Franscini, Monte Verità, Ascona, Switzerland
WG 2000	D. Wagner – Waldhaus Jakob, Konstanz, Germany
WG 2001	A. Brandstädt, Boltenhagen near Rostock, Germany
WG 2002	L. Kučera – Český Krumlov, Czech Republic
WG 2003	H. L. Bodlaender – Elspeet, The Netherlands
WG 2004	J. Hromkovič, M. Nagl – Bad Honnef, Germany
WG 2005	D. Kratsch – Île du Saulcy, Metz, France
WG 2006	F. V. Fomin – Sotra near Bergen, Norway
WG 2007	A. Brandstädt, D. Kratsch, H. Müller – Dornburg near Jena, Germany
WG 2008	H. Broersma, T. Erlebach – Durham, UK
WG 2009	C. Paul, M. Habib – Montpellier, France
WG 2010	D. M. Thilikos – Zarós, Crete, Greece

WG 2011	J. Kratochvíl – Teplá Monastery, West Bohemia, Czech Republic
WG 2012	M. C. Golumbic, G. Morgenstern, M. Stern, A. Levy – Ramat Rachel, Jerusalem, Israel
WG 2013	A. Brandstädt, K. Jansen, R. Reischuk – Lübeck, Germany
WG 2014	D. Kratsch, I. Todinca – Le Domaine de Chalès, Orléans, France
WG 2015	E. W. Mayr – Garching near Munich, Germany
WG 2016	P. Heggernes – Rumeli Hisarüstü, Istanbul, Turkey
WG 2017	H. L. Bodlaender, G. J. Woeginger – Heeze near Eindhoven, The Netherlands
WG 2018	A. Brandstädt, E. Köhler, K. Meer – Lübbenau near Cottbus, Germany
WG 2019	I. Sau, D. M. Thilikos – Vall de Núria, Catalunya, Spain
WG 2020	I. Adler, H. Müller – Leeds, UK

Contents

Combinatorial Bounds for Conflict-Free Coloring on Open Neighborhoods

Sriram Bhyravarapu and Subrahmanyam Kalyanasundaram$^{(\boxtimes)}$

Department of Computer Science and Engineering, IIT Hyderabad, Hyderabad, India
{cs16resch11001,subruk}@iith.ac.in

Abstract. In an undirected graph G, a conflict-free coloring with respect to open neighborhoods (denoted by CFON coloring) is an assignment of colors to the vertices such that every vertex has a uniquely colored vertex in its open neighborhood. The minimum number of colors required for a CFON coloring of G is the CFON chromatic number of G, denoted by $\chi_{ON}(G)$.

The decision problem that asks whether $\chi_{ON}(G) \leq k$ is NP-complete. Structural as well as algorithmic aspects of this problem have been well studied. We obtain the following results for $\chi_{ON}(G)$:

- Bodlaender, Kolay and Pieterse [WADS 2019] showed the upper bound $\chi_{ON}(G) \leq \mathsf{fvs}(G) + 3$, where $\mathsf{fvs}(G)$ denotes the size of a minimum feedback vertex set of G. We show the improved bound of $\chi_{ON}(G) \leq \mathsf{fvs}(G) + 2$, which is tight, thereby answering an open question in the above paper.
- We study the relation between $\chi_{ON}(G)$ and the pathwidth of the graph G, denoted $\mathsf{pw}(G)$. The above paper from WADS 2019 showed the upper bound $\chi_{ON}(G) \leq 2\mathsf{tw}(G) + 1$ where $\mathsf{tw}(G)$ stands for the treewidth of G. This implies an upper bound of $\chi_{ON}(G) \leq 2\mathsf{pw}(G) + 1$. We show an improved bound of $\chi_{ON}(G) \leq \lfloor \frac{5}{3}(\mathsf{pw}(G) + 1) \rfloor$.
- We prove new bounds for $\chi_{ON}(G)$ with respect to the structural parameters neighborhood diversity and distance to cluster, improving the existing results of Gargano and Rescigno [Theor. Comput. Sci. 2015] and Reddy [Theor. Comput. Sci. 2018], respectively. Furthermore, our techniques also yield improved bounds for the closed neighborhood variant of the problem.
- We also study the partial coloring variant of the CFON coloring problem, which allows vertices to be left uncolored. Let $\chi^*_{ON}(G)$ denote the minimum number of colors required to color G as per this variant. Abel et al. [SIDMA 2018] showed that $\chi^*_{ON}(G) \leq 8$ when G is planar. They asked if fewer colors would suffice for planar graphs. We answer this question by showing that $\chi^*_{ON}(G) \leq 5$ for all planar G. This approach also yields the bound $\chi^*_{ON}(G) \leq 4$ for all outerplanar G.

All our bounds are a result of constructive algorithmic procedures.

1 Introduction

A *proper coloring* of a graph is an assignment of a color to every vertex of the graph such that adjacent vertices receive distinct colors. Conflict-free coloring

© Springer Nature Switzerland AG 2020
I. Adler and H. Müller (Eds.): WG 2020, LNCS 12301, pp. 1–13, 2020.
https://doi.org/10.1007/978-3-030-60440-0_1

is a variant of the graph coloring problem. A conflict-free coloring of a graph G is a coloring such that for every vertex in G, there exists a uniquely colored vertex in its neighborhood. This problem was first introduced in 2002 by Even, Lotker, Ron and Smorodinsky [1]. This problem was originally motivated by wireless communication systems, where the base stations and clients have to communicate with each other. Each base station is assigned a frequency and if two base stations with the same frequency communicate with the same client, it leads to interference. So for each client, it is ideal to have a base station with a unique frequency. Since each frequency band is expensive, there is a need to minimize the number of frequencies used by the base stations.

Over the past two decades, this problem has been very well studied, see for instance the survey by Smorodinsky [2]. The conflict-free coloring problem has been studied with respect to the open neighborhood and the closed neighborhood. In this paper, we focus on the open neighborhood variant of the problem.

Definition 1 (Conflict-Free Coloring). *A CFON coloring of a graph $G = (V, E)$ using k colors is an assignment $C : V(G) \to \{1, 2, \ldots, k\}$ such that for every $v \in V(G)$, there exists an $i \in \{1, 2, \ldots, k\}$ such that $|N(v) \cap C^{-1}(i)| = 1$. The smallest number of colors required for a CFON coloring of G is called the* CFON chromatic number of G, denoted by $\chi_{ON}(G)$.

The closed neighborhood variant of the problem, CFCN coloring, *is obtained by replacing the open neighborhood $N(v)$ by the closed neighborhood $N[v]$ in the above. The corresponding chromatic number is denoted by $\chi_{CN}(G)$.*

The CFON coloring problem and many of its variants are known to be NP-complete [3,4]. It was further shown in [4] that the CFON coloring problem is hard to approximate within a factor of $n^{1/2-\varepsilon}$, unless P = NP. Since the problem is NP-hard, the parameterized aspects of the problem have been studied. The problems are fixed parameter tractable when parameterized by vertex cover number, neighborhood diversity [4], distance to cluster [5], and more recently, treewidth [6,7]. This problem has attracted special interest for graphs arising out of intersection of geometric objects, see for instance, [8–10].

The CFON coloring problem is considered as the harder of the open and closed neighborhood variants, see for instance, remarks in [8,11]. It is easy to construct example graphs G, for which $\chi_{CN}(G) = 2$ and $\chi_{ON}(G) = \Theta(\sqrt{n})$. Pach and Tardos [11] showed that for any graph G on n vertices, the closed neighborhood chromatic number $\chi_{CN}(G) = O(\log^2 n)$. The corresponding best bound [11,12] for open neighborhood is $\chi_{ON}(G) = O(\sqrt{n})$.

Another variant that has been studied [3] is the partial coloring variant:

Definition 2 (Partial Conflict-Free Coloring). *A partial conflict-free coloring on open neighborhood, denoted by CFON*, of a graph $G = (V, E)$ using k colors is an assignment $C : V(G) \to \{1, 2, \ldots, k, unassigned\}$ such that for every $v \in V(G)$, there exists an $i \in \{1, 2, \ldots, k\}$ such that $|N(v) \cap C^{-1}(i)| = 1$. The corresponding CFON* chromatic number is denoted $\chi_{ON}^*(G)$.*

The key difference between CFON* coloring and CFON coloring is that in the partial variant, we allow some vertices to be not assigned a color. If a graph can be CFON* colored using k colors, then all the uncolored vertices can be assigned the color $k + 1$, and thus is a CFON coloring using $k + 1$ colors.

1.1 Our Results and Discussion

In this paper, we obtain improved bounds for $\chi_{ON}(G)$ under different settings. More importantly, all our bounds are a result of constructive algorithmic procedures and hence can easily be converted into respective algorithms. We summarize our results below:

1. In Sect. 3, we show that $\chi_{ON}(G) \leq \lfloor \frac{5}{3}(\mathsf{pw}(G) + 1) \rfloor$ where $\mathsf{pw}(G)$ denotes the pathwidth of G. The previously best known bound in terms of $\mathsf{pw}(G)$ was $\chi_{ON}(G) \leq 2\mathsf{pw}(G) + 1$, implied by the results in [6].
   To the best of our knowledge, this is the first upper bound for $\chi_{ON}(G)$ in terms of pathwidth, which does not follow from treewidth. Our bound follows from an algorithmic procedure and uses an intricate analysis. We are unable to generalize our bound in terms of treewidth because we crucially use a fact (stated as Theorem 6) that applies to path decomposition, but does not seem to apply to tree decomposition. It will be of interest to see if this hurdle can be overcome to obtain an equivalent bound in terms of treewidth.
   There are graphs G for which $\chi_{ON}(G) = \mathsf{tw}(G) + 1 = \mathsf{pw}(G)$. It would be interesting to close the gaps between the respective upper and lower bounds.
2. In Sect. 4, we show that $\chi_{ON}(G) \leq \mathsf{fvs}(G) + 2$, where $\mathsf{fvs}(G)$ denotes the size of a minimum feedback vertex set of G. This bound is tight and is an improvement over the bound $\chi_{ON}(G) \leq \mathsf{fvs}(G) + 3$ by Bodlaender, Kolay and Pieterse [6].
3. In Sect. 5, we give improved bounds with respect to neighborhood diversity parameter. Gargano and Rescigno [4] showed that $\chi_{ON}(G) \leq \chi_{ON}(H) + cl(G) + 1$ and $\chi_{CN}(G) \leq \chi_{CN}(H) + ind(G) + 1$. Here H is the type graph of G, while $cl(G)$ and $ind(G)$ denote the number of cliques and independent sets respectively in the type partition of G. We present the improvements $\chi_{ON}(G) \leq \chi_{ON}(H) + cl(G)/2 + 2$ and $\chi_{CN}(G) \leq \chi_{CN}(H) + ind(G)/3 + 3$.
4. In Sect. 5, we show that $\chi_{ON}(G) \leq \mathsf{dc}(G) + 3$, where $\mathsf{dc}(G)$ is the distance to cluster parameter of G. This is an improvement over the previous bound [5] of $2\mathsf{dc}(G) + 3$. Our bound is nearly tight since there are graphs for which $\chi_{ON}(G) = \mathsf{dc}(G)$. Using a similar approach, we obtain the improved bound $\chi_{CN}(G) \leq \max\{3, \mathsf{dc}(G) + 1\}$.
   For the results in terms of parameters neighborhood diversity and distance to cluster, the obvious open questions are to improve the bounds and/or to provide tight examples.
5. When G is planar, we show that $\chi^*_{ON}(G) \leq 5$. This improves the previous best known bound by Abel et al. [3] of $\chi^*_{ON}(G) \leq 8$. The same approach helps us show that $\chi^*_{ON}(G) \leq 4$, when G is an outerplanar graph. These two results are discussed in Sect. 6.
   There are planar graphs G for which $\chi^*_{ON}(G) = 4$, which shows that our bound is nearly tight and leaves a gap of 1 between the upper and lower bounds. It will be of interest to close this gap.

6. For outerplanar graphs G, the bound $\chi^*_{ON}(G) \leq 4$ implies a bound of $\chi_{ON}(G) \leq 5$. We show a better bound of $\chi_{ON}(G) \leq 4$.

2   Preliminaries

In this paper, we consider only simple, finite, undirected and connected graphs. If the graph is not connected, we color each of the components independently. Also, we assume that the graphs do not have isolated vertices as they cannot be CFON colored. The graph induced by a set of vertices V' in G is denoted $G[V']$. For any two vertices $u, v \in V(G)$, the shortest distance between them is denoted $dist(u, v)$. The open neighborhood of v, denoted $N(v)$, is the set of vertices adjacent to v. The closed neighborhood of v, denoted $N[v]$, is $N[v] = N(v) \cup \{v\}$. The degree of a vertex v in the graph is denoted $\deg(v)$. The distance, degree and neighborhood restricted to a subgraph H is denoted $dist_H(u, v)$, $\deg_H(v)$ and $N_H(v)$ respectively.

We denote the set $\{1, 2, \ldots, q\}$ by $[q]$. Throughout this paper, we use the coloring functions $C : V \to [q]$ and $U : V \to [q]$ to denote the color assigned to a vertex and a unique color in its neighborhood, respectively. For a vertex $v \in V(G)$, if there exists a vertex $w \in N(v)$ such that $\{x \in N(v) \setminus \{w\}: C(x) = C(w)\} = \emptyset$, then w is called a uniquely colored neighbor of v.

For theorems marked $\star$, we provide the proofs in the full version of the paper [13].

3   Pathwidth

Theorem 3 (Main Pathwidth Result). *Let G be a graph and let $\mathsf{pw}(G)$ denote the pathwidth of G. Then there exists a CFON coloring of G using at most $\lfloor \frac{5}{3}(\mathsf{pw}(G) + 1) \rfloor$ colors.*

The proof of this theorem is a constructive procedure that assigns colors to the vertices of G from a set of size $5(\mathsf{pw}(G)+1)/3$. We first formally define pathwidth.

Definition 4 (Path decomposition [14]). *A path decomposition of a graph G is a sequence $\mathcal{P} = (X_1, X_2, \ldots, X_s)$ of bags such that, for every $p \in \{1, 2, \ldots, s\}$, we have $X_p \subseteq V(G)$ and the following hold:*

- *For each vertex $v \in V(G)$, there is a $p \in \{1, 2, \ldots, s\}$ such that $v \in X_p$.*
- *For each edge $\{u, v\} \in E(G)$, there is a $p \in \{1, 2, \ldots, s\}$ such that $u, v \in X_p$.*
- *If $v \in X_{p_1}$ and $v \in X_{p_2}$ for some $p_1 \leq p_2$, then $v \in X_p$ for all $p_1 \leq p \leq p_2$.*

The *width* of a path decomposition $(X_1, X_2, \ldots, X_s)$ is $\max_{1 \leq p \leq s}\{|X_p| - 1\}$. The *pathwidth* of a graph G, denoted $\mathsf{pw}(G)$, is the minimum width over all path decompositions of G. For the purposes of our algorithm, we need the path decomposition to satisfy certain additional properties too.

Definition 5 (Semi-nice Path Decomposition). *A path decomposition* $\mathcal{P} = (X_1, X_2, \ldots, X_s)$ *is called a* semi-nice path decomposition *if* $X_1 = X_s = \emptyset$ *and for all* $p \in \{2, \ldots, s\}$, *exactly one of the following hold:*

SN1. *There is a vertex* v *such that* $v \notin X_{p-1}$ *and* $X_p = X_{p-1} \cup \{v\}$. *In this case, we say that* X_p *introduces* v. *Further, when* X_p *introduces* v, $N(v) \cap X_p \neq \emptyset$.

SN2. *There is a vertex* v *such that* $v \in X_{p-1}$ *and* $X_p = X_{p-1} \setminus \{v\}$. *In this case, we say* X_p *forgets* v.

SN3. *There is a pair of vertices* $v, \widehat{v}$ *such that* $v, \widehat{v} \notin X_{p-1}$ *and* $X_p = X_{p-1} \cup \{v, \widehat{v}\}$. *We call such a bag* X_p *a* special bag *that introduces* v *and* $\widehat{v}$. *Further, in a special bag* X_p *that introduces* v *and* $\widehat{v}$, *it must be true that* $N(v) \cap X_p = \{\widehat{v}\}$ *and* $N(\widehat{v}) \cap X_p = \{v\}$.

We first note that the every graph without isolated vertices has a semi-nice path decomposition of width $\mathsf{pw}(G)$.

Theorem 6 ($\star$). *Let* G *be a graph that has no isolated vertices. Then it has a semi-nice path decomposition of width* $\mathsf{pw}(G)$.

Algorithm. We start with a semi-nice path decomposition $\mathcal{P} = (X_1, X_2, \ldots, X_q)$ of width $\mathsf{pw}(G)$. We process each bag in the order $X_1, X_2, \ldots, X_q$. As we encounter each bag, we assign to the vertices in the bag a color $C : V(G) \to [5(\mathsf{pw}(G) + 1)/3]$. We will also identify a unique color (from its neighborhood) for each vertex $U : V(G) \to [5(\mathsf{pw}(G) + 1)/3]$. We color the bags such that the below are satisfied:

Invariant 1. For any bag X, if $v, v' \in X$, then $C(v) \neq C(v')$.

Invariant 2. Suppose we have processed bags X_1 to X_p, where $p \geq 2$. At this point, the induced graph $G[\cup_{1 \leq j \leq p} X_j]$ is CFON colored.

Invariant 3. For every vertex v that appears in the bags processed, $U(v)$ is set as $C(w)$ for a neighbor w of v. Once $U(v)$ is assigned, it is ensured that for all "future" neighbors v' of v, $C(v') \neq U(v)$, thereby ensuring that $U(v)$ is retained as a unique color in $N(v)$.

Definitions Required for the Algorithm: For each bag X, we define the set of *free colors*, as $F(X) = \{U(x) : x \in X\} \setminus \{C(x) : x \in X\}$. That is, $F(X)$ is the set of colors that appear in X as unique colors of vertices in X, but not as colors of any vertex. Further, we partition $F(X)$ into two sets $F_1(X)$ and $F_{>1}(X)$. They are defined as $F_1(X) = \{c \in F(X) : |\{x \in X : U(x) = c\}| = 1\}$ and $F_{>1}(X) = \{c \in F(X) : |\{x \in X : U(x) = c\}| > 1\}$. A vertex v that appears in a bag X is called a *needy vertex* (or simply *needy*) in X, if $U(v) \in F(X)$. For a bag X, we say that a set $S \subseteq X$ is an *expensive subset* if $|\cup_{w \in S} \{C(w), U(w)\}| = 2|S|$.

When going through the sequence of bags in the semi-nice path decomposition, the bags X that forget a vertex only contain vertices that have already been assigned colors and hence no action needs to be taken. When we move from a bag X' to the next bag X that introduces either one vertex or two vertices, we need to handle the introduced vertices. Below, we explain the rules to assign

$C(v)$ and $U(v)$, when X is a bag that introduces one vertex v, and the rules to assign $C(v)$, $C(\widehat{v})$, $U(v)$ and $U(\widehat{v})$, when X is a special bag that introduces two vertices v and $\widehat{v}$.

For bags that introduce one vertex v

Rule 1 for assignment of $C(v)$:
- If there exists a color $c \in F_1(X') \setminus \{U(x) : x \in N(v) \cap X'\}$, then we assign $C(v) = c$. If there are more than one such color c, choose a c such that $|\{x : x \in X', C(U^{-1}(c)) = U(x)\}|$ is minimized. Note that for all $c \in F_1(X')$, there is a unique vertex $w \in X'$ such that $U(w) = c$, and hence $U^{-1}(c)$ is well defined.
- If $F_1(X') \setminus \{U(x) : x \in N(v) \cap X'\} = \emptyset$, we check if there exists a color $c \in F_{>1}(X') \setminus \{U(x) : x \in N(v) \cap X'\}$. If so, we assign $C(v) = c$. If there are multiple such c, then we choose one arbitrarily.
- If $F_1(X') \cup F_{>1}(X') \setminus \{U(x) : x \in N(v) \cap X'\} = \emptyset$, then there are no free colors that can be assigned as $C(v)$. We assign $C(v)$ to be a new color (a color not in $\cup_{x \in X'} \{C(x), U(x)\}$).

Rule 2 for assignment of $U(v)$: We assign $U(v) = C(y)$, where $y \in X'$ is a neighbor of v. Such a y exists by Theorem 6. If v has multiple neighbors, we follow the below priority order:
- If v has needy vertices in X' as neighbors, we choose y as a needy neighbor such that $|\{x : x \in X', U(y) = U(x)\}|$ is minimized.
- If v does not have needy vertices in X' as neighbors, then we choose $y \in X'$ arbitrarily from the set of neighbors of v.

For special bags that introduce two vertices v and $\widehat{v}$

For assignment of $C(v)$ and $C(\widehat{v})$: We select one of v and $\widehat{v}$ arbitrarily, say v, to be colored first. We use Rule 1 to assign $C(v)$ and then $C(\widehat{v})$, in that order. One point to note is that during the application of Rule 1 here, the part $\{U(x) : x \in N(v) \cap X'\}$ will not feature as neither v nor $\widehat{v}$ have neighbors in X'.

For assignment of $U(v)$ and $U(\widehat{v})$: Assign $U(v) = C(\widehat{v})$ and $U(\widehat{v}) = C(v)$.

It can easily be checked that the above rules maintain the invariants 1, 2 and 3 stated earlier and hence the algorithm results in a CFON coloring of G. What remains is to show that $5(\mathsf{pw}(G) + 1)/3$ colors are sufficient. We first prove a technical result.

Theorem 7 (Technical Pathwidth Result). *During the course of the algorithm, let k be the size of the largest expensive subset out of all the bags in the path decomposition. Then there must exist a bag of size at least $3k/2$.*

Proof. In the sequence of bags seen by the algorithm, let X be the first bag that has an expensive subset of size k. We show that $|X| \geq 3k/2$. Let $S = \{v_1, v_2, \ldots, v_k\} \subseteq X$ be an expensive subset of size k. For each v_i, let $C(v_i) = 2i - 1$ and $U(v_i) = 2i$.

Let X' be the bag that precedes X in the sequence. By the choice of X, no k-expensive subset is present in X'. It follows that $S \not\subseteq X'$. Hence the bag X must introduce a vertex[1] that belongs to S. Without loss of generality, let v_k be this vertex introduced in X. Further wlog, let $v_1, \ldots, v_r$ be the needy vertices (in X') of S for some $1 \leq r \leq k$. If none of the vertices in S are needy, then we have that $|X| \geq 2|S| = 2k$ and the theorem holds. So we can assume that $r \geq 1$.

Since the vertices $v_1, \ldots, v_r$ are needy in X', we have $\{2, 4, \ldots, 2r\} \subseteq F(X')$. The vertices $v_{r+1}, \ldots, v_k$ are not needy because there exist distinct vertices $Z = \{z_{r+1}, \ldots, z_k\}$ in the bag X such that $C(z_i) = U(v_i) = 2i$ for $r + 1 \leq i \leq k$. We have three cases. In Cases 1 and 2, X is a bag that introduces one vertex v_k. Case 1 is when none of the colors in $F(X')$ was eligible to be assigned as $C(v_k)$. Hence $C(v_k)$ is assigned from outside the set $\cup_{x \in X'}\{C(x), U(x)\}$. Case 2 is when there are eligible colors in $F(X')$, and $C(v_k)$ is chosen from $F(X') \subseteq \cup_{x \in X'}\{C(x), U(x)\}$. Case 3 is when X is a special bag that introduces two vertices.

Case 1: X is a bag that introduces one vertex v_k and $2k - 1 \notin \{U(x) : x \in X'\}$. There is no vertex $x \in X$ with $U(x) = 2k - 1$. To assign a color to v_k, the algorithm chose a new color. This means that $F(X') \setminus \{U(x) : x \in N(v_k) \cap X'\} = \emptyset$. In particular, for each $1 \leq i \leq r$, there[2] exists $v_i' \in N(v_k) \cap X'$ such that $U(v_i') = 2i$. Hence the colors $2i$, for $1 \leq i \leq r$ cannot be assigned as $C(v_k)$. By Rule 2, we must set $U(v_k)$ to be the $C(y)$ where y is a needy neighbor of v_k. Hence $C(y) = 2k$.

If $U(y) \notin \{1, 2, 3, \ldots, 2k - 2\}$, then $(S \cup \{y\}) \setminus \{v_k\}$ is a k-expensive subset in X', the predecessor of X. This contradicts the choice of X. Hence we can assume that $U(y) \in \{1, 2, 3, \ldots, 2k - 2\}$. Since y is needy in X', $U(y)$ is not $C(v)$ for any $v \in X'$ and hence we conclude $U(y) \notin \{1, 3, \ldots, 2k - 3\}$. Further, colors from $\{2(r+1), 2(r+2), \ldots, 2k - 2\}$ appear as $C(z)$ for the vertices $z \in Z$. Hence $U(y) \in \{2, 4, 6, \ldots, 2r\}$.

Let $U(y) = 2j$ for some $1 \leq j \leq r$. Notice that $U(v_j) = 2j$ as well, giving us $|\{x : x \in X', U(y) = U(x)\}| \geq 2$.

By Rule 2, we chose $U(v_k) = C(y)$, where y is the needy neighbor that minimizes $|\{x : x \in X', U(y) = U(x)\}|$. We chose y over other needy neighbors $v_1', \ldots, v_r'$ of v_k. Hence there exist r distinct vertices $Y = \{y_1, \ldots, y_r\}$ in the bag X, disjoint from S, such that $U(y_i) = U(v_i') = 2i$ for each $1 \leq i \leq r$.

Note that the set $Y \cup Z$ must be disjoint from S, but Y and Z may intersect with each other. Since $|Y| + |Z| = k$, we have $|Y \cup Z| \geq k/2$ and therefore $|X| \geq |S| + |Y \cup Z| \geq 3k/2$.

---

[1] In the case where X is a special bag that introduces two vertices, at most one of the two introduced vertices can be part of an expensive subset.

[2] The vertex v_i' may or may not be the same as v_i.

Case 2 ($\star$): X is a bag that introduces one vertex v_k and $2k - 1 \in \{U(x) : x \in X'\}$.

Case 3 ($\star$): X is a special bag that introduces v_k and $\widehat{v}_k$.

The proofs of Cases 2 and 3 are omitted due to space constraints and can be found in the full version of the paper [13]. □

Now we give a sketch of the proof of the main theorem of this section.

Proof (Proof Sketch of Theorem 3). Theorem 7 limits the size of largest expensive set in the bags during the course of the algorithm. Since the largest bag is of size $\mathsf{pw}(G) + 1$, the size of the largest expensive set is of size $2(\mathsf{pw}(G) + 1)/3$.

The reason for requiring more colors than the bag size is the presence of "extra colors", those that appear as $U(v)$, but not as $C(v)$. The extra colors can be used to form an expensive set, the size of which is bounded by the above argument. This helps us bring down the bound to $5(\mathsf{pw}(G) + 1)/3$. □

4   Feedback Vertex Set

Definition 8 (Feedback Vertex Set). *Let $G = (V, E)$ be an undirected graph. A feedback vertex set (FVS) is a set of vertices $S \subseteq V$, removal of which from the graph G makes the remaining graph ($G[V \setminus S]$) acyclic. The size of a smallest such set S is denoted as $\mathsf{fvs}(G)$.*

The main result of this section is the following. Though the full proof is omitted, we present the key ideas and a sketch in this section.

Theorem 9 ($\star$). $\chi_{ON}(G) \leq \mathsf{fvs}(G) + 2$.

The following graph (as observed in [6]), shows that the above theorem is tight. Let K_n^* be the graph obtained by starting with the clique on n vertices, and subdividing each edge with a vertex. Then K_n^* has an FVS of size $n - 2$, and it can be seen that $\chi_{ON}(K_n^*) = n$.

The proof of this theorem is through a constructive process to CFON color the vertices of the graph G, given a feedback vertex set F of G. By definition, $G[V \setminus F]$ is a collection of trees.

Each tree T in $G[V \setminus F]$ is rooted at an arbitrary vertex r_T. If $|V(T)| \geq 2$, we choose a neighbor of r_T and call it the *special vertex* in T, denoted by s_T. Let v be a vertex not in T. The *deepest neighbor* of v in T, denoted by $deep_T(v)$, is a vertex $w \in V(T) \cap N(v)$ such that $dist_T(r_T, w)$ is maximized. If there are multiple such vertices at the same distance, the deepest neighbor is chosen to be a vertex which is not the special vertex s_T.

Lemma 10. *Let T be a tree with $|V(T)| \geq 2$. Then $\chi_{ON}(T) \leq 2$.*

Proof. We assign colors $C : V(T) \to \{1, 2\}$ in the following manner.

– Assign $C(r_T) = 1$ and $C(s_T) = 2$.
– For each vertex $v \in N_T(r_T) \setminus \{s_T\}$, assign $C(v) = 1$.

- For the remaining vertices $v \in V(T)$, assign $C(v) = \{1, 2\} \setminus C(w)$, where w is the grandparent of v.

For each vertex $v \in V(T) \setminus \{r_T\}$, the uniquely colored neighbor is its parent. For r_T, the uniquely colored neighbor is s_T. This is a CFON 2-coloring of T. □

We first prove a special case of Theorem 9.

Lemma 11. *Let $G = (V, E)$ be a graph and $F \subseteq V$ be a feedback vertex set with $|F| = 1$. Then G can be CFON colored using 3 colors.*

Proof. Let $F = \{v\}$. First using Lemma 10, we color all the trees $T \subseteq G[V \setminus F]$ using the colors 2 and 3, whenever $|V(T)| \geq 2$. All the singleton components of $G[V \setminus F]$ are assigned the color 2. We assign $C(v) = 1$. Now all the vertices, except possibly v, have a uniquely colored neighbor. We explain how to fix this and obtain a CFON coloring.

- **Case 1: There exists a singleton component $\{w\} \subseteq G[V \setminus F]$.**
  Reassign $C(w) = 1$.
- **Case 2: Else, if there exists a component $T \subseteq G[V \setminus F]$, such that either (i) $deep_T(v) \neq s_T$ or (ii) $deep_T(v) = s_T$ and $\{r_T, v\} \notin E(G)$.**
  Reassign $C(deep_T(v)) = 1$.
- **Case 3: Else, for each component $T \subseteq G[V \setminus F]$, $N(v) \cap V(T) = \{r_T, s_T\}$.**
  If there exists a component $T \subseteq G[V \setminus F]$, such that $|V(T)| \geq 3$, choose a vertex $w \in V(T) \setminus \{r_T, s_T\}$ and set w as the new root of T. Reassign s_T and the colors of $V(T)$ accordingly. Doing so will ensure that $deep_T(v) \neq s_T$. We apply Case 2.
  Else, for all the components $T \subseteq G[V \setminus F]$, we have $|V(T)| = 2$. Choose a component $T' \subseteq G[V \setminus F]$. For all the other vertices $w \in V \setminus (\{v\} \cup V(T'))$, reassign $C(w) = 2$.

All the trees in $G[V \setminus F]$ are CFON colored as per the earlier described procedure. Even after reassigning some colors, they remain CFON colored. The vertex v sees another vertex w, with $C(w) = 1$ if in Case 1 or 2. In the last case, v sees a unique vertex that is colored 3. □

Proof (Proof Sketch of Theorem 9). When $|F| = 1$, three colors are sufficient to CFON color G by Lemma 11. We may assume that $|F| \geq 2$. The proof of Lemma 11 indicates the key ideas and the structure, however the proof of Theorem 9 will be longer and use more involved arguments. We start by coloring the trees in $G[V \setminus F]$ using two colors. We proceed in phases. We first handle the singleton components in $G[V \setminus F]$. Next, we consider trees in $G[V \setminus F]$ that interact with the uncolored vertices in F in certain specific ways. The final part of the argument shows that the collection of cases is comprehensive. If none of the cases apply, we show that the graph is not connected, which is a contradiction. □

5   Neighborhood Diversity and Distance to Cluster

In this section, we give improved bounds for $\chi_{ON}(G)$ and $\chi_{CN}(G)$ with respect to the parameters neighborhood diversity and distance to cluster.

Definition 12 (Neighborhood Diversity [4]). *Give a graph $G = (V, E)$, two vertices $v, w \in V$ have the same* type *if $N(v) \setminus \{w\} = N(w) \setminus \{v\}$. A graph G has* neighborhood diversity *at most t if $V(G)$ can be partitioned into t sets $V_1, V_2, \ldots, V_t$, such that all the vertices in each V_i, $1 \leq i \leq t$ have the same type. The partition $\{V_1, V_2, \ldots, V_t\}$ is called the* type partition *of G.*

It can be inferred from the above definition that all vertices in a V_i either form a clique or an independent set, $1 \leq i \leq t$. For two types V_i, V_j, either each vertex in V_i is neighbor to each vertex in V_j, or no vertex in V_i is neighbor to any vertex in V_j. This leads to the definition of the *type graph* $H = (\{1, 2, \ldots, t\}, E_H)$, where $E_H = \{\{i, j\} : 1 \leq i < j \leq t, \text{each vertex in } V_i \text{ is adjacent to each vertex in } V_j\}$.

In the above, $cl(G)$ and $ind(G)$ respectively denote the number of V_i's that form a clique and independent set in the type partition $\{V_1, V_2, \ldots, V_t\}$.

Theorem 13 ($\star$). $\chi_{ON}(G) \leq \chi_{ON}(H) + \frac{cl(G)}{2} + 2.$

Theorem 14 ($\star$). $\chi_{CN}(G) \leq \chi_{CN}(H) + \frac{ind(G)}{3} + 3.$

Definition 15 (Distance to Cluster). *Let $G = (V, E)$ be a graph. The distance to cluster of G, denoted $\mathsf{dc}(G)$, is the size of a smallest set $X \subseteq V$ such that $G[V \setminus X]$ is a disjoint union of cliques.*

Theorem 16 ($\star$). $\chi_{ON}(G) \leq \mathsf{dc}(G) + 3.$

Theorem 17 ($\star$). $\chi_{CN}(G) \leq \max\{3, \mathsf{dc}(G) + 1\}.$

For the subdivided clique K_n^*, we have $\chi_{ON}(K_n^*) = \mathsf{dc}(K_n^*) = n$. Hence Theorem 16 is nearly tight.

6   CFON* Coloring of Planar Graphs

Definition 18 (Planar and Outerplanar graphs). *A* planar graph *is a graph that can be drawn in $\mathbb{R}^2$ (a plane) such that the edges do not cross each other in the drawing. An* outerplanar graph *is a planar graph that has a drawing in a plane such that all the vertices of the graph belong to the outer face.*

Abel et al. showed [3] that eight colors are sufficient for CFON* coloring of a planar graph. In this section, we improve the bound to five colors.

We need the following definition:

Definition 19 (Maximal Distance-3 Set). *For a graph $G = (V, E)$, a maximal distance-3 set is a set $S \subseteq V(G)$ that satisfies the following:*

1. *For every pair of vertices* $w, w' \in S$, *we have* $dist(w, w') \geq 3$.
2. *For every vertex* $w \in S$, $\exists w' \in S$ *such that* $dist(w, w') = 3$.
3. *For every vertex* $x \notin S$, $\exists x' \in S$ *such that* $dist(x, x') < 3$.

The set S is constructed by initializing $S = \{v\}$ where v is an arbitrary vertex. We proceed in iterations. In each iteration, we add a vertex w to S if (1) for every v already in S, $dist(v, w) \geq 3$, and (2) there exists a vertex $w' \in S$ such that $dist(w, w') = 3$. We repeat this until no more vertices can be added.

The main component of the proof is the construction of an auxiliary graph G' from the given graph G.

Construction of G'**:** The first step is to pick a maximal distance-3 set V_0. Notice that any distance-3 set is an independent set by definition. We let V_1 denote the neighborhood of V_0. More formally, $V_1 = \{w : \{w, w'\} \in E(G), w' \in V_0\}$. Let V_2 denote the remaining vertices i.e., $V_2 = V \setminus (V_0 \cup V_1)$.

We note the following properties satisfied by the above partitioning of $V(G)$.

1. The set V_0 is an independent set.
2. For every vertex $w \in V_1$, there exists a unique vertex $w' \in V_0$ such that $\{w, w'\} \in E(G)$. This is because if there are two such vertices, this will violate the distance-3 property of V_0.
3. Every vertex in V_0 has a neighbor in V_1. If there exists $v \in V_0$ without a neighbor in V_1, then v is an isolated vertex. By assumption, G does not have isolated vertices.
4. There are no edges from V_0 to V_2.
5. Every vertex in V_2 has a neighbor in V_1, and is hence at distance 2 from some vertex in V_0. This is due to the maximality of the distance-3 set V_0.

Now we define $A = V_0 \cup V_2$. We first remove all the edges of $G[V_2]$ making A an independent set. For every vertex $v \in A$ we do the following: we identify an arbitrary neighbor $f(v) \in N(v) \subseteq V_1$. Then we contract the edge $\{v, f(v)\}$. That is, we first identify vertex v with $f(v)$. Then for every edge $\{v, v'\}$, we add an edge $\{f(v), v'\}$. The resulting graph is G'.

Theorem 20. *If G is a planar graph, $\chi^*_{ON}(G) \leq 5$.*

Proof. Let G be a planar graph. We first construct the graph G' as above. Since the steps for constructing G' involve only edge deletion and edge contraction, G' is also a planar graph. By the planar four-color theorem [15], there is an assignment $C : V(G') \rightarrow \{2, 3, 4, 5\}$ such that no two adjacent vertices of G' are assigned the same color. Now we have colored all the vertices in $V(G') = V_1$

Now, we extend C to get a CFON* coloring for G. For all vertices $v \in V_0$, we assign $C(v) = 1$. The vertices in V_2 are not assigned a color.

We will show that C is indeed a CFON* coloring of G. Consider a vertex $v \in A$ which is contracted to a neighbor $f(v) = w \in V_1$. The color assigned to w is distinct from all w's neighbors in G'. Hence the color assigned to w is the unique color among the neighbors of v in G.

For each vertex $w \in V_1$, w is a neighbor of exactly one vertex $v \in V_0$. Every vertex $v \in V_0$ is colored 1, which is different from all the colors assigned to the neighbors of w in G'. $\square$

Outerplanar graphs have a proper coloring using three colors. By an argument analogous to Theorem 20, we infer the following.

Corollary 21. *If G is an outerplanar graph, $\chi^*_{ON}(G) \leq 4$.*

For outerplanar graphs, a CFON* coloring using 4 colors implies a CFON coloring using 5 colors. However, we can show the following improved bound.

Theorem 22 (⋆). *If G is an outerplanar graph, $\chi_{ON}(G) \leq 4$.*

Acknowledgments. We would like to thank I. Vinod Reddy for suggesting the problem, Rogers Mathew and N. R. Aravind for helpful discussions and the anonymous reviewer who pointed out an issue with the proof of Theorem 6.

References

1. Even, G., Lotker, Z., Ron, D., Smorodinsky, S.: Conflict-free colorings of simple geometric regions with applications to frequency assignment in cellular networks. SIAM J. Comput. **33**(1), 94–136 (2004). https://doi.org/10.1137/S0097539702431840
2. Smorodinsky, S.: Conflict-free coloring and its applications. In: Bárány, I., Böröczky, K.J., Tóth, G.F., Pach, J. (eds.) Geometry—Intuitive, Discrete, and Convex. BSMS, vol. 24, pp. 331–389. Springer, Heidelberg (2013). https://doi.org/10.1007/978-3-642-41498-5_12
3. Abel, Z., et al.: Conflict-free coloring of graphs. SIAM J. Discret. Math. **32**(4), 2675–2702 (2018). https://doi.org/10.1137/17M1146579
4. Gargano, L., Rescigno, A.A.: Complexity of conflict-free colorings of graphs. Theor. Comput. Sci. **566**(C), 39–49 (2015). https://doi.org/10.1016/j.tcs.2014.11.029
5. Reddy, I.V.: Parameterized algorithms for conflict-free colorings of graphs. Theor. Comput. Sci. **745**, 53–62 (2018). https://doi.org/10.1016/j.tcs.2018.05.025
6. Bodlaender, H.L., Kolay, S., Pieterse, A.: Parameterized complexity of conflict-free graph coloring. In: Friggstad, Z., Sack, J.-R., Salavatipour, M.R. (eds.) WADS 2019. LNCS, vol. 11646, pp. 168–180. Springer, Cham (2019). https://doi.org/10.1007/978-3-030-24766-9_13
7. Agrawal, A., Ashok, P., Reddy, M.M., Saurabh, S., Yadav, D.: FPT algorithms for conflict-free coloring of graphs and chromatic terrain guarding. CoRR abs/1905.01822. arXiv:1905.01822 (2019)
8. Keller, C., Smorodinsky, S.: Conflict-free coloring of intersection graphs of geometric objects. Discret. Comput. Geom., 1–26 (2019). https://doi.org/10.1007/s00454-019-00097-8
9. Fekete, S.P., Keldenich, P.: Conflict-free coloring of intersection graphs. Int. J. Comput. Geom. Appl. **28**(03), 289–307 (2018). https://doi.org/10.1142/S0218195918500085
10. Chen, K.: Online conflict-free coloring for intervals. SIAM J. Comput. **36**(5), 1342–1359 (2006). https://doi.org/10.1137/S0097539704446682
11. Pach, J., Tardos, G.: Conflict-free colourings of graphs and hypergraphs. Comb. Probab. Comput. **18**(5), 819–834 (2009). https://doi.org/10.1017/S0963548309990290
12. Cheilaris, P.: Conflict-free coloring, Ph.D. thesis, New York, NY, USA (2009)

13. Bhyravarapu, S., Kalyanasundaram, S.: Combinatorial bounds for conflict-free coloring on open neighborhoods. CoRR abs/2007.05585. arXiv:2007.05585 (2020)
14. Cygan, M., et al.: Parameterized Algorithms, vol. 1. Springer, Cham (2015). https://doi.org/10.1007/978-3-319-21275-3
15. Robertson, N., Sanders, D., Seymour, P., Thomas, R.: The four-colour theorem. J. Comb. Theor. Ser. B **70**(1), 2–44 (1997). https://doi.org/10.1006/jctb.1997.1750

Guarding Quadrangulations and Stacked Triangulations with Edges

Paul Jungeblut$^{(\boxtimes)}$ and Torsten Ueckerdt

Karlsruhe Institute of Technology, Karlsruhe, Germany
{paul.jungeblut,torsten.ueckerdt}@kit.edu

Abstract. Let $G = (V, E)$ be a plane graph. A face f of G is *guarded* by an edge $vw \in E$ if at least one vertex from $\{v, w\}$ is on the boundary of f. For a planar graph class $\mathcal{G}$ we ask for the minimal number of edges needed to guard all faces of any n-vertex graph in $\mathcal{G}$. We prove that $\lfloor n/3 \rfloor$ edges are always sufficient for quadrangulations and give a construction where $\lfloor (n-2)/4 \rfloor$ edges are necessary. For 2-degenerate quadrangulations we improve this to a tight upper bound of $\lfloor n/4 \rfloor$ edges. We further prove that $\lfloor 2n/7 \rfloor$ edges are always sufficient for stacked triangulations (that are the 3-degenerate triangulations) and show that this is best possible up to a small additive constant.

Keywords: Edge guard sets · Art galleries · Quadrangulations · Stacked triangulations

1 Introduction

In 1975, Chvátal [5] laid the foundation for the widely studied field of *art gallery problems* by answering how many guards are needed to observe all interior points of any given n-sided polygon P. Here a guard is a point p in P and it can observe any other point q in P if the line segment pq is fully contained in P. He shows that $\lfloor n/3 \rfloor$ guards are sometimes necessary and always sufficient. Fisk [8] revisited Chvátal's Theorem in 1978 and gave a very short and elegant new proof by introducing diagonals into the polygon P to obtain a triangulated, outerplanar graph. Such graphs are 3-colorable and in each 3-coloring all faces are incident to vertices of all three colors, so the vertices of the smallest color class can be used as guard positions. Bose et al. [4] studied the problem to guard the faces of a plane graph instead of a polygon. A *plane graph* is a graph $G = (V, E)$ with an embedding in $\mathbb{R}^2$ with not necessarily straight edges and without crossings between any two edges. Here a face f is guarded by a vertex v, if v is on the boundary of f. They show that $\lfloor n/2 \rfloor$ vertices (so called *vertex guards*) are sometimes necessary and always sufficient for n-vertex plane graphs.

We consider a variant of this problem introduced by O'Rourke [13]. He shows that only $\lfloor n/4 \rfloor$ guards are necessary in Chvátal's original setting if each guard

---

A full version including all omitted proofs is available on arXiv [11].

© Springer Nature Switzerland AG 2020
I. Adler and H. Müller (Eds.): WG 2020, LNCS 12301, pp. 14–26, 2020.
https://doi.org/10.1007/978-3-030-60440-0_2

is assigned to an edge of the polygon that he can patrol along instead of being fixed to a single point. Considering plane graphs again, an *edge guard* is an edge $vw \in E$ and it guards all faces having v and/or w on their boundary. For a given planar graph class $\mathcal{G}$, we ask for the minimal number of edge guards needed to guard all faces of any n-vertex graph in $\mathcal{G}$. Here and in the following the outer face is treated just like any other face and must also be guarded.

General (not necessarily triangulated) n-vertex plane graphs might need at least $\lfloor n/3 \rfloor$ edge guards, even when requiring 2-connectedness [4]. The best known upper bounds have recently been presented by Biniaz et al. [1] and come in two different fashions: First, any n-vertex plane graph can be guarded by $\lfloor 3n/8 \rfloor$ edge guards found in an iterative process. Second, a coloring approach yields an upper bound of $\lfloor n/3 + \alpha/9 \rfloor$ edge guards where α counts the number of quadrangular faces in G. Looking at n-vertex triangulations, Bose et al. [4] give a construction for triangulations needing $\lfloor (4n - 8)/13 \rfloor$ edge guards.[1] A corresponding upper bound of $\lfloor n/3 \rfloor$ edge guards was published earlier in the same year by Everett and Rivera-Campo [7].

Preliminaries. All graphs considered throughout this paper are undirected and simple (unless explicitly stated otherwise). Let $G = (V, E)$ be a graph. For an edge $\{v, w\} \in E$ we use the shorter notation vw or wv and both mean the same. The *order* of G is its number of vertices and denoted by $|G|$. We say that G is k-regular if each vertex $v \in V$ has degree exactly k. Further G is called k-degenerate if every subgraph contains a vertex of degree at most k. For the subgraph induced by a subset $X \subseteq V$ of the vertices we write $G[X]$. Now assume that G is plane with face set F. The *dual graph* $G^* = (V^*, E^*)$ is defined by $V^* = \{f^* \mid f \in F\}$ and $E^* = \{f^* g^* \mid f, g \in F \wedge f, g$ share a boundary edge $vw \in E\}$. Note that G^* can be a multigraph. The dual graph G^* of a plane graph G is also planar and we assume below that a plane drawing of G^* is given that is inherited from the plane drawing of G as follows: Each dual vertex $f^* \in V^*$ is drawn inside face f, each dual edge $f^* g^* \in E^*$ crosses its primal edge vw exactly once and in its interior and no two dual edges cross.

Let $\Gamma \subseteq E$ be a set of edges. We write $V(\Gamma)$ for the set of endpoints of all edges in Γ. Further, Γ is an *edge guard set* if all faces $f \in F$ are guarded by at least one edge in Γ, i.e. each face f has a boundary vertex in $V(\Gamma)$.

Contribution. In Sect. 2 we consider the class of quadrangulations, i.e. plane graphs where every face is bounded by a 4-cycle. We describe a coloring based approach to improve the currently best known upper bound to $\lfloor n/3 \rfloor$ edge guards. In addition we also consider 2-degenerate quadrangulations and present an upper bound of $\lfloor n/4 \rfloor$ edge guards, which is best possible. Our motivation

---

[1] The authors of [4] actually claim that $\lfloor (4n - 4)/13 \rfloor$ edge guards are necessary, but this result is only valid for *near*-triangulations (this was noted first by Kaučič et al. [12] and later clarified by one of the original authors [2]). For proper triangulations an additional vertex is needed in the construction so only $\lfloor (4n - 8)/13 \rfloor$ edge guards are necessary.

to consider quadrangulations is that the coloring approaches developed earlier for general plane graphs [1,7] fail on quadrangular faces. As a second result in Sect. 3 we present a new upper bound of $\lfloor 2n/7 \rfloor$ edge guards for stacked triangulations and show that it is best possible. We saw above that no infinite family of triangulations is known that actually needs $\lfloor n/3 \rfloor$ edge guards. With the stacked triangulations we now know a non-trivial subclass of triangulations for which strictly fewer edge guards are necessary than for general plane graphs.

2   Quadrangulations

Quadrangulations are the maximal plane bipartite graphs and every face is bounded by exactly four edges. The currently best known upper bounds are the ones given by Biniaz et al. [1] for general plane graphs ($\lfloor 3n/8 \rfloor$ respectively $\lfloor n/3 + \alpha/9 \rfloor$, where α is the number of quadrilateral faces). For n-vertex quadrangulations we have $\alpha = n-2$, so $\lfloor n/3 + (n-2)/9 \rfloor = \lfloor (4n-2)/9 \rfloor > \lfloor 3n/8 \rfloor$ for $n \geq 4$. In this section we provide a better upper bound of $\lfloor n/3 \rfloor$ and a construction for quadrangulations needing $\lfloor (n-2)/4 \rfloor$ edge guards. Closing the gap remains an open problem.

Theorem 1. *For $k \in \mathbb{N}$ there exists a quadrangulation Q_k with $n = 4k + 2$ vertices needing $k = (n-2)/4$ edge guards.*

Proof. Define $Q_k = (V, E)$ with $V := \{s, t\} \cup \bigcup_{i=1}^{k} \{a_i, b_i, c_i, d_i\}$ and $E := \bigcup_{i=1}^{k} \{sa_i, sc_i, ta_i, tc_i, a_ib_i, a_id_i, c_ib_i, c_id_i\}$ as the union of k disjoint 4-cycles and two extra vertices s and t connecting them. Figure 1 shows this and a planar embedding. Now for any two distinct $i, j \in \{1, \ldots, k\}$ the two quadrilateral faces (a_i, b_i, c_i, d_i) and (a_j, b_j, c_j, d_j) are only connected via paths through s or t. Therefore, no edge can guard two or more of them and we need at least k edge guards for Q_k. On the other hand it is easy to see that $\{sa_1, \ldots, sa_k\}$ is an edge guard set of size k, so Q_k needs exactly k edge guards.   □

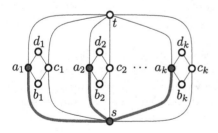

Fig. 1. A quadrangulation with $4k + 2$ vertices needing k edge guards (thick red edges). (Color figure online)

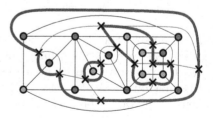

Fig. 2. A quadrangulation G (black edges) and its dual G^* (purple edges) with a 2-factor (thick edges). The vertex coloring in orange and green is a guard coloring. (Color figure online)

The following definition and lemma are from Bose et al. [3] and we cite it using the terminology of Biniaz et al. [1]. A *guard coloring* of a plane graph G is a non-proper 2-coloring of its vertex set, such that each face f of G has at least one boundary vertex of each color and at least one monochromatic edge (i.e. an edge where both endpoints receive the same color). They prove that a guard coloring exists for all graphs without any quadrangular faces.

Lemma 1 ([3, Lemma 3.1]). *If there is a guard coloring for an n-vertex plane graph G, then G can be guarded by $\lfloor n/3 \rfloor$ edge guards.*

Bose et al. [3] even present a linear time algorithm to compute a guard coloring for graphs without quadrangular faces. We extend their result by showing that plane graphs consisting of only quadrangular faces also have a guard coloring.

Theorem 2. *Every quadrangulation can be guarded by $\lfloor n/3 \rfloor$ edge guards.*

Proof. Let G be a quadrangulation. We show that there is a guard coloring for G, which is sufficient by Lemma 1. Consider the dual graph $G^* = (V^*, E^*)$ of G with its inherited plane embedding, so each vertex $f^* \in V^*$ is placed inside the face f of G corresponding to it. Since every face of G is bounded by a 4-cycle, its dual graph G^* is 4-regular. Using Petersen's 2-Factor Theorem [14][2] we get that G^* contains a 2-factor H (a spanning 2-regular subgraph). Therefore H is a set of vertex-disjoint cycles that might be nested inside each other. Now we define a 2-coloring col : $V \to \{0, 1\}$ for the vertices of G. For each $v \in V$ let c_v be the number of cycles $\mathcal{C}$ of H such that v belongs to the region of the embedding surrounded by $\mathcal{C}$. The color of v is determined by the parity of c_v as col$(v) := c_v$ mod 2.

We claim that this yields a guard coloring of G: Any edge $e = ab \in E$ has a corresponding dual edge e^*. If $e^* \in E(H)$, then e crosses exactly one cycle edge so $|c_a - c_b| = 1$ and therefore col$(a) \neq$ col(b). Otherwise $e \notin E(H)$ and its two endpoints are in the same cycles, thus col$(a) =$ col(b) and e is monochromatic. Because H is a 2-factor, each face has exactly two monochromatic edges.   □

Figure 2 shows an example quadrangulation and a 2-factor in its dual graph. By counting how many cycles each vertex lies inside, the vertices were colored in green and orange to obtain a guard coloring.

In order to bridge the gap between the construction needing $\lfloor (n - 2)/4 \rfloor$ edge guards and the upper bound of $\lfloor n/3 \rfloor$, we also consider the subclass of 2-degenerate quadrangulations in the master's thesis of the first author [10, Theorem 5.9]:

Theorem 3. *Every n-vertex 2-degenerate quadrangulation can be guarded by $\lfloor n/4 \rfloor$ edge guards.*

---

[2] Diestel [6, Corollary 2.1.5] gives a very short and elegant proof of this theorem in his book. He only considers simple graphs there, but all steps in the proof also work for multigraphs like G^* that have at most two edges between any pair of vertices.

Note that this bound is best possible, as the quadrangulations constructed in Theorem 1 are 2-degenerate. The proof of Theorem 3 follows the same lines as the one we present in the following section for stacked triangulations.[3]

3   Stacked Triangulations

The stacked triangulations (also known as Apollonian networks, maximal planar chordal graphs or planar 3-trees) are a subclass of the triangulations that can recursively be formed by the following two rules: (i) A triangle is a stacked triangulation and (ii) if G is a stacked triangulation and $f = (x, y, z)$ an inner face, then the graph obtained by placing a new vertex v into f and connecting it with all three boundary vertices is again a stacked triangulation.

Definition 1. *For a stacked triangulation G we define* height(G) *as*

$$
\text{height}(G) := \begin{cases} 0 & \text{if } |G| = 3 \\ 1 + \max\{\text{height}(G_1), \text{height}(G_2), \text{height}(G_3)\} & \text{otherwise} \end{cases}
$$

where G_1, G_2, G_3 are the stacked triangulations induced by (v, x, y), (v, y, z), (v, z, x) and their interior vertices, respectively.

The stacked triangulations are a non-trivial subclass of the triangulations and we shall prove that they need strictly less than $\lfloor n/3 \rfloor$ edge guards (which is the best known upper bound for general triangulations). To start, we present a family of stacked triangulations needing many edge guards allowing us to conclude that the upper bound presented later is tight.

Theorem 4. *For even $k \in \mathbb{N}$ there is a stacked triangulation G_k with $n = (7k + 4)/2$ vertices needing at least $k = (2n - 4)/7$ edge guards.*

Proof. Let S be a stacked triangulation with k faces and therefore $(k + 4)/2$ vertices (by Euler's formula). Insert three new vertices a_f, b_f, c_f into each face f of S such that the resulting graph is still a stacked triangulation and these three vertices form a new triangular face t_f, i.e. f and t_f do not share any boundary vertices. Figure 3 illustrates how the new vertices can be inserted into a single face f. Then G has $n = (k+4)/2 + 3k = (7k+4)/2$ vertices. For any two distinct faces f, g of S let P be a shortest path between any two boundary vertices of the new faces t_f and t_g. By our construction P has length at least 2, so no edge can guard both t_f and t_g. Therefore G needs at least k edge guards.   □

---

[3] For every n-vertex 2-degenerate quadrangulation G there is an $(n - k)$-vertex 2-degenerate quadrangulation G' ($k \geq 4$), such that an edge guard set Γ' for G' can be used to construct an edge guard set Γ for G with $|\Gamma| = |\Gamma'| + 1$. Obviously different cases need to be considered compared to stacked triangulations and analogous versions of the auxiliary lemmas are needed. But apart from that the proof strategy is the same.

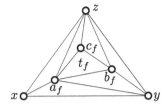

Fig. 3. Three new vertices a_f, b_f, c_f are stacked into face $f = (x, y, z)$ forming a new face t_f. Note that the graph remains a stacked triangulation.

Complementing this construction we state the following upper bound. The remaining part of this section states additional observations and introduces techniques that shall ultimately be combined into a proof.

Theorem 5. *Every n-vertex stacked triangulation with $n \geq 4$ can be guarded by $\lfloor 2n/7 \rfloor$ edge guards.*

Before going into detail, let us start with a high-level description of the proof strategy. We use induction on the number n of vertices. Given a stacked triangulation G we create a smaller stacked triangulation G' of size $|G'| = |G| - k$ for some $k \in \mathbb{N}$. Applying the induction hypothesis on G' yields an edge guard set Γ' of size $|\Gamma'| \leq \lfloor 2(n-k)/7 \rfloor$. Then we extend Γ' into an edge guard set Γ for G using ℓ additional edges. In each step we guarantee $\ell/k \leq 2/7$, such that $|\Gamma| = |\Gamma'| + \ell \leq \lfloor 2(n-k)/7 \rfloor + 2k/7 = \lfloor 2n/7 \rfloor$.

We create G' by choosing a triangle $\triangle$ and removing the set of vertices in its interior. Call this set V^-. Note that G' is still a stacked triangulation. Under all possible candidates we choose $\triangle$ such that V^- is of minimal cardinality but consists of at least four elements. By the choice of $\triangle$ we get that $G[\triangle \cup V^-]$ has at most ten inner vertices: The triangle $\triangle$ consists of three vertices x, y, z and there is a unique vertex v in its interior adjacent to all three of them. The remaining vertices of $G[\triangle \cup V^-]$ are distributed along the three triangles (v, x, y), (v, y, z) and (v, z, x). None of them can contain more than three vertices in its interior, otherwise it would be a triangle $\triangle'$ that would have been chosen instead of $\triangle$.

Now that we have a bound on the size of $G[\triangle \cup V^-]$, we systematically consider edge guard sets for small stacked triangulations.

Observation 1. Let $f = (x, y, z)$ be a face of a stacked triangulation and let Γ be an edge guard set. Now we add a new vertex v into f with edges to all x, y, z.

- If $|V(\Gamma) \cap \{x, y, z\}| \geq 2$, we say that f is *doubly* guarded. In this case the three new faces (v, x, y), (v, y, z) and (v, z, x) are all guarded.
- If $|V(\Gamma) \cap \{x, y, z\}| = 3$, we say that f is *triply* guarded. Furthermore the three new faces (v, x, y), (v, y, z) and (v, z, x) are all doubly guarded.

Observation 2. Let G be a stacked triangulation, and v be a vertex of degree 3 with neighbors x, y, z. Then for any edge guard set Γ we have $|\{v, x, y, z\} \cap V(\Gamma)| \geq 2$. If $v \notin V(\Gamma)$, then at least two of x, y, z must be in $V(\Gamma)$, because each

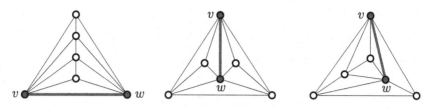

Fig. 4. The three different planar embeddings of the unique 6-vertex stacked triangulation. The thick red edge is the unique edge guard for all eight faces. (Color figure online)

of them is incident to only two of the three faces inside (x, y, z). But if $v \in V(\Gamma)$, it must be as part of an edge with one of its neighbors.

Lemma 2. *Let G be a 6-vertex stacked triangulation. Then G can be guarded by a unique edge guard. Further, if there is a vertex guard at an outer vertex x of G, then there is an edge ab guarding the remaining faces where $a \neq x$ is another outer vertex.*

Proof. There is only a single 6-vertex stacked triangulation and it has three substantially different planar embeddings, all shown in Fig. 4. We see that there is indeed only one possible edge guard vw in all three cases. Further both v and w are adjacent to all three outer vertices (or are one of them and adjacent to the other two). □

Lemma 3. *Let G be a 7-vertex stacked triangulation with a vertex guard at an outer vertex. Then one additional edge suffices to guard the remaining faces of G.*

The proof of Lemma 3 is by complete case enumeration and omitted here for space reasons. We refer the interested reader to the full version [11].
We can already see how Lemma 3 is used in our inductive step, namely in all cases where $|V^-| = 4$: After removing the vertices in V^- the triangle $\triangle$ from G is a face in G' and this face gets guarded by any edge guard set Γ' for G'. Using Lemma 3 we know that one additional edge is always enough to extend Γ' to an edge guard set Γ for G. However, for $|V^-| \geq 5$ the situation gets more complex; just removing V^- and applying the induction hypothesis might lead to an edge guard set Γ' for G' that cannot be extended to an edge guard set Γ for G with $\ell \leq 2k/7$ additional edges (remember that $k = |G| - |G'|$). See Fig. 5 as an example. To solve this we now describe how to extend G' with some new vertices and edges, such that there is always at least some edge guard set Γ' for G' of size $\lfloor 2|G'|/7 \rfloor$ that can be augmented into an edge guard set Γ for G of size $\lfloor 2|G|/7 \rfloor$.

Lemma 4. *Let $f = (x, y, z)$ be a face of a stacked triangulation. By adding two new vertices into f we can obtain a stacked triangulation G such that for each edge guard set Γ there is an edge guard set Γ' of equal size with $\{x, y\} \subseteq V(\Gamma')$.*

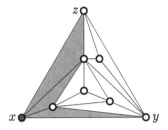

Fig. 5. Here $\triangle = (x, y, z)$ and G' was created by removing all interior vertices V^-. The induction hypothesis then provided an edge guard set Γ' for G' that guards face (x, y, z) of G' through $x \in V(\Gamma')$. After reinserting V^-, the faces shaded in red are already guarded. But none of the other edges is strong enough to guard the remaining faces. (Color figure online)

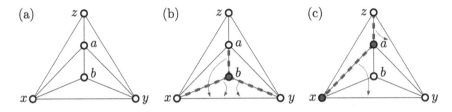

Fig. 6. (a) Add a, b as shown to get $x, y \subseteq V(\Gamma)$. (b) The thick dashed edges are all possible edge guards containing b as an endpoint. The little arrows indicate that each edge can be exchanged with edge xy. (c) The thick dashed edges are all possible edge guards containing a as an endpoint if $b, y \notin V(\Gamma)$. Again the little arrows indicate with which edge these edge guards can be exchanged.

Proof. Add vertex a with edges ax, ay, az and then vertex b with edges ab, bx, by to obtain G as shown in Fig. 6a. Now let Γ be an edge guard set for G with $|\{x, y\} \cap V(\Gamma)| \leq 1$. If $b \in V(\Gamma)$ as part of an edge bv, we can set $\Gamma' := (\Gamma \setminus \{bv\}) \cup \{xy\}$, see Fig. 6b. This is possible, because no matter what vertex v is, edge xy guards a superset of the faces that bv guards. If otherwise $b \notin V(\Gamma)$, we assume without loss of generality that $x \in V(\Gamma)$ so that face (x, y, b) is guarded. Face (a, b, y) can then only be guarded by edge av where $v \in \{x, z\}$. Since $N(a) \subseteq N(y)$ we can set $\Gamma' := (\Gamma \setminus \{av\}) \cup \{vy\}$, see Fig. 6c. In both cases $\{x, y\} \subseteq \Gamma'$ and $|\Gamma| = |\Gamma'|$. $\qquad\square$

With Lemma 4 at hand we can now consider the cases where $|V^-| \geq 5$, i.e. stacked triangulations on eight or more vertices.

Lemma 5. *Let G be an 8-vertex stacked triangulation with outer face (x, y, z), such that the following configuration applies (see Fig. 7a):*

- *Vertex v is the only vertex adjacent to all x, y, z.*
- *(v, x, y) and its interior vertices induce a 6-vertex stacked triangulation G_A.*
- *(v, z, x) and its interior vertex induce a 4-vertex stacked triangulation G_B.*

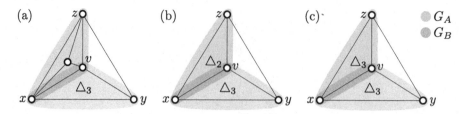

Fig. 7. Configurations as described in (a) Lemma 5, (b) Lemma 6 and (c) Lemma 7. Here $\triangle_k$ is a placeholder for k additional vertices inside the surrounding triangle (such that the graph is a stacked triangulation). The subgraphs induced by the vertices highlighted in green (blue) induce the stacked triangulation G_A (G_B). (Color figure online)

Then any edge guard set Γ' for the subgraph G' induced by $\{v, x, y, z\}$ can be extended by one edge e to an edge guard set Γ for G.

Proof. By Observation 2 we have $|\{v, x, y, z\} \cap V(\Gamma')| \geq 2$. Face (v, y, z) is then already guarded. If further $|\{v, x, z\} \cap V(\Gamma')| \geq 2$, then triangle (v, z, x) is doubly guarded, so all faces of G_B are guarded. In this case set e to be the unique edge guarding G_A, which exists by Lemma 2.

If otherwise $|\{v, x, z\} \cap V(\Gamma')| = 1$, we have $y \in V(\Gamma')$. By Lemma 2 an edge ab exists that guards the remaining faces of G_A with $a \in \{x, v\}$ but also $a \notin V(\Gamma')$. Then G_B is doubly guarded so all of its faces are guarded. ☐

Lemma 6. *Let G be a 9-vertex stacked triangulation with outer face (x, y, z), such that the following configuration applies (see Fig. 7b):*

- *Vertex v is the only vertex adjacent to all x, y, z.*
- *(v, x, y) and its interior vertices induce a 6-vertex stacked triangulation G_A.*
- *(v, z, x) and its interior vertices induce a 5-vertex stacked triangulation G_B.*

Then we can create a 5-vertex stacked triangulation G', such that any edge guard set Γ' for G' can be augmented into an edge guard set Γ for G with $|\Gamma| = |\Gamma'| + 1$.

Proof. Note that the 5-vertex stacked triangulation G_B can always be guarded by one of its outer edges vx, vz or xz. In all cases we first remove the interior vertices of (x, y, z) and add two new vertices a and b into (x, y, z) to get a stacked triangulation G'. By placing a and b appropriately, Lemma 4 allows us to force one of the three sets $\{x, y\}$, $\{x, z\}$, $\{y, z\}$ to be a subset of $V(\Gamma')$. Depending on which edge from $\{vx, vz, xz\}$ guards G_B, we force a different one of the three sets. In the following let $e = uw$ be the unique edge guarding G_A such that $u \in \{v, x, y\}$ by Lemma 2.

Case 1: xz guards G_B:
Place a and b such that $x, z \in V(\Gamma')$. Then all faces of G_B and face (v, y, z) are guarded. We set $\Gamma := \Gamma' \cup \{e\}$ to also guard all faces of G_A.

Case 2: xv guards G_B:

Place a and b such that $x, y \in V(\Gamma')$. Then face (v, y, z) is already guarded. If $u = v$, set $\Gamma := \Gamma' \cup \{e\}$ to guard all faces of G_B and G_A. Otherwise we can use $e' := vw$ instead by Lemma 2 and set $\Gamma := \Gamma' \cup \{e'\}$.

Case 3: vz guards G_B:

If $u = v$, we can place a and b to force $y, z \in V(\Gamma')$. Then $\Gamma := \Gamma' \cup \{e\}$ guards all faces of G_A and G_B. Otherwise, place a and b so that $u, z \in V(\Gamma')$ and set $e' := vw$ by Lemma 2. Then $\Gamma := \Gamma' \cup \{e'\}$ fulfills the requirements. $\square$

Lemma 7. *Let G be a 10-vertex stacked triangulation with outer face (x, y, z), such that the following configuration applies (see Fig. 7c):*

- *Vertex $v \in V$ is the unique vertex adjacent to all x, y and z.*
- *(v, x, y) and its interior vertices induce a 6-vertex stacked triangulation G_A.*
- *(v, z, x) and its interior vertices induce a 6-vertex stacked triangulation G_B.*

Then we can create a 6-vertex stacked triangulation G', such that any edge guard set Γ' for G' can be augmented to an edge guard set Γ for G with $|\Gamma| = |\Gamma'| + 1$.

The proof of Lemma 7 follows the same lines as the ones for Lemma 5 and Lemma 6. It is omitted here for space reasons and can be found in the full version [11]. Finally we are set up to prove Theorem 5 stating that $\lfloor 2n/7 \rfloor$ edge guards are always sufficient for any n-vertex stacked triangulation G.

Proof of Theorem 5. As described above, the proof is by induction on the number n of vertices. We find a smaller graph G' for which the induction hypothesis provides an edge guard set Γ' that we augment into an edge guard set Γ for G. By guaranteeing that $(|\Gamma| - |\Gamma'|)/(|G| - |G'|) \leq 2/7$ we hereby obtain an edge guard set for G of size at most $\lfloor 2n/7 \rfloor$.

For base case we note that if $n \leq 6$, we need a single edge guard by Lemma 2. So from now on assume $n \geq 7$. Let $\triangle = (x, y, z)$ be a triangle such that there are at least four vertices V^- inside $\triangle$ but among all candidates $|V^-|$ is minimal. Further let $v \in V^-$ be the unique vertex adjacent to all x, y, z. We consider the following cases in the order they are given: If a case applies, then all others before must not apply.

Case 1: $|V^-| = 4$:

Set $G' := G[V \setminus V^-]$ and use the induction hypothesis to get an arbitrary edge guard set Γ' for G'. Triangle $\triangle$ is a face in G' and as such guarded by Γ' through at least one of its boundary vertices. Together with the vertices in V^- it forms a 7-vertex stacked triangulation with at least one guarded outer vertex, so by Lemma 3 we can extend Γ' by one additional edge to an edge guard set Γ for G. We get $k = |G| - |G'| = 4$ and $\ell = 1$, so $\ell/k = 1/4 \leq 2/7$.

Case 2: $|V^-| \geq 5 \wedge \text{height}(G[\triangle \cup V^-]) \leq 3$:

Construct G' by removing all vertices from V^- except for v and let Γ' be an edge guard set for G' given by the induction hypothesis. By Observation 2 we have $|\{v, x, y, z\} \cap V(\Gamma')| \geq 2$ and we set Γ to be Γ' plus one additional edge, so that $\{v, x, y, z\} \subseteq V(\Gamma)$. This is always possible, because v, x, y, z induce a 4-clique in G. Because height$(G[\triangle \cup V^-]) \leq 3$, each face of G inside $\triangle$ is incident to at least one vertex in $\{v, x, y, z\}$, so all faces are guarded. We get $k = |G| - |G'| \geq 4$ and $\ell = 1$, so $\ell/k \leq 1/4 \leq 2/7$.

At this stage we finished all cases where height$(G[\triangle \cup V^-]) \leq 3$. The following cases all have height$(G[\triangle \cup V^-]) = 4$. (Note *equality* instead of *greater than or equal*. This is justified, because if one of the three triangles (v, x, y), (v, y, z) or (v, z, x) has height at least five, it would contain at least four vertices in its interior. This is impossible as $\triangle$ has a minimal number of vertices in its interior.)

Case 3: height$(G[\triangle \cup V^-]) = 4 \wedge [(v, x, y), (v, y, z)$ or (v, z, x) is a face]:
Without loss of generality we assume that (v, y, z) is a face. The other cases are symmetric. If $|V^-| = 5/6/7$, then $G[\triangle \cup V^-]$ induces an 8/9/10-vertex stacked triangulation fulfilling the conditions of Lemma 5/6/7, respectively. In all three cases, the lemma describes how G' is constructed and how an edge guard set Γ' obtained by applying the induction hypothesis can be extended to Γ. We always have $k = |G| - |G'| \geq 4$ and $\ell = 1$, so $\ell/k \leq 1/4 \leq 2/7$.

Case 4: height$(G[\triangle \cup V^-]) = 4$:
Partition V^- into $V^- = \{v\} \cup V_1^- \cup V_2^- \cup V_3^-$, where V_1^-, V_2^-, V_3^- are the vertices in the interior of (v, x, y), (v, y, z) and (v, z, x), respectively. At least one of them has cardinality three, because height$(G[\triangle \cup V^-]) = 4$. We assume without loss of generality that $|V_2^-| = 3$.

Assume first that $|V^-| \geq 7$. Remove V_2^- from G to get a graph $\widetilde{G}$. Then $\widetilde{G}$ fulfills the condition of either Case 1, Case 2 or Case 3 and can be treated as described there. In the corresponding case another $\widetilde{k} \geq 4$ vertices are removed from $\widetilde{G}$ to get G' and $\widetilde{\ell} = 1$ extra edge is needed to extend an edge guard set Γ' for G' to an edge guard set $\widetilde{\Gamma}$ for $\widetilde{G}$. After reinserting the vertices in V_2^- we need only one extra edge to extend $\widetilde{\Gamma}$ to an edge guard set Γ for G by Lemma 2, because $G[\{v, y, z\} \cup V_2^-]$ is a 6-vertex stacked triangulation. In total we get $k = \widetilde{k} + 3 \geq 7$ and $\ell = \widetilde{\ell} + 1 = 2$, so $\ell/k \leq 2/7$.

Now assume that $|V^-| \leq 6$. Since $|V_2^-| = 3$ it must be $|V_1^-| = |V_3^-| = 1$. Remove all vertices in V^- and add two new vertices using Lemma 4 to get a graph G', such that there is an edge guard set Γ' for G' with $y, z \in V(\Gamma')$. By Lemma 2 there is another edge e containing v as an endpoint such that all faces of $G[\{v, y, z\} \cup V_2^-]$ are guarded. Then $\Gamma := \Gamma' \cup \{e\}$ also doubly guards triangles (v, x, y) and (v, z, x), so all faces inside triangle (x, y, z) are guarded. In this case we get $k = 6 - 2 = 4$ and $\ell = 1$, so $\ell/k = 1/4 \leq 2/7$. $\square$

4   Conclusion and Open Problems

We proved new bounds on the size of edge guard sets for stacked triangulations and quadrangulations. Considering quadrangulations was motivated by the fact that previous coloring-based approaches for general plane graphs failed on quadrangular faces. Our upper bound of $\lfloor n/3 \rfloor$ as well as work from Biniaz et al. [1] about quadrangular faces that are far apart from each other suggests that the difficulty is not due to the quadrangular faces themselves. Instead the currently known methods seem to be not strong enough to capture the complexity introduced by a mix of quadrangular and non-quadrangular faces. Finding tight bounds remains an open question, as our construction needs only $\lfloor (n-2)/4 \rfloor$ edge guards. We proved that this is best possible for 2-degenerate quadrangulations and verified exhaustively and computer assisted that $\lfloor n/4 \rfloor$ is an upper bound for all quadrangulations with $n \leq 23$ (master's thesis of the first author [10]).

For stacked triangulations we proved tight bounds of $\lfloor 2n/7 \rfloor$. By this we identified a non-trivial subclass of triangulations needing strictly less than $\lfloor n/3 \rfloor$ edge guards. We hope that this can be used to improve the upper bound for general triangulations, for example by combining it with bounds for 4-connected triangulations along the lines of [9]. Lastly we want to highlight the open problem for general graphs, namely: Can every n-vertex plane graph be guarded by $\lfloor n/3 \rfloor$ edge guards?

Acknowledgements. We thank Kolja Knauer and Lukas Barth for interesting discussions on the topic.

References

1. Biniaz, A., Bose, P., Ooms, A., Verdonschot, S.: Improved bounds for guarding plane graphs with edges. Graphs and Combinatorics **35**(2), 437–450 (2019). https://doi.org/10.1007/s00373-018-02004-z
2. Bose, P.: A note on the lower bound of edge guards of polyhedral terrains. Int. J. Comput. Math. **86**(4), 577–583 (2009)
3. Bose, P., Kirkpatrick, D., Li, Z.: Worst-case-optimal algorithms for guarding planar graphs and polyhedral surfaces. Comput. Geom. **26**(3), 209–219 (2003). https://doi.org/10.1016/S0925-7721(03)00027-0
4. Bose, P., Shermer, T., Toussaint, G., Zhu, B.: Guarding polyhedral terrains. Comput. Geom. **7**(3), 173–185 (1997). https://doi.org/10.1016/0925-7721(95)00034-8
5. Chvátal, V.: A combinatorial theorem in plane geometry. J. Comb. Theory Ser. B **18**(1), 39–41 (1975)
6. Diestel, R.: Graph Theory. GTM, vol. 173, 5th edn. Springer, Heidelberg (2017). https://doi.org/10.1007/978-3-662-53622-3
7. Everett, H., Rivera-Campo, E.: Edge guarding polyhedral terrains. Comput. Geom. **7**(3), 201–203 (1997). https://doi.org/10.1016/0925-7721(95)00051-8
8. Fisk, S.: A short proof of Chvátal's Watchman theorem. J. Comb. Theory Ser. B **24**, 374 (1978). https://doi.org/10.1016/0095-8956(78)90059-X
9. Heldt, D., Knauer, K., Ueckerdt, T.: On the bend-number of planar and outerplanar graphs. Discrete Appl. Math. **179**, 109–119 (2014). https://doi.org/10.1016/j.dam.2014.07.015
10. Jungeblut, P.: Edge guarding plane graphs. Master's thesis, Karlsruhe Institute of Technology (2019). https://i11www.iti.kit.edu/extra/publications/j-egpg-19.pdf
11. Jungeblut, P., Ueckerdt, T.: Guarding quadrangulations and stacked triangulations with edges (2020). https://arxiv.org/pdf/2006.13722.pdf

12. Kaučič, B., Žalik, B., Novak, F.: On the lower bound of edge guards of polyhedral terrains. Int. J. Comput. Math. **80**(7), 811–814 (2003)
13. O'Rourke, J.: Galleries need fewer mobile guards: a variation on Chvátal's theorem. Geom. Dedicata. **14**(3), 273–283 (1983). https://doi.org/10.1007/BF00146907
14. Petersen, J.: Die Theorie der regulären graphs. Acta Mathematica **15**(1), 193–220 (1891)

Hamiltonian Cycle Parameterized by Treedepth in Single Exponential Time and Polynomial Space

Jesper Nederlof[1], Michał Pilipczuk[2], Céline M. F. Swennenhuis[3], and Karol Węgrzycki[2(✉)]

[1] Utrecht University, Utrecht, The Netherlands
j.nederlof@uu.nl
[2] Institute of Informatics, University of Warsaw, Warsaw, Poland
{michal.pilipczuk,k.wegrzycki}@mimuw.edu.pl
[3] Eindhoven University of Technology, Eindhoven, The Netherlands
c.m.f.swennenhuis@tue.nl

Abstract. For many algorithmic problems on graphs of treewidth t, a standard dynamic programming approach gives an algorithm with time and space complexity $2^{\mathcal{O}(t)} \cdot n^{\mathcal{O}(1)}$. It turns out that when one considers the more restrictive parameter treedepth, it is often the case that a variation of this technique can be used to reduce the space complexity to polynomial, while retaining time complexity of the form $2^{\mathcal{O}(d)} \cdot n^{\mathcal{O}(1)}$, where d is the treedepth. This transfer of methodology is, however, far from automatic. For instance, for problems with connectivity constraints, standard dynamic programming techniques give algorithms with time and space complexity $2^{\mathcal{O}(t \log t)} \cdot n^{\mathcal{O}(1)}$ on graphs of treewidth t, but it is not clear how to convert them into time-efficient polynomial space algorithms for graphs of low treedepth.

Cygan et al. (FOCS'11) introduced the Cut&Count technique and showed that a certain class of problems with connectivity constraints can be solved in time and space complexity $2^{\mathcal{O}(t)} \cdot n^{\mathcal{O}(1)}$. Recently, Hegerfeld and Kratsch (STACS'20) showed that, for some of those problems, the Cut&Count technique can be also applied in the setting of treedepth, and it gives algorithms with running time $2^{\mathcal{O}(d)} \cdot n^{\mathcal{O}(1)}$ and polynomial space usage. However, a number of important problems eluded such a treatment, with the most prominent examples being HAMILTONIAN CYCLE and LONGEST PATH.

In this paper we clarify the situation by showing that HAMILTONIAN CYCLE, HAMILTONIAN PATH, LONG CYCLE, LONG PATH, and MIN CYCLE COVER all admit $5^d \cdot n^{\mathcal{O}(1)}$-time and polynomial space algorithms on graphs of treedepth d. The algorithms are randomized Monte Carlo with only false negatives.

Keywords: Hamiltonian cycle · Connectivity · Polynomial space · Treedepth

© Springer Nature Switzerland AG 2020
I. Adler and H. Müller (Eds.): WG 2020, LNCS 12301, pp. 27–39, 2020.
https://doi.org/10.1007/978-3-030-60440-0_3

1   Introduction

It is widely believed that no NP-hard problem admits a polynomial time algorithm. However, actual instances of problems that we are interested in solving often admit much more structure than a general instance. This observation gave rise to the field of *parameterized complexity*, where the hardness of an instance does not depend exclusively on the input size. In the parameterized regime, we assume that each instance is equipped with an additional parameter k and the goal is to give a *fixed-parameter algorithm*: an algorithm with running time $f(k) \cdot n^{\mathcal{O}(1)}$, where f is a function independent of n. After settling that a problem admits such an algorithm, it is natural to lookfor one with function f as low as possible. We refer to [5, 10, 12] for an introduction to parameterized complexity.

One of the most widely used parameters is the *treewidth* t of the input graph. Usually, problems that involve only constraints of local nature admit an algorithm with running time of the form $2^{\mathcal{O}(t)} \cdot n^{\mathcal{O}(1)}$ [5]. For a long time, such algorithms remained out of reach for problems involving connectivity constraints, and for those only $2^{\mathcal{O}(t \log t)} \cdot n^{\mathcal{O}(1)}$-time algorithms were known. The breakthrough came with the Cut&Count technique, introduced by Cygan et al. in [7], that allows one to design randomized Monte-Carlo algorithms with running times of the form $2^{\mathcal{O}(t)} \cdot n^{\mathcal{O}(1)}$ for a wide range of connectivity problems, e.g., HAMILTONIAN PATH, CONNECTED VERTEX COVER, CONNECTED DOMINATING SET, etc. The technique was subsequently derandomized [4, 13].

One of the main issues with standard dynamic programming algorithms is that they tend to have prohibitively large space usage. The natural goal is therefore to reduce the space complexity while not sacrificing much on the time complexity. Unfortunately, Drucker et al. [11] and Pilipczuk and Wrochna [23] gave some complexity-theoretical evidence that for dynamic programming on graphs of bounded treewidth, such a reduction is probably impossible. For example, they showed that under plausible assumptions, there is no algorithm that works in time $2^{\mathcal{O}(t)} \cdot n^{\mathcal{O}(1)}$ and uses $2^{o(t)} \cdot n^{\mathcal{O}(1)}$ space for the 3-COLORING or INDEPENDENT SET problem.

Treedepth. The aforementioned issues motivate the research on a different, more restrictive parameterization, for which the reduction of space complexity would be possible. In this paper we will consider the parameterization by *treedepth*, defined as follows.

Definition 1. An *elimination forest* of a graph G is a rooted forest F on the same vertex set as G such that for every edge uv of G, either u is an ancestor of v in F or v is an ancestor of u in F. The *treedepth* of G is the minimum possible depth of an elimination forest of G.

The treedepth of a graph is never smaller than its treewidth, but it is also never larger than the treewidth times $\log n$. In many concrete cases, the two parameters have the same advantages. For example, planar graphs have treewidth $\mathcal{O}(\sqrt{n})$, but also *treedepth* $\mathcal{O}(\sqrt{n})$.

It has been recently realized that on graphs of treedepth d, many algorithmic problems indeed can be solved in time $2^{\mathcal{O}(d)} \cdot n^{\mathcal{O}(1)}$ and using only polynomial space.[1] For the most basic problems, such as 3-COLORING and INDEPENDENT SET, a simple branching algorithms achieves such complexity. However, in contrast to the treewidth parameterization, for many more complex problems it is highly non-trivial, yet possible to establish similar bounds. One technique that turns out to be useful here is the framework of algebraic transforms introduced by Loksthanov and Nederlof [18], who demonstrated how to reduce the space requirements of many dynamic programming algorithms to polynomial in the input size by reorganizing the computation using a suitable transform. Fürer and Yu [14] applied this framework to give $2^{\mathcal{O}(d)} \cdot n^{\mathcal{O}(1)}$-time and polynomial space algorithms on graphs of treedepth d for the DOMINATING SET problem and for the problem of counting the number of perfect matchings. Pilipczuk and Wrochna [23] considered algorithms with even more restricted space requirements: they showed that 3-COLORING, DOMINATING SET, and VERTEX COVER admit algorithms that work in $2^{\mathcal{O}(d)} \cdot n^{\mathcal{O}(1)}$ time and use $\mathcal{O}(d + \log n)$ space. For DOMINATING SET they avoided the explicit use of algebraization and instead provided a more combinatorial interpretation based on what one could call *inclusion-exclusion branching*. Later, Pilipczuk and Siebertz [22] used color-coding to give an $2^{\mathcal{O}(d \log d)} \cdot n^{\mathcal{O}(1)}$-time and polynomial space algorithm for the SUBGRAPH ISOMORPHISM problem. Recently, Belbasi and Fürer [1] presented an algorithm for counting Hamiltonian cycles in time $(4t)^d \cdot n^{\mathcal{O}(1)}$ and using polynomial space, where t is the width of a given tree decomposition and d is its (suitably defined) depth.

Treedepth and Cut&Count. Very recently, Hegerfeld and Kratsch [16] demonstrated that the Cut&Count technique can be also applied in the setting of the treedepth parameterization. Consequently, they gave randomized algorithms with running times $2^{\mathcal{O}(d)} \cdot n^{\mathcal{O}(1)}$ and polynomial space usage for a number of problems with connectivity constraints such as CONNECTED VERTEX COVER, CONNECTED DOMINATING SET, FEEDBACK VERTEX SET, or STEINER TREE. However, Hegerfeld and Kratsch found it problematic to apply the methodology to several important problems originally considered by Cygan et al. [7] in the context of Cut&Count. Specifically, these are problems based on selection of edges rather than vertices, such as HAMILTONIAN CYCLE or LONG CYCLE. For this reason, Hegerfeld and Kratsch explicitly asked in [16] whether HAMILTONIAN CYCLE, HAMILTONIAN PATH, LONG CYCLE, and MIN CYCLE COVER

---

[1] Throughout the introduction, when we speak about a graph of treedepth d, we mean a graph supplied with an elimination forest of depth d. While in the case of treewidth, a tree decomposition of approximately (up to a constant factor) optimum width can be computed in time $8^t \cdot n^{\mathcal{O}(1)}$ [5,26], the existence of such an approximation algorithm for treedepth is a notorious open problem.

also admit $2^{\mathcal{O}(d)} \cdot n^{\mathcal{O}(1)}$-time and polynomial space algorithms[2] on graphs of treedepth d (see the full version of this paper [21] for problem definitions[3]).

Our Contribution. In this paper we introduce additional techniques that allow us to extend the results of [16] and to answer the abovementioned open problem of Hegerfeld and Kratsch in the affirmative. More precisely, we prove the following theorem.

Theorem 1. *There is a randomized algorithm that given a graph G together with its elimination forest of depth d, and number $k \in \mathbb{N}$, solves* HAMILTONIAN CYCLE, HAMILTONIAN PATH, k-CYCLE, k-PATH *and* MIN CYCLE COVER *in time $5^d \cdot n^{\mathcal{O}(1)}$ and using polynomial space. The algorithm has a one-sided error: it may give false negatives with probability at most $\frac{1}{2}$.*

In fact, Theorem 1 is an easy corollary of the following result for a generalization of the considered problems. In the PARTIAL CYCLE COVER problem we are given an undirected graph G and integers k and ℓ, and we ask whether in G there is a family of at most k vertex-disjoint cycles that jointly visit exactly ℓ vertices. We will prove the following theorem.

Theorem 2. *There is a randomized algorithm that given a graph G together with its elimination forest of depth d, and numbers $k, \ell \in \mathbb{N}$, solves the* PARTIAL CYCLE COVER *problem for G, k, ℓ in time $5^d \cdot n^{\mathcal{O}(1)}$ and using polynomial space. The algorithm has a one-sided error: it may give false negatives with probability at most $\frac{1}{2}$.*

To see that Theorem 2 implies Theorem 1, note that HAMILTONIAN CYCLE, MIN CYCLE COVER and LONG CYCLE are special cases of the PARTIAL CYCLE COVER (for fixed parameters k and ℓ).

To solve LONG PATH, we can simply iterate through all pairs of non-adjacent vertices s, t and apply the LONG CYCLE algorithm to the graph G with edge st added; this increases the treedepth by at most 1 and the provided elimination forest can be easily adjusted. It is easy to see that then the original graph G contains a simple path on ℓ vertices if and only if for some choice of s and t, we find a cycle of length ℓ in G augmented with the edge st. Finally, HAMILTONIAN PATH is just LONG PATH applied for $\ell = |V(G)|$.

We remark that our algorithmic findings have concrete applications outside of the realm of structural parameterizations. For instance, Lokshtanov et al. [17] gave a $2^{\mathcal{O}(\sqrt{\ell} \log^2 \ell)} \cdot n^{\mathcal{O}(1)}$-time polynomial space algorithm for the LONG PATH problem on H-minor-free graphs, for every fixed H. In the full version of this paper [21] we present how using our results one can improve the running time to $2^{\mathcal{O}(\sqrt{\ell} \log \ell)} \cdot n^{\mathcal{O}(1)}$ while keeping the polynomial space complexity.

---

[2] Note, that graphs of treedepth at most k cannot contain a path of length 2^k. This leads to trivial FPT algorithms for these problems, however with doubly-exponential running time dependency on k.

[3] Note that when discussing the LONG PATH and the LONG CYCLE problems, we use the letter ℓ to denote the required length of a path, respectively of a cycle, instead of the letter k that is perhaps more traditionally used in this context.

Our Techniques. Similarly to Hegerfeld and Kratsch [16] we use the Cut&Count framework, but we apply a different view on the Count part, suited for problems based on edge selection. The main idea is that instead of counting cycle covers, as a standard application of Cut&Count would do, we count perfect matchings in an auxiliary graph, constructed by replacing every vertex with two adjacent copies. The number of such perfect matchings can be related to the number of cycle covers of the original graph. However, the considered perfect matchings can be counted within the claimed complexity by either employing the previous "algebraized" dynamic programming algorithm, or the algorithm based on inclusion-exclusion branching (our presentation chooses the latter).

Applying this approach naïvely would give us a polynomial space algorithm with running time $8^d \cdot n^{\mathcal{O}(1)}$. We improve the running time to $5^d \cdot n^{\mathcal{O}(1)}$ by employing several observations about the symmetries of recursive calls of our algorithms, in a similar way as in the algorithm for #k-MULTI-SET-COVER of Nederlof [20].

Organization of the Paper. The remainder of the paper is devoted to the proof of Theorem 2. In Sect. 2 we introduce the notation and present basic definitions. In Sect. 3 we discuss the Cut&Count technique in a self-contained manner and explain the Cut part. In Sect. 4 we reduce the Count part to counting perfect matchings in an auxiliary graph. In Sect. 5 we give an intuition behind counting such matchings. We conclude with several open questions in Sect. 6. Due to space restrictions, proofs of statements marked with $\Diamond$ are deferred to the full version of this paper [21]. The full version [21] also contains applications of our results to the LONG PATH problem in H-minor free graphs.

2   Preliminaries

Notation. For a graph G, by $\mathsf{cc}(G)$ we denote number of connected components of G. Let F be a subset of edges of G. By $\mathsf{cc}(F)$ we denote the number of connected components of the graph consisting of all the edges of F and vertices incident to them. For a vertex u, by $\deg_F(u)$ we mean the number of edges of F incident to u. Then F is a *matching* if $\deg_F(u) \in \{0,1\}$ for every vertex u, is a *perfect matching* if $\deg_F(u) = 1$ for every vertex u, and is a *partial cycle cover* if $\deg_F(u) \in \{0,2\}$ for every vertex u. Note that thus we treat partial cycle covers as sets of edges.

A *cut* of a set U is just an ordered partition of U into two sets, that is, a pair (L, R) such that $L \cap R = \emptyset$ and $L \cup R = U$. A cut (L, R) of the vertex set of a graph is *consistent* with a subset of edges F if there is no edge in F with one endpoint in L and second in R.

For a function f and elements x, y, where x is not in the domain of f, by $f[x \mapsto y]$ we denote the function obtained from f by extending its domain by x and setting $f(x) = y$.

We use the $\mathcal{O}^\star(\cdot)$ notation to hide factors polynomial in the input size. For convenience, throughout the paper we assume the RAM model: every integer takes a unit of space and arithmetic operations on integers have unit cost. However, it can be easily seen that all the numbers appearing during the computation have bit length bounded polynomially in the input size. Since we never specify the polynomial factors in the time or space complexity of our algorithms, without any influence on the claimed asymptotic bounds we may assume that the representation of any number takes polynomial space and arithmetic operations on the numbers take polynomial time.

Treedepth. A *rooted forest* is a directed acyclic graph T where every vertex has outdegree at most 1. The vertices of outdegree 0 in T are called the *roots*. Whenever a vertex u is reachable from a vertex v by a directed path in T, we say that u is an *ancestor* of v, and v is a *descendant* of u. Note that every vertex is its own ancestor as well as descendant. The *depth* of a rooted forest is the maximum number of vertices that can appear on a directed path in it.

We use the following notation from previous works [16,23]. For a vertex u of a rooted forest T, we denote:

$$\mathsf{subtree}[u] := \{v \colon u \text{ is ancestor of } v\}, \qquad \mathsf{subtree}(u) := \mathsf{subtree}[u] \setminus \{u\},$$
$$\mathsf{tail}[u] := \{v \colon v \text{ is ancestor of } u\}, \qquad \mathsf{tail}(u) := \mathsf{tail}[u] \setminus \{u\},$$
$$\mathsf{broom}[u] := \mathsf{tail}[u] \cup \mathsf{subtree}[u].$$

Additionally, $\mathsf{children}(u)$ denotes the set of children of u, whereas $\mathsf{parent}(u)$ is the *parent* of u, that is, the only outneighbor of u. If u is a root, we set $\mathsf{parent}(u) = \bot$.

For a graph G, an *elimination forest* of G is a rooted forest T on the same vertex set as G that satisfies the following property: whenever uv is an edge in G, then in T either u is an ancestor of v, or v is an ancestor of u. The *treedepth* of a graph is the minimum possible depth of an elimination forest of G.

Isolation Lemma. The only source of randomness in our algorithm is the Isolation Lemma of Mulmuley et al. [19]. Suppose U is a finite set and $\omega \colon U \to \mathbb{Z}$ is a weight function on U. We say that ω *isolates* a non-empty family of subsets $\mathcal{F} \subseteq 2^U$ if there is a unique $S \in \mathcal{F}$ such that

$$\omega(S) = \min_{X \in \mathcal{F}} \omega(X),$$

where $\omega(X) := \sum_{x \in X} \omega(x)$. Then the Isolation Lemma can be stated as follows.

Lemma 1 (Isolation Lemma [19]). *Let U be a finite set and $\mathcal{F} \subseteq 2^U$ be a non-empty family of subsets of U. Suppose for every $u \in U$ we choose its weight $\omega(u)$ uniformly and independently at random from the set $\{1, \ldots, N\}$, where $N \in \mathbb{N}$. Then ω isolates $\mathcal{F}$ with probability at least $1 - \frac{|U|}{N}$.*

3   The Cut Part

We now proceed to the proof of Theorem 2. Throughout the proof we fix the input graph $G = (V, E)$, its elimination forest T of depth d, and numbers $k, \ell \in \mathbb{N}$. We may assume that G is connected, as otherwise we may apply the algorithm to each connected component separately. Thus T has to be a tree, so we will call it an *elimination tree* to avoid confusion. Also, we denote $n := |V|$.

As mentioned before, we shall apply the Cut&Count technique of Cygan et al. [7]. This technique consists of two parts: the Cut part and the Count part. The idea is that in the first part, we relax the connectivity requirements and show that it is enough to count the number of relaxed solutions together with cuts consistent with them, as this number is congruent to the number of non-relaxed solutions modulo a power of 2. The Isolation Lemma is used here to ensure that with high probability, the number of solutions does not accidentally cancel out modulo this power of 2. More precisely, having drawn a weight function at random, for each possible total weight w we count the number of solutions of total weight w. Then the Isolation Lemma asserts that, with high probability, for some w there will be a unique solution of total weight w. Then comes the Count part, where the goal is to efficiently count the number of relaxed solutions together with cuts consistent with them.

We refer the reader to [7] for a more elaborate discussion of the Cut&Count technique, while now we apply it to the particular case of PARTIAL CYCLE COVER. A relaxed solution is just a partial cycle cover consisting of ℓ edges. Then a solution is a relaxed solution that spans at most k cycles. Formally, the sets of *solutions* ($\mathcal{S}$) and *relaxed solutions* ($\mathcal{R}$) are defined as follows:

$$\mathcal{R} := \{ F \subseteq E : |F| = \ell \text{ and } \deg_F(u) \subseteq \{0, 2\} \text{ for every } u \in V \};$$
$$\mathcal{S} := \{ F \in \mathcal{R} : \mathsf{cc}(F) \leqslant k \}.$$

Suppose now that the input graph G is supplied with a weight function on edges $\omega \colon E \to \mathbb{Z}$. Then we can stratify the families above using the total weight. That is, for every $w \in \mathbb{Z}$ we define:

$$\mathcal{R}_w := \{ F \in \mathcal{R} : \omega(F) = w \} \quad \text{and} \quad \mathcal{S}_w := \{ F \in \mathcal{S} : \omega(F) = w \}.$$

Now, let

$$\mathcal{C}_w := \{ (F, (L, R)) : F \in \mathcal{R}_w \text{ and } (L, R) \text{ is a cut of } V \text{ consistent with } F \}.$$

The following observation is the key idea in the Cut&Count technique.

Lemma 2 ($\Diamond$). *For every $w \in \mathbb{Z}$, we have*

$$|\mathcal{C}_w| \equiv \sum_{F \in \mathcal{S}_w} 2^{n - \ell + \mathsf{cc}(F)} \mod 2^{n - \ell + k + 1}.$$

In the next sections we will present the Count part of the technique, which boils down to proving the following lemma.

Lemma 3. *Given $w \in \mathbb{Z}$ and a weight function $\omega \colon E \to \{1, \ldots, N\}$, where $N = \mathcal{O}^\star(1)$, the number $|\mathcal{C}_w|$ can be computed in time $\mathcal{O}^\star(5^d)$ and space $\mathcal{O}^\star(1)$.*

In the full version of this paper [21] we show how to combine Lemma 2 with Lemma 3 to prove Theorem 2. Therefore, it remains to prove Lemma 3.

4   From Cycle Covers to Matchings

For the proof of Lemma 3, instead of counting the number of suitable partial cycle covers, we find it more convenient to count the number of perfect matchings in an auxiliary graph. Note, that this concept is natural when using *inclusion-exclusion branching* technique. A similar auxiliary graph arises in the algorithm for $\#k$-MULTI-SET-COVER [20].

We define a graph G' as follows. The vertex set V' of G' is $V' := \{u^0, u^1 : u \in V\}$. That is, we put two copies of each vertex of G into the vertex set of G'. The edge set E' of G' is the union of the following two sets:

$$E_0' := \{u^0 u^1 : u \in V\}, \qquad E_1' := \{u^0 v^0, u^0 v^1, u^1 v^0, u^1 v^1 : uv \in E\}.$$

In other words, for every vertex $u \in V$ we put an edge in E_0' connecting the two copies of u in V', while for every edge $uv \in E$ we put four different edges in E_1', each connecting a copy of u with a copy of v in V'.

Let $\pi : E_1' \to E$ be the natural projection from E_1' to E: for each $uv \in E$ and $s, t \in \{0, 1\}$, we set $\pi(u^s v^t) = uv$. We extend the mapping π to all subsets $F \subseteq E'$ by setting $\pi(F) := \pi(F \cap E_1')$. We also extend the weight function ω to the edges of E' by putting $\omega(e) = 0$ for each $e \in E_0'$ and $\omega(e) = \omega(\pi(e))$ for each $e \in E_1'$.

A set of edges F in G' shall be called *simple* if for every $e \in E$, we have

$$|F \cap \pi^{-1}(e)| \leqslant 1.$$

For now, we mainly focus on simple perfect matchings in G'. We observe that they are in correspondence with partial cycle covers in G, as explained next.

Lemma 4 ($\diamondsuit$). *For every simple perfect matching M in G', the set $\pi(M)$ is a partial cycle cover in G of size $|M \cap E_1'|$. Moreover, for every partial cycle cover F in G, there are exactly $2^{|F|}$ simple perfect matchings M in G' for which $F = \pi(M)$.*

Lemma 4 motivates introducing the following analogues of the sets $\mathcal{C}_w$. For $w \in \mathbb{Z}$, we define

$$\mathcal{M}_w \quad := \{(M, (L, R)) : M \text{ is a simple perfect matching in } G',$$
$$(L, R) \text{ is a cut of } V \text{ consistent with } \pi(M),$$
$$|M \cap E_1'| = \ell \text{ and } \omega(M) = w\}.$$

Since for every simple perfect matching M in G' we have $\omega(M) = \omega(\pi(M))$, from Lemma 4 we immediately obtain the following.

Corollary 1. *For every $w \in \mathbb{Z}$, we have $|\mathcal{M}_w| = 2^\ell \cdot |\mathcal{C}_w|$.*

Therefore, to prove Lemma 3 it suffices to apply the algorithm provided by the following lemma and divide the outcome by 2^ℓ.

Lemma 5 ($\diamondsuit$). *Given $w \in \mathbb{Z}$ and a weight function $\omega : E \to \{1, \ldots, N\}$, where $N = \mathcal{O}^\star(1)$, the number $|\mathcal{M}_w|$ can be computed in time $\mathcal{O}^\star(5^d)$ and space $\mathcal{O}^\star(1)$.*

5   The Count Part

Due to space restrictions we defer the formal proof of Lemma 5 to the full version of this paper [21]. In this section we discuss only the intuition behind the approach.

The basic idea is that we will compute the number $|\mathcal{M}_w|$ using bottom-up dynamic programming over the given elimination tree T. In order to achieve polynomial space complexity, this dynamic programming will be cast as a standard recursion, but for this to work, we need that the recurrence equations governing the dynamic programming have a specific form. In essence, whenever we compute an entry of the dynamic programming table at some vertex u, the value should be obtained as a simple aggregation of single entries from the tables of the children of u. The most straightforward approach to computing $|\mathcal{M}_w|$ would be to count partial perfect matchings and to remember, in the states corresponding to u, subsets of $\mathsf{tail}[u]$ consisting of vertices matched to $\mathsf{subtree}(u)$. This would yield a dynamic programming algorithm that is *not* of the form required for the space complexity reduction. However, we show that by counting different objects than partial perfect matchings, and using the inclusion-exclusion principle at every computation step, we can reorganize the computation so that the space reduction is possible.

We remark that even though at the end of the day our algorithm relies only on basic ideas such as branching and inclusion-exclusion, there is a deeper intuition behind the definitions of the computed values. In fact, from the right angle our algorithm can be seen as an application of the technique of *saving space by algebraization*, introduced by Lokshtanov and Nederlof [18], which boils down to applying the Fourier transform on the lattice of subsets in order to turn subset convolutions into pointwise products. We refer the reader to [1,14,16,23] for other applications of this technique in the context of treedepth-based algorithms.

6   Conclusion and Further Research

In this paper we answered the open question of Hegerfeld and Kratsch [16] by presenting an $\mathcal{O}^\star(5^d)$-time and polynomial space algorithm for HAMILTONIAN PATH, HAMILTONIAN CYCLE, LONGEST PATH, LONGEST CYCLE MIN CYCLE COVER, where d is the depth of a provided elimination forest of the input graph. However, there are still multiple open problems around time- and space-efficient algorithms on graphs of bounded treedepth. We list here a selection.

Approximation of Treedepth. Recall that the treewidth of a graph can be approximated up to a constant factor in fixed-parameter time. For instance, the classic algorithm of Robertson and Seymour [26] (see also [5]) takes on input a graph G and integer t, works in time $2^{\mathcal{O}(t)} \cdot n^{\mathcal{O}(1)}$ and in polynomial space, and either concludes that the treewidth of G is larger than t, or finds a tree decomposition of G of width at most $4t + 4$. This means that for the purpose of designing $2^{\mathcal{O}(t)} \cdot n^{\mathcal{O}(1)}$-time algorithms on graphs of treewidth t, we may assume that a tree decomposition of approximately optimum width is given, as it can be always computed from the input graph within the required complexity bounds. Unfortunately, no such approximation algorithm is known for the treedepth. Namely, it is known that the treedepth can be computed exactly in time and space $2^{\mathcal{O}(d^2)} \cdot n$ [25] and approximated up to factor $\mathcal{O}(t \log^{3/2} t)$ in polynomial time [8], where d and t are the values of the treedepth and the treewidth of the input graph, respectively. A piece of the theory that seems particularly missing is a constant-factor approximation algorithm for treedepth running in time $2^{\mathcal{O}(d)} \cdot n^{\mathcal{O}(1)}$; polynomial space usage would be also desired.

Faster Algorithms. The bases of the exponent of the running times of the algorithms given by Hegerfeld and Kratsch [16] for the treedepth parameterization match the ones obtained by Cygan et al. [7] for the treewidth parameterization. In the case of our results, the situation is different: while HAMILTONIAN CYCLE can be solved in time $4^t \cdot n^{\mathcal{O}(1)}$ in graphs of treewidth t [7] and in time $(2 + \sqrt{2})^p \cdot n^{\mathcal{O}(1)}$ in graphs of pathwidth p [6], we needed to increase the base of the exponent to 5 in order to achieve polynomial space complexity for the treedepth parameterization. As the treedepth of a graph is never smaller than its pathwidth, it is natural to ask whether there is an $(2 + \sqrt{2})^d \cdot n^{\mathcal{O}(1)}$-time polynomial-space algorithm for HAMILTONIAN CYCLE on graphs of treedepth d. In fact, reducing the base 5 to any $c < 5$ would be interesting.

Derandomization. Shortly after its introduction, the Cut&Count technique for the treewidth parameterization has been derandomized. Bodlaender et al. [4] presented two approaches for doing so. The first one, called the *rank-based approach*, boils down to maintaining a small set of representative partial solutions along the dynamic programming computation, and pruning irrelevant partial solutions on the fly using Gaussian elimination. Fomin et al. [13] later reinterpreted this technique in the language of matroids and extended it. The second approach, called *determinant-based*, uses the ideas behind Kirchoff's matrix-tree theorem to deliver a formula for counting suitable spanning trees of a graph, which can be efficiently evaluated by a dynamic programming over a tree decomposition.

It seems to us that none of these approaches applies in the context of the treedepth parameterization, where we additionally require polynomial space complexity. For the rank-based and matroid-based approaches, they are based on keeping track of a set of representative solutions, which in the worst case may have exponential size. In the determinant-based approach, when computing the formula for the number of spanning trees over a tree decomposition, the aggregation of dynamic programming tables is done using operations that are algebraically more involved, and which in particular are non-commutative. See the work of Włodarczyk [27] for a discussion. It is unclear whether this computation can be reorganized so that in the aggregation we use only pointwise product—which, in essence, is our current methodology from the algebraic perspective.

Hence, it is highly interesting whether our algorithm, or the algorithms of Hegerfeld and Kratsch [16], can be derandomized while keeping running time $2^{\mathcal{O}(d)} \cdot n^{\mathcal{O}(1)}$ and polynomial space usage.

Other Graph Parameters. Actually, Hegerfeld and Kratsch [16] were not the first to employ Cut&Count on structural graph parameters beyond treewidth. Pino et al. [24] used Cut&Count and rank-based approach to get single-exponential time algorithms for connectivity problems parametrized by *branchwidth*. Recently, Cut&Count was also applied in the context of *cliquewidth* [2], and of $\mathbb{Q}$-*rankwidth*, *rankwidth*, and *MIM-width* [3]. All these algorithms have exponential space complexity, as they follow the standard dynamic programming approach. One may expect that maybe for the depth-bounded counterparts of cliquewidth and rankwidth—*shrubdepth* [15] and *rankdepth* [9]—time-efficient polynomial-space algorithms can be designed, similarly as for treedepth.

Acknowledgements. We would like to thank anonymous reviewers and Petr Hliněný for their suggestions and comments. The second author would like to thank Marcin Wrochna for some early discussions on combining treedepth and Cut&Count. The work leading to the results presented in this paper was initiated during the Parameterized Retreat of the algorithms group of the University of Warsaw (PARUW), held in Karpacz in February 2019. This Retreat was financed by the project CUTACOMBS, which has received funding from the European Research Council (ERC) under the European Union's Horizon 2020 research and innovation programme (grant agreement No. 714704).

Jesper Nederlof is supported by the project CRACKNP that has received funding from the European Research Council (ERC) under the European Union's Horizon 2020 research and innovation programme (grant agreement No. 853234). Michał Pilipczuk is supported by the project TOTAL that has received funding from the European Research Council (ERC) under the European Union's Horizon 2020 research and innovation programme (grant agreement No 677651). Céline M. F. Swennenhuis is supported by the Netherlands Organization for Scientific Research under project no. 613.009.031b. Karol Węgrzycki is supported by the grants 2016/21/N/ST6/01468 and 2018/28/T/ST6/00084 of the Polish National Science Center and project TOTAL that has received funding from the European Research Council (ERC) under the European Union's Horizon 2020 research and innovation programme (grant agreement No. 677651).

References

1. Belbasi, M., Fürer, M.: A space-efficient parameterized algorithm for the hamiltonian cycle problem by dynamic algebraization. In: van Bevern, R., Kucherov, G. (eds.) CSR 2019. LNCS, vol. 11532, pp. 38–49. Springer, Cham (2019). https://doi.org/10.1007/978-3-030-19955-5_4
2. Bergougnoux, B., Kanté, M.M.: Fast exact algorithms for some connectivity problems parameterized by clique-width. Theor. Comput. Sci. **782**, 30–53 (2019)

3. Bergougnoux, B., Kanté, M.M.: More applications of the d-neighbor equivalence: connectivity and acyclicity constraints. In: 27th Annual European Symposium on Algorithms, ESA 2019. LIPIcs, vol. 144 , pp. 17:1–17:14. Schloss Dagstuhl – Leibniz-Zentrum für Informatik (2019)

4. Bodlaender, H.L., Cygan, M., Kratsch, S., Nederlof, J.: Deterministic single exponential time algorithms for connectivity problems parameterized by treewidth. Inf. Comput. **243**, 86–111 (2015)

5. Cygan, M.: Parameterized Algorithms. Springer, Cham (2015). https://doi.org/10.1007/978-3-319-21275-3

6. Cygan, M., Kratsch, S., Nederlof, J.: Fast Hamiltonicity checking via bases of perfect matchings. J. ACM **65**(3), 12:1–12:46 (2018)

7. Cygan, M., Nederlof, J., Pilipczuk, M., Pilipczuk, M., van Rooij, J.M.M., Wojtaszczyk, J.O.: Solving connectivity problems parameterized by treewidth in single exponential time. In: 52nd Annual Symposium on Foundations of Computer Science, FOCS 2011, pp. 150–159. IEEE (2011)

8. Czerwiński, W., Nadara, W., Pilipczuk, M.: Improved bounds for the excluded-minor approximation of treedepth. In: 27th Annual European Symposium on Algorithms, ESA 2019. LIPIcs, vol. 144, pp. 34:1–34:13. Schloss Dagstuhl – Leibniz-Zentrum für Informatik (2019)

9. DeVos, M., Kwon, O., Oum, S.: Branch-depth: Generalizing tree-depth of graphs. arXiv preprint, abs/1903.11988 (2019)

10. Downey, R.G., Fellows, M.R.: Fundamentals of Parameterized Complexity. TCS. Springer, London (2013). https://doi.org/10.1007/978-1-4471-5559-1

11. Drucker, A., Nederlof, J., Santhanam, R.: Exponential time paradigms through the polynomial time lens. In: 24th Annual European Symposium on Algorithms, ESA 2016. LIPIcs, vol. 57, pp. 36:1–36:14. Schloss Dagstuhl – Leibniz-Zentrum für Informatik (2016)

12. Flum, J., Grohe, M.: Parameterized Complexity Theory. TTCSAES. Springer, Heidelberg (2006). https://doi.org/10.1007/3-540-29953-X

13. Fomin, F.V., Lokshtanov, D., Panolan, F., Saurabh, S.: Efficient computation of representative families with applications in parameterized and exact algorithms. J. ACM **63**(4), 29:1–29:60 (2016)

14. Fürer, M., Yu, H.: Space saving by dynamic algebraization based on tree-depth. Theor. Comput. Syst. **61**(2), 283–304 (2017)

15. Ganian, R., Hliněný, P., Nešetřil, J., Obdržálek, J., de Mendez, P.O.: Shrub-depth: capturing height of dense graphs. Log. Meth. Comput. Sci. **15**(1), 71–725 (2019)

16. Hegerfeld, F., Kratsch, S.: Solving connectivity problems parameterized by treedepth in single-exponential time and polynomial space. arXiv preprint, abs/2001.05364 (2020). Accepted to STACS 2020

17. Lokshtanov, D., Mnich, M., Saurabh, S.: Planar k-path in subexponential time and polynomial space. In: Kolman, P., Kratochvíl, J. (eds.) WG 2011. LNCS, vol. 6986, pp. 262–270. Springer, Heidelberg (2011). https://doi.org/10.1007/978-3-642-25870-1_24

18. Lokshtanov, D., Nederlof, J.: Saving space by algebraization. In: 42nd ACM Symposium on Theory of Computing, STOC 2010, pp. 321–330. ACM (2010)

19. Mulmuley, K., Vazirani, U.V., Vazirani, V.V.: Matching is as easy as matrix inversion. Combinatorica **7**(1), 105–113 (1987)

20. Nederlof, J.: Inclusion exclusion for hard problems. Master's thesis, Utrecht University (2008)

21. Nederlof, J., Pilipczuk, M., Swennenhuis, C.M.F., Wegrzycki, K.: Hamiltonian cycle parameterized by treedepth in single exponential time and polynomial space. CoRR, abs/2002.04368 (2020)

22. Pilipczuk, M., Siebertz, S.: Polynomial bounds for centered colorings on proper minor-closed graph classes. In: 30th Annual ACM-SIAM Symposium on Discrete Algorithms, SODA 2019, pp. 1501–1520. SIAM (2019)

23. Pilipczuk, M., Wrochna, M.: On space efficiency of algorithms working on structural decompositions of graphs. ACM Trans. Comp. Theor. 9(4), 181–1836 (2018)

24. Pino, W.J.A., Bodlaender, H.L., van Rooij, J.M.M.: Cut and Count and representative sets on branch decompositions. In: 11th International Symposium on Parameterized and Exact Computation, IPEC 2016. LIPIcs, vol. 63, pp. 27:1–27:12. Schloss Dagstuhl – Leibniz-Zentrum für Informatik (2016)

25. Reidl, F., Rossmanith, P., Villaamil, F.S., Sikdar, S.: A faster parameterized algorithm for treedepth. In: Esparza, J., Fraigniaud, P., Husfeldt, T., Koutsoupias, E. (eds.) ICALP 2014. LNCS, vol. 8572, pp. 931–942. Springer, Heidelberg (2014). https://doi.org/10.1007/978-3-662-43948-7_77

26. Robertson, N., Seymour, P.D.: Graph Minors. XIII. The disjoint paths problem. J. Comb. Theor. Ser. B 63(1), 65–110 (1995)

27. Włodarczyk, M.: Clifford algebras meet tree decompositions. Algorithmica 81(2), 497–518 (2019)

Parameterized Inapproximability
of Independent Set in H-Free Graphs

Pavel Dvořák[1]([⊠]), Andreas Emil Feldmann[1], Ashutosh Rai[1],
and Paweł Rzążewski[2,3]

[1] Faculty of Mathematics and Physics, Charles University, Prague, Czech Republic
koblich@iuuk.mff.cuni.cz, feldmann.a.e@gmail.com, ashu.rai87@gmail.com
[2] Faculty of Mathematics and Information Science,
Warsaw University of Technology, Warsaw, Poland
p.rzazewski@mini.pw.edu.pl
[3] Institute of Informatics, University of Warsaw, Warsaw, Poland

Abstract. We study the INDEPENDENT SET problem in H-free graphs,
i.e., graphs excluding some fixed graph H as an induced subgraph. We
prove several inapproximability results both for polynomial-time and
parameterized algorithms.

Halldórsson [SODA 1995] showed that for every $\delta > 0$ the INDE-
PENDENT SET problem has a polynomial-time $(\frac{d-1}{2} + \delta)$-approximation
algorithm in $K_{1,d}$-free graphs. We extend this result by showing that
$K_{a,b}$-free graphs admit a polynomial-time $\mathcal{O}(\alpha(G)^{1-1/a})$-approximation,
where $\alpha(G)$ is the size of a maximum independent set in G. Furthermore, we complement the result of Halldórsson by showing that for some
$\gamma = \Theta(d/\log d)$, there is no polynomial-time γ-approximation algorithm
for these graphs, unless NP = ZPP.

Bonnet *et al.* [IPEC 2018] showed that INDEPENDENT SET parameterized by the size k of the independent set is W[1]-hard on graphs which do
not contain (1) a cycle of constant length at least 4, (2) the star $K_{1,4}$, and
(3) any tree with two vertices of degree at least 3 at constant distance.
We strengthen this result by proving three inapproximability results
under different complexity assumptions for almost the same class of
graphs (we weaken condition (2) that G does not contain $K_{1,5}$). First,
under the ETH, there is no $f(k) \cdot n^{o(k/\log k)}$ algorithm for any computable
function f. Then, under the deterministic Gap-ETH, there is a constant
$\delta > 0$ such that no δ-approximation can be computed in $f(k) \cdot n^{O(1)}$
time. Also, under the stronger randomized Gap-ETH there is no such
approximation algorithm with runtime $f(k) \cdot n^{o(k)}$.

Finally, we consider the parameterization by the excluded graph H,
and show that under the ETH, INDEPENDENT SET has no $n^{o(\alpha(H))}$

P. Dvořák and A.E. Feldmann—Supported by Czech Science Foundation GAČR (grant
#19-27871X).

A.E. Feldmann and A. Rai—Supported by Center for Foundations of Modern Computer
Science (Charles Univ. project UNCE/SCI/004).

P. Rzążewski—Supported by Polish National Science Centre grant no. 2018/31/D/
ST6/00062.

© Springer Nature Switzerland AG 2020
I. Adler and H. Müller (Eds.): WG 2020, LNCS 12301, pp. 40–53, 2020.
https://doi.org/10.1007/978-3-030-60440-0_4

algorithm in H-free graphs. Also, we prove that there is no $d/k^{o(1)}$-approximation algorithm for $K_{1,d}$-free graphs with runtime $f(d,k)\cdot n^{O(1)}$, under the deterministic Gap-ETH.

1   Introduction

The INDEPENDENT SET problem, which asks for a maximum sized set of pairwise non-adjacent vertices in a graph, is one of the most well-studied problems in algorithmic graph theory. It was among the first 21 problems that were proven to be NP-hard by Karp [22], and is also known to be hopelessly difficult to approximate in polynomial time: Håstad [21] proved that under standard assumptions from classical complexity theory the problem admits no $(n^{1-\varepsilon})$-approximation, for any $\varepsilon > 0$ (by n we always denote the number of vertices in the input graph). This was later strengthened by Khot and Ponnuswami [23], who were able to exclude any algorithm with approximation ratio $n/(\log n)^{3/4+\varepsilon}$, for any $\varepsilon > 0$. Let us point out that the currently best polynomial-time approximation algorithm for INDEPENDENT SET achieves the approximation ratio $\mathcal{O}(n\frac{(\log\log n)^2}{(\log n)^3})$ [17].

There are many possible ways of approaching such a difficult problem, in order to obtain some positive results. One could give up on generality, and ask for the complexity of the problem on restricted instances. For example, while the INDEPENDENT SET problem remains NP-hard in subcubic graphs [18], a straightforward greedy algorithm gives a 3-approximation.

H-Free Graphs. A large family of restricted instances, for which the INDEPENDENT SET problem has been well-studied, comes from forbidding certain induced subgraphs. For a (possibly infinite) family $\mathcal{H}$ of graphs, a graph G is $\mathcal{H}$-free if it does not contain any graph of $\mathcal{H}$ as an induced subgraph. If $\mathcal{H}$ consists of just one graph, say $\mathcal{H} = \{H\}$, then we say that G is H-free. The investigation of the complexity of INDEPENDENT SET in H-free graphs dates back to Alekseev, who proved the following.

Theorem 1 (Alekseev [2]). *Let $s \geq 3$ be a constant. The* INDEPENDENT SET *problem is* NP-*hard in graphs that do not contain any of the following induced subgraphs: (1) a cycle on at most s vertices, (2) the star $K_{1,4}$, and (3) any tree with two vertices of degree at least 3 at distance at most s.*

We can restate Theorem 1 as follows: the INDEPENDENT SET problem is NP-hard in H-free graphs, unless H is a subgraph of a subdivided claw (i.e., three paths which meet at one of their endpoints). The reduction also implies that for each such H the problem is APX-hard and cannot be solved in subexponential time, unless the Exponential Time Hypothesis (ETH) fails. On the other hand, polynomial-time algorithms are known only for very few cases. First let us consider the case when $H = P_t$, i.e., we forbid a path on t vertices. Note that the case of $t = 3$ is trivial, as every P_3-free graph is a disjoint union of cliques. Already in 1981 Corneil et al. [10] showed that INDEPENDENT SET is tractable for P_4-free graphs. For many years there was no improvement, until the breakthrough algorithm of Lokshtanov et al. [25] for P_5-free graphs. Their approach was recently extended to P_6-free graphs by Grzesik et al. [19]. We still do not know whether the problem is polynomial-time solvable in P_7-free graphs, and we do not know it to be NP-hard in P_t-free graphs, for any constant t.

Even less is known for the case if H is a subdivided claw. The problem can be solved in polynomial time in claw-free (i.e., $K_{1,3}$-free) graphs, see Sbihi [32] and Minty [31]. This was later extended to H-free graphs, where H is a claw with one edge once subdivided (see Alekseev [1] for the unweighted version and Lozin and Milanič [27] for the weighted one).

When it comes to approximations, Halldórsson [20] gave an elegant local search algorithm that finds a $(\frac{d-1}{2} + \delta)$-approximation of the maximum independent set in $K_{1,d}$-free graphs for any constant $\delta > 0$ in polynomial time. Very recently, Chudnovsky et al. [9] designed a QPTAS (quasi-polynomial-time approximation scheme) that works for *every* H, which is a subgraph of a subdivided claw (in particular, a path). Recall that on all other graphs H the problem is APX-hard.

Parameterized Complexity. Another approach that one could take is to look at the problem from the parameterized perspective: we no longer insist on finding the maximum independent set, but want to verify whether some independent set of size at least k exists. To be more precise, we are interested in knowing how the complexity of the problem depends on k. The best type of behavior we are hoping for is *fixed-parameter tractability* (FPT), i.e., the existence of an algorithm with running time $f(k) \cdot n^{\mathcal{O}(1)}$, for some function f (note that since the problem is NP-hard, we expect f to be super-polynomial).

It is known [11] that on general graphs the INDEPENDENT SET problem is W[1]-hard parameterized by k, which is a strong indication that it does not admit an FPT algorithm. Furthermore, it is even unlikely to admit any non-trivial *fixed-parameter approximation (FPA)*: a γ-FPA algorithm for the INDEPENDENT SET problem is an algorithm that takes as input a graph G and an integer k, and in time $f(k) \cdot n^{\mathcal{O}(1)}$ either correctly concludes that G has no independent set of size at least k, or outputs an independent set of size at least k/γ (note that γ does not have to be a constant). It was shown in [5] that on general graphs no $o(k)$-FPA exists for INDEPENDENT SET, unless the randomized Gap-ETH fails.

Parameterized Complexity in H-Free Graphs. As we pointed out, none of the discussed approaches, i.e., considering H-free graphs or considering parameterized algorithms, seems to make the INDEPENDENT SET problem more tractable. However, some positive results can be obtained by combining these two settings, i.e., considering the parameterized complexity of INDEPENDENT SET in H-free graphs. For example, the Ramsey theorem implies that any graph with $\Omega(4^p)$ vertices contains a clique or an independent set of size $\Omega(p)$. Since the proof actually tells us how to construct a clique or an independent set in polynomial time [16], we immediately obtain a very simple FPT algorithm for K_p-free graphs. Dabrowski [12] provided some positive and negative results for the complexity of the INDEPENDENT SET problem in H-free graphs, for various H.

The systematic study of the problem was initiated by Bonnet et al. [3] and continued by Bonnet et al. [4]. Among other results, Bonnet *et al.* [3] obtained the following analog of Theorem 1.

Theorem 2 (Bonnet *et al.* [3]). *Let $s \geq 3$ be a constant. The* INDEPENDENT SET *problem is* W[1]*-hard in graphs that do not contain any of the following induced subgraphs: (1) a cycle on at least 4 and at most s vertices, (2) the star $K_{1,4}$, and (3) any tree with two vertices of degree at least 3 at distance at most s.*

Note that, unlike in Theorem 1, we are not able to show hardness for C_3-free graphs: as already mentioned, the Ramsey theorem implies that INDEPENDENT SET is FPT in C_3-free graphs. Thus, graphs H for which there is hope for FPT algorithms in H-free graphs are essentially obtained from paths and subdivided claws (or their subgraphs) by replacing each vertex with a clique.

Let us point out that, even though it is not stated there explicitly, the reduction of Bonnet *et al.* [3] also excludes any algorithm solving the problem in time $f(k) \cdot n^{o(\sqrt{k})}$, unless the ETH fails.

Our Results. We study the approximation of the INDEPENDENT SET problem in H-free graphs, mostly focusing on approximation hardness. Our first two results are related to Halldórsson's [20] polynomial-time ($\frac{d-1}{2} + \delta$)-approximation algorithm for $K_{1,d}$-free graphs. We extend this result to $K_{a,b}$-free graphs, for any constants a, b. Moreover, we show that the approximation ratio of the algorithm of Halldórsson [20] is optimal, up to logarithmic factors.

Theorem 3. *Given a $K_{a,b}$-free graph G, an $\mathcal{O}\big((a+b)^{1/a} \cdot \alpha(G)^{1-1/a}\big)$-approximation can be computed in $n^{\mathcal{O}(a)}$ time.*

Theorem 4. *There is a function $\gamma = \Theta(d/\log d)$ such that the* INDEPENDENT SET *problem does not admit a polynomial time γ-approximation algorithm in $K_{1,d}$-free graphs, unless* ZPP $=$ NP.

The proofs of Theorems 3 and 4 can be find in the full version of the paper [15]. Note that the factor γ determining the approximation gap in Theorem 4 is expressed as an asymptotic function of d, i.e., for growing d. In our case however, it is an interesting question how small the degree d can be so that we obtain an inapproximability result. We prove Theorem 4 by a reduction from the LABEL COVER problem, and a corresponding inapproximability result by Laekhanukit [24]. By calculating the bounds given in [24] (which heavily depend on the constant of Chernoff bounds) it can be shown that an inapproximability gap exists for $d \geq 31$ in Theorem 4.

Then in Sect. 3 we study the existence of fixed-parameter approximation algorithms for the INDEPENDENT SET problem in H-free graphs. We show the following strengthening of Theorem 2, which also gives (almost) tight runtime lower bounds assuming the ETH or the randomized Gap-ETH (for more information about complexity assumptions used in Theorem 5 see Sect. 2).

Theorem 5. *Let $s \geq 4$ be a constant, and let $\mathcal{G}$ be the class of graphs that do not contain any of the following induced subgraphs: (1) a cycle on at least 5 and at most s vertices, (2) the star $K_{1,5}$, and (3) (i) the star $K_{1,4}$ or (ii) a cycle on 4 vertices and any tree with two vertices of degree at least 3 at distance at most s. The* INDEPENDENT SET *problem on $\mathcal{G}$ does not admit the following:*

(a) *an exact algorithm with runtime $f(k) \cdot n^{o(k/\log k)}$, for any computable function f, under the ETH,*

(b) *a γ-approximation algorithm with runtime $f(k) \cdot n^{\mathcal{O}(1)}$ for some constant $\gamma > 0$ and any computable function f, under the deterministic Gap-ETH,*

(c) *a γ-approximation algorithm with runtime $f(k) \cdot n^{o(k)}$ for some constant $\gamma > 0$ and any computable function f, under the randomized Gap-ETH.*

Finally, in Sect. 4 we study a slightly different setting, where the graph H is not considered to be fixed. As mentioned before, INDEPENDENT SET is known to be polynomial-time solvable in P_t-free graphs for $t \leq 6$. The algorithms for increasing values of t get significantly more complicated and their complexity increases. Thus it is natural to ask whether this is an inherent property of the problem and can be formalized by a runtime lower bound when parameterized by t.

We give an affirmative answer to this question, even if the forbidden family is not a family of paths: note that the independent set number $\alpha(P_t)$ of a path on t vertices is $\lceil t/2 \rceil$.

Proposition 1. *Let d be an integer and let $\mathcal{H}_d$ be a family of graphs, such that $\alpha(H) > d$ for every $H \in \mathcal{H}_d$. The* INDEPENDENT SET *problem in $\mathcal{H}_d$-free graphs is* W[1]*-hard parameterized by d and cannot be solved in $n^{o(d)}$ time, unless the ETH fails.*

The proof of Proposition 1 can be found in the full version of the paper. We also study the special case when $H = K_{1,d}$ and consider the inapproximability of the problem parameterized by both $\alpha(K_{1,d}) = d$ and k. Unfortunately, for the parameterized version we do not obtain a clear-cut statement as in Theorem 4, since in the following theorem d cannot be chosen independently of k in order to obtain an inapproximability gap.

Proposition 2. *Let $\varepsilon > 0$ be any constant and $\xi(k) = 2^{(\log k)^{1/2+\varepsilon}}$. The* INDEPENDENT SET *problem in $K_{1,d}$-free graphs has no $d/\xi(k)$-approximation algorithm with runtime $f(d,k) \cdot n^{\mathcal{O}(1)}$ for any computable function f, unless the deterministic Gap-ETH fails.*

Note that this in particular shows that if we allow d to grow as a polynomial k^ε for any constant $\frac{1}{2} > \varepsilon > 0$, then no k^δ-approximation is possible for any $\delta < \varepsilon$ (since $\xi(k) = k^{o(1)}$). This indicates that the $(\frac{d-1}{2} + \delta)$-approximation for $K_{1,d}$-free graphs [20] is likely to be best possible (up to sub-polynomial factors), even when parameterizing by k and d.

2   Preliminaries

All our hardness results for INDEPENDENT SET are obtained by reductions from some variant of the MAXIMUM COLORED SUBGRAPH ISOMORPHISM (MCSI) problem. This optimization problem has been widely studied in the literature, both to obtain polynomial-time and parameterized inapproximability results, but also in its decision version to obtain parameterized runtime lower bounds. We note that by applying standard transformations, MCSI contains the well-known problems LABEL COVER [24] and BINARY CSP [26]: for BINARY CSP the graph J is a complete graph, while for LABEL COVER J is usually bipartite.

MAXIMUM COLORED SUBGRAPH ISOMORPHISM (MCSI)
Input: A graph G, whose vertex set is partitioned into subsets $V_1, \ldots, V_\ell$, and a graph J on vertex set $\{1, \ldots, \ell\}$.
Goal: Find an assignment $\phi : V(J) \to V(G)$, where $\phi(i) \in V_i$ for every $i \in [\ell]$, that maximizes the number $S(\phi)$ of satisfied edges, i.e.,
$$S(\phi) := \big|\{ ij \in E(J) \mid \phi(i)\phi(j) \in E(G) \}\big|.$$

Given an instance $\Gamma = (G, V_1, \ldots, V_\ell, J)$ of MCSI, we refer to the number of vertices of G as the *size* of Γ. Any assignment $\phi : V(J) \to V(G)$, such that for every i it holds that $\phi(i) \in V_i$, is called a *solution of* Γ. The *value* of a solution ϕ is $\mathrm{val}(\phi) := S(\phi)/|E(J)|$, i.e., the fraction of satisfied edges. The value of the instance Γ, denoted by $\mathrm{val}(\Gamma)$, is the maximum value of any solution of Γ.

When considering the decision version of MCSI, i.e., determining whether $\mathrm{val}(\Gamma) = 1$ or $\mathrm{val}(\Gamma) < 1$, a result by Marx [29] gives a runtime lower bound for parameter ℓ under the *Exponential Time Hypothesis (ETH)*. That is, no $f(\ell) \cdot n^{o(\ell/\log \ell)}$ time algorithm can solve MCSI for any computable function f, assuming there is no deterministic $2^{o(n)}$ time algorithm to solve the 3-SAT problem. For the optimization version, an α-approximation is a solution ϕ with $\mathrm{val}(\phi) \geq 1/\alpha$. When J is a complete graph, a result by Dinur and Manurangsi [13,14] states that there is no $\ell/\xi(\ell)$-approximation algorithm (where $\xi(\ell) = 2^{(\log \ell)^{1/2+\varepsilon}}$ for any constant $\varepsilon > 0$) with runtime $f(\ell) \cdot n^{O(1)}$ for any computable function f, unless the *deterministic Gap-ETH* fails (see Theorem 9). This hypothesis assumes that there exists some constant $\delta > 0$ such that no deterministic $2^{o(n)}$ time algorithm for 3-SAT can decide whether all or at most a $(1-\delta)$-fraction of the clauses can be satisfied. A recent result by Manurangsi [28] uses an even stronger assumption, which also rules out randomized algorithms, and in turn obtains a better runtime lower bound at the expense of a worse approximation lower bound: he shows that, when J is a complete graph, there is no γ-approximation algorithm for MCSI with runtime $f(\ell) \cdot n^{o(\ell)}$ for any computable function f and any constant γ, under the *randomized Gap-ETH*. This assumes that there exists some constant $\delta > 0$ such that no randomized $2^{o(n)}$ time algorithm for 3-SAT can decide whether all or at most a $(1 - \delta)$-fraction of the clauses can be satisfied. (Note that the runtime lower bound under the stronger randomized Gap-ETH does not have the $\log(\ell)$ factor in the polynomial degree as the runtime lower bound under ETH does.)

For our results we will often need the special case of MCSI when the graph J has bounded degree. We define this problem in the following.

DEGREE-t MAXIMUM COLORED SUBGRAPH ISOMORPHISM (MCSI(t))
Input: A graph G, whose vertex set is partitioned into subsets $V_1, \ldots, V_\ell$, and a graph J on vertex set $\{1, \ldots, \ell\}$ and maximum degree t.
Goal: Find an assignment $\phi : V(J) \to V(G)$, where $\phi(i) \in V_i$ for every $i \in [\ell]$, that maximizes the number $S(\phi)$ of satisfied edges, i.e.,
$S(\phi) := \big|\{ij \in E(J) \mid \phi(i)\phi(j) \in E(G)\}\big|.$

The bounded degree case has been considered before, and we harness some of the known hardness results for MCSI(t) in our proofs. First, let us point out that the lower bound for exact algorithms holds even for the case when $t = 3$, as shown by Marx and Pilipczuk [30]. We also use a polynomial-time approximation lower bound given by Laekhanukit [24], where t can be set to any constant and the approximation gap depends on t. The complexity assumption of this algorithm is that NP-hard problems do not have polynomial time Las Vegas algorithms, i.e., NP $\neq$ ZPP. For parameterized approximations, we use a result by Lokshtanov et al. [26], who obtain a constant approximation gap for the case when $t = 3$ (see Theorem 6). It seems that this result for parameterized algorithms is not easily generalizable to arbitrary constants t so that the approximation gap would depend only on t, as in the result for polynomial-time algorithms provided by Laekhanukit [24]: neither the techniques found in [24] nor those of [26] seem to be usable to obtain an approximation gap that depends only on t but not the parameter ℓ. However, we develop a weaker parameterized inapproximability result for the case when $t \geq \xi(\ell) = \ell^{o(1)}$ (see Theorem 7 in Sect. 4), and use it to prove Proposition 2.

3 Parameterized Approximation for Fixed H

In this section we prove Theorem 5. Let us define an auxiliary family of classes of graphs: for integers $4 \leq a \leq b$ and $c \geq 3$, let $\mathcal{C}([a,b],c)$ denote the class of graphs that are $K_{1,c}$-free and C_p-free for any $p \in [a,b]$. Let $\mathcal{T}(b)$ be the class of trees with two vertices of degree at least 3 at distance at most b. Let $\mathcal{C}^*([a,b],c) \subseteq \mathcal{C}([a,b],c)$ be the set of those $G \in \mathcal{C}([a,b],c)$, which are are also $\mathcal{T}(\lceil\frac{b-1}{2}\rceil)$-free. Actually, we will prove a stronger statement that Theorem 5 holds for $G \in \mathcal{C}^*([4,z],5) \cup \mathcal{C}([5,z],4)$ for a constant $z \geq 5$.

The proof consists of two steps: first we will prove it for graphs in $\mathcal{C}^*([4,z],5)$, and then for graphs in $\mathcal{C}([5,z],4)$. In both proofs we will reduce from the MCSI(3) problem. Here, we sketch the proof for the class $\mathcal{C}^*([4,z],5)$. The rest of the proof can be found in the full version [15]. Let $\Gamma = (G, V_1, \ldots, V_\ell, H)$ be an instance of MCSI(3). For $ij \in E(H)$, by $E_{ij} = E_{ji}$ we denote the set of edges between V_i and V_j. Note that we may assume that H has no isolated vertices, each V_i is an independent set, and $E_{ij} \neq \emptyset$ if and only if $ij \in E(H)$.

Lokshtanov et al. [26] gave the following hardness result (the first statement actually follows from Marx [29] and Marx and Pilipczuk [30]). We note that Lokshtanov et al. [26] conditioned their result on the *Parameterized Inapproximability Hypothesis* (PIH) and W[1] $\neq$ FPT. Here we use stronger assumptions, i.e., the deterministic and randomized Gap-ETH, which are more standard in the area of parameterized approximation. The reduction in [26] yields the following theorem, when starting from [13,14] and [28], respectively (see also [7, Corollary 7.9]).

Theorem 6 (Lokshtanov et al. [26]). *Consider an arbitrary instance $\Gamma = (G, V_1, \ldots, V_\ell, H)$ of MCSI(3) with size n.*

1. *Assuming the ETH, for any computable function f, there is no $f(\ell) \cdot n^{o(\ell/\log \ell)}$ time algorithm that solves Γ.*
2. *Assuming the deterministic Gap-ETH there exists a constant $\gamma > 0$, such that for any computable function f, there is no $f(\ell) \cdot n^{\mathcal{O}(1)}$ time algorithm that can distinguish between the two cases: (YES-case) $\mathrm{val}(\Gamma) = 1$, and (NO-case) $\mathrm{val}(\Gamma) < 1 - \gamma$.*
3. *Assuming the randomized Gap-ETH there exists a constant $\gamma > 0$, such that for any computable function f, there is no $f(\ell) \cdot n^{o(\ell)}$ time algorithm that can distinguish between the two cases: (YES-case) $\mathrm{val}(\Gamma) = 1$, and (NO-case) $\mathrm{val}(\Gamma) < 1 - \gamma$.*

Let $\Gamma = (G, V_1, \ldots, V_\ell, H)$ be an instance of MCSI(3). We aim to build an instance (G', k) of INDEPENDENT SET, such that the graph $G' \in \mathcal{C}^*([4, z], 5)$.

For each $ij \in E(H)$, we introduce a clique C_{ij} of size $|E_{ij}|$, whose every vertex *represents* a different edge from E_{ij}. The cliques constructed at this step will be called *primary cliques*, note that their number is $|E(H)|$. Choosing a vertex v from C_{ij} to an independent set of G' will correspond to mapping i and j to the appropriate endvertices of the edge from E_{ij}, corresponding to v.

Now we need to ensure that the choices in primary cliques corresponding to edges of G are consistent. Consider $i \in V(H)$ and suppose it has three neighbors j_1, j_2, j_3 (the cases if i has fewer neighbors are dealt with analogously). We will connect the cliques $C_{ij_1}, C_{ij_2}, C_{ij_3}$ using a gadget called a *vertex-cycle*, whose construction we describe below. For each $a \in \{1, 2, 3\}$, we introduce s copies of C_{ij_a} and denote them by $D_{ij_a}^1, D_{ij_a}^2, \ldots, D_{ij_a}^s$, respectively. Let us call these copies *secondary cliques*. The vertices of secondary cliques represent the edges from E_{ij_a} analogously as the ones of C_{ij_a}. We call primary and secondary cliques as *base cliques*. We connect the base cliques corresponding to the vertex $i \in V(H)$ into *vertex-cycle* $\mathcal{C}_i$. Imagine that secondary cliques, along with primary cliques $C_{ij_1}, C_{ij_2}, C_{ij_3}$, are arranged in a cycle-like fashion, as follows:

$C_{ij_1}, D_{ij_1}^1, D_{ij_1}^2, \ldots, D_{ij_1}^s, C_{ij_2}, D_{ij_2}^1, D_{ij_2}^2, \ldots, D_{ij_2}^s, C_{ij_3}, D_{ij_3}^1, D_{ij_3}^2, \ldots, D_{ij_3}^s, C_{ij_1}.$

This cyclic ordering of cliques constitutes the vertex-cycle, let us point out that we treat this cycle as a directed one. As we describe below we put some edges between two base cliques D_1 and D_2 only if they belong to some vertex-cycle $\mathcal{C}_i$. See Fig. 1 for an example of how we connect base cliques.

Now, we describe how we connect the consecutive cliques in $\mathcal{C}_i$. Recall that each vertex v of each clique represents exactly one edge uw of G, whose exactly one vertex, say u, is in V_i. We extend the notion of representing and say that v *represents* u, and denote it by $r_i(v) = u$.

Let us fix an arbitrary ordering $\prec_i$ on V_i. Now, consider two consecutive cliques of the vertex-cycle. Let v be a vertex of the first clique and v' be a vertex from the second clique, and let u and u' be the vertices of V_i represented by v and v', respectively. The edge vv' exists in G' if and only if $u \prec_i u'$. See Fig. 2 how we connect two consecutive base cliques in a vertex-cycle. This finishes the construction of $\mathcal{C}_i$. We introduce a vertex-cycle $\mathcal{C}_i$ for every vertex i of H, note that each primary clique C_{ij} is in exactly two vertex-cycles: $\mathcal{C}_i$ and $\mathcal{C}_j$. The number of all base cliques is

$$k := \underbrace{|E(H)|}_{\substack{\text{primary} \\ \text{cliques}}} + \underbrace{\sum_{i \in V(H)} \deg_H(i) \cdot s}_{\text{secondary cliques}} = |E(H)| \cdot \left(1 + \frac{s}{2}\right) \leq \frac{3\ell}{2} \cdot \left(1 + \frac{s}{2}\right) = \mathcal{O}(\ell).$$

This concludes the construction of (G', k). Since $V(G')$ is partitioned into k base cliques, k is an upper bound on the size of any independent set in G', and a solution of size k contains exactly one vertex from each base clique.

We claim that the graph G' is in the class $\mathcal{C}^*([4, z], 5)$. Moreover, if $\text{val}(\Gamma) = 1$, then the graph G' has an independent set of size k and if the graph G' has an independent set of size at least $(1 - \gamma') \cdot k$ for $\gamma' = \frac{\gamma}{6+3s}$, then $\text{val}(\Gamma) \geq 1 - \gamma$.

4   Parameterized Approximation with H as a Parameter

Now let us consider the INDEPENDENT SET problem in $K_{1,d}$-free graphs, parameterized by *both* k and d. In this case we are able to give parameterized approximation lower bounds based on the following sparsification of MCSI. Recall that $\xi(\ell) = 2^{(\log \ell)^{1/2+\varepsilon}} = \ell^{o(1)}$ for any constant $\frac{1}{2} > \varepsilon > 0$, i.e., the term grows slower than any polynomial (but faster than any polylogarithm).

Theorem 7. *Consider an instance $\Gamma = (G, V_1, \ldots, V_\ell, J)$ of MCSI(t) with size n and $t > \xi(\ell)$. Assuming the deterministic Gap-ETH, for any computable*

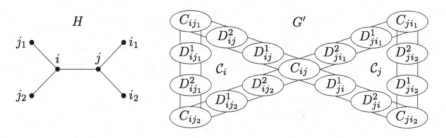

Fig. 1. A part of the construction of G' for $s = 2$. Cliques C_{ab} representing edge sets $E_{ab} \subseteq E(G)$ are connected through secondary cliques D^p_{ab}.

function f, there is no $f(\ell) \cdot n^{O(1)}$ time algorithm that can distinguish between the two cases: (YES-case) $\mathrm{val}(\Gamma) = 1$, *and* (NO-case) $\mathrm{val}(\Gamma) \leq \xi(\ell)/t$.

Proposition 2 follows from Theorem 7 by easy reduction (see the full version [15]). To prove Theorem 7 we need two facts. The first is the Erdős-Gallai theorem on *degree sequences*, which are sequences of non-negative integers $d_1, \ldots, d_n$, for each of which there exists a simple graph on n vertices such that vertex $i \in [n]$ has degree d_i. We use the following constructive formulation due to Choudum [8].

Theorem 8 (Erdős-Gallai theorem [8]). *A sequence of non-negative integers $d_1 \geq \cdots \geq d_n$ is a degree sequence of a simple graph on n vertices if $d_1 + \cdots + d_n$ is even and for every $1 \leq k \leq n$ the following inequality holds: $\sum_{i=1}^{k} d_i \leq k(k-1) + \sum_{i=k+1}^{n} \min(d_i, k)$. Moreover, given such a degree sequence, a corresponding graph can be constructed in polynomial time.*

We also need a parameterized approximation lower bound for MCSI, as given by Dinur and Manurangsi [13].

Theorem 9 (Dinur and Manurangsi [13]). *Consider an instance $\Gamma = (G, V_1, \ldots, V_\ell, J)$ of MCSI with size n and J a complete graph. Assuming the deterministic Gap-ETH, there is no $f(\ell) \cdot n^{O(1)}$ time algorithm for any computable function f, that can distinguish between the following two cases:* (YES-case) $\mathrm{val}(\Gamma) = 1$, *and* (NO-case) $\mathrm{val}(\Gamma) \leq \xi(\ell)/\ell$.

Proof of Theorem 7. Let $t < \ell$ and let $\Gamma = (G, V_1, \ldots, V_\ell, J)$ be an instance of MCSI where J is a complete graph. To find an instance of MCSI(t), we want to construct a graph J' on ℓ vertices with maximum degree t, for which we use the Erdős-Gallai theorem. By Theorem 8 it is easy to verify that a t-regular graph on ℓ vertices exists if $t\ell$ is even. However, if $t\ell$ is odd, there is a graph with $\ell - 1$ vertices of degree t and one vertex of degree $t - 1$. Moreover, the proof of Theorem 8 by Choudum [8] is constructive, and gives a polynomial time algorithm. Hence we can compute a graph J' with maximum degree t and $|E(J')| \geq (t\ell - 1)/2$. Since J is a complete graph, J' is a subgraph of J.

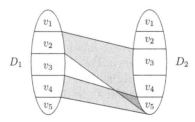

Fig. 2. Edges between two consecutive cliques D_1 and D_2 in a vertex-cycle $\mathcal{C}_i$, where $V_i = \{v_1, \ldots, v_5\}$. We show only edges incident to $u \in V(D_1)$ such that $r_i(u) \in \{v_2, v_4\}$.

We create a graph G' by removing edges from G according to J': we remove all edges between sets V_i and V_j of G if and only if $ij \notin E(J')$, and call the resulting graph G'. Thus, we get new instance $\Gamma' = (G', V_1, \ldots, V_\ell, J')$ of MCSI(t).

It is easy to see that if val(Γ) = 1, then val(Γ') = 1 as well: we just use the optimal solution for Γ. Now suppose that val(Γ) $\leq \nu$, which means that each solution ϕ satisfies at most a ν-fraction of edges of J. Let ϕ be an arbitrary solution of Γ', which is also a solution for Γ because $V(G) = V(G')$ and $V(J) = V(J')$. By our assumption we know that it satisfies at most $\nu \cdot E(J)$ edges of J. Thus, the solution ϕ satisfies at most $\nu \cdot E(J)$ edges of J' as well. Hence we obtain

$$\text{val}(\Gamma') \leq \frac{\nu \cdot E(J)}{E(J')} = \nu \cdot \frac{\ell(\ell-1)}{t\ell - 1} \leq \nu \cdot \frac{\ell}{t - 1/\ell}.$$

Now, by Theorem 9 we know that under the deterministic Gap-ETH, no $f(\ell) \cdot n^{O(1)}$ time algorithm can distinguish between val(Γ) = 1 and val(Γ) $\leq \xi(\ell)/\ell$ given Γ. By the above calculations, for Γ' we obtain that no such algorithm can distinguish between val(Γ') = 1 and val(Γ') $\leq \frac{\xi(\ell)}{t-1/\ell}$ by setting $\nu = \xi(\ell)/\ell$. Recall that $\xi(\ell) = 2^{(\log \ell)^{1/2+\varepsilon}}$ where ε can be set to any positive constant in Theorem 9. Given any constant $\varepsilon' > 0$, we choose ε such that $2^{(\log \ell)^{1/2+\varepsilon}}/2^{(\log \ell)^{1/2+\varepsilon'}} \leq (t - 1/\ell)/t$. It can be verified that such a constant $\varepsilon > 0$ always exists, assuming w.l.o.g. that ℓ is larger than some sufficiently large constant. This implies that val(Γ') $\leq \frac{\xi(\ell)}{t-1/\ell} \leq 2^{(\log \ell)^{1/2+\varepsilon'}}/t$. Note that val($\Gamma'$) < 1 if $t > 2^{(\log \ell)^{1/2+\varepsilon'}}$, and so we obtain Theorem 7 (for $\xi(\ell) := 2^{(\log \ell)^{1/2+\varepsilon'}}$). □

5   Conclusion and Open Problems

Our parameterized inapproximability results of Theorem 5 suggest that the INDEPENDENT SET problem is hard to approximate to within some constant, whenever it is W[1]-hard to solve on H-free graphs, according to Theorem 2. In most cases it is unclear though whether any approximation can be computed (either in polynomial time or by exploiting the parameter k), which beats the strong lower bounds for polynomial-time algorithms for general graphs. The only known exceptions to this are the $K_{1,d}$-free case, where a polynomial-time $(\frac{d-1}{2} + \delta)$-approximation algorithm was shown by Halldórsson [20], and the $K_{a,b}$-free case, for which we showed a polynomial-time $\mathcal{O}((a + b)^{1/a} \cdot \alpha(G)^{1-1/a})$-approximation algorithm in Theorem 3. For $K_{1,d}$-free graphs, we were also able to show an almost asymptotically tight lower bound for polynomial-time algorithms in Theorem 4. For parameterized algorithms, our lower bound of Proposition 2 for $K_{1,d}$-free graphs does not give a tight bound, but seems to suggest that parameterizing by k does not help to obtain an improvement. For P_t-free graphs, for which the INDEPENDENT SET problem is conjectured to be polynomial-time solvable, we showed in Proposition 1 that the complexity of any such algorithm must grow with the length t of the excluded path.

Settling the question whether H-free graphs admit better approximations to INDEPENDENT SET than general graphs, remains a challenging open problem, both for polynomial-time algorithms and algorithms exploiting the parameter k.

Let us point out one more, concrete open question. Recall from Theorem 2 Bonnet *et al.* [3] were able to show W[1]-hardness for graphs which *simultaneously* exclude $K_{1,4}$ and all induced cycles of length in $[4, z]$, for any constant $z \geq 5$. On the other hand, we presented two separate reductions, one for $(K_{1,5}, C_4, \ldots, C_z)$-free graphs, and another one for $(K_{1,4}, C_5, \ldots, C_z)$-free graphs. It would be nice to provide a uniform reduction, i.e., prove hardness for parameterized approximation in $(K_{1,4}, C_4, \ldots, C_z)$-free graphs.

References

1. Alekseev, V.: Polynomial algorithm for finding the largest independent sets in graphs without forks. Discrete Appl. Math. **135**(1), 3–16 (2004). Russian Translations II
2. Alekseev, V.E.: The effect of local constraints on the complexity of determination of the graph independence number. In: Combinatorial-Algebraic Methods in Applied Mathematics, pp. 3–13 (1982)
3. Bonnet, É., Bousquet, N., Charbit, P., Thomassé, S., Watrigant, R.: Parameterized complexity of independent set in H-free graphs. In: Paul, C., Pilipczuk, M. (eds.) 13th International Symposium on Parameterized and Exact Computation, IPEC 2018, Helsinki, Finland, 20–24 August 2018. LIPIcs, vol. 115, pp. 17:1–17:13. Schloss Dagstuhl - Leibniz-Zentrum für Informatik (2018)
4. Bonnet, É., Bousquet, N., Thomassé, S., Watrigant, R.: When maximum stable set can be solved in FPT time. In: Lu, P., Zhang, G. (eds.) 30th International Symposium on Algorithms and Computation, ISAAC 2019, Shanghai University of Finance and Economics, Shanghai, China, 8–11 December 2019. LIPIcs, vol. 149, pp. 49:1–49:22. Schloss Dagstuhl - Leibniz-Zentrum für Informatik (2019)
5. Chalermsook, P., et al.: From gap-ETH to FPT-inapproximability: clique, dominating set, and more. In: 58th IEEE Annual Symposium on Foundations of Computer Science, FOCS 2017, Berkeley, CA, USA, 15–17 October 2017, pp. 743–754 (2017)
6. Chawla, S. (ed.): Proceedings of the 2020 ACM-SIAM Symposium on Discrete Algorithms, SODA 2020, Salt Lake City, UT, USA, 5–8 January 2020. SIAM (2020)
7. Chitnis, R., Feldmann, A.E., Manurangsi, P.: Parameterized approximation algorithms for bidirected Steiner Network problems (2017)
8. Choudum, S.: A simple proof of the Erdős-Gallai theorem on graph sequences. Bull. Aust. Math. Soc. **33**(1), 67–70 (1986)
9. Chudnovsky, M., Pilipczuk, M., Pilipczuk, M., Thomassé, S.: Quasi-polynomial time approximation schemes for the maximum weight independent set problem in H-free graphs. In: Chawla [6], pp. 2260–2278 (2020)
10. Corneil, D., Lerchs, H., Burlingham, L.: Complement reducible graphs. Discrete Appl. Math. **3**(3), 163–174 (1981)
11. Cygan, M., et al.: Parameterized Algorithms. Springer, Cham (2015). https://doi.org/10.1007/978-3-319-21275-3
12. Dabrowski, K., Lozin, V., Müller, H., Rautenbach, D.: Parameterized algorithms for the independent set problem in some hereditary graph classes. In: Iliopoulos, C.S., Smyth, W.F. (eds.) IWOCA 2010. LNCS, vol. 6460, pp. 1–9. Springer, Heidelberg (2011). https://doi.org/10.1007/978-3-642-19222-7_1

13. Dinur, I., Manurangsi, P.: ETH-hardness of approximating 2-CSPs and Directed Steiner Network. In: Karlin, A.R. (ed.) 9th Innovations in Theoretical Computer Science Conference, ITCS 2018, Cambridge, MA, USA, 11–14 January 2018. LIPIcs, vol. 94, pp. 36:1–36:20. Schloss Dagstuhl - Leibniz-Zentrum für Informatik (2018)

14. Dinur, I., Manurangsi, P.: ETH-hardness of approximating 2-CSPs and Directed Steiner Network. CoRR, abs/1805.03867 (2018)

15. Dvořák, P., Feldmann, A.E., Rai, A., Rzążewski, P.: Parameterized inapproximability of independent set in H-free graphs. CoRR, abs/2006.10444 (2020)

16. Erdős, P., Szekeres, G.: A Combinatorial Problem in Geometry, pp. 49–56. Birkhäuser Boston, Boston (1987)

17. Feige, U.: Approximating maximum clique by removing subgraphs. SIAM J. Discrete Math. **18**(2), 219–225 (2004)

18. Garey, M., Johnson, D., Stockmeyer, L.: Some simplified NP-complete graph problems. Theoret. Comput. Sci. **1**(3), 237–267 (1976)

19. Grzesik, A., Klimosova, T., Pilipczuk, M., Pilipczuk, M.: Polynomial-time algorithm for maximum weight independent set on P6-free graphs. In: Chan, T.M. (ed.) Proceedings of the Thirtieth Annual ACM-SIAM Symposium on Discrete Algorithms, SODA 2019, San Diego, California, USA, 6–9 January 2019, pp. 1257–1271. SIAM (2019)

20. Halldórsson, M.M.: Approximating discrete collections via local improvements. In: Clarkson, K.L. (ed.) Proceedings of the Sixth Annual ACM-SIAM Symposium on Discrete Algorithms, San Francisco, California, USA, 22–24 January 1995, pp. 160–169. ACM/SIAM (1995)

21. Håstad, J.: Clique is hard to approximate within $n^{(1-\varepsilon)}$. In: Acta Mathematica, pp. 627–636 (1996)

22. Karp, R.M.: Reducibility among combinatorial problems. In: Miller, R.E., Thatcher, J.W., Bohlinger, J.D. (eds.) Complexity of Computer Computations. IRSS, pp. 85–103. Springer, Boston (1972). https://doi.org/10.1007/978-1-4684-2001-2_9

23. Khot, S., Ponnuswami, A.K.: Better inapproximability results for MaxClique, chromatic number and Min-3Lin-Deletion. In: Bugliesi, M., Preneel, B., Sassone, V., Wegener, I. (eds.) ICALP 2006. LNCS, vol. 4051, pp. 226–237. Springer, Heidelberg (2006). https://doi.org/10.1007/11786986_21

24. Laekhanukit, B.: Parameters of two-prover-one-round game and the hardness of connectivity problems. In: Proceedings of the Twenty-Fifth Annual ACM-SIAM Symposium on Discrete Algorithms (SODA), pp. 1626–1643. SIAM (2014)

25. Lokshantov, D., Vatshelle, M., Villanger, Y.: Independent set in P5-free graphs in polynomial time. In: Chekuri, C. (ed.) Proceedings of the Twenty-Fifth Annual ACM-SIAM Symposium on Discrete Algorithms, SODA 2014, Portland, Oregon, USA, 5–7 January 2014, pp. 570–581. SIAM (2014)

26. Lokshtanov, D., Ramanujan, M.S., Saurabh, S., Zehavi, M.: Parameterized complexity and approximability of directed odd cycle transversal. CoRR, abs/1704.04249 (2017)

27. Lozin, V.V., Milanic, M.: A polynomial algorithm to find an independent set of maximum weight in a fork-free graph. J. Discrete Algorithms **6**(4), 595–604 (2008)

28. Manurangsi, P.: Tight running time lower bounds for strong inapproximability of maximum k-coverage, unique set cover and related problems (via t-wise agreement testing theorem). In: Chawla [6], pp. 62–81 (2020)

29. Marx, D.: Can you beat treewidth? Theory Comput. **6**(1), 85–112 (2010)

30. Marx, D., Pilipczuk, M.: Optimal parameterized algorithms for planar facility location problems using Voronoi diagrams. CoRR, abs/1504.05476 (2015)
31. Minty, G.J.: On maximal independent sets of vertices in claw-free graphs. J. Comb. Theory Ser. B **28**(3), 284–304 (1980)
32. Sbihi, N.: Algorithme de recherche d'un stable de cardinalite maximum dans un graphe sans etoile. Discrete Math. **29**(1), 53–76 (1980)

Clique-Width of Point Configurations

Onur Çağırıcı[1], Petr Hliněný[1(✉)], Filip Pokrývka[1],
and Abhisekh Sankaran[2]

[1] Faculty of Informatics, Masaryk University, Brno, Czech Republic
`onur@mail.muni.cz`, {`hlineny,xpokryvk`}`@fi.muni.cz`
[2] Department of Computer Science and Technology, University of Cambridge,
Cambridge, UK
`abhisekh.sankaran@cl.cam.ac.uk`

Abstract. While structural width parameters (of the input) belong to
the standard toolbox of graph algorithms, it is not the usual case in computational geometry. As a case study we propose a natural extension of
the structural graph parameter of *clique-width* to geometric point configurations represented by their *order type*. We study basic properties
of this clique-width notion, and relate it to the monadic second-order
logic of point configurations. As an application, we provide several linear
FPT time algorithms for geometric point problems which are NP-hard
in general, in the special case that the input point set is of bounded
clique-width and the clique-width expression is also given.

Keywords: Point configuration · Order type · Fixed-parameter
tractability · Relational structure · Clique-width

1 Introduction

An order type is a useful means to characterize the combinatorial properties of
a finite point configuration in the plane. As introduced in Goodman and Pollack [17,18], the *order type* of a given set P of points assigns, to each ordered
triple $(a, b, c) \in P^3$ of points, the orientation (either clockwise or counter-
clockwise) of the triangle abc in the plane. More generally, if the point set P
is not in a general position, the triple (a, b, c) may also be collinear (as the
natural third option).

Knowing the order type of a point set P is sufficient to determine some useful
combinatorial properties of the geometric set P, such as the convex hull of P and
other. For example, problems of finding convex holes in P or dealing with the
intersection pattern of straight line segments with ends in P, can be solved by
looking only at the order type of P and not on its geometric properties. That is

O. Çağırıcı, P. Hliněný and F. Pokrývka have been supported by the Czech Science
Foundation, project no. 20-04567S. A. Sankaran has been supported by the Leverhulme
Trust through a Research Project Grant on 'Logical Fractals'.

I. Adler and H. Müller (Eds.): WG 2020, LNCS 12301, pp. 54–66, 2020.
https://doi.org/10.1007/978-3-030-60440-0_5

why order types of points sets are commonly studied from various perspectives in the field of computational geometry, e.g., [2–6,16,19,27].

On the other hand, knowing the order type of P is obviously not sufficient to answer questions involving truly "geometric" aspects of P, e.g., distances in P (straight-line or geodesic), or angles between the lines or the area of polygons within P. Nevertheless, even in such geometry-based problems, a more efficient subroutine computing with the order type of P might speed-up the overall computation, which can be a promising direction for future research.

Unlike in the area of graphs and graph algorithms, where structural width parameters are very common for many years, at least since the 90's, no similar effort can be seen in combinatorial and computational geometry. We would like to introduce, in this paper, possible combinatorial handling of "structural complexity" of a given point configuration P through defining its "width" (which we would assume to be small for the studied inputs).

Inspired by graph structure parameters, the obvious first attempt could be to extend the traditional notion of *tree-width* [26]. Such an extension is technically possible (cf. tree-width of the Gaifman graph of a relational structure), but the huge problem is that for the tree-width to be upper-bounded, the underlying structure must be "sparse" – in particular, it can only have a linear number of edges/tuples. This is clearly not satisfied for the order type in which about half of all triples are of each orientation.

A better option comes with another traditional, but not so well-known, notion of *clique-width* [12]. Clique-width can be bounded even on dense graphs, such as on cliques, and, similarly to the case of Courcelle's theorem [9] for tree-width, clique-width also enjoys some nice metaalgorithmic properties, e.g. [11,15]. This includes solving any decision (and some optimization as well) problems formulated in the monadic second-order (MSO) logic in linear time. Hence, alongside the (Sect. 2) proposed definition of the clique-width of point configurations, we will introduce the MSO language of their order types and discuss which problems can be formulated in this language (and hence solved in linear time if a point set with a decomposition of bounded clique-width is given on the input).

Arguments which were omitted due to space restrictions can be found in the full paper preprint arXiv:2004.02282.

2   Order Types and Clique-Width

We now recall the notion of an order type in a formal setting, and propose a definition of the clique-width of (the order type of) a point configuration, based on a natural specialization of the very general concept of clique-width of relational structures. A *relational structure* $S = (U, R_1^S, \ldots, R_a^S)$ of the signature $\sigma = \{R_1, \ldots, R_a\}$ consists of a universe (a finite set) U and a (finite) list of relations $R_1^S, \ldots, R_a^S$ over U. For instance, for graphs, $U = V(G)$ is the vertex set and $R_1^G = E(G)$ is the binary symmetric relation of edges of G.

For a set of points P, here *always* considered in the plane, consider a map $\omega : P^3 \to \{+, -, 0\}$ where $\omega(a, b, c) = 0$ if the triple of points[1] a, b, c is collinear, $\omega(a, b, c) = +$ if abc forms a counter-clockwise oriented triangle, and $\omega(a, b, c) = -$ otherwise. Then ω is traditionally called the order type of P, but we, for technical reasons, prefer defining the *order type* of P as the ternary relation $\Omega \subseteq P^3$ such that $(a, b, c) \in \Omega$ iff $\omega(a, b, c) = +$. Hence we have formally got a relational structure (P, Ω) of the signature consisting of one ternary symbol. We will also write $\Omega(P)$ to emphasize that Ω is the order type of the point set P.

Observe that $\omega(a, b, c) = -$ iff $(b, a, c) \in \Omega$, and $\omega(a, b, c) = 0$ iff $(a, b, c) \notin \Omega$ and $(b, a, c) \notin \Omega$. Hence, the relation Ω fully determines the usual order type of P. Furthermore, $(a, b, c) \in \Omega$ implies $(b, c, a) \in \Omega$ and $(c, a, b) \in \Omega$, and so we call the set of triples $\{(a, b, c), (b, c, a), (c, a, b)\}$ the *cyclic closure* of $(a, b, c) \in \Omega$.

Unary Clique-Width. We start with the definition of ordinary graph clique-width. Let an *ℓ-expression* be an algebraic expression using the following four operations on vertex-labelled graphs using ℓ labels:

(u1) create a new vertex with single label i;
(u2) take the disjoint union of two labelled graphs;
(u3) add all edges between the vertices of label i and label j $(i \neq j)$; and
(u4) relabel all vertices with label i to label j.

The *clique-width* $\mathrm{cw}(G)$ of a graph G equals the minimum ℓ such that (some labelling of) G is the value of an ℓ-expression.

The idea behind this definition is that the edge set of a graph G can be constructed with "bounded amount of information"; this is since we have only a fixed number of distinct labels and vertices of the same label are, intuitively speaking, further indistinguishable by the expression.

This definition has an immediate generalization to the *unary clique-width* of an order type $\Omega(P)$ of a point set P (the adjective referring to the fact that labels occur as unary predicates in the definition): replace (u3) with

(u3') add to Ω the cyclic closures of all triples (a, b, c) of distinct elements such that a is labelled i, b is labelled j and c is labelled k.

Unfortunately, although being very simple, this definition is generally not satisfactory due to problems discussed, e.g., in [1] and specifically illustrated for order types in our Proposition 2.

Multi-ary Clique-Width. While in the case of graphs (whose edge relation is binary) it is sufficient to consider clique-width expressions with unary labels, for the ternary order-type relation (as well as for other relational structures of higher arity) it is generally necessary to allow creation of "intermediate" binary labels, which are labelled pairs of points of P.

This generalization, which is in agreement with the treatment by Blumensath and Courcelle [7], leads to the proposed new definition:

---

[1] Note that if any two of a, b, c are not distinct, then we automatically get $\omega(a, b, c) = 0$.

Definition 1 (Clique-width of a point configuration). Consider an algebraic expression $\mathcal{E}$ using the following five operations on labelled relational structures (of arity 3 in this case) over point sets:

(w1) create a new point with single label i;

(w2) take the disjoint union of two point sets;

(w3) for every two points, point a of label i and point b of label j ($i \neq j$), give the ordered pair (a, b) binary label k;[2]

(w4) for every three pairwise distinct points, a, b and c such that c is of (unary) label i, and the pair (a, b) is of (binary) label k, add to the structure the cyclic closure of the ordered triple (a, b, c);

(w4') under the same conditions as in (w4), add the cyclic closure of (b, a, c);

(w5) relabel all tuples (singletons or pairs) with label i to label j of equal arity.

The *value* of such expression $\mathcal{E}$ is the ternary relational structure on the points created by (w1) and consisting of the triples added by (w4) and (w4'). The auxiliary labels introduced in $\mathcal{E}$ are no longer relevant after the evaluation of $\mathcal{E}$.

The *width* of an expression $\mathcal{E}$ constructed as in (w1)–(w5) equals the sum of arities of the labels occuring in $\mathcal{E}$.[3] The *clique-width* $cw(P)$ of a point configuration P equals the minimum ℓ such that the order type $\Omega(P)$ of P is the value of an expression of width at most ℓ.

Note that, although the clique-width is a concrete natural number, we will not be interested in the exact value of it, but instead study whether the clique-width is bounded or unbounded on a given class of point configurations.

For a closer explanation of this concept, we present a basic example:

Proposition 2. *Let P be an arbitrary finite set of points in a strictly convex position.[4] Then the clique-width of P is bounded by a constant, while the unary clique-width of P is unbounded.*

Proof Outline. Let the points of P be $p_1, p_2, \ldots, p_n$ in the counter-clockwise order (starting arbitrarily). We start with p_1 and stepwise add p_2, p_3 etc., changing previous points to label 1 and the added point created with unique label 2. See Fig. 1. Along the steps, after the creation of p_j, we add the binary label 3 to all pairs labelled 1 and 2, i.e., to (p_i, p_j) for all $i < j$, and create the order triples $(p_i, p_{i'}, p_j)$ of three distinct points over all pairs $(p_i, p_{i'})$ of label 3 and p_j of label 2. This construction witnesses that the clique-width of P is at most 4.

On the other hand, take unary clique-width with ℓ labels, and $|P| \geq 2\ell + 1$. An arbitrary ℓ-expression for $\Omega(P)$ must involve a union operation (the "last" one) over two sets such that one has more than ℓ points. Then, at the time of taking the union, there are two points a, b of the same label in the set by the pigeon-hole principle. Let c be any point from the other set. Then there is no

---

[2] After this operation, (a, b) may hold more than one binary label, which is ok.

[3] Note that this 'sum of arities' measure directly generalizes the number ℓ of unary labels in the expression of (u1)–(u4).

[4] That is, in the convex hull of P every point of P is a vertex.

way, based on the labels, to distinguish between the triples (a, b, c) and (b, a, c), which must have the opposite orientations in $\Omega(P)$. Therefore, the clique-with of P must be at least $\ell + 1$.

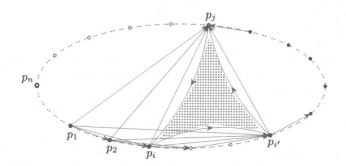

Fig. 1. An illustration of the expression (width 4) in Proposition 2. Unary labels 1 are blue (on $p_1, \ldots, p_{j-1}$), the unique label 2 is orange (on p_j just added), and the binary labels 3 are with green arrows. We are just creating the red triple(s) $(p_i, p_{i'}, p_j)$. (Color figure online)

Annotated Point Configurations. In some situations, it may be useful to consider a point configuration P with additional information (or structure) on the points or selected pairs of them. An exemplary use case for such annotations is to study polygons, with P as the vertex set, for which case we are considering an order type $\Omega(P)$ together with a directed Hamiltonian cycle on P representing the counter-clockwise boundary of P.

Formally, we simply consider relational structures (over P) with the signature consisting of the ternary order type and arbitrary binary or unary symbols. The *clique-width* of such an *annotated point configuration* P is, naturally, as in Definition 1 with additional rules that some of the auxiliary unary and binary labels are at the end turned into the desired unary and binary relations on P.

3   MSO Logic of Order Types

The beginning of this section is devoted to a short introduction of the *monadic second-order (MSO) logic* of relational structures. Recall a *relational structure* $S = (U, R_1^S, \ldots, R_q^S)$ of the signature $\sigma = \{R_1, \ldots, R_q\}$.

The language of MSO logic (of the signature σ) then consists of the standard propositional logic, quantifiers $\forall, \exists$ ranging over elements and subsets of the universe U, and the relational symbols $R_1, \ldots, R_q$ with the following meaning: for R_i of arity a, we have $S \models R_i(x_1, \ldots, x_a)$ if and only if $(x_1, \ldots, x_a) \in R_i^S$.

In our specific case of order types $\Omega(P)$ of point sets P, we use the relational symbol $ordccw(x_1, x_2, x_3)$ for Ω within MSO logic. For example, we can express that a point y lies strictly in the convex hull of points x_1, x_2, x_3 as follows

$$\left[ordccw(x_1, x_2, x_3) \wedge \bigwedge\nolimits_{i=1,2,3} ordccw(x_i, x_{i+1}, y)\right] \vee \qquad (1)$$

$$\left[ordccw(x_3, x_2, x_1) \wedge \bigwedge\nolimits_{i=1,2,3} ordccw(x_{i+1}, x_i, y)\right],$$

where x_4 is taken as x_1.

More generally, we can express that a point $y \in P$ belongs to the convex hull (not necessarily strictly now) of a set $X \subset P$ with the following formula:

$$convhull(X, y) \equiv y \in X \vee \forall x, x' \in X \qquad (2)$$
$$\left[(x \neq x' \wedge \forall z \in X \neg ordccw(x', x, z)) \rightarrow \neg ordccw(x', x, y)\right]$$

Then we may express, for example, that a set $X \subseteq P$ is a convex hole (i.e., no point outside of X belongs to the convex hull of X, and no point of X belongs to the convex hull of the rest of X) with the following:

$$\forall y \notin X \, (\neg convhull(X, y)) \wedge \forall Y \subseteq X \forall z \in X (convhull(Y, z) \rightarrow z \in Y) \qquad (3)$$

Further similar examples are easy to come up with.

Interpretations and Transductions. We sketch the concept of "translating" between relational structures. Consider relational signatures $\sigma = \{R_1, \ldots, R_q\}$ and $\tau = \{R'_1, \ldots, R'_t\}$. A (simple) *MSO interpretation of τ-structures in σ-structures* is a t-tuple of MSO formulas $\Psi = (\psi_i : 1 \leq i \leq t)$ of the signature σ, where the number of free variables of ψ_i equals the arity a_i of R'_i. A τ-structure T is *interpreted* in a σ-structure S via Ψ if T and S share the same ground set U and, for each $1 \leq i \leq t$, we have $(x_1, \ldots, x_{a_i}) \in R'^T_i \iff S \models \psi_i(x_1, \ldots, x_{a_i})$.

As a short example, consider a point set P and its mirror image P'. Then the order type $\Omega(P')$ can be interpreted in $\Omega(P)$ simply by taking $\psi_1(a, b, c) \equiv ordccw(b, a, c)$. The true power of interpretations will show up in the following.

There is a more general concept of a *transduction* from a σ-structure S to a set of τ-structures which, before taking an (MSO) interpretation, has abilities (in this order of application): (i) to equip S with a fixed number of arbitrary parameters given as unary labels (because of this, the result of a transduction is not deterministic, but a set of τ-structures), (ii) to "amplify" the ground set of S by taking a bounded number of disjoint copies of S, and (iii) to subsequently restrict the ground set by an MSO formula with one free variable. See Courcelle and Engelfriet [10] for more technical details on transductions.

Considering a transduction Ψ (as described above) and a σ-structure S, let $\Psi(S)$ denote the set of τ-structures which result from S under the transduction Ψ. For a class of relational structures $\mathcal{S}$, the image under a transduction Ψ of the class $\mathcal{S}$ is the union of all transduction results, precisely, $\Psi(\mathcal{S}) := \bigcup_{S \in \mathcal{S}} \Psi(S)$.

Note that one can come up with various notions of clique-width of relational structures (also giving distinct numbers for the same structure), but the underlying essence is always "the same". In order to smoothen out marginal technical differences between the various definitions, we consider the following. We say that a class $\mathcal{S}$ is of *bounded clique-width* if there exists a constant h such that the clique-width of every $S \in \mathcal{S}$ is at most h. On such abstract level, we then have the following *crucial characterization* (essentially a metadefinition):

Theorem 3 (Blumensath and Courcelle [7, Proposition 27]**).** *A class S of finite relational structures (of the same signature) is of bounded clique-width, if and only if S is contained in the image of the class of finite trees under an MSO transduction.*

For a *very informal* explanation of the meaning of this statement, we remark that a tree which is the preimage of the mentioned transduction gives a hierarchical structure to the clique-width expression in Definition 1. The arbitrary transduction parameters then determine particular operations (and labelling) used within the expression, and the formula(s) of a final interpretation roughly encodes Definition 1 itself. No copying ("amplification") is necessary there.

Since the concept of a transduction is transitive, Theorem 3 implies:

Corollary 4. *If a class S of order types (of points) is of bounded clique-width, then the image of S under an MSO transduction is also of bounded clique-width.*

Deciding MSO Properties. Perhaps the most important application of bounded clique-width of point configurations P could be in faster deciding of MSO-definable properties (and, in greater generality, of some optimization and counting properties as well, see examples in [11]) of the order type of P.

Theorem 5 (Courcelle et al. [11]**, via Theorem** 3**).** *Consider a class S of finite relational structures of signature σ and of bounded clique-width. For any MSO sentence φ of signature σ, if a structure $S \in S$ is given on the input alongside with a clique-width expression of bounded width, then we can decide in linear time whether $S \models \varphi$ (i.e., whether S has the property φ).*

Furthermore, under the same assumptions for S and for an MSO formula $\varphi(X)$ with a free set variable X, we can find in linear time a minimum- or maximum-cardinality set X such that $S \models \varphi(X)$, and we can enumerate all sets X such that $S \models \varphi(X)$ in time which is linear in the input plus output size.

4   Assorted Examples

First, to give readers a better feeling about how big the clique-width of "nicely looking" point sets in the plane can be, we show the following:

Theorem 6. *Let P be a point configuration, $P_0 \subseteq P$ and $d = |P \setminus P_0|$.*

(a) *If all points of P_0 are collinear, then the clique-width of P is in $\mathcal{O}(d)$.*
(b) *Assume the points of P_0 are in a strictly convex position. If $d \leq 1$, then the clique-width of P is bounded (by a constant). On the other hand, there exist examples already with $d = 2$ and unbounded clique-width of P.*

Proof Outline. In case (a), we first create the d points of $P \setminus P_0$, each with its unique label, and their counter-clockwise order triples. See Fig. 2(a). Then we stepwise create the collinear points of P_0, ordered from left to right. During the steps, we add binary labels on P_0 between each pair from left to right, and we also in the right order create the needed order triples having one point in P_0 and two points in $P \setminus P_0$. At the end, we easily create from the binary labels on P_0 the remaining order triples having two points in P_0 and one in $P \setminus P_0$.

In case (b), if $d = 1$, we construct an expression similarly as in Proposition 2, but we simultaneously proceed in two subsequences of the counter-clockwise perimeter of P_0, "opposite" to each other. This process of construction allows us to create also the counter-clockwise order triples involving the sole point of $P \setminus P_0$ (in "the middle").

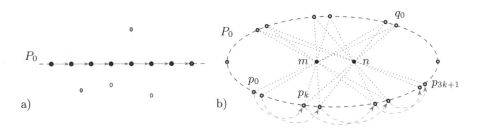

Fig. 2. Illustrations of the two parts of Theorem 6. (a) Labelling for an expression of bounded width. (b) A sketch of interpreting a large grid within the point configuration. (Color figure online)

In case (b) with $d \geq 2$, we present a construction informally shown in Fig. 2(b). The underlying idea is to use the points of $P \setminus P_0$ to "mutually relate" opposite points of P_0, such as the depicted collinear triples p_0, m, q_0 and q_0, n, p_k. Collinear triples are easy to detect within the order type, hence we can this way interpret the binary relation between p_0 and p_k, analogously between subsequent p_1 and p_{k+1}, and so on (see the green dashed arrows in the picture). Together with a description of neighbouring points of P_0 – see x and x' in (2), we can interpret an arbitrarily large square grid graph on the points $p_0, p_1, \ldots, p_k, p_{k+1}, \ldots$ Since the square grid is a folklore basic example of unbounded clique-width [10], Corollary 4 implies that the clique-width of such configurations P (with $d = 2$) is unbounded. A similar construction, albeit more complicated, with "doubling" the points q_i, can show the same result without having collinear triples in P.

Some NP-Hard Problems of Point Configurations

As already mentioned, perhaps the most interesting computing application of clique-width of point sets could be in designing algorithms which run in parameterized polynomial, or even linear, time with respect to the clique-width as the parameter. This is especially relevant for problems for which no such algorithms are believed to exist in general, such as for NP-hard problems.

A parameterized problem has an *FPT algorithm* if the algorithm runs in time $\mathcal{O}(f(d) \cdot n^c)$ where f is an arbitrary computable function of the (fixed) parameter d, and c is a constant. If $c = 1$, then we speak about a linear FPT algorithm (e.g., this is the complete case of Theorem 5).

Since, except the binary case such as that of graphs, there is no known FPT algorithm (even approximation one) for finding a clique-width expression of relational structures of bounded clique-width, we must assume that an expression of bounded width is given alongside with the input point configuration. Notice that for the above presented examples of small clique-width, the relevant expressions are very natural and easy to come with.

General Position Subset. This problem asks whether, for a given point set P and integer k, there exists a subset $Q \subseteq P$ such that no three points of Q are collinear and $|Q| > k$. This problem is NP-hard and APX-hard by [14].

Theorem 7. *Assume a point set P is given alongside with a clique-width expression (for $\Omega(P)$) of width d. Then the* GENERAL POSITION SUBSET *problem of P is solvable in linear FPT time with respect to the parameter d.*

Proof. We write the MSO formula

$$\varphi(X) \equiv \forall x, y, z \in X \left[x \neq y \neq z \neq x \to \left(\mathrm{ordccw}(x, y, z) \vee \mathrm{ordccw}(y, x, z) \right) \right]$$

to say that no three points in X are collinear, and then compute using Theorem 5 the value $\max_{\Omega(P) \models \varphi(X)} |X|$ and compare to k. □

A very similar simple approach works also for the NP-hard problem HITTING SET FOR INDUCED LINES [25], which asks for a minimum-cardinality subset $H \subseteq P$ such that the lines between each pair of points of P all contain a point of H.

Minimum Convex Partition. Consider a given point set P and an integer k. The objective of this problem [13] is to decide whether the convex hull $conv(P)$ of P can be partitioned into $\leq k$ convex faces. By a *convex face* in this situation we mean the convex hull of a subset $Q \subseteq P$ which is a convex hole of P (recall (3)). Note that in our definition Q must be strictly convex, but we may as well apply a non-strict variant in which some points of Q (possibly) are not vertices of $conv(Q)$ but lie on the boundary of $conv(Q)$; the arguments would be similar.

This problem has been recently claimed NP-hard [20]. Unfortunately, inherent limitations of MSO logic do not allow us to directly formulate the MINIMUM CONVEX PARTITION as an MSO optimization problem (one is not allowed to quantify set families), but we can handle it if we take k as an additional parameter.

Theorem 8. *Assume a point set P given alongside with a clique-width expression of width d. The* MINIMUM CONVEX PARTITION *problem of P into $\leq k$ convex faces is solvable in linear FPT time with respect to the parameter $d + k$.*

Proof Outline. Let *convhole*(X) denote the MSO formula (3). We may now write

$$\exists X_1, \ldots, X_k \left[\bigwedge_{1 \leq i \leq k} \text{convhole}(X_i) \; \wedge \; \text{convpartition}(X_1, \ldots, X_k) \right]$$

where the subformula *convpartition* checks whether the convex hulls of the sets X_i partition *conv*(P). At this point, we know that each X_i is a convex hole in P, and we further test for set inclusion and the following two conditions:

- the boundaries of *conv*(X_i) and *conv*(X_j) ($1 \leq i < j \leq k$) do not cross, and
- every boundary edge of *conv*(X_i) is, at the same time, a boundary edge of exactly one of *conv*(X_j) ($i \neq j$) or of *conv*(P).

Both conditions can be, although not easily, stated in MSO over order types.

Terrain Guarding. Another NP-hard problem formulated on point sets [22] is that of guarding an x-monotone polygonal line L with the given vertex set P. The objective of guarding is to find a minimum-cardinality vertex guard set $G \subseteq P$ such that every point ℓ on L is seen by some point $g \in G$ "from above the terrain", that is, the straight line segment from g to ℓ is never strictly below L.

Note that the point set P (no two points of the same x-coordinate) uniquely determines the terrain L, with the vertices ordered by their x-coordinates as $P = (p_1, p_2, \ldots, p_n)$. However, the order type $\Omega(P)$ does not (unless we would add an auxiliary point "at infinity" in the y-axis direction). That is why we assume the terrain L given as a relational structure consisting of ternary $\Omega(P)$ and the binary successor relation consisting of the pairs (p_1, p_{i+1}) for $1 \leq i < n$.

Fig. 3. Guarding a terrain: the two black square vertices guard the whole terrain, but the bottom horizontal segment is not seen by any single one of them. To turn this (pair of guards) into a valid segmented terrain guarding instance with a solution, we may subdivide the bottom segment into two segments with a new vertex (the hollow dot) of the terrain – each guard would then see an entire one of the two segments.

There is one further complication in regard of the order type $\Omega(P)$ of the terrain in this problem: if, in an instance, some edge of L is seen together by two guards, but no one sees the full edge, then knowing only $\Omega(P)$ is not sufficient to verify validity of such a solution (see Fig. 3). That is why we define here the SEGMENTED TERRAIN GUARDING variant as follows: for every segment s of L there must exist a vertex guard g seeing entire s and, moreover, there is a dedicated subset $P_1 \subseteq P$ such that the guards g are selected from P_1. By a natural subdivision of terrains in the hard instances of terrain guarding [22] we immediately get that also SEGMENTED TERRAIN GUARDING is NP-hard.

Theorem 9. *Assume a polygonal terrain L given alongside with a clique-width expression of width d (defining both the successor relation and the order type of the vertices, cf. end of Sect. 2). The* SEGMENTED TERRAIN GUARDING *problem of L is solvable in linear FPT time with respect to the parameter d.*

Proof Outline. We show a formula *seguard*(X) stating that every segment of the terrain L is seen by one point of X. Then, we verify that, for every successive pair of vertices (p_i, p_{i+1}) of L, there exists $x \in X$ such that;

- the triple (x, p_i, p_{i+1}) is oriented counter-clockwise (for x to see the segment $\overline{p_i p_{i+1}}$ "from above"), and
- no "peak" z on L between $\overline{p_i, p_{i+1}}$ and x is oriented clockwise from (x, p_{i+1}) (if z is to the left of x) or counter-clockwise from (x, p_i) (z to the right of x).

This suffices since L is x-monotone. Then Theorem 5 finishes the argument.

We can similarly handle the orthogonal terrain guarding problem which is also NP-hard [8]. Another possible extension is to minimize the sum of weighted guards, using a weighted variant of Theorem 5 (as in [11]). However, our approach to terrain guarding cannot be directly extended to the traditional and more general *Art gallery* (guarding) problem [23], not even in the adjusted case when each edge of the polygon is seen by a single vertex guard. This is due to possible presence of "blind spots" in the interior of the polygon which cannot be determined knowing just the order type $\Omega(P)$ and the boundary edges of the polygon on P. Interested readers may find more in the full paper.

Polygon Visibility Graph. As we have mentioned the Art gallery problem, we briefly add that people are also studying problems related to the *visibility graph* of a given polygon Q. The visibility graph of Q has the same vertex set as Q and the edges are those line segments with ends in the vertices of Q which are disjoint from the complement of the polygon. We give the following toolbox:

Theorem 10. *Assume a polygon Q with vertex set P given as a relational structure consisting of the order type $\Omega(P)$ and the counter-clockwise Hamiltonian cycle of edges of Q. Then the visibility graph of Q has an MSO interpretation in Q.*

5 Conclusions

We managed to show, in this limited space, only few example applications of bounding the clique-width in efficient parameterized algorithms for geometric point problems. More examples of similar kind could be added but, as a future work, we would especially like to investigate possible applications to "metric" problems. Of course, MSO logic of order types cannot express metric properties of a point set, but it could be possible that in some problems the enumerative part of Theorem 5 provided us with a relatively short list of small subconfigurations which would then be processed even by brute force, resulting in a faster algorithm. For instance, we suggest to investigate in this manner the problem of a minimum area triangle on a given point set, which is in general 3SUM-hard (that is, not believed to have a subquadratic algorithm).

Another possible extension would be to consider order types in dimension 3 (or higher), but then even a strictly convex point set could easily have unbounded clique-width – the quaternary relational structures of such order types just seem to be too complex even in very simple cases.

Lastly, we mention another very natural question; can the clique-width of a point configuration be at least approximated by an FPT algorithm with the width as the fixed parameter? Such an approximation is possible in the case of graph clique-width [21,24], thanks to the close relation of graph clique-width to rank-width and to binary matroids. Perhaps the natural correspondence of order types to oriented matroids could be of some help in this research direction.

Acknowledgments. We would like to thank to Achim Blumensath and Bruno Courcelle for discussions about the clique-width of relational structures.

References

1. Adler, H., Adler, I.: A note on clique-width and tree-width for structures. CoRR abs/0806.0103 (2008). http://arxiv.org/abs/0806.0103
2. Aichholzer, O., Aurenhammer, F., Krasser, H.: Enumerating order types for small point sets with applications. Order **19**(3), 265–281 (2002)
3. Aichholzer, O., et al.: Minimal representations of order types by geometric graphs. In: Archambault, D., Tóth, C.D. (eds.) GD 2019. LNCS, vol. 11904, pp. 101–113. Springer, Cham (2019). https://doi.org/10.1007/978-3-030-35802-0_8
4. Aichholzer, O., Krasser, H.: Abstract order type extension and new results on the rectilinear crossing number. Comput. Geom. **36**(1), 2–15 (2007)
5. Aichholzer, O., Kusters, V., Mulzer, W., Pilz, A., Wettstein, M.: An optimal algorithm for reconstructing point set order types from radial orderings. Int. J. Comput. Geom. Appl. **27**(1–2), 57–84 (2017)
6. Aloupis, G., Iacono, J., Langerman, S., Özkan, Ö., Wuhrer, S.: The complexity of order type isomorphism. In: SODA, pp. 405–415. SIAM (2014)
7. Blumensath, A., Courcelle, B.: Recognizability, hypergraph operations, and logical types. Inf. Comput. **204**(6), 853–919 (2006)
8. Bonnet, É., Giannopoulos, P.: Orthogonal terrain guarding is NP-complete. JoCG **10**(2), 21–44 (2019)
9. Courcelle, B.: The monadic second order logic of graphs I: recognizable sets of finite graphs. Inf. Comput. **85**, 12–75 (1990)
10. Courcelle, B., Engelfriet, J.: Graph Structure and Monadic Second-Order Logic: A Language-Theoretic Approach, Encyclopedia of Mathematics and Its Applications, vol. 138. Cambridge University Press, Cambridge (2012)
11. Courcelle, B., Makowsky, J.A., Rotics, U.: Linear time solvable optimization problems on graphs of bounded clique-width. Theory Comput. Syst. **33**(2), 125–150 (2000)
12. Courcelle, B., Engelfriet, J., Rozenberg, G.: Context-free handle-rewriting hypergraph grammars. In: Ehrig, H., Kreowski, H.-J., Rozenberg, G. (eds.) Graph Grammars 1990. LNCS, vol. 532, pp. 253–268. Springer, Heidelberg (1991). https://doi.org/10.1007/BFb0017394
13. Demaine, E., Fekete, S., Keldenich, P., Krupke, D., Mitchell, J.: Geometric optimization challenge, part of CG Week in Zurich, Switzerland, 22–26 June 2020 (2020). https://cgshop.ibr.cs.tu-bs.de/competition/cg-shop-2020/
14. Froese, V., Kanj, I.A., Nichterlein, A., Niedermeier, R.: Finding points in general position. Int. J. Comput. Geom. Appl. **27**(4), 277–296 (2017)

15. Ganian, R., Hliněný, P.: On parse trees and Myhill-Nerode-type tools for handling graphs of bounded rank-width. Discrete Appl. Math. **158**(7), 851–867 (2010)
16. Goaoc, X., Hubard, A., de Joannis de Verclos, R., Sereni, J., Volec, J.: Limits of order types. In: Symposium on Computational Geometry. LIPIcs, vol. 34, pp. 300–314. Schloss Dagstuhl - Leibniz-Zentrum für Informatik (2015)
17. Goodman, J.E., Pollack, R.: Multidimensional sorting. SIAM J. Comput. **12**(3), 484–507 (1983)
18. Goodman, J.E., Pollack, R.: Upper bounds for configurations and polytopes in R^d. Discrete Comput. Geom. **1**, 219–227 (1986)
19. Goodman, J.E., Pollack, R., Sturmfels, B.: Coordinate representation of order types requires exponential storage. In: Proceedings of the 21st Annual ACM Symposium on Theory of Computing, pp. 405–410. ACM (1989)
20. Grelier, N.: Minimum convex partition of point sets is NP-hard. CoRR abs/1911.07697 (2019). http://arxiv.org/abs/1911.07697
21. Hliněný, P., Oum, S.: Finding branch-decomposition and rank-decomposition. SIAM J. Comput. **38**, 1012–1032 (2008)
22. King, J., Krohn, E.: Terrain guarding is NP-hard. SIAM J. Comput. **40**(5), 1316–1339 (2011)
23. Lee, D.T., Lin, A.K.: Computational complexity of art gallery problems. IEEE Trans. Inf. Theory **32**(2), 276–282 (1986)
24. Oum, S., Seymour, P.D.: Approximating clique-width and branch-width. J. Comb. Theory Ser. B **96**(4), 514–528 (2006)
25. Rajgopal, N., Ashok, P., Govindarajan, S., Khopkar, A., Misra, N.: Hitting and piercing rectangles induced by a point set. In: Du, D.-Z., Zhang, G. (eds.) COCOON 2013. LNCS, vol. 7936, pp. 221–232. Springer, Heidelberg (2013). https://doi.org/10.1007/978-3-642-38768-5_21
26. Robertson, N., Seymour, P.D.: Graph minors. II. Algorithmic aspects of tree-width. J. Algorithms **7**(3), 309–322 (1986)
27. Roy, B.: Point visibility graph recognition is NP-hard. Int. J. Comput. Geom. Appl. **26**(1), 1–32 (2016)

On the Complexity of Finding Large Odd Induced Subgraphs and Odd Colorings

Rémy Belmonte[1] and Ignasi Sau[2($\boxtimes$)]

[1] University of Electro-Communications, Chofu, Japan
remy.belmonte@gmail.com
[2] LIRMM, Université de Montpellier, CNRS, Montpellier, France
ignasi.sau@lirmm.fr

Abstract. We study the complexity of the problems of finding, given a graph G, a largest induced subgraph of G with all degrees odd (called an *odd* subgraph), and the smallest number of odd subgraphs that partition $V(G)$. We call these parameters $\mathsf{mos}(G)$ and $\chi_{\mathsf{odd}}(G)$, respectively. We prove that deciding whether $\chi_{\mathsf{odd}}(G) \leq q$ is polynomial-time solvable if $q \leq 2$, and **NP**-complete otherwise. We provide algorithms in time $2^{\mathcal{O}(\mathsf{rw})} \cdot n^{\mathcal{O}(1)}$ and $2^{\mathcal{O}(q \cdot \mathsf{rw})} \cdot n^{\mathcal{O}(1)}$ to compute $\mathsf{mos}(G)$ and to decide whether $\chi_{\mathsf{odd}}(G) \leq q$ on n-vertex graphs of rank-width at most rw, respectively, and we prove that the dependency on rank-width is asymptotically optimal under the **ETH**. Finally, we give some tight bounds for these parameters on restricted graph classes or in relation to other parameters.

Keywords: Odd subgraph · Odd coloring · Rank-width · Parameterized complexity · Single-exponential algorithm · Exponential Time Hypothesis

1 Introduction

Gallai proved, around 60 years ago, that the vertex set of every graph can be partitioned (in polynomial time) into two sets, each of them inducing a subgraph in which all vertices have even degree (cf. [26, Exercise 5.19]). Let us call such a subgraph an *even* subgraph, and an *odd* subgraph is defined similarly. Hence, every graph G contains an even induced subgraph with at least $|V(G)|/2$ vertices. The analogous properties for odd subgraphs seem to be more elusive. For a graph G, let $\mathsf{mos}(G)$ and $\chi_{\mathsf{odd}}(G)$ be the order of a largest odd induced subgraph of G and the minimum number of odd induced subgraphs of G that partition $V(G)$, respectively. Note that for $\chi_{\mathsf{odd}}(G)$ to be well-defined, each connected component of G must have even order.

Concerning the former parameter, the following long-standing –and still open– conjecture is cited as "part of the graph theory folklore" by Caro [7]:

Work supported by French projects DEMOGRAPH (ANR-16-CE40-0028) and ESIGMA (ANR-17-CE23-0010), the program "Exploration Japon 2017" of the French embassy in Japan, and the JSPS KAKENHI grant number JP18K11157.

I. Adler and H. Müller (Eds.): WG 2020, LNCS 12301, pp. 67–79, 2020.
https://doi.org/10.1007/978-3-030-60440-0_6

there exists a positive constant c such that every graph G without isolated vertices satisfies $\mathsf{mos}(G) \geq c \cdot |V(G)|$. In the following discussion we only consider graphs without isolated vertices. Caro [7] proved that $\mathsf{mos}(G) \geq (1 - o(1))\sqrt{n/6}$ where $n = |V(G)|$, and Scott [33] improved this bound to $\frac{cn}{\log n}$ for some $c > 0$. The conjecture has been proved for particular graph classes, such as trees [30], graphs of bounded chromatic number [33], graphs of maximum degree three [2], and graphs of tree-width at most two [20], also obtaining best possible constants.

As for the complexity of computing $\mathsf{mos}(G)$, Cai and Yang [6] studied, among other problems, two parameterized versions of this problem, and their reductions imply that it is NP-hard. They also prove the NP-hardness of computing the largest size of an even induced subgraph of a graph G, denoted $\mathsf{mes}(G)$. As a follow-up of [6], related problems were studied by Cygan et al. [9] and Goyal et al. [19].

The parameter χ_{odd}, which we call the *odd chromatic number*, has attracted much less interest in the literature. To the best of our knowledge, it has only been considered by Scott [34], who defined it (using a different notation) and proved that the necessary condition discussed above for $\chi_{\mathsf{odd}}(G)$ to be well-defined is also sufficient. He also provided lower and upper bounds on the maximum value of $\chi_{\mathsf{odd}}(G)$ over all n-vertex graphs. In particular, there are graphs G for which $\chi_{\mathsf{odd}}(G) = \Omega(\sqrt{n})$.

Our Contribution. In this article we mostly focus on computational aspects of the parameters mos and χ_{odd}. Note that, given a graph G, deciding whether $\chi_{\mathsf{odd}}(G) \leq 1$ is trivial. We prove that deciding whether $\chi_{\mathsf{odd}}(G) \leq q$ is NP-complete for every $q \geq 3$ using a reduction from q-COLORING. We obtain a dichotomy on the complexity of computing χ_{odd} by showing that deciding whether $\chi_{\mathsf{odd}}(G) \leq 2$ can be solved in polynomial time, through a reduction to the existence of a feasible solution to a system of linear equations over GF[2].

Given the NP-hardness of computing both parameters, we are interested in its parameterized complexity [8,11], namely in identifying relevant parameters k that allow for FPT algorithms, that is, algorithms running in time $f(k) \cdot n^{\mathcal{O}(1)}$ for some computable function f. Since the natural parameter, that is, the solution size, for mos has been studied by Cai and Yang [6] (and its dual as well), and for χ_{odd} the problem is para-NP-hard by our hardness results, we rather focus on *structural parameters*. Two of the most successful ones are definitely tree-width and clique-width, or its parametrically equivalent parameter *rank-width* introduced by Oum and Seymour [29]. This latter parameter is *stronger* than tree-width, in the sense that graph classes of bounded tree-width also have bounded rank-width. We present algorithms running in time $2^{\mathcal{O}(\mathsf{rw})} \cdot n^{\mathcal{O}(1)}$ for computing $\mathsf{mes}(G)$ and $\mathsf{mos}(G)$ for an n-vertex graph G given along with a decomposition tree of width at most rw, and an algorithm in time $2^{\mathcal{O}(q \cdot \mathsf{rw})} \cdot n^{\mathcal{O}(1)}$ for deciding whether $\chi_{\mathsf{odd}}(G) \leq q$. These algorithms are inspired by the ones of Bui-Xuan et al. [3,4] to solve MAXIMUM INDEPENDENT SET parameterized by rank-width and boolean-width, respectively. To the best of our knowledge, our algorithms are the first ones parameterized by rank-width for an NP-hard problem running in time $2^{o(\mathsf{rw}^2)} \cdot n^{\mathcal{O}(1)}$ [1,3,17,18,28].

We also show that the dependency on rank-width of the above algorithms is asymptotically optimal under the Exponential Time Hypothesis (ETH) of Impagliazzo et al. [21,22]. For this, it suffices to obtain a *linear* NP-hardness reduction from a problem for which a subexponential algorithm does not exist under the ETH. While our reduction to decide whether $\chi_{\mathsf{odd}}(G) \leq q$ already satisfies this property, the NP-hardness proof of Cai and Yang [6] for computing $\mathsf{mes}(G)$ and $\mathsf{mos}(G)$, which is from the EXACT ODD SET problem [12], has a *quadratic* blow-up, so only a lower bound of $2^{o(\sqrt{n})}$ can be deduced from it. Motivated by this, we present linear NP-hardness reductions from 2IN3-SAT to the problems of computing $\mathsf{mes}(G)$ and $\mathsf{mos}(G)$. The reduction itself is not very complicated, but the correctness proof requires some non-trivial arguments[1].

Finally, motivated by the complexity of computing these parameters, we obtain two tight bounds on their values. We first prove that for every graph G with all components of even order, $\chi_{\mathsf{odd}}(G) \leq \mathsf{tw}(G)+1$, where $\mathsf{tw}(G)$ denotes the tree-width of G. This result improves the best known lower bound on a parameter defined by Hou et al. [20] (cf. Sect. 5 for the details). On the other hand, we prove that, for every n-vertex graph G such that $V(G)$ can be partitioned into two non-empty sets that are complete to each other (i.e., a *join*), $\mathsf{mos}(G) \geq 2 \cdot \left\lceil \frac{n-2}{4} \right\rceil$. In particular, this proves the conjecture about the linear size of an odd induced subgraph for *cographs*, which are the graphs of clique-width two. This adds another graph class to the previous ones for which the conjecture is known to be true [2,20,30,33]. It is interesting to mention that our proof implies that, for a cograph G, $\chi_{\mathsf{odd}}(G) \leq 3$, and this bound is also tight. While for cographs, or equivalently P_4-free graphs, we have proved that the odd chromatic number is bounded, we also show that it is unbounded for P_5-free graphs.

Organization. We start with some preliminaries in Sect. 2. In Sect. 3 we provide the linear NP-hardness reductions and the polynomial-time algorithm for deciding whether $\chi_{\mathsf{odd}}(G) \leq 2$. The FPT algorithms by rank-width are presented in Sect. 4, and the tight bounds in Sect. 5. We conclude the article in Sect. 6 with a number of open problems and research directions. Additional results for related problems can be found in the full version, available at https://arxiv.org/abs/2002.06078. Due to space limitations, the proofs of the results marked with '($\star$)' can be found in the full version.

2   Preliminaries

Graphs. We use standard graph-theoretic notation, and we refer the reader to [10] for any undefined notation. Let $G = (V, E)$ be a graph, $S \subseteq V$, and H be a subgraph of G. We denote an edge between u and v by uv. The *order* of G is $|V|$. The *degree* (resp. *open neighborhood*, *closed neighborhood*) of a vertex $v \in V$

---

[1] We would like to mention that another NP-hardness proof for computing $\mathsf{mes}(G)$ has very recently appeared online [32]. The proof uses a chain of reductions from MAXIMUM CUT and, although it also involves a quadratic blow-up, it can be avoided by starting from MAXIMUM CUT restricted to graphs of bounded degree.

is denoted by $\mathsf{deg}(v)$ (resp. $N(v)$, $N[v]$), and we let $\mathsf{deg}_H(v) = |N(v) \cap V(H)|$. We use the notation $G - S = G[V(G) \setminus S]$. The *maximum* and *minimum degree* of G are denoted by $\Delta(G)$ and $\delta(G)$, respectively. We denote by P_i the path on i vertices. For two graphs G_1 and G_2, with $V(G_2) \subseteq V(G_1)$, the *union* of G_1 and G_2 is the graph $(V(G_1), E(G_1) \cup E(G_2))$. The operation of *contracting* an edge uv consists in deleting both u and v and adding a new vertex w with neighbors $N(u) \cup N(v) \setminus \{u, v\}$. A graph M is a *minor* of G if it can be obtained from a subgraph of G by a sequence of edge contractions. For a positive integer $k \geq 3$, the *k-wheel* is the graph obtained from a cycle C on k vertices by adding a new vertex v adjacent to all the vertices of C. A *join* in a graph G is a partition of $V(G)$ into two non-empty sets V_1 and V_2 such that every vertex in V_1 is adjacent to every vertex in V_2. For a positive integer i, we denote by $[i]$ the set containing every integer j such that $1 \leq j \leq i$.

Parameterized Complexity. We refer the reader to [8,11,14,27] for basic background on parameterized complexity, and we recall here only some basic definitions. A *parameterized problem* is a decision problem whose instances are pairs $(x, k) \in \Sigma^* \times \mathbb{N}$, where k is called the *parameter*. A parameterized problem is *fixed-parameter tractable* (FPT) if there exists an algorithm $\mathcal{A}$, a computable function f, and a constant c such that given an instance $I = (x, k)$, $\mathcal{A}$ (called an FPT *algorithm*) correctly decides whether $I \in L$ in time $f(k) \cdot |I|^c$. A parameterized problem is *slice-wise polynomial* (XP) if there exists an algorithm $\mathcal{A}$ and two computable functions f, g such that given an instance $I = (x, k)$, $\mathcal{A}$ (called an XP *algorithm*) correctly decides whether $I \in L$ in time $f(k) \cdot |I|^{g(k)}$.

The *Exponential Time Hypothesis* (ETH) of Impagliazzo et al. [21,22] implies that the 3-SAT problem on n variables cannot be solved in time $2^{o(n)}$. We say that a polynomial reduction from a problem Π_1 to a problem Π_2, generating an input of size n_2 from an input of size n_1, is *linear* if $n_2 = \mathcal{O}(n_1)$. Clearly, if Π_1 cannot be solved, under the ETH, in time $2^{o(n)}$ on inputs of size n, and there exists a linear reduction from Π_1 to Π_2, then Π_2 cannot either.

Width Parameters. In this article we mention several width parameters of graphs, such as tree-width, rank-width, clique-width, or boolean-width. However, since we only deal with rank-width in our algorithms (cf. Sect. 4), we give only the definition of this parameter here.

A *decomposition tree* of a graph G is a pair (T, δ) where T is a full binary tree (i.e., T is rooted and every non-leaf node has two children) and δ a bijection between the leaf set of T and the vertex set of G. For a node w of T, we denote by V_w the subset of $V(G)$ in bijection –via δ– with the leaves of the subtree of T rooted at w. We say that the decomposition defines the *cut* $\left(V_w, \overline{V_w}\right)$. The *rank-width* of a decomposition tree (T, δ) of a graph G, denoted by $\mathsf{rw}(T, \delta)$, is the maximum over all $w \in V(T)$ of the rank of the adjacency matrix of the bipartite graph $G[V_w, \overline{V_w}]$. The *rank-width of G*, denoted by $\mathsf{rw}(G)$, is the minimum $\mathsf{rw}(T, \delta)$ over all decomposition trees (T, δ) of G.

Definition of the Problems. A graph is called *odd* (resp. *even*) if every vertex has odd (resp. even) degree. The MAXIMUM ODD SUBGRAPH (resp. MAXIMUM

EVEN SUBGRAPH problem consists in, given a graph G, determining the maximum order of an odd (resp. even) induced subgraph of G, that is, $\mathsf{mos}(G)$ (resp. $\mathsf{mes}(G)$). An *odd q-coloring* of a graph $G = (V, E)$ is a set of q odd induced subgraphs $H_1, \ldots, H_q$ of G such that $V(H_1) \uplus \cdots \uplus V(H_q)$ is a partition of V. The ODD q-COLORING problem consists in determining whether an input graph G admits an odd q-coloring. In the ODD CHROMATIC NUMBER problem, the objective is to determine the smallest integer q such that an input graph G admits an odd q-coloring.

3   Linear Reductions and a Polynomial-Time Algorithm

We first present the linear reductions for MAXIMUM EVEN SUBGRAPH and MAXIMUM ODD SUBGRAPH, and then for ODD q-COLORING for $q \geq 3$.

Theorem 1 ($\star$). *The* MAXIMUM EVEN SUBGRAPH *and* MAXIMUM ODD SUBGRAPH *problems are* NP-*hard. Moreover, none of them can be solved in time* $2^{o(n)}$ *on n-vertex graphs unless the* ETH *fails.*

Theorem 2. *For every integer $q \geq 3$, given a graph G on n vertices, determining whether $\chi_{\mathsf{odd}}(G) \leq q$ is* NP-*complete and, moreover, cannot be solved in time* $2^{o(n)}$ *unless the* ETH *fails.*

Proof: Membership in NP is clear. For every integer $q \geq 3$, we present a linear reduction from the q-COLORING problem, which is well-known to be NP-hard and not solvable in time $2^{o(n)}$ on n-vertex graphs unless the ETH fails [21,22]. We will use the fact that any graph $G = (V, E)$ such that $|V| + |E|$ is even admits an orientation of E such that, in the resulting digraph, all the vertex in-degrees are odd; we call such an orientation an *odd orientation*. Moreover, an odd orientation can be found in polynomial time (for a proof, see for instance [16]).

Given an instance $G = (V, E)$ of q-COLORING, we build from G an instance $G^{\bullet}$ of ODD q-COLORING as follows. First, if $|V| + |E|$ is odd, we arbitrarily select a vertex $v \in V$ and add a triangle on three new vertices v_1, v_2, v_3 and the edge vv_1. Note that the resulting graph $G' = (V', E')$ is q-colorable for $q \geq 3$ if and only if G is, and that $|V'| + |E'|$ is even. Hence, E' admits an odd orientation ϕ. We let $G^{\bullet}$ be the graph obtained from G' by subdividing every edge once. Note that the size of $G^{\bullet}$ depends linearly on the size of G, as required. We claim that $\chi(G) \leq q$ if and only if $\chi_{\mathsf{odd}}(G^{\bullet}) \leq q$.

Assume first that we are given a proper q-coloring $c : V \to [q]$, which can trivially be extended to a proper q-coloring of G'. We define an odd q-coloring c_{odd} of $G^{\bullet}$ as follows. If $v \in V(G^{\bullet})$ is an original vertex of V', we set $c_{\mathsf{odd}}(v) = c(v)$. Otherwise, if v is a subdivision vertex between two vertices u and w of V', we set $c_{\mathsf{odd}}(v) = c(u)$ if edge uw is oriented toward u in ϕ, and $c_{\mathsf{odd}}(v) = c(w)$ otherwise. It can be easily verified that c_{odd} is indeed an odd q-coloring of $G^{\bullet}$.

Conversely, let $c_{\mathsf{odd}} : V(G^{\bullet}) \to [q]$ be an odd q-coloring of $G^{\bullet}$, let uw be an edge of G', and let v be the subdivision vertex in $G^{\bullet}$ between u and w. If follows that $c_{\mathsf{odd}}(u) \neq c_{\mathsf{odd}}(w)$, as otherwise vertex v would have degree zero or two in its color class. Therefore, letting $c(v) = c_{odd}(v)$ for every vertex $v \in V(G)$ defines a proper q-coloring of G, and the theorem follows.   □

Theorem 2 establishes the NP-hardness of ODD q-COLORING for every $q \geq 3$. On the other hand, the ODD 1-COLORING is trivial, as for any graph G, $\chi_{odd}(G) \leq 1$ if and only if G is an odd graph itself. Therefore, the only remaining case is ODD 2-COLORING. In the next theorem we prove that this problem can be solved in polynomial time.

Theorem 3. *The* ODD 2-COLORING *problem can be solved in polynomial time.*

Proof: We will express the ODD 2-COLORING problem as the existence of a feasible solution to a system of linear equations over the binary field, which can be determined in polynomial time using, for instance, Gaussian elimination. Given an instance $G = (V, E)$ of ODD 2-COLORING, let its vertices be labeled $v_1, \ldots, v_n$. For every vertex $v_i \in V$ we create a binary variable x_i, and for every edge $v_i v_j \in E$, we create a binary variable $x_{i,j}$. The interpretation of these two types of variables is quite different. Namely, for a vertex variable x_i, its value corresponds to the color (either 0 or 1) assigned to vertex v_i. On the other hand, the value of an edge variable corresponds the whether this edge belongs to a monochromatic subgraph, that is, to whether both its endvertices get the same color. In this case, its value is 1, and 0 otherwise. We guarantee this latter property by adding the following set of linear equations:

$$x_i + x_j + x_{i,j} \equiv 1 \quad \text{for every edge } v_i v_j \in E. \tag{1}$$

To guarantee that the degree of every vertex in each of the two monochromatic subgraphs is odd, we add the following set of linear equations (for an edge variable $x_{i,j}$, to simplify the notation we interpret $x_{j,i} = x_{i,j}$):

$$\sum_{j:v_j \in N(v_i)} x_{i,j} \equiv 1 \quad \text{for every vertex } v_i \in V. \tag{2}$$

Note that by Eq. (1), only monochromatic edges contribute to the sum of Eq. (2). Therefore, the above discussion implies that $\chi_{odd}(G) \leq 2$ if and only if the system of linear equations given by Eqs. (1) and (2) admits a feasible solution, and the theorem follows.   □

Note that the EVEN 2-COLORING problem could be formulated in a similar way, just by replacing Eq. (2) with $\sum_{j:v_j \in N(v_i)} x_{i,j} \equiv 0$. However, this is not that interesting, since all the instances of EVEN 2-COLORING are positive [26].

4   Dynamic Programming Algorithms

In this section, we present FPT algorithms for MAXIMUM ODD/EVEN SUB-GRAPH and ODD q-COLORING, parameterized by the rank-width of the input graph. The algorithms are similar to those of Bui-Xuan et al. [3, 4] for MAXIMUM

INDEPENDENT SET parameterized by rank-width and boolean-width, respectively, and also to the one by Bui-Xuan et al. [5] for so-called locally checkable vertex partitioning problems. There are however two key differences with our algorithms. First, while partial solutions for MAXIMUM INDEPENDENT SET are, themselves, independent sets, this is not true in general for odd subgraphs, where partial solutions may consist in a subgraph some vertices of which have even degree. Those vertices will impose some extra constraints on the remainder of the solution. The second difference is that, while the equivalence classes of [3] and [4] are based on neighborhoods of vertex sets, those for MAXIMUM ODD SUBGRAPH only require "neighborhoods modulo 2". This will allow us to consider only $2^{\mathcal{O}(\mathsf{rw})}$ equivalence classes, compared to $2^{\mathcal{O}(\mathsf{rw}^2)}$ classes used in [3] for MAXIMUM INDEPENDENT SET.

Throughout this section, we will rely on the notion of "neighborhood modulo 2" of a set of vertices, defined as follows. Given a graph G and $X \subseteq V(G)$, the *neighborhood of X modulo 2*, denoted by $N_2(X)$, is the set $\triangle_{u \in X}(N(u))$, where the operator $\triangle$ denotes the symmetric difference. Note that $N_2(X)$ is exactly the set of vertices in $V(G) \setminus X$ that have an odd number of neighbors in X. The results in this section are stated using the $\mathcal{O}^*$ notation, which hides polynomial factors in the input size.

Theorem 4. *Given a graph G along with a decomposition tree of rank-width* rw, *the* MAXIMUM ODD SUBGRAPH *problem can be solved in time* $\mathcal{O}^*(2^{3\mathsf{rw}})$.

Proof: We give a dynamic programming over the given decomposition tree (T, L). Recall that there is a bijection between the leaves of T and $V(G)$, and that each edge of T corresponds to a cut $(A, \overline{A})$ of G. We begin by defining the equivalence relation over subsets of A, given a cut $(A, \overline{A})$: two sets $X, Y \subseteq V(G)$ are *odd neighborhood equivalent* with regard to A, denoted by $X \equiv_2^A Y$, if $N_2(X) \setminus A = N_2(Y) \setminus A$. Then, given a row basis $\mathcal{B}$ of the adjacency matrix of $(A, \overline{A})$ over $\mathsf{GF}[2]$, where we interpret a vertex set as the vector corresponding to its vertices, we define the *representative* of a set $X \subseteq A$ as the unique set of vertices $R_A(X) \subseteq A$ such that $R_A(X) \subseteq \mathcal{B}$ and $X \equiv_2^A R_A(X)$. Observe that since $(A, \overline{A})$ is a cut of (T, L), its adjacency matrix has rank at most $\mathsf{rw}(G)$, and therefore $|R_A(X)| \leq \mathsf{rw}(G)$. This implies, in particular, that there are at most $2^{\mathsf{rw}(G)}$ distinct representatives for subsets of a given set A.

We are now ready to define the tables of our algorithm. Given an edge e of (T, L) and its associated cut $(A, \overline{A})$ of G, we store in table T_A, for every pair R, R' of representatives of subsets of A and $\overline{A}$, respectively, a largest set $S \subseteq A$ such that S is odd neighborhood equivalent to R, and all the vertices that have even degree in $G[S]$ is exactly the set $N_2(R') \cap S$. More formally:

(✖) $T_A[R, R'] = \underset{S \subseteq A}{\mathrm{maxset}}\{S \equiv_2^A R \wedge \{v \in S : |N(v) \cap S| \text{ is even}\} = N_2(R') \cap S\}$,

where the notation 'maxset' indicates a largest set that satisfies the conditions. In cases where edge e is incident with a leaf, the cut associated with e is of the form $(\{u\}, V(G) \setminus \{u\})$. We set $T_{\{u\}}[\emptyset, \emptyset] = T_{\{u\}}[\emptyset, \{v\}] = \emptyset$, and $T_{\{u\}}[\{u\}, \{v\}] =$

$\{u\}$, where v is the unique vertex of a basis of the adjacency matrix of the cut $(V(G) \setminus \{u\}, \{u\})$, which is the only non-empty choice for R'. The entry $T_{\{u\}}[\{u\}, \emptyset]$ is left empty, due to there being no subgraph of $G[\{u\}]$ with the same neighborhood as $\{u\}$ in $G - \{u\}$, all vertices of which that have even degree lying in $N_2(\emptyset)$.

Given an edge e of (T, L) such that the tables of both edges incident with one endvertex of e, say f, f', have been computed, we compute the table of e as follows. Let us denote by $(A, \overline{A}), (X, \overline{X})$, and $(Y, \overline{Y})$ the cuts associated with e, f, and f', respectively. For each pair of representatives $R_A, R_{\overline{A}}$ of the cut $(A, \overline{A})$, the value of $T_A[R_A, R_{\overline{A}}]$ is the largest $T_X[R_X, R_{\overline{X}}] \cup T_Y[R_Y, R_{\overline{Y}}]$, such that $R_X, R_{\overline{X}}, R_Y$, and $R_{\overline{Y}}$ satisfy the following conditions with regard to R_A and $R_{\overline{A}}$:

(i) $R_A \equiv_2^A R_X \triangle R_Y$, (ii) $R_{\overline{X}} \equiv_2^{\overline{X}} R_{\overline{A}} \triangle R_Y$, and (ii') $R_{\overline{Y}} \equiv_2^{\overline{Y}} R_{\overline{A}} \triangle R_X$.

We proceed with this computation, starting from the leaves, in a bottom-up manner, having previously rooted T by choosing an arbitrary edge, subdividing it, and making the newly created vertex the root of T. Observe that in the final stage of the algorithm, when the tables of both edges f, f' incident with the root have been computed, we compute the table for the root node as described above, with $\overline{A} = \emptyset$, since $X \cup Y = V(G)$ in this case. Of the three conditions described above, condition (i) becomes trivial, since $R_A = \emptyset$, and conditions (ii) and (ii') simplify to $R_{\overline{X}} \equiv_2^{\overline{X}} R_Y$, and $R_{\overline{Y}} \equiv_2^{\overline{Y}} R_X$, respectively.

We first observe that since, as noted above, there are at most 2^{rw} representatives on each side of each cut, and the choices of R_X, R_Y and $R_{\overline{A}}$ uniquely determines $R_{\overline{X}}, R_{\overline{Y}}$ and R_A through equations (i), (ii), and (ii'), and computing new tables can be carried out in time $\mathcal{O}^*(2^{3\mathrm{rw}})$, as desired. It now remains to prove that the algorithm correctly computes an optimal solution. The correctness of the tables for the leaves of T follows from their description. We now prove by induction that the tables are correct for internal edges of T as well. Let us assume T_X and T_Y have been fully and correctly computed for all possible representatives $R_X, R_{\overline{X}}, R_Y$, and $R_{\overline{Y}}$ as per the description above. We first argue that the tables' description is correct, i.e., given an optimal solution OPT (that is, an induced subgraph of G achieving $\mathsf{mos}(G)$) and a cut $(A, \overline{A})$, $S = \mathsf{OPT} \cap A$ is a largest set that satisfies (✠) for some pair R, R' of representatives. Indeed, assume for contradiction that there exists $S^* \subseteq A$ such that $S^* \equiv_2^A S$, $\{v \in S^* : |N(v) \cap S^*| \text{ is even}\} = S^* \cap N_2(\mathsf{OPT} \cap \overline{A})$, and $|S| < |S^*|$. Then, $\mathsf{OPT}^* = (\mathsf{OPT} \setminus S) \cup S^*$ induces an odd subgraph of G and $|\mathsf{OPT}^*| > |\mathsf{OPT}|$, contradicting the optimality of OPT.

Finally, we argue that if T_X and T_Y are computed correctly, then so is T_A, i.e., given any two representatives R_A and $R_{\overline{A}}$ of A and $\overline{A}$, respectively, there exist representatives $R_X, R_{\overline{X}}, R_Y$, and $R_{\overline{Y}}$ of $X, \overline{X}, Y$, and $\overline{Y}$, respectively, that satisfy conditions (i), (ii), and (ii'), and such that $T_X[R_X, R_{\overline{X}}] \cup T_Y[R_Y, R_{\overline{Y}}]$ is a largest set that satisfies (✠) with respect to $(R_A, R_{\overline{A}})$. Let $R_X, R_{\overline{X}}, R_Y$, and $R_{\overline{Y}}$ be representatives such that $T_A[R_A, R_{\overline{A}}] = T_X[R_X, R_{\overline{X}}] \cup T_Y[R_Y, R_{\overline{Y}}] = S$, and let S_X and S_Y denote $S \cap X$ and $S \cap Y$, respectively. Note that, since X and Y form a partition of A, S_X and S_Y form a partition of S, which implies

$S = S_X \cup S_Y = S_X \triangle S_Y$. We first show that S indeed satisfies (✠) with respect to $(A, \overline{A})$, i.e., $S \equiv_2^A R_A$ and $\{v \in S : |N(v) \cap S| \text{ is even}\} = N_2(R_{\overline{A}}) \cap S$. For the first of those two conditions, combining it with the fact that $S = S_X \cup S_Y = S_X \triangle S_Y$, we only need to prove that $S_X \triangle S_Y \equiv_2^A R_X \triangle R_Y$. Observe first that, since X and Y form a partition of A, we have that for every vertex $v \in \overline{A}, |N(v) \cap A| = |N(v) \cap X| + |N(v) \cap Y|$. Therefore, for every sets $X', X'' \subseteq X$ and $Y', Y'' \subseteq Y$, it holds that if $X' \equiv_2^X X''$ and $Y' \equiv_2^Y Y''$, then $X' \triangle Y' \equiv_2^A X'' \triangle Y''$. From the definition of representative we obtain that $S \equiv_2^A S_X \triangle S_Y \equiv_2^A R_X \triangle R_Y$, as desired.

Let us now consider the second condition, i.e., $\{v \in S : |N(v) \cap S| \text{ is even}\} = N_2(R_{\overline{A}}) \cap S$. Let us assume first that $v \in S_X$. If $|N(v) \cap S|$ is even, then at least one of the following cases holds:

- $|N(v) \cap S_X|$ is even and $v \notin N_2(S_Y)$. Since $|N(v) \cap S_X|$ is even, we obtain from (✠) in T_X that $v \in N_2(R_{\overline{X}})$, which when combined with (ii) implies $v \in N_2(\overline{A}) \triangle N_2(S_Y)$. Since $v \notin N_2(S_Y)$, it follows that $v \in N_2(\overline{A})$, as desired.
- $|N(v) \cap S_X|$ is odd and $v \in N_2(S_Y)$. Symmetrically to the case above, we have that $v \notin N_2(R_{\overline{X}})$, hence $v \notin N_2(\overline{A}) \triangle N_2(S_Y)$ from (ii), and since $v \in N_2(S_Y)$, it follows that $v \in N_2(\overline{A})$, as desired.

The case where $v \in S_Y$ is proved similarly, replacing condition (ii) with (ii'). Therefore, $\{v \in S : |N(v) \cap S| \text{ is even}\} \subseteq S \cap N_2(R_{\overline{A}})$. Let us now assume that $v \in S_X \cap N_2(R_{\overline{A}})$. From (ii), we obtain that $v \in N_2(S \cap \overline{X})$ if and only if $v \notin N_2(S_Y)$. Since T_X satisfies (✠), it holds that $v \in N_2(S \cap \overline{X})$ if and only if $|N(v) \cap S_X|$ is even, and therefore $v \notin N_2(S_Y)$ if and only if $|N(v) \cap S_X|$ is even. Therefore, $|N(v) \cap S| = |N(v) \cap S_X| + |N(v) \cap S_Y|$ is even, as desired. As above, the case where $v \in S_Y$ is proved similarly, replacing condition (ii) with (ii'). Therefore, $\{v \in S : |N(v) \cap S| \text{ is even}\} = S \cap N_2(R_{\overline{A}})$.

Finally, we prove the maximality of S among all those sets that satisfy (✠) with respect to $(R_A, R_{\overline{A}})$. Let us assume for a contradiction that there exists S^* that satisfies (✠) with respect to $(R_A, R_{\overline{A}})$ and such that $|S^*| > |S|$. Let S_X^* and S_Y^* denote $S^* \cap X$ and $S^* \cap Y$, respectively. Observe that S_X^* and S_Y^* satisfy (✠) with respect to some pairs of representatives $(R_X, R_{\overline{X}})$ and $(R_Y, R_{\overline{Y}})$, respectively. In addition, observe that since S satisfies (✠) with respect to $(R_A, R_{\overline{A}})$, it follows that S, S_X, and S_Y satisfy conditions (i), (ii), and (ii') with respect to $(R_A, R_{\overline{A}})$, contradicting the assumption that T_X and T_Y were computed correctly. □

Small variations of Theorem 4 allow us to prove the following two theorems.

Theorem 5 (⋆). *Given a graph G along with a decomposition tree of rank-width* rw, *the* Maximum Even Subgraph *problem can be solved in time* $\mathcal{O}^*(2^{3\text{rw}})$.

Theorem 6 (⋆). *Given a graph G along with a decomposition tree of rank-width* w, *the* Odd q-Coloring *problem can be solved in time* $\mathcal{O}^*(2^{\mathcal{O}(q \cdot \text{rw})})$.

5   Tight Bounds

In this section we provide two tight bounds concerning odd induced subgraphs and odd colorings. Namely, we first provide in Theorem 7 a tight upper bound on the odd chromatic number in terms of tree-width, and then we provide in Theorem 8 a tight lower bound on the size of a maximum odd induced subgraph for graphs that admit a join.

Theorem 7. *For every graph G with all components of even order we have that $\chi_{\mathsf{odd}}(G) \leq \mathsf{tw}(G) + 1$, and this bound is tight.*

Proof: Scott proved [34, Corollary 3] that every graph G with all components of even order admits a vertex partition such that every vertex class induces a tree with all degrees odd. Consider such a vertex partition, and let G' be the graph obtained from G by contracting each of the trees to a single vertex. Since G' is a minor of G, we have that $\mathsf{tw}(G') \leq \mathsf{tw}(G)$. Now note that every proper vertex coloring of G' using q colors can be lifted to a partition of $V(G)$ into q odd induced subgraphs (in fact, odd induced forests). Indeed, with every color i of a proper q-coloring of $V(G')$ we associate an induced forest of G defined by the union of the trees whose corresponding vertex in G' is colored i. Therefore,

$$\chi_{\mathsf{odd}}(G) \leq \chi(G') \leq \mathsf{tw}(G') + 1 \leq \mathsf{tw}(G) + 1,$$

where we have used the well-known fact that the chromatic number of a graph is at most its tree-width plus one [23].

To see that this bound it tight, consider a subdivided clique $K_n^{\bullet}$, that is, the graph obtained from a clique on n vertices, with $n \equiv 0, 3 \pmod 4$, by subdividing every edge once. Since no pair of original vertices of the clique can get the same color, we have that $\chi_{\mathsf{odd}}(K_n^{\bullet}) = n = \mathsf{tw}(K_n^{\bullet}) + 1$.   □

Let us mention some consequences of Theorem 7. Hou et al. [20] define the following parameter. Let $\mathcal{G}_k$ be the set of all graphs of treewidth at most k without isolated vertices, and let $c_k = \min_{G \in \mathcal{G}_k} \frac{\mathsf{mos}(G)}{|V(G)|}$. In [20] the authors prove that $c_2 = 2/5$ and say that the best general lower bound is $c_k \geq \frac{1}{2k+2}$, which follows from a result of Scott [33]. As an immediate corollary of Theorem 7 it follows that $c_k \geq \frac{1}{k+1}$, which improves the lower bound by a factor two. As it is known [20] that, for $k \in [4]$, $c_k \leq \frac{2}{k+3}$, our lower bound implies that $1/4 \leq c_3 \leq 1/3$ and $1/5 \leq c_4 \leq 2/7$.

We now provide a lower bound on $\mathsf{mos}(G)$ for every graph that admits a join.

Theorem 8 ($\star$). *For every n-vertex graph G that admits a join we have*

$$\mathsf{mos}(G) \geq 2 \cdot \left\lceil \frac{n-2}{4} \right\rceil, \text{ and this bound is tight even for cographs.}$$

Determining a tight lower bound for cographs that are not necessarily connected remains open. The proof of Cases 1 and 2 of Theorem 7 together with the fact that $\chi_{\mathsf{odd}}(K_{2,2,2}) = 3$ (since $\mathsf{mos}(K_{2,2,2}) = 2$) yield the following corollary.

Corollary 1. *Let G be a cograph with every connected component of even order. Then $\chi_{\mathsf{odd}}(G) \leq 3$. Moreover, this bound is tight.*

Note that cographs can be equivalently defined as P_4-free graphs. It is interesting to note that, in contrast to Corollary 1, P_5-free graphs have unbounded odd chromatic number. Indeed, let H_n be the graph obtained from the subdivided clique $K_n^{\bullet}$, with $n \equiv 0, 3 \pmod 4$, by adding an edge between each pair of original vertices of the clique. It can be checked that $\chi_{\mathsf{odd}}(H_n) \geq n$ and, in fact, the proof of Theorem 2 implies that $\chi_{\mathsf{odd}}(H_n) = n$. Note that H_n is a split graph, hence split graphs have unbounded odd chromatic number.

6   Further Research

We considered computational aspects of the MAXIMUM ODD SUBGRAPH and ODD q-COLORING problems. A number of interesting questions remain open.

We gave in Theorem 6 an algorithm that solves ODD q-COLORING in time $\mathcal{O}^*(2^{\mathcal{O}(q \cdot \mathsf{rw})})$. Is the ODD CHROMATIC NUMBER problem FPT or W[1]-hard parameterized by rank-width? A strongly related question is how the odd chromatic number depends on rank-width. We proved in Theorem 7 that $\chi_{\mathsf{odd}}(G) \leq \mathsf{tw}(G) + 1$, but we do not know whether $\chi_{\mathsf{odd}}(G) \leq f(\mathsf{rw}(G))$ for some function f. Note that this would not only yield an FPT algorithm for ODD CHROMATIC NUMBER by rank-width, but would also prove the conjecture about the linear size of a largest odd induced subgraph [7] for all graphs of bounded rank-width. As a first step in this direction, we proved in Corollary 1 that cographs, which have rank-width at most one, have odd chromatic number at most three. It would be interesting to prove an upper bound for distance-hereditary graphs, which can be equivalently defined as graphs of rank-width one.

In fact, we do not even know whether ODD CHROMATIC NUMBER by rank-width is in XP. In view of the algorithm of Theorem 6, a sufficient condition for this would be that there exists a function f such that $\chi_{\mathsf{odd}}(G) \leq f(\mathsf{rw}(G)) \cdot \log |V(G)|$ for every graph G with all components of even order. Another promising strategy would be to generalize the XP algorithms of Rao [31] to *counting* monadic second-odder logic. Toward an eventual W[1]-hardness proof, a natural strategy is to try to adapt the reduction given by Fomin et al. [15] to prove that CHROMATIC NUMBER is W[1]-hard by clique-width (hence, rank-width). This reduction is from EQUITABLE COLORING parameterized by the number of colors plus tree-width, proved to be W[1]-hard by Fellows et al. [13]. By appropriately modifying the chain of reductions given in [13], we have only managed to prove that the naturally defined ODD EQUITABLE COLORING problem is W[1]-hard by tree-width, but not if we add the number of colors as a parameter.

Concerning ODD q-COLORING parameterized by tree-width, a straightforward dynamic programming algorithm that guesses, for every vertex, its color class and the parity of its degree within that class, runs in time $\mathcal{O}^*((2q)^{\mathsf{tw}})$. Note that this algorithm together with Theorem 7 yield an algorithm for ODD

CHROMATIC NUMBER in time $\mathcal{O}^*((2\mathsf{tw}+2)^{\mathsf{tw}})$. By the lower bound under the ETH of Lokshtanov et al. [25] for CHROMATIC NUMBER by tree-width and the fact that our reduction of Theorem 2 preserves tree-width, it follows that the dependency on tree-width of this algorithm is asymptotically optimal under the ETH. It would be interesting to prove lower bounds under the *Strong Exponential Time Hypothesis* (SETH). Note that our reduction of Theorem 2 together with the lower bound under the SETH of Lokshtanov et al. [24] for q-COLORING by tree-width yield a lower bound for ODD q-COLORING of $\mathcal{O}^*((q-\varepsilon)^{\mathsf{tw}})$ under the SETH.

References

1. Bergougnoux, B., Kanté, M.M.: Rank based approach on graphs with structured neighborhood. CoRR, abs/1805.11275 (2018)
2. Berman, D.M., Wang, H., Wargo, L.: Odd induced subgraphs in graphs of maximum degree three. Australas. J. Comb. **15**, 81–86 (1997)
3. Bui-Xuan, B., Telle, J.A., Vatshelle, M.: H-join decomposable graphs and algorithms with runtime single exponential in rankwidth. Discrete Appl. Math. **158**(7), 809–819 (2010)
4. Bui-Xuan, B., Telle, J.A., Vatshelle, M.: Boolean-width of graphs. Theoret. Comput. Sci. **412**(39), 5187–5204 (2011)
5. Bui-Xuan, B., Telle, J.A., Vatshelle, M.: Fast dynamic programming for locally checkable vertex subset and vertex partitioning problems. Theoret. Comput. Sci. **511**, 66–76 (2013)
6. Cai, L., Yang, B.: Parameterized complexity of even/odd subgraph problems. J. Discrete Algorithms **9**(3), 231–240 (2011)
7. Caro, Y.: On induced subgraphs with odd degrees. Discrete Math. **132**(1–3), 23–28 (1994)
8. Cygan, M., et al.: Parameterized Algorithms. Springer, Cham (2015). https://doi.org/10.1007/978-3-319-21275-3
9. Cygan, M., Marx, D., Pilipczuk, M., Pilipczuk, M., Schlotter, I.: Parameterized complexity of Eulerian deletion problems. Algorithmica **68**(1), 41–61 (2012). https://doi.org/10.1007/s00453-012-9667-x
10. Diestel, R.: Graph Theory. Graduate Texts in Mathematics, vol. 173, 4th edn. Springer, Heidelberg (2012)
11. Downey, R.G., Fellows, M.R.: Fundamentals of Parameterized Complexity. TCS. Springer, London (2013). https://doi.org/10.1007/978-1-4471-5559-1
12. Downey, R.G., Fellows, M.R., Vardy, A., Whittle, G.: The parametrized complexity of some fundamental problems in coding theory. SIAM J. Comput. **29**(2), 545–570 (1999)
13. Fellows, M.R., et al.: On the complexity of some colorful problems parameterized by treewidth. Inf. Comput. **209**(2), 143–153 (2011)
14. Flum, J., Grohe, M.: Parameterized Complexity Theory. TTCSAES. Springer, Heidelberg (2006). https://doi.org/10.1007/3-540-29953-X
15. Fomin, F.V., Golovach, P.A., Lokshtanov, D., Saurabh, S.: Intractability of clique-width parameterizations. SIAM J. Comput. **39**(5), 1941–1956 (2010)
16. Frank, A., Jordán, T., Szigeti, Z.: An orientation theorem with parity conditions. In: Cornuéjols, G., Burkard, R.E., Woeginger, G.J. (eds.) IPCO 1999. LNCS, vol. 1610, pp. 183–190. Springer, Heidelberg (1999). https://doi.org/10.1007/3-540-48777-8_14

17. Ganian, R., Hlinený, P., Obdržálek, J.: Better algorithms for satisfiability problems for formulas of bounded rank-width. Fundamenta Informaticae **123**(1), 59–76 (2013)
18. Ganian, R., Hlinený, P., Obdržálek, J.: A unified approach to polynomial algorithms on graphs of bounded (bi-)rank-width. Eur. J. Comb. **34**(3), 680–701 (2013)
19. Goyal, P., Misra, P., Panolan, F., Philip, G., Saurabh, S.: Finding even subgraphs even faster. J. Comput. Syst. Sci. **97**, 1–13 (2018)
20. Hou, X., Yu, L., Li, J., Liu, B.: Odd induced subgraphs in graphs with treewidth at most two. Graphs Comb. **34**(4), 535–544 (2018)
21. Impagliazzo, R., Paturi, R.: On the complexity of k-SAT. J. Comput. Syst. Sci. **62**(2), 367–375 (2001)
22. Impagliazzo, R., Paturi, R., Zane, F.: Which problems have strongly exponential complexity? J. Comput. Syst. Sci. **63**(4), 512–530 (2001)
23. Kloks, T.: Treewidth. Computations and Approximations. LNCS, vol. 842. Springer, Heidelberg (1994). https://doi.org/10.1007/BFb0045375
24. Lokshtanov, D., Marx, D., Saurabh, S.: Known algorithms on graphs of bounded treewidth are probably optimal. ACM Trans. Algorithms **14**(2), 13:1–13:30 (2018)
25. Lokshtanov, D., Marx, D., Saurabh, S.: Slightly superexponential parameterized problems. SIAM J. Comput. **47**(3), 675–702 (2018)
26. Lovász, L.: Combinatorial Problems and Exercises. North-Holland, Amsterdam (1979)
27. Niedermeier, R.: Invitation to Fixed-Parameter Algorithms. Oxford University Press, Oxford (2006)
28. Oum, S.: Rank-width: algorithmic and structural results. Discrete Appl. Math. **231**, 15–24 (2017)
29. Oum, S., Seymour, P.D.: Approximating clique-width and branch-width. J. Comb. Theory Ser. B **96**(4), 514–528 (2006)
30. Radcliffe, A.J., Scott, A.D.: Every tree contains a large induced subgraph with all degrees odd. Discrete Math. **140**(1–3), 275–279 (1995)
31. Rao, M.: MSOL partitioning problems on graphs of bounded treewidth and clique-width. Theoret. Comput. Sci. **377**(1–3), 260–267 (2007)
32. Röyskö, A.: https://cstheory.stackexchange.com/questions/45885/complexity-of-finding-the-largest-induced-subgraph-with-all-even-degrees
33. Scott, A.D.: Large induced subgraphs with all degrees odd. Comb. Probab. Comput. **1**, 335–349 (1992)
34. Scott, A.D.: On induced subgraphs with all degrees odd. Graphs Comb. **17**(3), 539–553 (2001)

Knot Diagrams of Treewidth Two

Hans L. Bodlaender[1(✉)], Benjamin Burton[2], Fedor V. Fomin[3],
and Alexander Grigoriev[4]

[1] Department of Information and Computing Sciences, Utrecht University,
P.O. Box 80.089, 3508 TB Utrecht, The Netherlands
h.l.bodlaender@uu.nl
[2] School of Mathematics and Physics, The University of Queensland,
Brisbane, QLD 4072, Australia
[3] Department of Informatics, University of Bergen, 5020 Bergen, Norway
[4] Maastricht University School of Business and Economics,
Maastricht, The Netherlands

Abstract. In this paper, we study knot diagrams for which the underlying graph has treewidth two. We give a linear time algorithm for the following problem: given a knot diagram of treewidth two, does it represent the trivial knot? We also show that for a link diagram of treewidth two we can test in linear time if it represents the unlink. From the algorithm, it follows that a diagram of the trivial knot of treewidth 2 can always be reduced to the trivial diagram with at most n untwist and unpoke Reidemeister moves.

Keywords: Knot diagrams · Knot theory · Graph algorithms · Treewidth · Series parallel graphs

1   Introduction

A *knot* is a piecewise linear closed curve S^1 embedded into the 3-sphere S^3 (or the three-dimensional Euclidean space $\mathbb{R}^3$). Two knots are said to be *equivalent* if there is an ambient isotopy between them. In other words, two knots are equivalent if it is possible to distort one knot into the other without breaking it. The basic problem of knot theory is the following unknotting problem: given a knot, determine whether it is equivalent to a knot that bounds an embedded disk in S^3. Such a knot is called the *trivial knot* or simply the *unknot*.

Despite a significant progress, the computational complexity of the unknotting problem remains open. Even the existence of *any* algorithm for this problem

A large part of this research was done during the workshop *Fixed Parameter Computational Geometry* at the Lorentz Center in Leiden. The research of the first author was partially supported by the *Networks* project, supported by the Netherlands Organization for Scientific Research N.W.O. The second author was supported by the Australian Research Council under the Discovery Projects scheme (DP150104108). The third author was supported by the Research Council of Norway via the project "MULTIVAL".

I. Adler and H. Müller (Eds.): WG 2020, LNCS 12301, pp. 80–91, 2020.
https://doi.org/10.1007/978-3-030-60440-0_7

Fig. 1. A vertex of degree four representing a crossing of two strings

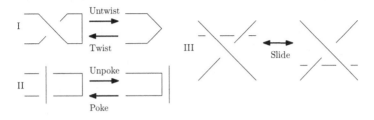

Fig. 2. Reidemeister moves

is a highly non-trivial question. As was stated by Turing in 1954 in [17], "No systematic method is yet known by which one can tell whether two knots are the same." The first algorithm resolving this problem is due to Haken [6]. By the celebrated result of Hass, Lagarias, and Pippenger [8], unknot recognition is in NP. The problem is also in co-NP, see Lackenby [12]. However, no polynomial algorithm for the unknotting problem is known.

It was understood already in 1920s that the question about equivalence of knots in $\mathbb{R}^3$ is reducible to a combinatorial question about knot diagrams [1,14]. Knot diagrams are labeled planar graphs representing a projection of the knot onto a plane. Thus every vertex of the graph in knot diagram is of degree 4 and edges are marked as overcrossing and undercrossing, see Fig. 1 and Sect. 2.

It is one of the most fundamental theorems in knot theory from 1920s that any two diagrams of a knot or link in $\mathbb{R}^3$ differ by a sequence of Reidemeister moves [14], illustrated in Fig. 2. We refer to these moves as (I) twist moves, (II) poke moves, and (III) slide moves, with the reverse operation of a twist move the untwist, and the reverse operation of a poke the unpoke.

With help of Reidemeister moves, see Fig. 2, we obtain an equivalence relation on knot diagrams: if a diagram can be obtained from another by zero or more Reidemeister moves, then these diagrams are equivalent. In our paper, we allow subdivision vertices (i.e., vertices of degree two), and extend the notion of equivalence in the following trivial way: diagrams are equivalent if they can be obtained from another by zero or more Reidemeister moves and additions or removals of subdivisions.

In particular, the diagram of every unknot can be reduced to the trivial diagram (a circle) by performing Reidemeister moves. While each of the Reidemeister moves can be performed in polynomial time, it is very unclear how many of these moves are required to transform an unknot to the trivial diagram. The problem is that sometimes a successful unknotting sequence of Reidemeister moves is not monotone, that is, it has to increase the number of crossings (vertices) in the knot diagram, see e.g. [9]. Bounding the number of required Reidemeister moves by any function on the number of vertices in the knot diagram was a long-standing open question in the area. The answer to this question was given by Hass and Lagarias [7] who gave the first (exponential) upper bound on the number of Reidemeister moves. Later Lackenby in [11] improved the bound significantly by showing that any diagram of the unknot with n crossings may be reduced to the trivial diagram using at most $(236n)^{11}$ Reidemeister moves. Let us note that this also implies that the unknotting problem is in NP.

In this work we consider the unknotting problem when the given knot diagram has treewidth at most 2. Our main algorithmic result is Theorem 1.

Theorem 1. *Deciding whether any diagram with n crossings and treewidth at most 2 is a diagram of the unknot can be decided in time $O(n)$.*

Our proof yields also the following combinatorial result about the number of Reidemeister moves. It is interesting to note that in Theorem 2 we do not use any slide moves.

Theorem 2. *Any diagram of treewidth 2 of the unknot with n crossings may be reduced to the trivial diagram using at most n untwist and unpoke Reidemeister moves.*

Actually, the techniques developed to prove Theorems 1 and 2 can be used to solve a slightly more general problems about links with diagrams of treewidth 2.

Related Work. To the best of our knowledge, the question whether the unknotting problem with diagrams of bounded treewidth can be resolved in polynomial time is open. Makowsky and Mariño in [13] studied the parametrized complexity of the knot (and link) polynomials known as Jones polynomials, Kauffman polynomials and HOMFLY polynomials on graphs of bounded treewidth. For the Jones and HOMFLY polynomials no example of a non-trivial knot with trivial polynomial is known [4]. Therefore, if e.g. the Jones polynomial recognizes the unknot, then the algorithm from [13] also recognizes the unknot in time FPT in the treewidth.

Rué et al. [16] studied the class of link-types that admit a K_4-minor-free diagram (which is of treewidth at most 2). They obtain counting formulas and asymptotic estimates for the connected K_4-minor-free link and unknot diagrams. While Rué et al. [16] do not discuss algorithms in their work, the combinatorial tools developed in their paper can also be used to obtain Theorem 1. Our approach is more direct, and gives a fairly simple algorithm which is very straightforward to implement. We believe that the notion of double edge is an interesting concept of separate interest. Also, our work was done independently from the work by Rué et al. [16].

Approach. Our main approach is the following. We introduce the notion of *generalized knot diagrams*—these extend knot diagrams with the notion of *double edges* (see Sect. 2). The algorithm starts making the graph simple by adding subdivision vertices. By repeatedly applying *safe reduction rules* (see Sect. 3), we obtain a series of generalized knot diagrams, that all are equivalent to the input knot diagram, have treewidth at most two, and are simple. This continues till we have classified the (generalized) diagram as knot, unknot, link or unlink. The safe reduction rules come from a well known insight of graphs of treewidth two (Theorem 3), and a case analysis for different types of vertices and edges in the generalized knot diagram. The main algorithm is explained in Sect. 4.

2   Graphs and (Generalized) Knot Diagrams

2.1   Graphs

A *subdivision* in a graph $G = (V, E)$ is a vertex of degree two. The operation to *add a subdivision* is the following: take an edge $\{v, w\}$, and replace this edge by edges $\{v, x\}$ and $\{x, w\}$ with x a new vertex. The operation to *remove* a subdivision is the following: take a vertex of degree 2, add an edge between its neighbors and then remove the vertex and its incident edges.

We do not need to give the definition of treewidth [15], but instead rely on the following well known results on treewidth.

Theorem 3 (Folklore, see e.g., Theorem 33 and Lemma 90 in [2]).

(i) If G has treewidth at most two and is not the empty graph, then G has a vertex of degree at most two.

(ii) If G has treewidth at most two, and G' is obtained from G by removing a vertex, removing an edge, adding a subdivision or removing a subdivision, then the treewidth of G is also at most two.

2.2   Generalized Knot Diagrams

Our algorithm is based upon a generalization of knot diagrams, which we call *generalized knot diagrams*. The main ingredient is a new type of edges, which are created in the course of the algorithm. While a *single* edge in a knot diagram represents a piece of a single string, a *double* edge in a generalized knot diagram represents two pieces of strings between two pairs of vertices of degree two. Specifically, consider any two pieces of strings not intersected by any other piece of string. Let the two pieces of strings have the (four) endpoints at vertices of degree two. Moreover, let the strings alternate at every two consecutive crossings with respect to over- and under-crossing, i.e., if string s is over-crossing string s' at a crossing, then at the next (consecutive) crossing s' is over-crossing s. In accordance with Reidemeister terminology, we refer to these alternating crossings as *twists*. Such two strings with twists between two pairs of vertices of degree two are referred as double edges.

For each double edge we create an integer label that gives the number of twists/crossings in the double edge. If the two pieces of string do not cross, the label is zero. With labelings of endpoints of the strings with u (up) and d (down),

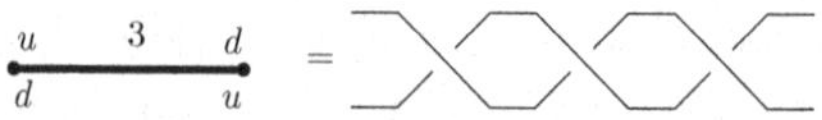

Fig. 3. A three-twist double edge and its string representation

we can distinguish between overcrossings and undercrossings; details are given later in this section.

See Fig. 3 for an illustration how a double edge represents two pieces of string with three twists.

In the generalized knot diagram we identify a pair of degree two vertices associated with an endpoint of a double edge as one *double vertex*, thus creating a new simple graph with a mix of knot diagram (*single*) vertices, double vertices, single and double edges, where double edges are labeled with numbers of twists.

Types of vertices. During the algorithm, we maintain that at each vertex of the diagram, either two or four pieces of string meet. Thus, we have the following types of vertices:

- A vertex of degree one, incident to one double edge (Type 1)
- A vertex of degree two, incident to two double edges (Type 2D)
- A vertex of degree two, incident to two single edges (Type 2S)
- A vertex of degree four, incident to four single edges (Type 4)
- A vertex of degree three, incident to one double edge and two single edges (Type 3)

Type 1, 2D and 3 vertices are called *double vertices*. Each of these is incident to a double edge. It is important to note that we do not have a crossing *at* a double vertex, i.e., all crossings either are at Type 4 vertices or at double edges with a non-zero label.

Each double edge whose integer label is non-zero has u and d-labelings attached to it, that determine the over- and undercrossings for the part of the diagram modelled by this edge. Each endpoint of the edge has a pair consisting of a u and a d attached to it; one of these comes clockwise directly before the edge, and one directly clockwise after the edge. (This can be represented by one bit.)

Thus, the two endpoints at a double vertex of the strings represented by a double edge have labels u and d, respectively. This models that the string labeled by u starts with an overcrossing and the string labeled by d starts with an undercrossing. See Fig. 4 at vertices v, w and x.

In this way, to each generalized knot diagram we can associate a knot diagram: we replace each double edge by a subgraph. If the double edge has integer label i, then we have i vertices of degree four. The u and d labels at these vertices are determined by the d and u labels at the endpoints of the double edge, as explained above. We add where necessary subdivision vertices of degree two to ensure that the graph is simple. We say that two generalized knot diagrams are equivalent when their associated knot diagrams are equivalent.

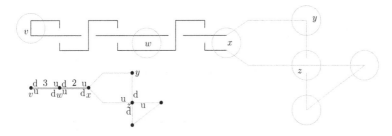

Fig. 4. A knot diagram and a corresponding generalized knot diagram. v is Type 1; w is Type 2D; x is Type 3; y is Type 2S; z is Type 4.

3   Safe Reduction Rules

In this section we introduce a number of reduction rules for generalized knot diagrams. The result of a rule is always again a generalized knot diagram. Note that we always remove one (single or double) vertex of degree at most 2, and possibly add an edge between the neighbors of a removed vertex of degree 2. Thus, when any of these rules is applied, the size of the generalized knot diagram is decreased by at least one vertex.

A rule is *safe*, if whenever we obtain the generalized knot diagram G' from G by applying the rule, we have that:

– The knot diagram associated with G is equivalent to the knot diagram associated with G'. (I.e., it can be obtained from the other by applying Reidemeister moves and adding or removing subdivision vertices.)
– If the treewidth of G is at most two, then the treewidth of G' is at most two.

Note that safeness of a rule implies that application of the rule preserves the ambient isotopy of the original and the resulting knots.

We will show that for all vertices of degree at most two, when G has at least three vertices, we have a safe rule that decreases the number of vertices by at least one, or we can resolve the problem. We have seven cases: vertices of Type 1, 2S, and 2D, where for the latter, their neighbors can be non-adjacent, adjacent by a single edge, or adjacent by a double edge.

Several of the rules have a straightforward case analysis for the markings of double edges. Many of these details can be found in the full version, see [3].

3.1   Vertices of Type 1

The first case is a vertex of Type 1: a vertex incident to one double edge. We can remove this vertex with its incident edge. Safeness follows by observing that the twists on this double edge can be removed with Reidemeister untwists. By removing subdivision vertices, we obtain a knot diagram represented by the diagram obtained by removing v and its incident double edge (Fig. 5).

Fig. 5. Removing a Type 1 vertex.

Rule 1. *Let v be of Type 1, incident to double edge $\{v, w\}$. Remove v and its incident edge.*

3.2   Vertices of Type 2S

For a vertex of Type 2S, we have three cases, depending on whether its neighbors are not adjacent, adjacent by a single edge, or adjacent by a double edge. The next Rule 2 is trivially safe as we just remove an edge subdivision. See Fig. 6.

Rule 2. *Let v be incident to two single edges $\{v, w\}$ and $\{v, x\}$, where w and x are not adjacent. Remove v and the edges $\{v, w\}$ and $\{v, x\}$, and add a single edge $\{w, x\}$.*

Fig. 6. Removing a vertex with two single edges and non-adjacent neighbors.

The second case is when the neighbors of v are connected by a single edge. This case has a number of different subcases. All are illustrated in Fig. 7.

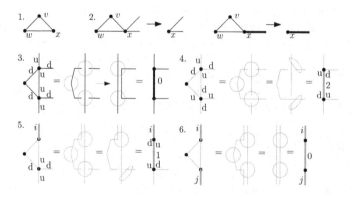

Fig. 7. The cases of Rule 3.

Rule 3. *Let v a Type 2S vertex with neighbors w and x, and suppose $\{w, x\}$ is a single edge.*

1. *If w and x are also Type 2S vertices, then G is a simple cycle of length three, and we do not apply any further rules as the diagram represents the unknot.*
2. *If w is of Type 2S and w of Type 3 or 4, then delete vertices v and w together with their incident edges.*
3. *If w and x are both Type 4 and the $\{u, d\}$-labels of the edge $\{w, x\}$ at vertices w and x are the same, we delete vertex v together with edges $\{v, w\}, \{v, x\}$ and $\{w, x\}$, and we create a double edge $\{w, x\}$ with label 0.*
4. *If w and x are both Type 4 and the the $\{u, d\}$-labels of the edge $\{w, x\}$ at vertices w and x are different, we delete vertex v together with edges $\{v, w\}, \{v, x\}$ and $\{w, x\}$, and we create a double edge $\{w, x\}$ of label 2.*
5. *If w is of Type 4 and x is of Type 3, we delete vertex v together with edges $\{v, w\}, \{v, x\}$ and $\{w, x\}$, and create a double edge $\{w.x\}$ with label 1.*
6. *If both x and w are of Type 3, we delete vertex v together with edges $\{v, w\}, \{v, x\}$ and $\{w, x\}$, and we create a double edge $\{w, x\}$ with label 0.*

Equivalence of the diagrams before and after the reduction step is easy to see. In Case 2, when w is of Type 4, we do one untwist removing the crossing at w; in Case 3, we do one unpoke. In the other cases, we do not perform Reidemeister moves, but by removing subdivisions, we can replace the generalized knot diagram by one with fewer vertices that represents the same knot diagram.

We now look at the third case. Suppose v is adjacent by two single edges to two double vertices, w and x, and there is a double edge between w and x with label i, i.e., having i twists.

Rule 4. *Suppose v is of Type 2S with neighbors w and x, and there is a double edge between w and x with i twists.*

1. *If $i = 0$, then the generalized knot diagram represents an unlink. We recurse on the generalized diagram obtained by removing v and incident edges, and making the edge $\{w, x\}$ a single edge.*
2. *If $i \neq 1$ is odd, the generalized knot diagram represents a non-trivial knot;*
3. *If $i \neq 0$ is even, the generalized knot diagram represents a non-trivial link.*
4. *If $i = 1$, then delete vertex v together with adjacent edges and delete double edge $\{w, x\}$, make w and x single vertices adjacent by a single edge.*

The cases are illustrated in Fig. 8. Correctness of the first three cases is evident. Safeness of the fourth case follows as this step represents a single Reidemeister untwist with subsequent contraction of subdivision.

3.3   Vertices of Type 2D

For vertices of Type 2D (incident to two double edges), we have again three cases: the neighbors are not adjacent, the neighbors are adjacent by a single edge, or the neighbors are adjacent by a double edge.

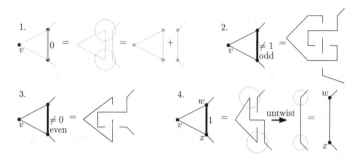

Fig. 8. The cases of Rule 4.

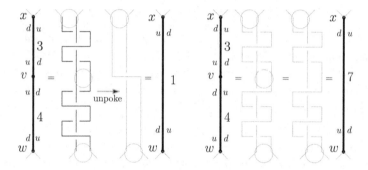

Fig. 9. Rule 5 with agreeing (left) and disagreeing (right) labels at v.

Rule 5. *Let v be of Type 2D where neighbors w and x are not adjacent. Suppose $\{v, w\}$ has label i, and $\{v, x\}$ has label j. Remove v and the edges $\{v, w\}$ and $\{v, x\}$, and add a double edge $\{w, x\}$.*

1. *If $i = 0$ ($j = 0$), the double edge $\{w, x\}$ gets label j (i), and the $\{u, d\}$-labels are as for the edge $\{v, x\}$ ($\{v, w\}$).*
2. *If the $\{u, d\}$-labels at v are at both sides equal (the left case in Fig. 9), then the double edge $\{w, x\}$ gets label $|i - j|$. If $i \neq j$, keep the $\{u, d\}$-labels at the side of the larger number i or j, and switch them at the other side.*
3. *If the $\{u, d\}$-labels at v differ at each of the sides (the right case in Fig. 9), then the double edge gets label $i + j$. Set the labels of the new double edge at w and x in the same way as the original double edges of these vertices to v.*

Lemma 1. *Rule 5 is safe.*

Proof. In the case of agreement on labels, see Fig. 9 (left), we proceed with $\min\{i, j\}$ Reidemeister unpoke moves followed by removing the subdivision. In case of label disagreement, see Fig. 9 (right), we keep exactly the same knot diagram, but simplify the generalized knot diagram by removing the subdivision on the double edge. In the latter case the number of twists on the new double edge is exactly the sum $i + j$ of the numbers of twists on the two original double edges. □

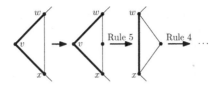

Fig. 10. Rule 6 is reduced to rule Rule 4

Rule 6. *Let v be of Type 2D, where neighbors w and x are adjacent by a single edge. This case can be handled with help of earlier cases. First, we add a subdivision of $\{w, x\}$. Then, we apply Rule 5 arriving in and applying one of the cases of Rule 4, see Fig. 10. Therefore, we either classify the diagram or safely reduce the graph.*

Rule 7. *Consider three double vertices v, w and x with three double edges, $\{v, w\}$, $\{x, v\}$ and $\{w, x\}$ of labels i, j and k, respectively. Apply Rule 5 to w, but instead of removing w, we set the new double edge (with label $i + k$ or $|i - k|$, say ℓ) to be $\{v, w\}$, and let $\{w, x\}$ be a double edge with label 0. Now, apply Rule 5 to v, but instead of removing v, we set the new double edge (with label $j + \ell$ or $|j - \ell|$, say m) to be $\{v, w\}$, and let $\{v, x\}$ be a double edge with label 0. Depending on the value of m, we can classify the knot diagram.*

1. *If $m = 0$, then the generalized knot diagram represents the unlink.*
2. *If $m \neq 1$ is odd, the generalized knot diagram represents the $(m, 2)$-torus knot, for definition and notations see [18].*
3. *If $m \neq 0$ is even, the generalized knot diagram represents the $(m, 2)$-torus link.*
4. *If $m = 1$, then the generalized knot diagram represents the unknot as a single Reidemeister untwist turns the diagram to a circle (see Fig 11, right.)*

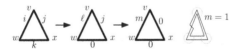

Fig. 11. Illustration to Rule 7

4   Main Algorithm

The main algorithm starts by subdividing where necessary edges to obtain a simple graph. Then, while there are at least three vertices, we repeatedly take a vertex that is incident to at most two edges, and apply a safe rule. Each rule application decreases the number of vertices, so after $O(n)$ such steps we resolve the problem. Standard techniques (for details see [3]) give a linear time implementation. Note that the safe rules only execute untwists and unpokes, but no other Reidemeister moves. Each untwist and unpoke decreases the number of twists by at least one. Thus we have shown Theorems 1 and 2. Next to these main results we have the following two straightforward corollaries.

Corollary 1. *Any knot/link having a diagram of treewidth 2 is a knot sum of* $(\cdot, 2)$-*torus knots/links.*

Corollary 2 (See [3]**).** *Given two knot diagrams of treewidth 2, the equivalence of the knots is verifiable in linear time (by simply comparing the resulting torus knots in Rules 4 and 7).*

5   Conclusions

We conclude with the following questions.

– We gave a linear time algorithm deciding whether any diagram with n crossings and treewidth 2 is a diagram of the unknot. The question arises: Is it possible to extend our result to graphs of treewidth $t \geq 3$? Even the existence of a polynomial time algorithm for $t = 3$ is open. Extension of our results to the graphs of treewidth $t = 3$ requires new arguments and techniques: Our algorithm monotonically decreases the number of crossings in a treewidth 2 diagram as only untwist and unpoke Reidemeister moves are performed, while there are unknot diagrams of treewidth 3 requiring increase of the number of crossings for unknotting, e.g., the *Culprit*, the *Goeritz unknot* and some other small but hard unknots [9].
– Koenig and Tsvietkova [10] conjectured and de Mesmay et al. [5] proved that deciding if a diagram of the unknot can be untangled using at most k Reidemeister moves (where k is part of the input) is NP-hard. Could this problem be solved in polynomial time on knots with diagrams of treewidth 2?

References

1. Alexander, J.W., Briggs, G.B.: On types of knotted curves. Ann. Math. (2) **28**(1–4), 562–586 (1926). https://doi.org/10.2307/1968399
2. Bodlaender, H.L.: A partial k-arboretum of graphs with bounded treewidth. Theoret. Comput. Sci. **209**, 1–45 (1998)
3. Bodlaender, H.L., Burton, B.A., Fomin, F.V., Grigoriev, A.: Knot diagrams of treewidth two (2019). http://arxiv.org/abs/1904.03117
4. Dasbach, O.T., Hougardy, S.: Does the Jones polynomial detect unknottedness? Exp. Math. **6**(1), 51–56 (1997)
5. de Mesmay, A., Rieck, Y.A., Sedgwick, E., Tancer, M.: The unbearable hardness of unknotting. In: 35th International Symposium on Computational Geometry, SoCG 2019, LIPIcs, vol. 129, pp. 49:1–49:19. Schloss Dagstuhl - Leibniz-Zentrum für Informatik (2019)

6. Haken, W.: Theorie der Normalflächen. Acta Math. **105**, 245–375 (1961). https://doi.org/10.1007/BF02559591

7. Hass, J., Lagarias, J.C.: The number of Reidemeister moves needed for unknotting. J. Am. Math. Soc. **14**(2), 399–428 (2001). https://doi.org/10.1090/S0894-0347-01-00358-7

8. Hass, J., Lagarias, J.C., Pippenger, N.: The computational complexity of knot and link problems. J. ACM **46**(2), 185–211 (1999). https://doi.org/10.1145/301970.301971

9. Henrich, A., Kauffman, L.H.: Unknotting unknots. arXiv preprint arXiv:1006.4176 (2010)

10. Koenig, D., Tsvietkova, A.: NP-hard problems naturally arising in knot theory. arXiv preprint arXiv:1809.10334 (2018)

11. Lackenby, M.: A polynomial upper bound on Reidemeister moves. Ann. Math. **182**(2), 491–564 (2015). https://doi.org/10.4007/annals.2015.182.2.3

12. Lackenby, M.: The efficient certification of knottedness and thurston norm. arXiv preprint arXiv:1604.00290 (2016)

13. Makowsky, J.A., Mariño, J.P.: The parameterized complexity of knot polynomials. J. Comput. Syst. Sci. **67**, 742–756 (2003)

14. Reidemeister, K.: Knoten und Gruppen. Abh. Math. Sem. Univ. Hamburg **5**(1), 7–23 (1927). https://doi.org/10.1007/BF02952506

15. Robertson, N., Seymour, P.D.: Graph minors. II. Algorithmic aspects of tree-width. J. Algorithms **7**, 309–322 (1986)

16. Rué, J., Thilikos, D.M., Velona, V.: Structure and enumeration of K_4-minor-free links and link diagrams. Electron. Notes Discret. Math. **68**, 119–124 (2018). https://doi.org/10.1016/j.endm.2018.06.021

17. Turing, A.M.: Solvable and Unsolvable Problems. Penguin Books, London (1954)

18. Torus knot (2019). http://mathworld.wolfram.com/TorusKnot.html

Treewidth Versus Clique Number in Graph Classes with a Forbidden Structure

Clément Dallard[1,2], Martin Milanič[1,2(✉)], and Kenny Štorgel[1,3]

[1] FAMNIT, University of Primorska, Koper, Slovenia
{clement.dallard,martin.milanic}@famnit.upr.si,
kennystorgel.research@gmail.com
[2] IAM, University of Primorska, Koper, Slovenia
[3] Faculty of Information Studies, Novo mesto, Slovenia

Abstract. Treewidth is an important graph invariant, relevant for both structural and algorithmic reasons. A necessary condition for a graph class to have bounded treewidth is the absence of large cliques. We study graph classes in which this condition is also sufficient, which we call (tw, ω)-bounded. Such graph classes are known to have useful algorithmic applications related to variants of the clique and k-coloring problems. We consider six well-known graph containment relations: the minor, topological minor, subgraph, induced minor, induced topological minor, and induced subgraph relations. For each of them, we give a complete characterization of the graphs H for which the class of graphs excluding H is (tw, ω)-bounded. Our results imply that the class of 1-perfectly orientable graphs is (tw, ω)-bounded, answering a question of Brešar, Hartinger, Kos, and Milanič from 2018. We also reveal some further algorithmic implications of (tw, ω)-boundedness related to list k-coloring and clique problems.

Keywords: Graph class · Treewidth · Clique number

1 Introduction

1.1 Background and Motivation

The treewidth of a graph measures, roughly speaking, how similar the graph is to a tree. This invariant played a crucial role in the theory of graph minors due to Robertson and Seymour (see, e.g., [34]), and many decision and optimization problems that are generally NP-hard are solvable in linear time for graph classes of bounded treewidth [4,6,12]. A necessary condition for bounded treewidth is the absence of large cliques. When is this condition also sufficient? We say that a graph class $\mathcal{G}$ is (tw, ω)-*bounded* if there exists a function $f : \mathbb{N} \to \mathbb{N}$

This research was funded in part by the Slovenian Research Agency (I0-0035, research program P1-0285, research projects J1-9110, N1-0102, and N1-0160, and a Young Researchers Grant).

I. Adler and H. Müller (Eds.): WG 2020, LNCS 12301, pp. 92–105, 2020.
https://doi.org/10.1007/978-3-030-60440-0_8

such that $tw(G) \leq f(\omega(G))$ for all graphs $G \in \mathcal{G}$, where $tw(G)$ and $\omega(G)$ denote the treewidth and the clique number of G, respectively. Such a function f is called a (tw, ω)-*binding function* for the class $\mathcal{G}$. Many graph classes studied in the literature are known to be (tw, ω)-bounded. For every positive integer t, the class of intersection graphs of connected subgraphs of graphs with treewidth at most t is (tw, ω)-bounded (see [3,40]). This includes the classes of chordal graphs and circular-arc graphs. Further examples include graph classes of bounded treewidth, classes of graphs in which all minimal separators are of bounded size [45], and, as a consequence of Ramsey's theorem, classes of graphs of bounded independence number.

There are multiple motivations for the study of (tw, ω)-bounded graph classes, from both algorithmic and structural points of view. The k-CLIQUE problem asks whether the input graph contains a clique of size k; the problem is known to be W[1]-hard (see [16]). Given a graph G and a list of available colors from the set $\{1, \ldots, k\}$ for each vertex, the LIST k-COLORING problem asks whether G can be properly vertex-colored by assigning to each vertex a color from its list. This is a generalization of the classical k-coloring problem and is thus NP-hard for all $k \geq 3$. Chaplick and Zeman gave fixed-parameter tractable algorithms for k-CLIQUE and LIST k-COLORING in any (tw, ω)-bounded classes of graphs with a computable binding function f [9]. From the structural point of view, identifying new (tw, ω)-bounded graph classes directly addresses a recent question of Weißauer [46] asking for which classes can we force large cliques by assuming large treewidth. Weißauer distinguishes graph parameters as being either *global* or *local* (see [46] for precise definitions). In this terminology, (tw, ω)-boundedness of a graph class is a sufficient condition for treewidth to become a local parameter.

1.2   Our Results

The main aim of this paper is to further the knowledge of (tw, ω)-bounded graph classes. We consider six well-known graph containment relations and for each of them, give a complete characterization of the graphs H for which the class of graphs excluding H (with respect to the relation) is (tw, ω)-bounded. These six relations are the minor relation, the topological minor relation, the subgraph relation, and their induced variants: the induced minor relation, the induced topological minor relation, and the induced subgraph relation. (Precise definitions will be given in Sect. 2.) To explain our results, we need to introduce some notation. We denote by $\subseteq_{is}$ the induced subgraph relation. By $K_{p,q}$ we denote the complete bipartite graph with parts of size p and q; if $p = q$, then, the complete bipartite graph is said to be *balanced*. The *claw* is the complete bipartite graph $K_{1,3}$. A *subdivided claw* is the graph obtained from the claw by replacing each edge with a path of length at least one. We denote by $\mathcal{S}$ the class of graphs in which every connected component is either a path or a subdivided claw. For $q \geq 1$, we denote by $K_{2,q}^+$ the graph obtained from $K_{2,q}$ by adding an additional edge between the two vertices in the part of size 2. Similarly, we denote by K_q^- the graph obtained from the complete graph K_q by removing an edge. Note that the graph K_4^- is sometimes called the *diamond*. The graph C_ℓ is the cycle on ℓ vertices, and the 4-*wheel*, also denoted by W_4, is the graph obtained from the C_4 by adding a new vertex adjacent to all vertices of the C_4. A graph is *subcubic* if every vertex is incident with at most three edges.

Our characterizations are summarized in Table 1 where each entry corresponds to one of the six containment relations and contains a description of necessary and sufficient conditions for a graph H such that the class of graphs excluding H with respect to the relation considered in the entry is (tw, ω)-bounded.

Table 1. Summary of (tw, ω)-bounded graph classes excluding a fixed graph H for six graph containment relations.

	General	Induced
Subgraph	$H \in \mathcal{S}$	$H \subseteq_{is} P_3$ or H is edgeless
Topological minor	H is subcubic and planar	$H \subseteq_{is} C_3$, $H \subseteq_{is} C_4$, $H \cong K_4^-$, or H is edgeless
Minor	H is planar	$H \subseteq_{is} W_4$, $H \subseteq_{is} K_5^-$, $H \subseteq_{is} K_{2,q}$ or $H \subseteq_{is} K_{2,q}^+$, for some $q \in \mathbb{N}$

To the best of our knowledge, these six dichotomies represent the first set of results towards a systematic study of the problem of classifying (tw, ω)-bounded graph classes. Furthermore, the (tw, ω)-boundedness of the class of $K_{2,3}$-induced-minor-free graphs, implies that the class of 1-perfectly orientable graphs is (tw, ω)-bounded. This answers a question raised by Brešar et al. [7].

From the algorithmic point of view, we observe that for any fixed positive integer k, the approach of Chaplick and Zeman from [9] can be adapted to obtain a robust polynomial-time algorithm for LIST k-COLORING in any graph class with a computable (tw, ω)-binding function. We also show how to approximate the clique number to within a factor of $\mathsf{opt}^{1-1/\mathcal{O}(1)}$ in graph classes with a polynomially bounded (tw, ω)-binding function, where opt is the clique number of the input graph.

Our techniques combine the development and applications of structural properties of graphs in restricted classes, connections with Hadwiger number and with minimal separators, as well as applications of Ramsey's theorem and known results on treewidth and graph minors. Results given by Table 1 are derived in Sects. 3 to 5. The algorithmic results are presented in Sect. 6.

Due to lack of space, proofs of results marked by ♦ are omitted.

1.3 Related Work

The (tw, ω)-boundedness property for the class of even-hole-free graphs was studied in [8,43,44]. Dichotomy studies similar to ours exist for many other properties of graph classes, including boundedness of the clique-width [13,14], well-quasi-ordering [2,15,31], and polynomial-time solvability of various graph problems [20,27,35,41].

The concept of a (tw, ω)-bounded graph class is part of the following more general framework. An (integer) *graph invariant* is a mapping from the class of all graphs to the set of non-negative integers $\mathbb{N}$ that does not distinguish between isomorphic graphs. Given two graph invariants ρ and σ and a graph class $\mathcal{G}$, we say that $\mathcal{G}$ is (ρ, σ)-*bounded* if there exists a (ρ, σ)-*binding function* for $\mathcal{G}$, that is, a function $f : \mathbb{N} \to \mathbb{N}$ such that $\rho(G) \le f(\sigma(G))$ for all graphs $G \in \mathcal{G}$. Probably the most well-known and well-studied case of (ρ, σ)-bounded graph classes corresponds to the pair $(\rho, \sigma) = (\chi, \omega)$, where $\chi(G)$ denotes the chromatic number of G. Such graph classes are called simply χ-*bounded*. They were introduced by Gyárfás in the late 1980s to generalize perfection [23] and studied extensively in the literature (see [42] for a survey). Note that every (tw, ω)-bounded graph class is also χ-bounded but not vice versa. Furthermore, (tw, ω)-boundedness generalizes chordality in the same way that χ-boundedness generalizes perfection. Several other variants of (ρ, σ)-bounded graph classes were studied in the literature, though not to the same extent as the χ-bounded ones (see, e.g., [5,24,28,30,36,48]).

2 Preliminaries

We now define the six graph containment relations studied in this paper. If a graph H can be obtained from a graph G by only deleting vertices, then H is an *induced subgraph* of G, and we write $H \subseteq_{is} G$. If H is obtained from G by deleting vertices and edges, then H is a *subgraph* of G, and we write $H \subseteq_s G$. Note that if $H \subseteq_{is} G$, then $H \subseteq_s G$. A subdivision of a graph H is a graph obtained from H by a sequence of edge subdivisions. The graph H is said to be a *topological minor* (or *topological subgraph*) of a graph G if G contains a subdivision of H as a subgraph, and we write $H \subseteq_{tm} G$. Similarly, H is an *induced topological minor* of G if G contains a subdivision of H as an induced subgraph, and we write $H \subseteq_{itm} G$. Again, if $H \subseteq_{itm} G$, then $H \subseteq_{tm} G$. An edge contraction is the operation of deleting a pair of adjacent vertices and replacing them with a new vertex whose neighborhood is the union of the neighborhoods of the two original vertices. We say that G contains H as *induced minor* if H can be obtained from G by a sequence of vertex deletions and edge contractions, and we write $H \subseteq_{im} G$. Finally, if H can be obtained from G by a sequence of vertex deletions, edge deletions, and edge contractions, then H is said to be a *minor* of G, and we write $H \subseteq_m G$. Here also, if $H \subseteq_{im} G$, then $H \subseteq_m G$. Besides the already observed implications, one can notice that

$$H \subseteq_s G \implies H \subseteq_{tm} G \implies H \subseteq_m G \quad \text{and}$$
$$H \subseteq_{is} G \implies H \subseteq_{itm} G \implies H \subseteq_{im} G.$$

If G does not contain an induced subgraph isomorphic to H, then we say that G is H-free. Analogously, we may also say that G is H-subgraph-free, H-

topological-minor-free, H-induced-topological-minor-free, H-minor-free, or H-induced-minor-free, respectively, for the other five relations. It is well known that G contains H as a minor if and only if there exists a *minor model* of H in G, that is, a collection $(X_u : u \in V(H))$ of pairwise disjoint subsets of $V(G)$ called *bags* such that each X_u induces a connected subgraph of G and for every two adjacent vertices $u, v \in V(H)$, there is an edge in G between a vertex of X_u and a vertex of X_v. Similarly, G contains H as an induced minor if and only if there exists an *induced minor model* of H in G, which is defined similarly as a minor model, except that for every two distinct vertices $u, v \in V(H)$, there is an edge in G between a vertex of X_u and a vertex of X_v if and only if $uv \in E(H)$.

Given a set $S \subseteq V(G)$, we denote by $G - S$ the graph obtained from G by removing all vertices S and by $G[S]$ the *subgraph of G induced by S*, that is, the graph $G - (V(G) \setminus S)$. For $u \in V$, $N(u) = \{v \in V : uv \in E\}$ is the *neighborhood* of u and $N[u] = N(u) \cup \{u\}$ is the *closed neighborhood* of u. The *degree* of u in G is the cardinality of its neighborhood. The *clique number* of a graph G, denoted by $\omega(G)$, is the maximum size of a clique in G. A *tree decomposition* of a graph G is a pair $(T, \{X_t : t \in V(T)\})$, where T is a tree and each $t \in V(T)$ is associated with a vertex subset $X_t \subseteq V(G)$ such that $\bigcup_{t \in V(T)} X_t = V$, for each edge $uv \in E(G)$ there exists some $t \in V(T)$ such that $u, v \in X_t$, and for every $u \in V(G)$, the set $T_u = \{t \in V(T) : u \in X_t\}$ induces a connected subtree of T. The *width* of a tree decomposition equals $\max_{t \in V(T)} |X_t| - 1$, and the *treewidth* of a graph G, denoted by $\mathrm{tw}(G)$, is the minimum possible width of a tree decomposition of G. A graph class $\mathcal{G}$ is said to be of *bounded treewidth* if there exists a constant c such that $\mathrm{tw}(G) \leq c$ for all $G \in \mathcal{G}$; otherwise, $\mathcal{G}$ is of *unbounded treewidth*. A *hole* in a graph G is an induced subgraph of G isomorphic to a cycle of length at least four. A graph is said to be *chordal* if it does not contain any hole.

Observation 1. *A graph G is chordal if and only if G is C_4-induced-minor-free, if and only if G is C_4-induced-topological-minor-free.*

Some of our proofs will make use of the following classical result due to Ramsey [38].

Ramsey's theorem. *For every two positive integers k and ℓ, there exists a least positive integer $R(k, \ell) \leq \binom{k+\ell-2}{k-1}$ such that every graph with at least $R(k, \ell)$ vertices contains either a clique of size k or an independent set of size ℓ.*

Using Ramsey's theorem, we can already derive the following.

♦ **Lemma 1.** *Let H be an edgeless graph. Then the class of H-free graphs is (tw, ω)-bounded, with a binding function $f(k) = R(k + 1, |V(H)|) - 2$.*

A graph class that is not (tw, ω)-bounded is said to be (tw, ω)-*unbounded*. Lemma 2 is about some specific (tw, ω)-unbounded graph classes, which will play a crucial role in our proofs. The *line graph* of a graph G, denoted by $L(G)$, is the graph with vertex set $E(G)$ where two vertices are adjacent if and only if the corresponding edges intersect. For the definition of an *elementary wall*, we refer to [11]. For a non-negative integer q, we say that a graph is a *q-subdivided-wall* if it can be obtained from an elementary wall by subdividing each edge q times.

♦ **Lemma 2.** *The class of balanced complete bipartite graphs and, for all $q \geq 0$, the class of q-subdivided walls and the class of their line graphs are (tw, ω)-unbounded.*

3   Forbidding a Subgraph, a Topological Minor, or a Minor

We use known results on treewidth and graph minors to derive characterizations of (tw, ω)-bounded graph classes excluding a single graph as either a subgraph, a topological minor, or a minor.

Theorem 1 (Robertson and Seymour [39]). *For every planar graph H, the class of H-minor-free graphs has bounded treewidth.*

Lemma 3 (Golovach et al. [21]). *For every $H \in \mathcal{S}$, a graph G is H-subgraph-free if and only if it is H-minor-free.*

Theorem 2. *Let H be a graph. Then, the class of H-subgraph-free graphs is (tw, ω)-bounded if and only if $H \in \mathcal{S}$.*

Proof. Suppose that $H \in \mathcal{S}$. Then following Lemma 3 every H-subgraph-free graph is also H-minor-free. Hence, by Theorem 1, the class of H-subgraph-free graphs has bounded treewidth. In particular, it is (tw, ω)-bounded.

Suppose now that the class of H-subgraph-free graphs is (tw, ω)-bounded. Note that H must be a subgraph of an elementary wall, since otherwise the class of H-subgraph-free graphs would contain the class of elementary walls, which, following Lemma 2, would contradict the assumption that the class is (tw, ω)-bounded. It follows that H is subcubic. Suppose that H contains a connected component with two vertices u and v of degree 3 and let ℓ be the distance between u and v. Then the class of ℓ-subdivided walls is a subclass of the class of H-subgraph-free graphs. Following Lemma 2, this contradicts the assumption that the class of H-subgraph-free graphs is (tw, ω)-bounded. Thus, every connected component of H has at most one vertex of degree 3. Using a similar reasoning, we can conclude that H is acyclic, and thus $H \in \mathcal{S}$.   □

A similar approach can be used to prove Theorems 3 and 4.

♦ **Theorem 3.** *Let H be a graph. Then, the class of H-topological-minor-free graphs is (tw, ω)-bounded if and only if H is subcubic and planar.*

♦ **Theorem 4.** *Let H be a graph. Then, the class of H-minor-free graphs is (tw, ω)-bounded if and only if H is planar.*

The proofs actually show that when H is forbidden as a subgraph, topological minor, or minor, (tw, ω)-boundedness is equivalent to bounded treewidth.

4   Forbidding an Induced Subgraph or an Induced Topological Minor

The following characterization of (tw, ω)-bounded graph classes excluding a single forbidden induced subgraph is derived using Lemmas 1 and 2.

♦ **Theorem 5.** *Let H be a graph. Then, the class of H-free graphs is (tw, ω)-bounded if and only if $H \subseteq_{is} P_3$ or H is edgeless.*

A *cut-vertex* in a connected graph G is a vertex whose removal disconnects the graph. A *block* of a graph is a maximal connected subgraph without cut-vertices. A *block-cactus graph* is a graph every block of which is a cycle or a complete graph. In her PhD thesis [25], Hartinger proved that a graph is K_4^--induced-minor-free if and only if G is a block-cactus graph. In fact, the same approach shows the following stronger claim.

♦ **Lemma 4.** *A graph G is a block-cactus graph, if and only if G is K_4^--induced-minor-free, if and only if G is K_4^--induced-topological-minor-free.*

♦ **Lemma 5.** *The class of block-cactus graphs is (tw, ω)-bounded, with a binding function $f(k) = \max\{k - 1, 2\}$.*

Graphs H such that the class of graphs excluding H as an induced topological minor are (tw, ω)-bounded are characterized as follows.

Theorem 6. *Let H be a graph. Then, the class of H-induced-topological-minor-free graphs is (tw, ω)-bounded if and only if $H \subseteq_{is} C_4$, $H \subseteq_{is} K_4^-$, or H is edgeless.*

Proof. If H is edgeless, then Lemma 1 applies. If $H \subseteq_{is} C_4$, then $H \subseteq_{itm} C_4$. Hence, by Observation 1, the class of H-induced-topological-minor-free graphs is a subclass of the class of chordal graphs, and thus (tw, ω)-bounded. If $H \subseteq_{is} K_4^-$, then according to Lemma 4 the class of H-induced-topological-minor-free graphs is a subclass of the class of block-cactus graphs, which is (tw, ω)-bounded by Lemma 5.

For the converse direction, suppose that $H \nsubseteq_{is} C_4$, $H \nsubseteq_{is} K_4^-$, H is not edgeless, and that the class of H-induced-topological-minor-free graphs is (tw, ω)-bounded. By Lemma 2, the class of line graph of 1-subdivided walls is (tw, ω)-unbounded. It follows that H must be an induced topological minor of the line graph of some 1-subdivided wall. Since the line graph of every 1-subdivided wall is planar, subcubic, and claw-free, this implies that H is planar, subcubic, and claw-free. Similarly, since the class of complete bipartite graphs is (tw, ω)-unbounded, H must be an induced topological minor of some complete bipartite graph. Since H is planar, subcubic, and not edgeless, it follows that $H \subseteq_{itm} K_{2,3}$. We obtain that $H \in \{P_2, P_3, C_3, C_4, K_4^-\}$, a contradiction. □

5   Forbidding an Induced Minor

Finally, we consider graph classes excluding a single graph H as an induced minor. Given a graph G, we denote by $\eta(G)$ the *Hadwiger number of G*, defined as the largest value of p such that K_p is a minor of G (see [33]). We first develop some sufficient conditions for when sufficiently large Hadwiger number implies large clique number and then apply these results to characterize the graphs H such that the class of H-induced-minor-free graphs is (tw, ω)-bounded.

5.1   A Detour: Hadwiger Number Versus Clique Number

Theorems 7 and 8 show that excluding either a complete graph minus an edge or a 4-wheel as an induced minor results in an (η, ω)-bounded graph class, with a linear binding function.

Theorem 7. *For each $p \geq 2$, the class of K_p^--induced-minor-free graphs is (η, ω)-bounded, with a binding function $f(k) = \max\{2p - 4, k\}$.*

Proof. Fix $p \geq 2$ and $k \in \mathbb{N}$, and let G be a K_p^--induced-minor-free graph with $\omega(G) = k$. Let $q = \max\{2p - 4, k\} + 1$. We want to show that G contains no K_q as a minor. Suppose for a contradiction that G contains K_q as a minor. Fix a minor model $M = (X_u : u \in V(K_q))$ of K_q in G such that the total number of vertices in the bags, that is, the sum $\sum_{u \in V(K_q)} |X_u|$, is minimized.

If for all $u \in V(K_q)$ we have $|X_u| = 1$, then the set $\bigcup_{u \in V(K_q)} X_u$ is a clique in G, implying that $\omega(G) \geq |V(K_q)| - q \geq k + 1$, a contradiction. Therefore, there exists some $u \in V(K_q)$ such that $|X_u| \geq 2$. Furthermore, note that for every vertex $y \in X_u$ there exists a vertex $v(y)$ of $K_q - u$ such that y has no neighbors in $X_{v(y)}$, since otherwise replacing the bag X_u with $\{y\}$ would result in a minor model of K_q smaller than M. Since $|X_u| \geq 2$ and the subgraph of G induced by X_u is connected, there exists a vertex $x \in X_u$ such that the subgraph of G induced by $X_u \setminus \{x\}$ is connected. (For example, take x to be a leaf of a spanning tree of $G[X_u]$.)

Let Z be the set of vertices $z \in V(K_q) \setminus \{u\}$ such that x has a neighbor in X_z. Suppose first that $|Z| \geq (q - 1)/2$. Recall that $X_{v(x)}$ is a bag in which x has no neighbor. In particular, $v(x) \neq u$ and $v(x) \notin Z$. Then, the bags from $(X_z : z \in Z)$ along with $\{x\}$ and $X_{v(x)}$ form an induced minor model of $K_{|Z|+2}^-$. Since $|Z| + 2 \geq (q - 1)/2 + 2 \geq (2p - 4)/2 + 2 = p$, we obtain a contradiction with the fact that G is K_p^--induced-minor-free.

Finally, suppose that $|Z| < (q - 1)/2$. The minimality of M implies that for some $w \in Z$ we have $\bigcup_{v \in X_w} N(v) \cap X_u = \{x\}$. Let $Z' = V(K_q) \setminus (Z \cup \{u\})$. Note that for every vertex $z \in Z'$ there exists an edge from X_z to $X_u \setminus \{x\}$. Since $|Z| + |Z'| = q - 1$ and $|Z| < (q - 1)/2$, we have $|Z'| \geq (q - 1)/2$. Thus, the bags from $(X_z : z \in Z')$ along with $X_u \setminus \{x\}$ and X_w form an induced minor model of $K_{|Z'|+2}^-$, leading again to a contradiction with the fact that G is K_p^--induced-minor-free. □

◆ **Theorem 8.** *The class of W_4-induced-minor-free graphs is (η, ω)-bounded, with a binding function $f(k) = k + 5$.*

5.2   Back to Treewidth

As explained by Belmonte *et al.* [1] (and observed also in [7]), the following fact can be derived from the proof of Theorem 9 in [29].

Theorem 9. *For every graph F and every planar graph H, the class of graphs that are both F-minor-free and H-induced-minor-free has bounded treewidth.*

Since excluding a complete graph as a minor is the same as excluding it as an induced minor, Theorem 9 implies the following.

Corollary 1. *For every positive integer p and every planar graph H, the class of $\{K_p, H\}$-induced-minor-free graphs has bounded treewidth.*

Observe that no graph G contains $K_{\eta(G)+1}$ as an (induced) minor.

♦ **Corollary 2.** *Let H be a planar graph. The class of H-induced-minor-free graphs is (η, ω)-bounded if and only if it is (tw, ω)-bounded.*

From Theorem 7 we obtain that the class of K_5^--induced-minor-free graphs is (η, ω)-bounded. Since K_5^- is planar, a direct application of Corollary 2 implies the following result.

Corollary 3. *The class of K_5^--induced-minor-free graphs is (tw, ω)-bounded.*

Similarly, since W_4 is planar, we can directly apply Theorem 8 and Corollary 2 and obtain the following result.

Corollary 4. *The class of W_4-induced-minor-free graphs is (tw, ω)-bounded.*

Our next result makes use of minimal separators. Given two non-adjacent vertices u and v in a graph G, a *u,v-separator* in G is a set S of vertices such that u and v are in different connected components of $G - S$. A u,v-separator is *minimal* if it does not contain any other u,v-separator. A *minimal separator* in a graph G is a minimal u,v-separator for some non-adjacent vertex pair u, v.

Theorem 10 (Skodinis [45]). *Let s be a positive integer and let $\mathcal{G}$ be the class of graphs in which all minimal separators have size at most s. Then, $\mathcal{G}$ is (tw, ω)-bounded, with a binding function $f(k) = \max\{k, 2s\} - 1$.*

Using Theorem 10, we infer our next result.

Lemma 6. *For every $q \in \mathbb{N}$, the class of $K_{2,q}$-induced-minor-free graphs is (tw, ω)-bounded, with a binding function $f(k) = \max\{k, 2R(k+1, q) - 2\} - 1$.*

Proof. Fix two positive integers q and k, and let G be a $K_{2,q}$-induced-minor-free graph with $\omega(G) = k$. We claim that every minimal separator in G has size at most $R(k+1, q) - 1$. Suppose this is not the case, and let u and v be two non-adjacent vertices in G such that $|S| \geq R(k+1, q)$ for some minimal u,v-separator S in G. Since $|S| \geq R(k+1, q)$, Ramsey's theorem implies that

$G[S]$ contains either a clique of size $k+1$ or an independent set of size q. Since $\omega(G[S]) \leq \omega(G) = k$, we infer that $G[S]$ contains an independent set I of size q. Let C_u and C_v denote the connected components of $G - S$ containing u and v, respectively. By the minimality of S, every vertex in S has a neighbor in C_u and a neighbor in C_v (see, e.g., [22]). But now, the sets $V(C_u)$, $V(C_v)$, and $\{x\}$ for all $x \in I$ form the bags of an induced minor model of $K_{2,q}$ in G, a contradiction. Therefore, every minimal separator in G has size at most $R(k+1, q) - 1$. Using Theorem 10, we obtain that $\mathrm{tw}(G) \leq \max\{k, 2R(k+1, q) - 2\} - 1$.   □

A graph G is said to be 1-*perfectly orientable* if it has an orientation D such that for every vertex $v \in V(G)$, the out-neighborhood of v in D is a clique in G. The class of 1-perfectly orientable graphs is a common generalization of the classes of chordal graphs and circular-arc graphs. Brešar et al. showed in [7] that the treewidth of every 1-perfectly orientable planar graph is at most 21 and asked whether the class of 1-perfectly orientable graphs is (tw, ω)-bounded. Since every 1-perfectly orientable graph excludes $K_{2,3}$ as an induced minor (see [26]), Lemma 6 answers their question in the affirmative.

Corollary 5. *The class of 1-perfectly orientable graphs is* (tw, ω)-*bounded with a binding function* $f(k) = \max\{k, 2R(k+1, 3) - 2\} - 1$.

Lemma 2 and Corollaries 3 and 4 lead to the following characterization.

◆ **Theorem 11.** *Let H be a graph. Then, the class of H-induced-minor-free graphs is* (tw, ω)-*bounded if and only if one of the following conditions holds:* $H \subseteq_{is} W_4$, $H \subseteq_{is} K_5^-$, $H \subseteq_{is} K_{2,q}$, *or* $H \subseteq_{is} K_{2,q}^+$, *for some* $q \in \mathbb{N}$.

6   Algorithmic Implications of (tw, ω)-Boundedness

As explained in the introduction, the (tw, ω)-bounded classes having a computable binding function possess some algorithmically useful properties for variants of the clique and coloring problems. All the (tw, ω)-bounded graph classes identified in this work have a computable binding function. For the (tw, ω)-bounded graph classes discussed in Sect. 3, a result of Chekuri and Chuzhoy applies stating that there is a universal constant $c \leq 100$ such that, if G excludes a planar graph H as a minor, then the treewidth of G is $\mathcal{O}(|V(H)|^c)$ [10]. The (tw, ω)-boundedness of graph classes discussed in Sects. 4 and 5 is derived either using the structure of graphs in the resulting class (Theorem 5 and Lemma 5), Ramsey's theorem (Theorem 5 and Lemma 6), or graph minors theory (Corollaries 3 and 4). In the former two cases, the binding functions are explicit polynomials. In the case of applications of graph minors theory, the key result to deriving those bounds is Theorem 9, the proof of which relies on results of Fomin *et al.* [18]. As explained in [19, Section 9], recent developments in the area of graph minors imply that these bounds are computable, too.

As shown by Chaplick and Zeman, for every (tw, ω)-bounded class $\mathcal{G}$ with a computable binding function and for every fixed k, LIST k-COLORING is solvable in linear time for graphs in $\mathcal{G}$ [9]. If we are satisfied with polynomial running time, we can extend their approach to obtain an algorithm for LIST k-COLORING that is *robust* in the sense of Raghavan and Spinrad [37]: it either solves the problem or determines that the input graph is not in $\mathcal{G}$.

♦ **Theorem 12.** *Let $\mathcal{G}$ be a (tw, ω)-bounded graph class having a computable (tw, ω)-binding function f. Then, for every positive integer k there exists a robust polynomial-time algorithm for the LIST k-COLORING problem on graphs in $\mathcal{G}$.*

We next discuss some possible implications of (tw, ω)-boundedness for improved approximations for the MAXIMUM CLIQUE problem. For general graphs, this problem is notoriously difficult to approximate: for every $\varepsilon > 0$, there is no polynomial-time algorithm for approximating the maximum clique in an n-vertex graph to within a factor of $n^{1-\varepsilon}$ unless $\mathsf{P} = \mathsf{NP}$ [49]. For (tw, ω)-bounded graph classes with a polynomial binding function, known approximation algorithms for treewidth (see, e.g., [17]) lead to an improved approximability bound, provided that we allow the algorithm to output only a number approximating the value of the optimal solution and not the approximate solution itself. We denote by opt the optimal solution value of the maximum clique problem on the input graph G, that is, $\omega(G)$.

♦ **Theorem 13.** *Let $\mathcal{G}$ be a graph class having a polynomial (tw, ω)-binding function $f(k) = \mathcal{O}(k^c)$ for some constant c. Then, for all $\varepsilon > 0$ the clique number can be approximated for graphs in $\mathcal{G}$ in polynomial time to within a factor of $\mathsf{opt}^{1-1/(c+\varepsilon)}$.*

For graph classes having a linear (tw, ω)-binding function, Theorem 13 implies an $\mathsf{opt}^{\varepsilon}$ approximation for all $\varepsilon > 0$. Note that this result cannot be improved to a polynomial-time approximation scheme for the maximum clique problem, unless $\mathsf{P} = \mathsf{NP}$, as there exist graph classes with a linear (tw, ω)-binding function in which the clique number is APX-hard to compute, see [9].

7   Open Problems

We obtained a first set of results aimed towards classifying (tw, ω)-bounded graph classes, by considering six well-known graph containment relations and for each of them, characterizing the graphs H for which the class of graphs excluding H is (tw, ω)-bounded. In conclusion, we pose the following open problems:

1. Which graph classes defined by larger finite sets of forbidden structures (with respect to various graph containment relations) are (tw, ω)-bounded?
2. All the (tw, ω)-bounded graph classes identified in this paper have a polynomial (tw, ω)-binding function. A natural question arises: Does every hereditary (tw, ω)-bounded graph class have a polynomial (tw, ω)-binding function? A positive answer to this question would answer the analogous question by Esperet on χ-boundedness (see [32]) for the case of (tw, ω)-bounded graph classes.

3. Which graph classes have a linear (tw, ω)-binding function?
4. It is a well-known open problem whether treewidth can be approximated within a constant factor (see, e.g., [17,47]). It seems plausible that the problem could be easier for (tw, ω)-bounded classes, especially if an additional constraint is imposed on the binding function, for example that it is polynomial or linear. Note that for graph classes with a linear (tw, ω)-binding function, a constant factor approximation for treewidth would also imply a constant factor approximation for the clique number.

References

1. Belmonte, R., Otachi, Y., Schweitzer, P.: Induced minor free graphs: isomorphism and clique-width. Algorithmica **80**(1), 29–47 (2016). https://doi.org/10.1007/s00453-016-0234-8

2. Błasiok, J., Kamiński, M., Raymond, J.F., Trunck, T.: Induced minors and well-quasi-ordering. J. Comb. Theor. Ser. B **134**, 110–142 (2019). https://doi.org/10.1016/j.jctb.2018.05.005

3. Bodlaender, H., Gustedt, J., Telle, J.A.: Linear-time register allocation for a fixed number of registers. In: Proceedings of the Ninth Annual ACM-SIAM Symposium on Discrete Algorithms, San Francisco, CA, pp. 574–583. ACM, New York (1998)

4. Bodlaender, H.L.: A linear-time algorithm for finding tree-decompositions of small treewidth. SIAM J. Comput. **25**(6), 1305–1317 (1996). https://doi.org/10.1137/S0097539793251219

5. Bodlaender, H.L., Ono, H., Otachi, Y.: Degree-constrained orientation of maximum satisfaction: graph classes and parameterized complexity. In: 27th International Symposium on Algorithms and Computation, LIPIcs. LeibnizInt. Proc. Inform., vol. 64, pp. Art. No. 20, 12. Schloss Dagstuhl. Leibniz-Zent. Inform., Wadern (2016). https://doi.org/10.1007/s00453-017-0399-9

6. Borie, R.B., Parker, R.G., Tovey, C.A.: Automatic generation of linear-time algorithms from predicate calculus descriptions of problems on recursively constructed graph families. Algorithmica **7**(5–6), 555–581 (1992). https://doi.org/10.1007/BF01758777

7. Brešar, B., Hartinger, T.R., Kos, T., Milanič, M.: 1-perfectly orientable K_4-minor-free and outerplanar graphs. Discret. Appl. Math. **248**, 33–45 (2018). https://doi.org/10.1016/j.dam.2017.09.017

8. Cameron, K., Chaplick, S., Hoàng, C.T.: On the structure of (pan, even hole)-free graphs. J. Graph Theor. **87**(1), 108–129 (2018). https://doi.org/10.1002/jgt.22146

9. Chaplick, S., Zeman, P.: Combinatorial problems on H-graphs. Electron. Notes Discret. Math. **61**, 223–229 (2017). https://doi.org/10.1016/j.endm.2017.06.042

10. Chekuri, C., Chuzhoy, J.: Polynomial bounds for the grid-minor theorem. J. ACM **63**(5), 65 (2016). https://doi.org/10.1145/2820609. Art. 40

11. Chuzhoy, J.: Improved bounds for the flat wall theorem. In: Proceedings of the Twenty-Sixth Annual ACM-SIAM Symposium on Discrete Algorithms, SIAM, Philadelphia, PA, pp. 256–275 (2015). https://doi.org/10.1137/1.9781611973730.20

12. Courcelle, B.: The monadic second-order logic of graphs. I. Recognizable sets of finite graphs. Inf. Comput. **85**(1), 12–75 (1990). https://doi.org/10.1016/0890-5401(90)90043-H

13. Dabrowski, K.K., Johnson, M., Paulusma, D.: Clique-width for hereditary graph classes. In: Surveys in Combinatorics 2019, London Math. Soc. Lecture Note Series, vol. 456, pp. 1–56. Cambridge University Press, Cambridge (2019). https://doi.org/10.1017/9781108649094

14. Dabrowski, K.K., Lozin, V.V., Paulusma, D.: Clique-width and well-quasi-ordering of triangle-free graph classes. J. Comput. Syst. Sci. **108**, 64–91 (2020). https://doi.org/10.1016/j.jcss.2019.09.001

15. Ding, G.: Subgraphs and well-quasi-ordering. J. Graph Theor. **16**(5), 489–502 (1992). https://doi.org/10.1002/jgt.3190160509

16. Downey, R.G., Fellows, M.R.: Fixed-parameter tractability and completeness. II. On completeness for $W[1]$. Theoret. Comput. Sci. **141**(1–2), 109–131 (1995). https://doi.org/10.1016/0304-3975(94)00097-3

17. Feige, U., Hajiaghayi, M., Lee, J.R.: Improved approximation algorithms for minimum weight vertex separators. SIAM J. Comput. **38**(2), 629–657 (2008). https://doi.org/10.1137/05064299X

18. Fomin, F.V., Golovach, P., Thilikos, D.M.: Contraction obstructions for treewidth. J. Comb. Theor. Ser. B **101**(5), 302–314 (2011). https://doi.org/10.1016/j.jctb.2011.02.008

19. Garnero, V., Paul, C., Sau, I., Thilikos, D.M.: Explicit linear kernels for packing problems. Algorithmica **81**(4), 1615–1656 (2018). https://doi.org/10.1007/s00453-018-0495-5

20. Golovach, P.A., Johnson, M., Paulusma, D., Song, J.: A survey on the computational complexity of coloring graphs with forbidden subgraphs. J. Graph Theor. **84**(4), 331–363 (2017). https://doi.org/10.1002/jgt.22028

21. Golovach, P.A., Paulusma, D., Ries, B.: Coloring graphs characterized by a forbidden subgraph. Discret. Appl. Math. **180**, 101–110 (2015). https://doi.org/10.1016/j.dam.2014.08.008

22. Golumbic, M.C.: Algorithmic Graph Theory and Perfect Graphs, Annals of Discrete Mathematics, vol. 57, 2nd edn. Elsevier, Amsterdam (2004)

23. Gyárfás, A.: Problems from the world surrounding perfect graphs. Zastos. Mat. **19**(3–4), 413–441 (1987)

24. Gyárfás, A., Zaker, M.: On (δ, χ)-bounded families of graphs. Electron. J. Comb. **18**(1), 108 (2011). https://doi.org/10.37236/595

25. Hartinger, T.R.: New Characterizations in Structural Graph Theory: 1-Perfectly Orientable Graphs, Graph Products, and the Price of Connectivity. Ph.D. Thesis. University of Primorska (2017)

26. Hartinger, T.R., Milanič, M.: Partial characterizations of 1-perfectly orientable graphs. J. Graph Theor. **85**(2), 378–394 (2017). https://doi.org/10.1002/jgt.22067

27. Hell, P., Nešetřil, J.: On the complexity of H-coloring. J. Comb. Theor. Ser. B **48**(1), 92–110 (1990). https://doi.org/10.1016/0095-8956(90)90132-J

28. Hermelin, D., Mestre, J., Rawitz, D.: Optimization problems in dotted interval graphs. Discret. Appl. Math. **174**, 66–72 (2014). https://doi.org/10.1016/j.dam.2014.04.014

29. van't Hof, P., Kamiński, M., Paulusma, D., Szeider, S., Thilikos, D.M.: On graph contractions and induced minors. Discret. Appl. Math. 160(6), 799–809 (2012). https://doi.org/10.1016/j.dam.2010.05.005

30. Jensen, T.R., Toft, B.: Graph Coloring Problems. Wiley, New York (1995). Wiley-Interscience Series in Discrete Mathematics and Optimization

31. Kamiński, M., Raymond, J.F., Trunck, T.: Well-quasi-ordering H-contraction-free graphs. Discret. Appl. Math. **248**, 18–27 (2018). https://doi.org/10.1016/j.dam.2017.02.018

32. Karthick, T., Maffray, F.: Vizing bound for the chromatic number on some graph classes. Graphs Comb. **32**(4), 1447–1460 (2015). https://doi.org/10.1007/s00373-015-1651-1

33. Kostochka, A.V.: Lower bound of the Hadwiger number of graphs by their average degree. Combinatorica **4**(4), 307–316 (1984). https://doi.org/10.1007/BF02579141

34. Lovász, L.: Graph minor theory. Bull. Am. Math. Soc. (N.S.) **43**(1), 75–86 (2006). https://doi.org/10.1090/S0273-0979-05-01088-8

35. Malyshev, D.S.: A complexity dichotomy and a new boundary class for the dominating set problem. J. Comb. Optim. **32**(1), 226–243 (2016). https://doi.org/10.1007/s10878-015-9872-z

36. Markossian, S.E., Gasparian, G.S., Reed, B.A.: β-perfect graphs. J. Comb. Theor. Ser. B **67**(1), 1–11 (1996). https://doi.org/10.1006/jctb.1996.0030

37. Raghavan, V., Spinrad, J.: Robust algorithms for restricted domains. J. Algorithms **48**(1), 160–172 (2003). https://doi.org/10.1016/S0196-6774(03)00048-8

38. Ramsey, F.P.: On a problem of formal logic. Proc. London Math. Soc. (2) **30**(4), 264–286 (1929). https://doi.org/10.1112/plms/s2-30.1.264

39. Robertson, N., Seymour, P.D.: Graph minors. V. Excluding a planar graph. J. Comb. Theor. Ser. B **41**(1), 92–114 (1986). https://doi.org/10.1016/0095-8956(86)90030-4

40. Scheffler, P.: What graphs have bounded tree-width? In: Proceedings of the 7th Fischland Colloquium, III (Wustrow, 1988), pp. 31–38, no. 41 (1990)

41. Schweitzer, P.: Towards an isomorphism dichotomy for hereditary graph classes. Theo. Comput. Syst. **61**(4), 1084–1127 (2017). https://doi.org/10.1007/s00224-017-9775-8

42. Scott, A., Seymour, P.: A survey of χ-boundedness. arXiv:1812.07500 [math.CO] (2018)

43. Silva, A., da Silva, A.A., Sales, C.L.: A bound on the treewidth of planar even-hole-free graphs. Discret. Appl. Math. **158**(12), 1229–1239 (2010). https://doi.org/10.1016/j.dam.2009.07.010

44. Sintiari, N.L.D., Trotignon, N.: (Theta, triangle)-free and (even hole, K_4)-free graphs. Part 1 : Layered wheels. arXiv:1906.10998 [cs.DM] (2019)

45. Skodinis, K.: Efficient analysis of graphs with small minimal separators. In: Widmayer, P., Neyer, G., Eidenbenz, S. (eds.) WG 1999. LNCS, vol. 1665, pp. 155–166. Springer, Heidelberg (1999). https://doi.org/10.1007/3-540-46784-X_16

46. Weißauer, D.: In absence of long chordless cycles, large tree-width becomes a local phenomenon. J. Comb. Theor. Ser. B **139**, 342–352 (2019). https://doi.org/10.1016/j.jctb.2019.04.004

47. Wu, Y., Austrin, P., Pitassi, T., Liu, D.: Inapproximability of treewidth and related problems. J. Artif. Intell. Res. **49**, 569–600 (2014). https://doi.org/10.1613/jair.4030

48. Zaker, M.: On lower bounds for the chromatic number in terms of vertex degree. Discret. Math. **311**(14), 1365–1370 (2011). https://doi.org/10.1016/j.disc.2011.03.025

49. Zuckerman, D.: Linear degree extractors and the inapproximability of max clique and chromatic number. Theor. Comput. **3**, 103–128 (2007). https://doi.org/10.4086/toc.2007.v003a006

Graph Isomorphism Restricted by Lists

Pavel Klavík[1,2], Dušan Knop[3], and Peter Zeman[4(✉)]

[1] Department of Mathematics, Faculty of Applied Sciences,
University of West Bohemia, Pilsen, Czech Republic
[2] OrgPad, Prague, Czech Republic
klavik@orgpad.com
[3] Department of Theoretical Computer Science, Faculty of Information Technology,
Czech Technical University in Prague, Prague, Czech Republic
knopdusa@fit.cvut.cz
[4] Department of Applied Mathematics, Faculty of Mathematics and Physics,
Charles University, Prague, Czech Republic
zeman@kam.mff.cuni.cz
https://www.orgpad.com

Abstract. The complexity of *graph isomorphism* (GRAPHISO) is a famous problem in computer science. For graphs G and H, it asks whether they are the same up to a relabeling of vertices. In 1981, Lubiw proved that *list restricted graph isomorphism* (LISTISO) is NP-complete: for each $u \in V(G)$, we are given a list $\mathfrak{L}(u) \subseteq V(H)$ of possible images of u. After 35 years, we revive the study of this problem and consider which results for GRAPHISO can be modified to solve LISTISO.

We prove: 1) Under certain conditions, GI-completeness of a class of graphs implies NP-completeness of LISTISO. 2) Several combinatorial algorithms for GRAPHISO can be modified to solve LISTISO: for trees, planar graphs, interval graphs, circle graphs, permutation graphs, and bounded treewidth graphs. 3) LISTISO is NP-complete for cubic colored graphs with sizes of color classes bounded by 8.

Keywords: Graph isomorphism · Restricted computational problem · NP-completeness

1 Introduction

For graphs G and H, a bijection $\pi : G \to H$ is called an *isomorphism* if $uv \in E(G) \iff \pi(u)\pi(v) \in E(H)$. The *graph isomorphism problem* (GRAPHISO) asks whether there exists an isomorphism from G to H. It obviously belongs to NP, and no polynomial-time algorithm is known, and it is unlikely NP-complete. It is a prime candidate for an intermediate problem with complexity between

Pavel Klavík and Peter Zeman were supported by GAČR 20-15576S. Peter Zeman was further supported by GAUK 1224120. An OrgPage summarizing the results of this paper is available at https://orgpad.com/o/ef3f3616-0661-49ac-871a-cb458078f083?presentation=c12c1eea-f6e8-4d11-a87c-016d8b6f4ced.

© Springer Nature Switzerland AG 2020
I. Adler and H. Müller (Eds.): WG 2020, LNCS 12301, pp. 106–118, 2020.
https://doi.org/10.1007/978-3-030-60440-0_9

P and NP-complete. The graph isomorphism problem is solved efficiently for various restricted graph classes and parameters; see Fig. 1.

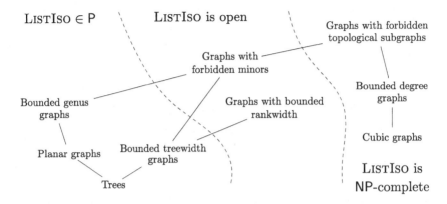

Fig. 1. Important graph classes for which the graph isomorphism problem is in P. Our complexity results for the list restricted graph isomorphism problem are depicted.

Combinatorial Algorithms. A prime example is the linear-time algorithm for testing graph isomorphism of trees. It is a bottom-up procedure comparing subtrees. For other graph classes, graph isomorphism reduces to graph isomorphism of labeled trees: for planar graphs [17], interval graphs [25], circle graphs [20], and permutation graphs [5]. Involved combinatorial arguments are used to solve graph isomorphism for bounded genus graphs [21] and bounded treewidth graphs [23].

Algorithms Based on Group Theory. The graph isomorphism problem is closely related to group theory, in particular to computing generators of automorphism groups of graphs. Assuming that G and H are connected, we can test $G \cong H$ by computing generators of $\text{Aut}(G \,\dot\cup\, H)$ and checking whether there exists a generator which swaps G and H. For the converse relation, Mathon [28] proved that generators of the automorphism group can be computed using $\mathcal{O}(n^3)$ instances of graph isomorphism.

Therefore, GRAPHISO can be attacked by techniques of group theory. The seminal result of Luks [26] uses group theory to solve GRAPHISO for graphs of bounded degree in polynomial time. If G has bounded degree, its automorphism group $\text{Aut}(G)$ may be arbitrary, but the stabilizer $\text{Aut}_e(G)$ of an edge e is restricted. Luks' algorithm tests GRAPHISO by an iterative process which determines $\text{Aut}_e(G)$ in steps, by adding layers around e.

Group theory can be used to solve GRAPHISO of colored graphs with bounded sizes of color classes [12] and of graphs with bounded eigenvalue multiplicity [2]. Miller [29] solved GRAPHISO of k-contractible graphs (which generalize both bounded degree and bounded genus graphs), and his results are used by Ponomarenko [30] to show that GRAPHISO can be decided in polynomial time for

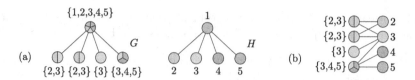

Fig. 2. (a) Two isomorphic graphs G and H with no list-compatible isomorphism. (b) It does not exist because there is no perfect matching between the lists of the leaves of G and of H.

graphs with excluded minors. Luks' algorithm [26] for bounded degree graphs is also used by Grohe and Marx [14] as a subroutine to solve GRAPHISO on graphs with excluded topological subgraphs. The recent breakthrough of Babai [1] heavily uses group theory to solve the graph isomorphism problem in quasipolynomial time.

Is Group Theory Needed? One of the fundamental problems for understanding the graph isomorphism problem is to understand in which cases group theory is really needed, and in which cases it can be avoided.[1] For instance, for which graph classes can GRAPHISO be decided by the classical combinatorial algorithm called k-dimensional Weisfieler-Leman refinement (k-WL)? (See the full version for details.)

Ponomarenko [30] used group theory to solve GRAPHISO in polynomial time on graphs with excluded minors. Robertson and Seymour [31] proved that a graph G with an excluded minor can be decomposed into pieces which are "almost embeddable" to a surface of genus g, where g depends on this minor. Recently, Grohe [13] generalized this to show that for G, there exists a treelike decomposition into almost embeddable pieces which is *automorphism-invariant* (every automorphism of G induces an automorphism of the treelike decomposition). Using this decomposition, it is possible to solve graph isomorphism in polynomial time and to avoid group theory techniques. In particular, k-WL can decide graph isomorphism on graphs with excluded minors where k depends on the minor.

It is a long-standing open problem whether the graph isomorphism problem for bounded degree graphs, and in particular for cubic graphs, can be solved in polynomial time without group theory. It is known that k-WL, for any fixed k, cannot decide graph isomorphism on cubic graphs [4]. Very recently, fixed parameter tractable algorithms for graphs of bounded treewidth [23] and for graphs of bounded genus [21] were constructed. On the other hand, the best known parameterized algorithms for graphs of bounded degree are XP algorithms [15,26], and it is a major open problem whether an FPT algorithm exists.

List Restricted Graph Isomorphism. In 1981, Lubiw [24] introduced the following computational problems. Let G and H be graphs, and the vertices of G be equipped with lists: each vertex $u \in V(G)$ has a list $\mathfrak{L}(u) \subseteq V(H)$. We say

---

[1] Ilya Ponomarenko in personal communication.

that an isomorphism $\pi : G \to H$ is *list-compatible* if, for all vertices $u \in V(G)$, we have $\pi(u) \in \mathfrak{L}(u)$; see Fig. 2a. A list-compatible isomorphism $\pi : G \to G$ is called a *list-compatible automorphism*.

Problem:	List restricted graph isomorphism – LISTISO
Input:	Graphs G and H, and the vertices of G are equipped by lists $\mathfrak{L}(u) \subseteq V(H)$.
Output:	Is there a list-compatible isomorphism $\pi : G \to H$?

Problem:	List restricted graph automorphism – LISTAUT
Input:	A graph G with vertices equipped with lists $\mathfrak{L}(u) \subseteq V(G)$.
Output:	Is there a list-compatible automorphism $\pi : G \to G$?

These two problems are polynomially equivalent (see Lemma 3). Lubiw [24] proved the following surprising result:

Theorem 1 (Lubiw [24]). *The problems* LISTISO *is* NP-*complete and the problem* LISTAUT *is* NP-*complete.*

Independently, LISTISO was rediscovered in [9,11]. Given two graphs G and H, we say that G *regularly covers* H if there exists a semiregular subgroup $\Gamma \leq \mathrm{Aut}(G)$ such that $G/\Gamma \cong H$. The list restricted graph isomorphism problem was used as a subroutine in [9,11] for 3-connected planar and projective graphs to test regular covering when G is a planar graph. The key idea is that a planar graph G can be reduced to a 3-connected planar graph G_r, for which $\mathrm{Aut}(G_r)$ is a spherical group. Therefore, we can compute all regular quotients G_r/Γ_r. Next, we reduce H towards G_r/Γ_r. The problem is that subgraphs of H may correspond to several different parts in G, so we compute lists of all possibilities. This leads to LISTISO of 3-connected planar and projective planar graphs.

Our Results. We revive the study of list restricted graph isomorphism. The goal is to determine which techniques for GRAPHISO can be modified to solve LISTISO. We believe that LISTISO is a very natural computational problem, as evidenced by its application in [9,11]. Moreover, the study of list-restricted problems is natural in computer science and discrete mathematics. These include for example the list homomorphism problem [16], and the list coloring problem [8]. We informally state the results proved in this paper; see Fig. 1 for an overview:

Result 1. GI-*completeness results for* GRAPHISO *with polynomial-time reductions using vertex-gadgets imply* NP-*completeness for* LISTISO.

For many classes $\mathcal{C}$ of graphs, it is known that GRAPHISO is equally hard for them as for general graphs, i.e., it is GI-complete. For instance, GRAPHISO is GI-complete for bipartite graphs, split and chordal graphs [25], chordal bipartite and strongly chordal graphs [36], trapezoid graphs [33], comparability graphs of dimension 4 [22], grid intersection graphs [35], line graphs [37], and self-complementary graphs [6].

The polynomial-time reductions are often done in a way that all graphs are encoded into $\mathcal{C}$, by replacing each vertex with a small vertex-gadget. We prove in Theorem 2 that such reductions can be modified for LISTISO: they imply NP-completeness of LISTISO for $\mathcal{C}$. For instance, LISTISO is NP-complete for all graph classes mentioned above (Corollary 2).

Result 2. *The problem* LISTISO *can be solved in polynomial-time for trees, planar graphs, interval graphs, circle graphs, permutation graphs, and bounded treewidth graphs.*

These algorithms are modifications of combinatorial techniques for GRAPHISO. As a by-product, our paper gives an overview of these techniques which can be modified to solve LISTISO in a straightforward way. Moreover, we can describe them more naturally with lists.

For example, the bottom-up linear-time algorithm for testing graph isomorphism of (rooted) trees can be modified to solve LISTISO in Theorem 4, since it captures all possible isomorphisms. The key difference is that the algorithm for LISTISO finds perfect matchings in bipartite graphs, in order to decide whether lists of several subtrees are simultaneously compatible; see Fig. 2b. We use the algorithm of Hopcroft and Karp [19], running in time $\mathcal{O}(\sqrt{n}m)$.

The algorithms for graph isomorphism of planar, interval, permutation and circle graphs based on tree decompositions and can be modified to solve LISTISO, as we show in Theorems 9, 5, 6 and 7. The algorithm for graph isomorphism of bounded treewidth graphs can be modified to solve LISTISO, as we show in Theorem 10. The complexity for graphs with bounded rankwidth and graphs with excluded minors remains open, see Conclusions for details.

Result 3. *The problem* LISTISO *is* NP-*complete for 3-regular colored graphs with all color classes of size at most 8 and with all lists of size at most 3.*

This result contrasts with two fundamental results using group theory techniques to solve graph isomorphism in polynomial time for graphs of bounded degrees [26] and bounded color classes [12]. In general, our impression is that group theory techniques probably cannot be modified in a straightforward way to solve LISTISO since list-compatible automorphisms of a graph G do not form a subgroup of $\mathrm{Aut}(G)$. In Theorem 3, we prove Result 3 by describing a non-trivial modification of the original NP-hardness reduction of Lubiw [24].

2   Preliminaries

Let G be an input graph of LISTISO or LISTAUT. We denote the set of its vertices by $V(G)$ and the set of its edges by $E(G)$. Let $n = |V(G)|$, $m = |E(G)|$ and ℓ be the total size of all lists. To make the problem non-trivial, we can assume that $\ell \geq n$.

Bipartite Perfect Matchings. We frequently use *bipartite perfect matching*:

Lemma 1 (Hopcroft and Karp [19]). *The bipartite perfect matching problem can be solved in time $\mathcal{O}(\sqrt{n}m)$, where n is the number of vertices and m is the number of edges.*

We repeatedly use this subroutine to solve LISTISO for many graph classes. Therefore, the running time of many of our algorithms $\mathcal{O}(\sqrt{n}\ell)$ while the input size is $\Omega(n + \ell)$. This cannot be avoided for the following reason (see the full version for a proof):

Lemma 2. *There exists a linear-time reduction from the bipartite perfect matching problem for n vertices and m edges to LISTISO of two independent sets with n vertices and $\ell = m$.*

Similar reductions work for trees, etc. Therefore, finding bipartite perfect matchings is the bottleneck in many of our algorithms and cannot be avoided.

Basic Results. We state some basic results concerning the complexity of LISTISO and LISTAUT, all proofs are in the full version.

Lemma 3. *Both problems LISTAUT and LISTISO are polynomially equivalent.*

Lemma 4. *The problem LISTISO can be solved in $\mathcal{O}(n + m)$ if $\ell \le 2n$.*

Lemma 5. *Let $G_1, \ldots, G_k$ be the components of G and $H_1, \ldots, H_k$ be the components of H. If we can decide LISTISO in polynomial time for all pairs G_i and H_j, then we can solve LISTISO for G and H in polynomial time.*

Lemma 6. *The problem LISTISO can be solved for cycles in time $\mathcal{O}(\ell)$.*

Lemma 7. *The problem LISTISO can be solved for graphs of maximum degree 2 in $\mathcal{O}(\sqrt{n}\ell)$.*

3   GI-Completeness Implies NP-Completeness

We want to show that in most cases, the GI-completeness reductions can be modified to show NP-completeness of LISTISO for $\mathcal{C}'$.

Vertex-Gadget Reductions. Suppose that GRAPHISO is GI-complete for a class $\mathcal{C}$. To show that GRAPHISO is GI-complete for another class $\mathcal{C}'$, one builds a polynomial-time reduction ψ from GRAPHISO of $\mathcal{C}$: given graphs $G, H \in \mathcal{C}$, we construct graphs $G', H' \in \mathcal{C}'$ in polynomial time such that $G \cong H$ if and only if $G' \cong H'$. Such reductions were described for certain graph classes (e.g., chordal graphs [25]) and they were systematically studied in [7].

We say that ψ uses *vertex-gadgets*, if to every vertex $u \in V(G)$ (resp. $u \in V(H)$), it assigns a *vertex-gadget* $\mathcal{V}_u$, and these gadgets are subgraphs of G' (resp. of H'), and satisfy the following two conditions:

1. Every isomorphism $\pi : G \to H$ induces an isomorphism $\pi' : G' \to H'$ such that $\pi(u) = v$ implies $\pi'(\mathcal{V}_u) = \mathcal{V}_v$.

2. Every isomorphism $\pi' : G' \to H'$ maps vertex-gadgets to vertex-gadgets and induces an isomorphism $\pi : G \to H$ such that $\pi'(\mathcal{V}_u) = \mathcal{V}_v$ implies $\pi(u) = v$.

Theorem 2. *Let $\mathcal{C}$ and $\mathcal{C}'$ be graph classes. Suppose that there exists a polynomial-time reduction ψ using vertex-gadgets from* GRAPHISO *of $\mathcal{C}$ to* GRAPHISO *of $\mathcal{C}'$. Then there exists a polynomial-time reduction from* LISTISO *of $\mathcal{C}$ to* LISTISO *of $\mathcal{C}'$.*

Corollary 1. *Let $\mathcal{C}$ be a class of graphs with* NP-*complete* LISTISO. *Suppose that there exists a reduction ψ using vertex-gadgets from* GRAPHISO *of $\mathcal{C}$ to* GRAPHISO *of $\mathcal{C}'$. Then* LISTISO *is* NP-*complete for $\mathcal{C}'$.*

Corollary 2. *The problem* LISTISO *is* NP-*complete for bipartite graphs, split and chordal graphs, chordal bipartite graphs, strongly chordal graphs, trapezoid graphs, comparability graphs of dimension 4, grid intersection graphs, line graphs, and self-complementary graphs.*

We are not aware of any polynomial-time reduction for graph isomorphism used in the literature which cannot be easily modified to use vertex-gadgets. The reason is that most of the reductions use the following operations: (1) taking the complement of the graph, (2) replacing all vertices by small disjoint isomorphic gadgets, (3) replacing all edges by small disjoint isomorphic gadgets, (4) taking disjoint copies of the graph or its complement, (5) adding a universal vertex, (6) adding a complete subgraph on some vertices or a complete bipartite graph between two sets of vertices. For instance, all reductions described in [7] can be easily modified to use vertex-gadgets.

4   NP-Completeness for 3-Regular Colored Graphs

Using group theory techniques, graph isomorphism can be solved in polynomial time for graphs of bounded degree [26] and for colored graphs with color classes of bounded sizes [12]. We modify the reduction of Lubiw [24] to show that LISTISO remains NP-complete even for 3-regular colored graphs with color classes of size at most 8 and each list of size at most 3. We use LISTAUT to solve 1-IN-3 SAT [32]: all literals are positive, each clause is of size 3 and a satisfying assignment has exactly one true literal in each clause.

Variable Gadget. For each variable u_i, we construct the *variable gadget* H_i which consists of two isolated vertices $u_i(0)$ and $u_i(1)$. We assign $\mathcal{L}(u_i(0)) = \mathcal{L}(u_i(1)) = \{u_i(0), u_i(1)\}$. There exist two list-compatible automorphisms of H_i: the transposition α_i swapping $u_i(0)$ and $u_i(1)$ and the identity β_i fixing both $u_i(0)$ and $u_i(1)$.

Clause Gadget. Let c_j be a clause with the literals q_j, r_j, and s_j. For every such clause c_j, the *clause gadget* G_j consists of the isolated vertices $c_j(0), \ldots, c_j(7)$. For every $k = 0, \ldots, 7$, we consider its binary representation $k = abc_2$, for

$a, b, c \in \{0, 1\}$. The vertex $c_j(k)$ has three neighbors $q_j(a)$, $r_j(b)$, and $s_j(c)$ belonging to the variable gadgets of its literals. We assign the list

$$\mathfrak{L}(c_j(k)) = \{c_j(k \oplus 100_2), c_j(k \oplus 010_2), c_j(k \oplus 001_2)\},$$

where $\oplus$ denotes the bitwise XOR; i.e., $\mathfrak{L}(c_j(k))$ contains all $c_j(k')$ in which k' differs from k in exactly one bit. Let G be the resulting graph consisting of all variable and clause gadgets.

Lemma 8. *Let π' be a partial automorphism of G obtained by choosing α_i or β_i on each H_i. There exists a unique automorphism π extending π' such that $\pi(G_j) = G_j$.*

Proof. Let c_j be a clause with the literals q_j, r_j, and s_j. We claim that $\pi(c_j(k))$ is determined by the images of its neighbors. Recall that β_i preserves the vertices of H_i, but α_i swaps them. Therefore, one neighbor of $\pi(c_j(k))$ is different from the corresponding neighbor of $c_j(k)$ for every application of α_i on q_j, r_j and s_j. Let $p = abc_2$ such that $a = 1$, $b = 1$ and $c = 1$ if and only if α_i is applied on the variable gadget of q_j, r_j, and s_j, respectively. Then $\pi(c_j(k)) = c_j(k \oplus p)$; otherwise π would not be an automorphism.

Lemma 9. *The 1-IN-3 SAT formula is satisfiable if and only if there exists a list-compatible automorphism of G.*

Proof. Let T be a truth value assignment satisfying the input formula. We construct a list-compatible automorphism π of G. If $T(u_i) = 1$, we put $\pi|_{H_i} = \alpha_i$, and if $T(u_i) = 0$, we put $\pi|_{H_i} = \beta_i$. By Lemma 8, this partial isomorphism has a unique extension to an automorphism π of G. It is list-compatible since T satisfies the 1-in-3 condition, so $\pi(c_j(k)) = c_j(k \oplus p)$, for $p \in \{100_2, 010_2, 001_2\}$.

For the other implication, let π be a list-compatible automorphism. Then $\pi|_{H_i}$ is either equal α_i, or β_i, which gives the values $T(u_i)$. By Lemma 8, $\pi(c_j(k)) = c_j(k \oplus p)$ and since π is a list-compatible isomorphism, we have $p \in \{100_2, 010_2, 001_2\}$. Therefore, exactly one literal in each clause is true, so all clauses are satisfied in T.

The described reduction clearly runs in polynomial-time, so we have established a proof of Theorem 1. For colored graphs, we require that automorphisms preserve colors. By altering the above reduction, we get the following:

Theorem 3. *The problem* LISTISO *is* NP*-complete for 3-regular colored graphs for which each color class is of size at most 8 and each list is of size at most 3.*

Proof. We modify the graph G to a 3-regular graph. For a clause gadget G_j representing c_j, every vertex $c_j(k)$ already has degree 3. Suppose that a variable u_i has o literals in the formula. Then both vertices of H_i have degrees $4o$, so we modify the variable gadgets.

We replace H_i by two cycles of length o, with the vertices $u_{i,1}(0), \ldots, u_{i,o}(0)$ and $u_{i,1}(1), \ldots, u_{i,o}(1)$, respectively. To each of these vertices, we attach a small gadget depicted in Fig. 3a. We have $\mathcal{L}(u_{i,t}(0)) = \mathcal{L}(u_{i,t}(1)) = \{u_{i,t}(0), u_{i,t}(1)\}$. There are two list-compatible automorphisms: α_i exchanging these two cycles by swapping $u_{i,t}(0)$ with $u_{i,t}(1)$, and β_i which is the identity fixing all $2o$ vertices. We note that when $o \le 2$, we get parallel edges or loops; to avoid this, we replace edges of two cycle by some 3-regular subgraphs.

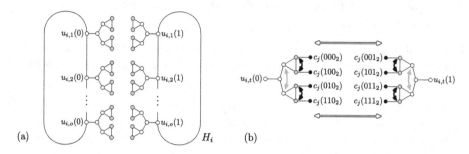

Fig. 3. (a) The variable gadget H_i. (b) The connection between H_i and G_j. Suppose that the variable u_i has a literal in the clause c_j, so $k = yzx_2$. We connect H_i with G_j as depicted. Suppose that an automorphism π maps $c_j(k)$ to $c_j(k \oplus p)$. We show the action of π on the vertices of H_i when $p = 001$ (in white), $p = 010$ (in gray), and $p = 100$ (in black).

Consider the attached gadgets to the vertices $u_{i,t}(0)$ and $u_{i,t}(1)$ corresponding to one literal of a clause c_j. Each vertex depicted in gray is adjacent to exactly one $c_j(k)$ of G_j, as depicted in Fig. 3b. Each k consists of three bits, denoted x, y and z (in some order). The bit x corresponds to this literal of u_i (i.e, x is the first bit for q_j being a literal of u_i, and so on). The gray vertices of gadgets attached to $u_{i,t}(j)$ are adjacent to $c_j(k)$ with $x = j$. Adjacent pairs of gray vertices are connected to $c_j(k)$ where k differs in the bit y. Non-adjacent pairs of gray vertices in one gadget are connected to $c_j(k)$ where k differs in the bit z.

In Fig. 3b, the action of $\mathbb{Z}_2^3$ is depicted. Lemma 8 translated to the modified definitions of variable gadgets which implies correctness of the reduction. The lists for the vertices of the attached gadgets are created as images of three depicted automorphisms; they clearly are of size at most 3. The constructed graph G is 3-regular and all lists of G are of size at most 3. We color the vertices by the orbits of all list-compatible automorphisms and their compositions. Notice that each color class is of size at most 8.

With Lemma 7, we get a dichotomy for the maximum degree: LISTISO can be solved in time $\mathcal{O}(\sqrt{n}\ell)$ for the maximum degree 2, and it is NP-complete for the maximum degree 3. Similarly, Lemma 4 implies a dichotomy for the list sizes: LISTISO can be solved in time $\mathcal{O}(n + m)$ where all lists are of size 2, and it is NP-complete for lists of size at most 3. For the last parameter, the maximum size of color classes, there is a gap. Lemma 4 implies that LISTISO can be solved in time $\mathcal{O}(n + m)$ when all color classes are of size 2 while it is NP-complete for size at most 8.

5   Polynomial-Time Algorithms

We start by modifying the standard algorithm for tree isomorphism to solve list restricted isomorphism of trees. We may assume that both trees G and H are rooted, otherwise we root them by their centers (and possibly subdivide the central edges). The algorithm for GRAPHISO processes both trees from bottom to the top. Using dynamic programming, it computes for every vertex possible images using possible images of its children. This algorithm can be modified to LISTISO.

Theorem 4. *The problem* LISTISO *can be solved for trees in time* $\mathcal{O}(\sqrt{n}\ell)$.

Proof. We apply the same dynamic algorithm with lists and update these lists as we go from bottom to the top. After processing a vertex u, we compute an updated list $\mathcal{L}'(u)$ which contains all elements of $\mathcal{L}(u)$ to which u can be mapped compatibly with its descendants. To initiate, each leaf u of G has $\mathcal{L}'(u) = \{w : w$ is a leaf and $w \in \mathcal{L}(u)\}$.

Next, we want to compute $\mathcal{L}'(u)$ and we know $\mathcal{L}'(u_i)$ of all children $U = \{u_1, \ldots, u_k\}$ of u. For each $w \in \mathcal{L}(u)$ with k children $w_1, \ldots, w_k$, we want to decide whether to put $w \in \mathcal{L}'(u)$. Let $W = \{w_1, \ldots, w_k\}$. Each u_i can be mapped to all vertices in $\mathcal{L}'(u_i) \cap W$. We need to decide whether all u_i's can be mapped simultaneously. Therefore, we form a bipartite graph $B(U, W)$ between U and W: we put an edge $u_i w_j$ if and only if $w_j \in \mathcal{L}'(u_i)$. Simultaneous mapping is possible if and only if there exists a perfect matching in this bipartite graph.

Let r be the root of G and r' be the root of H. We claim that there is a list-compatible isomorphism $\pi : G \to H$, if and only if $\mathcal{L}'(r) = \{r'\}$. Suppose that π exists. When $\pi(u) = w$, its children U are mapped to W. Since this mapping is compatible with the lists, $w \in \mathcal{L}(u)$, and the mapping of $u_1, \ldots, u_k$ gives a perfect matching in $B(U, W)$. Therefore, $w \in \mathcal{L}'(u)$, and by induction $r' \in \mathcal{L}'(r)$. On the other hand, we can construct π from the top to the bottom. We start by putting $\pi(r) = r'$. When $\pi(u) = w$, we map its children U to W according to some perfect matching in $B(U, W)$ which exists from the fact that $w \in \mathcal{L}'(u)$.

It remains to argue details of the complexity. We process the tree which takes time $\mathcal{O}(\ell)$ (assuming $n \leq \ell$) and we process each list constantly many times which takes $\mathcal{O}(\ell)$. To compute $\mathcal{L}'(u)$, we consider all vertices $w^1, \ldots, w^p \in \mathcal{L}(u)$, and let W^j be the children of w^j. We go through all lists of $\mathcal{L}'(u_1), \ldots, \mathcal{L}'(u_k)$ in linear time, and split them into sublists $\mathcal{L}'(u_i^j)$ of vertices whose parent is w^j. Only these sublists are used in the construction of the bipartite graph $B(U, W^j)$. Using Lemma 1, we decide existence of a perfect matching in time $\mathcal{O}(\sqrt{k}\ell_j)$ which is at most $\mathcal{O}(\sqrt{n}\ell_j)$, where ℓ_j is the total size of all sublists $\mathcal{L}'(u_i^j)$. When we sum this complexity for all vertices u, we get the total running time $\mathcal{O}(\sqrt{n}\ell)$. $\square$

Interval, Permutation and Circle Graphs. Standard algorithms solving GRAPHISO on interval, circle and permutation graphs can be modified to solve LISTISO. The structure of these graph classes can be captured by graph-labeled trees which are unique up to isomorphism and which capture the structure of all automorphisms; see [22] and the references therein.

For interval graphs, circle graphs, and permutation graphs we use MPQ-trees, split trees, and modular trees, respectively. We apply a bottom-up procedure similarly as in the proof of Theorem 4. The key difference is that nodes correspond to either prime, or degenerate graphs. Degenerate graphs are simple and lead to perfect matchings in bipartite graphs. Prime graphs have a small number of automorphisms [22], so all of them can be tested.

Theorem 5. *The problem* LISTISO *can be solved for interval graphs in* $\mathcal{O}(\sqrt{n}\ell + m)$.

Theorem 6. *The problem* LISTISO *can be solved for permutation graphs in* $\mathcal{O}(\sqrt{n}\ell + m)$.

Theorem 7. *The problem* LISTISO *can be solved for circle graphs in* $\mathcal{O}(\sqrt{n}\ell + m \cdot \alpha(m))$, *where* α *is the inverse Ackermann function.*

Planar Graphs. We sketch how to solve LISTISO on planar graphs. It is done in two steps. When G is a 3-connected planar graph, it has a unique embedding into the sphere (up to the reflection). It implies that $\text{Aut}(G)$ is a spherical group, so it is small and we can test all automorphisms whether they are list-compatible.

Lemma 10. *The problem* LISTISO *(with lists on both vertices and darts) can be solved for 3-connected planar graphs in time* $\mathcal{O}(\ell)$.

Seminal papers by Mac Lane [27] and Trakhtenbrot [34] introduced reduction which decomposes a graph into its *3-connected components*. We use an augmentation described in [9–11] (and the references therein) which behaves well with respect to automorphism groups. We get a rooted reduction tree whose nodes are either (essentially) 3-connected graphs or simple graphs (paths, cycles, dipoles). This tree can be computed in linear time [18].

Theorem 8. *Let* $\mathcal{C}$ *be a class of connected graphs closed under contractions and taking connected subgraphs. Suppose that* LISTISO *with lists on both vertices and darts can be solved for 3-connected graphs in* $\mathcal{C}$ *in time* $\varphi(n, m, \ell)$. *We can solve* LISTISO *on* $\mathcal{C}$ *in time* $\mathcal{O}(\sqrt{m}\ell + m + \varphi(n, m, \ell))$.

Theorem 9. *The problem* LISTISO *can be solved for planar graphs in time* $\mathcal{O}(\sqrt{n}\ell)$.

Bounded Treewidth Graphs. We prove that LISTISO can be solved in FPT with respect to the parameter treewidth $\text{tw}(G)$. We follow the approach of Bodlaender [3] which describes the following XP algorithm for GRAPHISO of bounded treewidth graphs, running in time $n^{\mathcal{O}(\text{tw}(G))}$. It is a dynamic programming algorithm running over all potential bags of all tree decompositions, and it can be modified with lists. The recent FPT-time algorithm by Lokshtanov et al. [23] can also be modified to solve LISTISO.

Theorem 10. LISTISO *can be solved in* FPT-*time* $2^{\mathcal{O}(\text{tw}(G)^5 \log \text{tw}(G))} n^5$.

References

1. Babai, L.: Graph isomorphism in quasipolynomial time. In: STOC (2016)
2. Babai, L., Grigoryev, D., Mount, D.: Isomorphism of graphs with bounded eigen-value multiplicity. In: STOC 1982, pp. 310–324. ACM (1982)
3. Bodlaender, H.L.: Polynomial algorithms for graph isomorphism and chromatic index on partial k-trees. J. Algorithms 11(4), 631–643 (1990)
4. Cai, J., Fürer, M., Immerman, N.: An optimal lower bound on the number of variables for graph identification. Combinatorica 12(4), 389–410 (1992). https://doi.org/10.1007/BF01305232
5. Colbourn, C.J.: On testing isomorphism of permutation graphs. Networks 11(1), 13–21 (1981)
6. Colbourn, C.J., Colbourn, M.J.: Isomorphism problems involving self-complementary graphs and tournaments. In: Proceedings of the Eighth Manitoba Conference on Numerical Mathematics and Computing, Congr. Numer. vol. 22, pp. 153–164 (1978)
7. Corneil, D.G., Kirkpatrick, D.G.: A theoretical analysis of various heuristics for the graph isomorphism problem. SIAM J. Comput. 9(2), 281–297 (1980)
8. Diestel, R.: Graph Theory. GTM, vol. 173, 3rd edn. Springer, Heidelberg (2017). https://doi.org/10.1007/978-3-662-53622-3
9. Fiala, J., Klavík, P., Kratochvíl, J., Nedela, R.: Algorithmic aspects of regular graph covers with applications to planar graphs. In: Esparza, J., Fraigniaud, P., Husfeldt, T., Koutsoupias, E. (eds.) ICALP 2014. LNCS, vol. 8572, pp. 489–501. Springer, Heidelberg (2014). https://doi.org/10.1007/978-3-662-43948-7_41
10. Fiala, J., Klavík, P., Kratochvíl, J., Nedela, R.: 3-connected reduction for regular graph covers. CoRR abs/1503.06556 (2016)
11. Fiala, J., Klavík, P., Kratochvíl, J., Nedela, R.: Algorithmic aspects of regular graph covers. CoRR abs/1609.03013 (2016)
12. Furst, M., Hopcroft, J., Luks, E.: Polynomial-time algorithms for permutation groups. In: FOCS, pp. 36–41. IEEE (1980)
13. Grohe, M.: Descriptive Complexity, Canonisation, and Definable Graph Structure Theory. Cambridge University Press, Cambridge (2012)
14. Grohe, M., Marx, D.: Structure theorem and isomorphism test for graphs with excluded topological subgraphs. In: Proceedings of the Forty-fourth Annual ACM Symposium on Theory of Computing, STOC 2012, pp. 173–192 (2012)
15. Grohe, M., Neuen, D., Schweitzer, P.: A faster isomorphism test for graphs of small degree. SIAM J. Comput. (2020)
16. Hell, P., Nesetril, J.: Graphs and Homomorphisms. Oxford University Press, Oxford (2004)
17. Hopcroft, J.E., Tarjan, R.E.: Isomorphism of planar graphs. In: Miller, R.E., Thatcher, J.W., Bohlinger, J.D. (eds.) Complexity of computer computations, pp. 131–152. Springer, Heidelberg (1972). https://doi.org/10.1007/978-1-4684-2001-2_13
18. Hopcroft, J.E., Tarjan, R.E.: Dividing a graph into triconnected components. SIAM J. Comput. 2(3), 135–158 (1973)
19. Hopcroft, J., Karp, R.: An $n^{5/2}$ algorithm for maximum matchings in bipartite graphs. SIAM J. Comput. 2(4), 225–231 (1973)
20. Hsu, W.L.: $\mathcal{O}(M \cdot N)$ algorithms for the recognition and isomorphism problems on circular-arc graphs. SIAM J. Comput. 24(3), 411–439 (1995)

21. Kawarabayashi, K.: Graph isomorphism for bounded genus graphs in linear time. CoRR abs/1511.02460 (2015)
22. Klavík, P., Zeman, P.: Automorphism groups of geometrically represented graphs. In: STACS 2015, LIPIcs, vol. 30, pp. 540–553 (2015)
23. Lokshtanov, D., Pilipczuk, M., Pilipczuk, M., Saurabh, S.: Fixed-parameter tractable canonization and isomorphism test for graphs of bounded treewidth. In: FOCS, pp. 186–195 (2014)
24. Lubiw, A.: Some NP-complete problems similar to graph isomorphism. SIAM J. Comput. **10**(1), 11–21 (1981)
25. Lueker, G.S., Booth, K.S.: A linear time algorithm for deciding interval graph isomorphism. J. ACM (JACM) **26**(2), 183–195 (1979)
26. Luks, E.M.: Isomorphism of graphs of bounded valence can be tested in polynomial time. J. Comput. Syst. Sci. **25**(1), 42–65 (1982)
27. Mac Lane, S.: A structural characterization of planar combinatorial graphs. Duke Math. J. **3**(3), 460–472 (1937)
28. Mathon, R.: A note on the graph isomorphism counting problem. Inf. Process. Lett. **8**(3), 131–136 (1979)
29. Miller, G.L.: Isomorphism testing and canonical forms for k-contractable graphs (A generalization of bounded valence and bounded genus). In: Karpinski, M. (ed.) FCT 1983. LNCS, vol. 158, pp. 310–327. Springer, Heidelberg (1983). https://doi.org/10.1007/3-540-12689-9_114
30. Ponomarenko, I.: The isomorphism problem for classes of graphs closed under contraction. J. Sov. Math. **55**(2), 1621–1643 (1991)
31. Robertson, N., Seymour, P.: Graph minors XVI. excluding a non-planar graph. J. Comb. Theor. Ser. B **77**, 1–27 (1999)
32. Schaefer, T.J.: The complexity of satisfiability problems. In: Proceedings of the Tenth Annual ACM Symposium on Theory of Computing, STOC 1978, pp. 216–226. ACM (1978)
33. Takaoka, A.: Graph isomorphism completeness for trapezoid graphs. IEICE Trans. Fundam. Electron. **98**(8), 1838–1840 (2015)
34. Trakhtenbrot, B.: Towards a theory of non-repeating contact schemes. Trudi Mat. Inst. Akad. Nauk SSSR **51**, 226–269 (1958)
35. Uehara, R.: Tractabilities and intractabilities on geometric intersection graphs. Algorithms **6**(1), 60–83 (2013)
36. Uehara, R., Toda, S., Nagoya, T.: Graph isomorphism completeness for chordal bipartite graphs and strongly chordal graphs. Discret. Appl. Math. **145**(3), 479–482 (2005)
37. Whitney, H.: Congruent graphs and the connectivity of graphs. Am. J. Math. **54**(1), 150–168 (1932)

Clique-Width: Harnessing the Power of Atoms

Konrad K. Dabrowski[1](✉)[iD], Tomáš Masařík[2,3,4][iD], Jana Novotná[2,3][iD], Daniël Paulusma[1][iD], and Paweł Rzążewski[3,5][iD]

[1] Department of Computer Science, Durham University, Durham, UK
{konrad.dabrowski,daniel.paulusma}@durham.ac.uk
[2] Faculty of Mathematics and Physics, Charles University, Prague, Czech Republic
{masarik,janca}@kam.mff.cuni.cz
[3] Institute of Informatics, University of Warsaw, Warsaw, Poland
[4] Department of Mathematics, Simon Fraser University, Burnaby, Canada
[5] Faculty of Mathematics and Information Science,
Warsaw University of Technology, Warsaw, Poland
p.rzazewski@mini.pw.edu.pl

Abstract. Many NP-complete graph problems are polynomial-time solvable on graph classes of bounded clique-width. Several of these problems are polynomial-time solvable on a hereditary graph class $\mathcal{G}$ if they are so on the atoms (graphs with no clique cut-set) of $\mathcal{G}$. Hence, we initiate a systematic study into boundedness of clique-width of atoms of hereditary graph classes. A graph G is *H-free* if H is not an induced subgraph of G, and it is (H_1, H_2)-*free* if it is both H_1-free and H_2-free. A class of H-free graphs has bounded clique-width if and only if its atoms have this property. This is no longer true for (H_1, H_2)-free graphs, as evidenced by one known example. We prove the existence of another such pair (H_1, H_2) and classify the boundedness of clique-width on (H_1, H_2)-free atoms for all but 18 cases.

1 Introduction

Many hard graph problems become tractable when restricting the input to some graph class. The two central questions are "for which graph classes does a graph problem become tractable" and "for which graph classes does it stay computationally hard?" Ideally, we wish to answer these questions for a large set of problems simultaneously instead of considering individual problems one by one.

Graph width parameters [26, 39, 41, 45, 54] make such results possible. A graph class has *bounded* width if there is a constant c such that the width of all its

---

The research in this paper received support from the Leverhulme Trust (RPG-2016-258). Masařík and Novotná were supported by Charles University student grants (SVV-2017-260452 and GAUK 1277018) and GAČR project (17-09142S). The last author was supported by Polish National Science Centre grant no. 2018/31/D/ST6/00062. A preprint of the full version of this paper is available from arXiv [29].

© Springer Nature Switzerland AG 2020
I. Adler and H. Müller (Eds.): WG 2020, LNCS 12301, pp. 119–133, 2020.
https://doi.org/10.1007/978-3-030-60440-0_10

members is at most c. There are several meta-theorems that provide sufficient conditions for a problem to be tractable on a graph class of bounded width.

Two popular width parameters are treewidth (tw) and clique-width (cw). For every graph G the inequality $cw(G) \leq 3 \cdot 2^{tw(G)-1}$ holds [19]. Hence, every problem that is polynomial-time solvable on graphs of bounded clique-width is also polynomial-time solvable on graphs of bounded treewidth. However, the converse statement does not hold: there exist graph problems, such as LIST COLOURING, which are polynomial-time solvable on graphs of bounded treewidth [44], but NP-complete on graphs of bounded clique-width [23]. Thus, the trade-off between treewidth and clique-width is that the former can be used to solve more problems, but the latter is *more powerful* in the sense that it can be used to solve problems for larger graph classes.

Courcelle [20] proved that every graph problem definable in MSO_2 is linear-time solvable on graphs of bounded treewidth. Courcelle, Makowsky and Rotics [22] showed that every graph problem definable in the more restricted logic MSO_1 is polynomial time solvable even for graphs of bounded clique-width (see [21] for details on MSO_1 and MSO_2). Since then, several clique-width meta-theorems for graph problems not definable in MSO_1 have been developed [32,36,46,51].

All of the above meta-theorems require a constant-width decomposition of the graph. We can compute such a decomposition in polynomial time for treewidth [4] and clique-width [50], but not for all parameters. For instance, unless $NP = ZPP$, this is not possible for mim-width [52], another well-known graph parameter, which is even more powerful than clique-width [54]. Hence, meta-theorems for mim-width [2,16] require an appropriate constant-width decomposition as part of the input (which may still be found in polynomial time for some graph classes).

Our Focus. In our paper we concentrate on *clique-width*[1] in an attempt to find *larger* graph classes for which certain NP-complete graph problems become tractable without the requirement of an appropriate decomposition as part of the input. The type of graph classes we consider all have the natural property that they are closed under vertex deletion. Such graph classes are said to be *hereditary* and there is a long-standing study on boundedness of clique-width for hereditary graph classes (see, for example, [3,6–8,10–13,24,25,27,28,30,31,39,45,48]).

Besides capturing many well-known classes, the framework of hereditary graph classes also enables us to perform a *systematic* study of a width parameter or graph problem. This is because every hereditary graph class $\mathcal{G}$ is readily seen to be uniquely characterized by a minimal (but not necessarily finite) set $\mathcal{F}_\mathcal{G}$ of forbidden induced subgraphs. If $|\mathcal{F}_\mathcal{G}| = 1$ or $|\mathcal{F}_\mathcal{G}| = 2$, then $\mathcal{G}$ is said to be *monogenic* or *bigenic*, respectively. Monogenic and bigenic graph classes already have a rich structure, and studying their properties has led to deep insights into the complexity of bounding graph parameters and solving graph problems; see e.g. [18,26,37,40] for extensive algorithmic and structural studies and surveys.

---

[1] See Sect. 2 for a definition of clique-width and other terminology used in Sect. 1.

It is well known (see e.g. [31]) that a monogenic class of graphs has bounded clique-width if and only if it is a subclass of the class $\mathcal{G}$ with $\mathcal{F}_\mathcal{G} = \{P_4\}$. The survey [26] gives a state-of-the-art theorem on the boundedness and unboundedness of clique-width of bigenic graph classes. Unlike treewidth, for which a complete dichotomy is known [5], and mim-width, for which there is an infinite number of open cases [15], this state-of-the-art theorem shows that there are still five open cases (up to an equivalence relation). From the same theorem we observe that many graph classes are of unbounded clique-width. However, if a graph class has unbounded clique-width, then this does not mean that a graph problem must be NP-hard on this class. For example, COLOURING is polynomial-time solvable on the (bigenic) class of (C_4, P_6)-free graphs [35], which contains the class of split graphs and thus has unbounded clique-width [48]. In this case it turns out that the *atoms* (graphs with no clique cut-set) in the class of (C_4, P_6)-free graphs *do* have bounded clique-width. This immediately gives us an algorithm for the whole class of (C_4, P_6)-free graphs due to Tarjan's decomposition theorem [53].

In fact, Tarjan's result holds not only for COLOURING, but also for many other graph problems. For instance, several other classical graph problems, such as MINIMUM FILL-IN, MAXIMUM CLIQUE, MAXIMUM WEIGHTED INDEPENDENT SET [53] (see [1] for the unweighted variant) and MAXIMUM INDUCED MATCHING [14] are polynomial-time solvable on a hereditary graph class $\mathcal{G}$ if and only if this is the case on the atoms of $\mathcal{G}$. Hence, we aim to investigate, in a systematic way, the following natural research question:

Which hereditary graph classes of unbounded *clique-width have the property that their atoms have* bounded *clique-width?*

Known Results. For monogenic graph classes, the restriction to atoms does not yield any algorithmic advantages, as shown by Gaspers et al. [35].

Theorem 1 ([35]). *Let H be a graph. The class of H-free atoms has bounded clique-width if and only if the class of H-free graphs has bounded clique-width (so, if and only if H is an induced subgraph of P_4).*

The result for (C_4, P_6)-free graphs [35] shows that the situation is different for bigenic classes. We are aware of two more hereditary graph classes $\mathcal{G}$ with this property, but in both cases $|\mathcal{F}_\mathcal{G}| > 2$. Split graphs, or equivalently, $(C_4, C_5, 2P_2)$-free graphs have unbounded clique-width [48], but split atoms are complete graphs and have clique-width at most 2. Cameron et al. [17] proved that (cap, C_4)-free odd-signable atoms have clique-width at most 48, whereas the class of all (cap, C_4)-free odd-signable graphs contains the class of split graphs and thus has unbounded clique-width. See [33,34] for algorithms for COLOURING on hereditary graph classes that rely on boundedness of clique-width of atoms of subclasses.

Our Results. Due to Theorem 1, and motivated by algorithmic applications, we focus on the atoms of bigenic graph classes. Recall that the class of (C_4, P_6)-free graphs has unbounded clique-width but its atoms have bounded clique-width [35]. This also holds, for instance, for its subclass of $(C_4, 2P_2)$-free graphs and thus for (C_4, P_5)-free graphs and $(C_4, P_2 + P_3)$-free graphs. We determine a new, incomparable case where we forbid $2P_2$ and $\overline{P_2 + P_3}$ (also known as the *paraglider* [43]); see Fig. 1 for illustrations of these forbidden induced subgraphs.

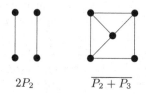

$2P_2$ $\overline{P_2 + P_3}$

Fig. 1. The two forbidden induced subgraphs from Theorem 2.

Theorem 2. *The class of $(2P_2, \overline{P_2 + P_3})$-free atoms has bounded clique-width (whereas the class of $(2P_2, \overline{P_2 + P_3})$-free graphs has unbounded clique-width).*

We sketch the proof of Theorem 2 in Sect. 3 after first giving an outline. Our approach shares some similarities with the approach Malyshev and Lobanova [49] used to show that (WEIGHTED) COLOURING is polynomial-time solvable on $(P_5, \overline{P_2 + P_3})$-free graphs. We explain the differences between both approaches and the new ingredients of our proof in detail in Sect. 3. Here, we only discuss a complication that makes proving boundedness of clique-width of atoms more difficult in general. Namely, when working with atoms, we need to be careful with performing complementation operations. In particular, a class of (H_1, H_2)-free graphs has bounded clique-width if only if the class of $(\overline{H_1}, \overline{H_2})$-free graphs has bounded clique-width. However, this equivalence relation holds for classes of (H_1, H_2)-free atoms. For example, (C_4, P_5)-free (and even (C_4, P_6)-free) atoms have bounded clique-width [35], but we prove that $(\overline{C_4}, \overline{P_5})$-free atoms have unbounded clique-width.

We also identify a number of new bigenic graph classes whose atoms already have unbounded clique-width. We prove this by modifying existing graph constructions for proving unbounded clique-width of the whole class (proofs omitted due to space restrictions). Combining these constructions with Theorem 2 and the state-of-art theorem on clique-width from [26] yields the following summary.

Theorem 3. *For graphs H_1 and H_2, let $\mathcal{G}$ be the class of (H_1, H_2)-free graphs.*

1. *The class of atoms in $\mathcal{G}$ has bounded clique-width if*
   (i) *H_1 or $H_2 \subseteq_i P_4$*
   (ii) *$H_1 = $ paw or K_s and $H_2 = P_1 + P_3$ or tP_1 for some $s, t \geq 1$*
   (iii) *$H_1 \subseteq_i$ paw and $H_2 \subseteq_i K_{1,3} + 3P_1$, $K_{1,3} + P_2$, $P_1 + P_2 + P_3$, $P_1 + P_5$, $P_1 + S_{1,1,2}$, $P_2 + P_4$, P_6, $S_{1,1,3}$ or $S_{1,2,2}$*
   (iv) *$H_1 \subseteq_i P_1 + P_3$ and $H_2 \subseteq_i \overline{K_{1,3} + 3P_1}$, $\overline{K_{1,3} + P_2}$, $\overline{P_1 + P_2 + P_3}$, $\overline{P_1 + P_5}$, $\overline{P_1 + S_{1,1,2}}$, $\overline{P_2 + P_4}$, $\overline{P_6}$, $\overline{S_{1,1,3}}$ or $\overline{S_{1,2,2}}$*
   (v) *$H_1 \subseteq_i$ diamond and $H_2 \subseteq_i P_1 + 2P_2$, $3P_1 + P_2$ or $P_2 + P_3$*

    (vi) $H_1 \subseteq_i 2P_1 + P_2$ and $H_2 \subseteq_i \overline{P_1 + 2P_2}$, $\overline{3P_1 + P_2}$ or $\overline{P_2 + P_3}$
    (vii) $H_1 \subseteq_i$ gem and $H_2 \subseteq_i P_1 + P_4$ or P_5
    (viii) $H_1 \subseteq_i P_1 + P_4$ and $H_2 \subseteq_i \overline{P_5}$
    (ix) $H_1 \subseteq_i K_3 + P_1$ and $H_2 \subseteq_i K_{1,3}$,
    (x) $H_1 \subseteq_i \overline{2P_1 + P_3}$ and $H_2 \subseteq_i 2P_1 + P_3$
    (xi) $H_1 \subseteq_i P_6$ and $H_2 \subseteq_i C_4$, or
    (xii) $H_1 \subseteq_i 2P_2$ and $H_2 \subseteq_i \overline{P_2 + P_3}$.

2. *The class of atoms in $\mathcal{G}$ has unbounded clique-width if*
    (i) $H_1 \notin \mathcal{S}$ and $H_2 \notin \mathcal{S}$
    (ii) $H_1 \notin \overline{\mathcal{S}}$ and $H_2 \notin \overline{\mathcal{S}}$
    (iii) $H_1 \supseteq_i K_3 + P_1$ and $H_2 \supseteq_i 4P_1$ or $2P_2$
    (iv) $H_1 \supseteq_i K_{1,3}$ and $H_2 \supseteq_i K_4$ or C_4
    (v) $H_1 \supseteq_i$ diamond and $H_2 \supseteq_i K_{1,3}, 5P_1, P_2 + P_4$ or $P_1 + P_6$
    (vi) $H_1 \supseteq_i 2P_1 + P_2$ and $H_2 \supseteq_i K_3 + P_1, K_5, \overline{P_2 + P_4}$ or $\overline{P_6}$
    (vii) $H_1 \supseteq_i K_3$ and $H_2 \supseteq_i 2P_1 + 2P_2, 2P_1 + P_4, 4P_1 + P_2, 3P_2$ or $2P_3$
    (viii) $H_1 \supseteq_i 3P_1$ and $H_2 \supseteq_i \overline{2P_1 + 2P_2}, \overline{2P_1 + P_4}, \overline{4P_1 + P_2}, \overline{3P_2}$ or $\overline{2P_3}$
    (ix) $H_1 \supseteq_i K_4$ and $H_2 \supseteq_i P_1 + P_4, 3P_1 + P_2$ or $2P_2$
    (x) $H_1 \supseteq_i 4P_1$ and $H_2 \supseteq_i$ gem, $\overline{3P_1 + P_2}$ or C_4
    (xi) $H_1 \supseteq_i$ gem, $\overline{P_1 + 2P_2}$ or $\overline{P_2 + P_3}$ and $H_2 \supseteq_i P_1 + 2P_2$ or P_6
    (xii) $H_1 \supseteq_i P_1 + P_4$ and $H_2 \supseteq_i \overline{P_1 + 2P_2}$, or
    (xiii) $H_1 \supseteq_i 2P_2$ and $H_2 \supseteq_i \overline{P_2 + P_4}, \overline{3P_2}$ or $\overline{P_5}$.

Due to Theorem 3, we are left with 18 open cases, listed in Sect. 4, where we discuss directions for future work.

2 Preliminaries

Let G be a graph. For a subset $S \subseteq V(G)$, the subgraph of G *induced by S* is the graph $G[S]$, which has vertex set S and edge set $\{uv \mid uv \in E(G), u, v \in S\}$. If $S = \{s_1, \ldots, s_r\}$, we may write $G[s_1, \ldots, s_r]$ instead of $G[\{s_1, \ldots, s_r\}]$. We write $F \subseteq_i G$ to denote that F is an induced subgraph of G. We say that G is *H-free* if G does not contain H as an induced subgraph, and that G is *$(H_1, \ldots, H_p)$-free* if it is H_i-free for all $i \in \{1, \ldots, p\}$. A *(connected) component* of G is a maximal connected subgraph of G. A clique $K \subseteq V(G)$ is a *clique cut-set* of G if $G \setminus K = G[V(G) \setminus K]$ is disconnected. A graph with no clique cut-sets is an *atom*; note that such graphs are connected. The *complement* $\overline{G}$ of G has vertex set $V(\overline{G}) = V(G)$ and edge set $E(\overline{G}) = \{uv \mid u, v \in V(G), u \neq v, uv \notin E(G)\}$. The *neighbourhood* of a vertex $u \in V(G)$ is the set $N(u) = \{v \in V(G) \mid uv \in E(G)\}$. Let X and Y be two disjoint vertex subsets of G. A vertex $x \in V(G) \setminus Y$ is *(anti-)complete* to Y if it is (non-)adjacent to every vertex in Y. Similarly, X is *complete* to Y if every vertex of X is complete to Y and *anti-complete* to Y if every vertex of X is anti-complete to Y.

    The graph $G_1 + G_2$ is the *disjoint union* of two vertex-disjoint graphs G_1 and G_2 and has vertex set $V(G_1) \cup V(G_2)$ and edge set $E(G_1) \cup E(G_2)$. The graph rG is the disjoint union of r copies of a graph G. The graphs C_t, K_t, and P_t denote the cycle, complete graph, and path on t vertices, respectively.

The paw is the graph $\overline{P_1 + P_3}$, the diamond is the graph $\overline{2P_1 + P_2}$, and the gem is the graph $\overline{P_1 + P_4}$. The *subdivided claw* $S_{h,i,j}$, for $1 \le h \le i \le j$ is the tree with one vertex x of degree 3 and exactly three leaves, which are of distance h, i and j from x, respectively. We let S denote the class of graphs every connected component of which is either a subdivided claw or a path on at least one vertex. Note that $S_{1,1,1} = K_{1,3}$.

The *clique-width* of a graph G, denoted by $\mathrm{cw}(G)$, is the minimum number of labels needed to construct G using the following four operations:

1. create a new graph consisting of a single vertex v with label i;
2. take the disjoint union of two labelled graphs G_1 and G_2;
3. add an edge between every vertex with label i and every vertex with label j $(i \ne j)$;
4. relabel every vertex with label i to have label j.

A class of graphs $\mathcal{G}$ has *bounded* clique-width if there is a constant c such that $\mathrm{cw}(G) \le c$ for every $G \in \mathcal{G}$; otherwise the clique-width of $\mathcal{G}$ is *unbounded*.

For an induced subgraph G' of a graph G, the *subgraph complementation* acting on G with respect to G' replaces every edge of G' by a non-edge, and vice versa. Hence, the resulting graph has vertex set $V(G)$ and edge set $(E(G) \setminus E(G')) \cup E(\overline{G'})$. For two disjoint vertex subsets S and T in G, the *bipartite complementation* acting on G with respect to S and T replaces every edge with one end-vertex in S and the other in T by a non-edge and vice versa.

For a constant $k \ge 0$ and a graph operation γ, a graph class $\mathcal{G}'$ is (k, γ)-*obtained* from a graph class $\mathcal{G}$ if (i) every graph in $\mathcal{G}'$ is obtained from a graph in $\mathcal{G}$ by performing γ at most k times, and (ii) for every $G \in \mathcal{G}$, there exists at least one graph in $\mathcal{G}'$ obtained from G by performing γ at most k times. Then γ *preserves* boundedness of clique-width if for every constant k and every graph class $\mathcal{G}$, every graph class $\mathcal{G}'$ that is (k, γ)-obtained from $\mathcal{G}$ has bounded clique-width if and only if $\mathcal{G}$ has bounded clique-width.

Fact 1. Vertex deletion preserves boundedness of clique-width [47].
Fact 2. Subgraph complementation preserves boundedness of clique-width [45].
Fact 3. Bipartite complementation preserves boundedness of clique-width [45].

A graph is *split* if its vertex set can be partitioned into a clique K and an independent set I. Note that if there is a vertex $v \in I$ with $N(v) \subsetneq K$, then $N(v)$ is a clique cut-set. Furthermore, if $|I| > 1$ then K is a clique cut-set. It follows that split atoms are complete graphs. Since complete graphs have clique-width at most 2, this means that split atoms have bounded clique-width.

3   The Proof of Theorem 2

Here, we prove Theorem 2, namely that the class of $(2P_2, \overline{P_2 + P_3})$-free atoms has bounded clique-width. Our approach is based on the following three claims:

(i) $(2P_2, \overline{P_2 + P_3})$-free atoms with an induced C_5 have bounded clique-width.
(ii) $(2P_2, \overline{P_2 + P_3})$-free atoms with an induced C_4 have bounded clique-width.
(iii) $(C_4, C_5, 2P_2, \overline{P_2 + P_3})$-free atoms have bounded clique-width.

We prove Claims (i) and (ii) in Lemmas 4 and 5, respectively, whereas Claim (iii) follows from the fact that $(C_4, C_5, 2P_2)$-free graphs are split graphs and so the atoms in this class are complete graphs, which therefore have clique-width at most 2. We partition the vertex set of an arbitrary $(2P_2, \overline{P_2 + P_3})$-free atom G into a number of different subsets with according to their neighbourhoods in an induced C_5 in Lemma 4 or an induced C_4 in Lemma 5. We then analyse the properties of these different subsets of $V(G)$ and how they are connected to each other, and use this knowledge to apply a number of appropriate vertex deletions, subgraph complementations and bipartite complementations. These operations will modify G into a graph G' that is a disjoint union of a number of smaller "easy" graphs known to have "small" clique-width. We then use Facts 1–3 to conclude that G also has small clique-width.

This approach works, as we will:

- apply the vertex deletions, subgraph complementations, and bipartite complementations only a constant number of times;
- not use the properties of being an atom or being $(2P_2, \overline{P_2 + P_3})$-free once we "leave the graph class" due to applying the above graph operations.

Our approach is similar to the approach used by Malyshev and Lobanova [49] for showing that COLOURING is polynomial-time solvable on the superclass of $(P_5, \overline{P_2 + P_3})$-free graphs. However, we note the following two differences:

1. Prime atoms restriction: OK for COLOURING, but not for clique-width. A set $X \subseteq V(G)$ is said to be a *module* if all vertices in X have the same set of neighbours in $V(G) \setminus X$. A module X in a graph G is *trivial* if it contains either all or at most one vertex of G. A graph G is *prime* if it has no non-trivial modules. To solve COLOURING in polynomial time on some hereditary graph class $\mathcal{G}$, one may restrict to prime atoms from $\mathcal{G}$ [42]. Malyshev and Lobanova proved that $(P_5, \overline{P_2 + P_3})$-free prime atoms with an induced C_5 are $3P_1$-free or have a bounded number of vertices. In both cases, COLOURING can be solved in polynomial time. We cannot make the pre-assumption that our atoms are prime. To see this, let G be a split graph. Add two new non-adjacent vertices to G and make them complete to the rest of $V(G)$. Let $\mathcal{G}$ be the (hereditary) graph class that consists of all these "enhanced" split graphs and their induced subgraphs. These enhanced split graphs are atoms, which have unbounded clique-width due to Fact 1 and the fact that split graphs have unbounded clique-width [48]. However, the prime atoms of $\mathcal{G}$ are the complete graphs, which have clique-width at most 2.

2. Perfect graphs restriction: OK for COLOURING, but not for clique-width. Malyshev and Lobanova observed that $(P_5, \overline{P_2 + P_3}, C_5)$-free graphs are perfect. Hence, COLOURING can be solved in polynomial time on such graphs [38]. However, being perfect does not imply boundedness of clique-width (for instance, split graphs are perfect graphs with unbounded clique-width).

We omit the proof of the next lemma.

Lemma 4. *The class of $(2P_2, \overline{P_2 + P_3})$-free atoms that contain an induced C_5 has bounded clique-width.*

Lemma 5. *The class of $(2P_2, \overline{P_2 + P_3})$-free atoms that contain an induced C_4 has bounded clique-width.*

Proof. Suppose G is a $(2P_2, \overline{P_2 + P_3})$-free atom containing an induced cycle C on four vertices, say $v_1, \ldots, v_4$ in that order. By Lemma 4, we may assume that G is C_5-free. For $S \subseteq \{1, \ldots, 4\}$, let V_S be the set of vertices $x \in V(G) \setminus V(C)$ such that $N(x) \cap V(C) = \{v_i \mid i \in S\}$.

To simplify notation, in the following claims, subscripts on vertices and vertex sets should be interpreted modulo 4 and whenever possible we will write V_i instead of $V_{\{i\}}$, write $V_{i,j}$ instead of $V_{\{i,j\}}$, and so on.

Claim 1. *For $i \in \{1, \ldots, 4\}$, $V_{i,i+1,i+2}$ is empty.*

Proof of Claim. Suppose, for contradiction, that $x \in V_{1,2,3}$. Then $G[v_1, v_3, v_2, v_4, x]$ is a $\overline{P_2 + P_3}$, a contradiction. The claim follows by symmetry.   ◇

Claim 2. *For $i \in \{1, \ldots, 4\}$, $V_\emptyset \cup V_i \cup V_{i+1} \cup V_{i,i+1}$ is an independent set.*

Proof of Claim. Suppose, for contradiction, that $x, y \in V_\emptyset \cup V_1 \cup V_2 \cup V_{1,2}$ are adjacent. Then $G[x, y, v_3, v_4]$ is a $2P_2$, a contradiction. The claim follows by symmetry.   ◇

Claim 3. *For $i \in \{1, \ldots, 4\}$, $V_{i,i+1} \cup V_{i,i+2}$ and $V_{i,i+1} \cup V_{i+1,i+3}$ are independent sets.*

Proof of Claim. Suppose, for contradiction, that $x, y \in V_{1,2} \cup V_{1,3}$ are adjacent. By Claim 2, x and y cannot both be in $V_{1,2}$, so assume without loss of generality that $x \in V_{1,3}$. Now $G[x, v_2, v_1, v_3, y]$ or $G[v_1, v_3, x, v_2, y]$ is a $\overline{P_2 + P_3}$ if $y \in V_{1,2}$ or $y \in V_{1,3}$, respectively, a contradiction. The claim follows by symmetry.   ◇

Claim 4. *$G[V_{1,2,3,4}]$ is $(P_1 + P_2)$-free and so it has bounded clique-width.*

Proof of Claim. Suppose, for contradiction, that $x, y, y' \in V_{1,2,3,4}$ induce a $P_1 + P_2$ in G. Then $G[v_1, v_3, y, x, y']$ is a $\overline{P_2 + P_3}$, a contradiction. Therefore $G[V_{1,2,3,4}]$ is $(P_1 + P_2)$-free and so P_4-free, so it has bounded clique-width by Theorem 1.   ◇

Claim 5. *For $i \in \{1, 2\}$, $V_{i,i+2}$ is complete to $V_{1,2,3,4}$.*

Proof of Claim. Suppose, for contradiction, that $x \in V_{1,3}$ is non-adjacent to $y \in V_{1,2,3,4}$. Then $G[v_1, v_3, v_2, x, y]$ is a $\overline{P_2 + P_3}$, a contradiction. The claim follows by symmetry.   ◇

Claim 6. *For $i \in \{1, 2, 3, 4\}$ either $V_{i-1} \cup V_{i-1,i}$ or $V_{i,i+1} \cup V_{i+1}$ is empty.*

Proof of Claim. Suppose, for contradiction, that $x \in V_1 \cup V_{1,2}$ and $y \in V_{2,3} \cup V_3$. Then $G[v_1, x, y, v_3, v_4]$ is a C_5 or $G[x, v_1, y, v_3]$ is a $2P_2$ if x is adjacent or non-adjacent to y, respectively, a contradiction. The claim follows by symmetry.   ◇

Claim 7. *If $x \in V_\emptyset$ then x has at least two neighbours in one of $V_{1,3}$ and $V_{2,4}$ and is anti-complete to the other. Furthermore, in this case x is complete to $V_{1,2,3,4}$.*

Proof of Claim. Suppose $x \in V_\emptyset$. Since G is not an atom, $N(x)$ cannot be a clique, and so must contain two non-adjacent vertices y, y'. By Claims 1 and 2, and the definition of $V_\emptyset$, it follows that $y, y' \in V_{1,3} \cup V_{2,4} \cup V_{1,2,3,4}$. If $y, y' \in V_{1,2,3,4}$, then $G[y, y', v_1, x, v_2]$ is a $\overline{P_2 + P_3}$, a contradiction. By Claim 5, $V_{1,2,3,4}$ is complete to $V_{1,3} \cup V_{2,4}$, so it follows that $y, y' \in V_{1,3} \cup V_{2,4}$. If $y \in V_{1,3}$ and $y' \in V_{2,4}$, then $G[v_1, v_2, y', x, y']$ is a C_5, a contradiction. It follows that $y, y' \in V_{1,3}$ or $y, y' \in V_{2,4}$.

Suppose $y, y' \in V_{1,3}$. If $z \in V_{2,4}$ is a neighbour of x, then z must be adjacent to y and y' (since, as shown above, x cannot have a pair of non-adjacent neighbours one of which is in $V_{1,3}$ and the other of which is in $V_{2,4}$), in which case $G[y, y', x, v_1, z]$ is a $\overline{P_2 + P_3}$, a contradiction. Therefore x cannot have a neighbour in $V_{2,4}$. If $z \in V_{1,2,3,4}$ is a non-neighbour of x, then z must be adjacent to y and y' by Claim 5, so $G[y, y', v_1, x, z]$ is a $\overline{P_2 + P_3}$, a contradiction. Therefore x is complete to $V_{1,2,3,4}$. The claim follows by symmetry.   ◇

Claim 8. *For $i \in \{1, 2\}$, $|V_{i,i+1} \cup V_{i+2,i+3}| \le 2$.*

Proof of Claim. Suppose, for contradiction, that $|V_{1,2} \cup V_{3,4}| \ge 3$. First note that if $x \in V_{1,2}$, $y \in V_{3,4}$ are non-adjacent, then $G[v_1, x, v_3, y]$ is a $2P_2$, a contradiction. Therefore $V_{1,2}$ is complete to $V_{3,4}$. By Claim 2, both $V_{1,2}$ and $V_{3,4}$ are independent sets. If $x \in V_{1,2}$ and $y, y' \in V_{3,4}$, then $G[y, y', v_3, x, v_4]$ is a $\overline{P_2 + P_3}$, a contradiction. By symmetry, we conclude that either $V_{1,2}$ or $V_{3,4}$ is empty. Suppose $V_{3,4}$ is empty, so $V_{1,2}$ contains at least three vertices and let $x \in V_{1,2}$ be such a vertex. Since G is an atom, $N(x)$ cannot be a clique, so x must have two neighbours y, y' that are non-adjacent. By Claims 1, 2, 3 and 6, and the definition of $V_{1,2}$, every neighbour of $x \in V_{1,2}$ lies in $\{v_1, v_2\} \cup V_{1,2,3,4}$. Since v_1 is complete to $\{v_2\} \cup V_{1,2,3,4}$ and v_2 is complete to $\{v_1\} \cup V_{1,2,3,4}$, it follows that $y, y' \in V_{1,2,3,4}$. Now $G[y, y', v_1, v_3, x]$ is a $\overline{P_2 + P_3}$, a contradiction. The claim follows by symmetry.   ◇

Claim 9. *For $i \in \{1, 2, 3, 4\}$, V_i is complete to $V_{1,2,3,4}$ and at most one vertex of $V_{i,i+2}$ has neighbours in V_i.*

Proof of Claim. Suppose $x \in V_1$. Since G is an atom, x must have two neighbours y, y' that are non-adjacent. By Claims 1, 2 and 6, and the definition of V_1, every neighbour of x lies in $\{v_1\} \cup V_{1,3} \cup V_{2,4} \cup V_{1,2,3,4}$. If $y, y' \in V_{1,3} \cup V_{1,2,3,4}$, then $G[y, y', v_1, v_3, x]$ is a $\overline{P_2 + P_3}$, a contradiction. The vertex v_1 is complete to $V_{1,3} \cup V_{1,2,3,4}$. Therefore without loss of generality, we may assume $y \in V_{2,4}$. Furthermore, note that $V_{1,3}$ is an independent set by Claim 3, so x has at most one neighbour in $V_{1,3}$. Since V_1 is an independent set by Claim 2, it follows that $G[V_1 \cup V_{1,3}]$ is a bipartite graph with parts V_1 and $V_{1,3}$. Since G is $2P_2$-free, it follows that no two vertices in V_1 can have different neighbours in $V_{1,3}$. Therefore at most one vertex of $V_{1,3}$ has a neighbour in V_1. Now if $z \in V_{1,2,3,4}$, then z is adjacent to y by Claim 5. If x is non-adjacent to z, then $G[v_1, y, v_2, x, z]$ is a $\overline{P_2 + P_3}$, a contradiction. We conclude that V_1 is complete to $V_{1,2,3,4}$. The claim follows by symmetry.   ◇

We now proceed as follows. By Claim 1, the set $V_{1,2,3} \cup V_{2,3,4} \cup V_{1,3,4} \cup V_{1,2,4}$ is empty. By Claims 6 and 8, there are at most two vertices in $V_{1,2} \cup V_{2,3} \cup V_{3,4} \cup V_{1,4}$, so after doing at most two vertex deletions, we may assume these sets are empty (note that the resulting graph may no longer be an atom). Applying four further vertex deletions, we can remove the cycle C from G. By Claim 6, we may assume without loss of generality that V_3 and V_4 are empty. The remaining vertices of G all lie in $V_\emptyset \cup V_1 \cup V_2 \cup V_{1,3} \cup V_{2,4} \cup V_{1,2,3,4}$ and by Fact 1, it suffices to show that this modified graph has bounded clique-width. By Claims 5, 7 and 9, $V_{1,2,3,4}$ is complete to $V_\emptyset \cup V_1 \cup V_2 \cup V_{1,3} \cup V_{2,4}$, and so applying a bipartite complementation between these two sets disconnects $G[V_{1,2,3,4}]$ from the rest of the graph. By Claim 4, $G[V_{1,2,3,4}]$ has bounded clique-width, so by Fact 3, we may assume $V_{1,2,3,4}$ is empty. By Claim 9, at most one vertex of $V_{1,3}$ (resp. $V_{2,4}$) has a neighbour in V_1 (resp. V_2). Applying at most two further vertex deletions, we may assume that $V_{1,3}$ is anti-complete to V_1 and $V_{2,4}$ is anti-complete to V_2. By Claim 7, we can partition $V_\emptyset$ into the set $V_\emptyset^{1,3}$ of vertices that have neighbours in $V_{1,3}$ and the set $V_\emptyset^{2,4}$ of vertices that have neighbours in $V_{2,4}$. Now Claims 2 and 3 imply that $V_\emptyset^{2,4} \cup V_1 \cup V_{1,3}$ and $V_\emptyset^{1,3} \cup V_2 \cup V_{2,4}$ are independent sets, and so $G[V_\emptyset \cup V_1 \cup V_2 \cup V_{1,3} \cup V_{2,4}]$ is a $2P_2$-free bipartite graph. Such graphs are also known as bipartite chain graphs and are well known to have bounded clique-width (see e.g. [30, Theorem 2]). By Fact 1, this completes the proof.   □

The class of split graphs is the class of $(C_4, C_5, 2P_2)$-free graphs. Since split graphs therefore form a subclass of the class of $(2P_2, \overline{P_2 + P_3})$-free graphs, and split graphs have unbounded clique-width, it follows that $(2P_2, \overline{P_2 + P_3})$-free graphs also have unbounded clique-width. Recall that split atoms are complete graphs, which therefore have clique-width at most 2. The $(2P_2, \overline{P_2 + P_3})$-free atoms that are not split must therefore contain an induced C_4 or C_5. Applying Lemmas 4 and 5, we obtain Theorem 2, which we restate below.

Theorem 2 (restated). *The class of $(2P_2, \overline{P_2 + P_3})$-free atoms has bounded clique-width (whereas the class of $(2P_2, \overline{P_2 + P_3})$-free graphs has unbounded clique-width).*

4   Conclusions

Motivated by algorithmic applications, we determined a new class of (H_1, H_2)-free graphs of unbounded clique-width whose atoms have *bounded* clique-width, namely when $(H_1, H_2) = (2P_2, \overline{P_2 + P_3})$. We also identified a number of classes of (H_1, H_2)-free graphs of unbounded clique-width whose atoms still have *unbounded* clique-width. The latter results show that boundedness of clique-width of (H_1, H_2)-free atoms does not necessarily imply boundedness of clique-width of $(\overline{H_1}, \overline{H_2})$-free atoms. For example, (C_4, P_5)-free atoms have bounded clique-width [35], but we proved that $(\overline{C_4}, \overline{P_5})$-free atoms have unbounded clique-width (Theorem 3). Note however that while it is not known whether the class of $(K_3, S_{1,2,3})$-free graphs has bounded clique-width, we can show that the class of $(K_3, S_{1,2,3})$-free atoms has bounded clique-width if and only if the class of $(3P_1, \overline{S_{1,2,3}})$-free atoms has bounded clique-width (proof omitted).

We also presented a summary theorem (Theorem 3), from which we can deduce the following list of **18** open cases. The cases marked with a * are those for which even the boundedness of clique-width of the whole class of (H_1, H_2)-free graphs is unknown.

Open Problem 6. *Does the class of (H_1, H_2)-free atoms have bounded clique-width if*

- (i) $H_1 = $ diamond *and* $H_2 = P_6$
- (ii) $H_1 = C_4$ *and* $H_2 \in \{P_1 + 2P_2, P_2 + P_4, 3P_2\}$
- (iii) $H_1 = \overline{P_1 + 2P_2}$ *and* $H_2 \in \{2P_2, P_2 + P_3, P_5\}$
- (iv) $H_1 = \overline{P_2 + P_3}$ *and* $H_2 \in \{P_2 + P_3, P_5\}$
- *(v) $H_1 = K_3$ *and* $H_2 \in \{P_1 + S_{1,1,3}, S_{1,2,3}\}$
- *(vi) $H_1 = 3P_1$ *and* $H_2 = \overline{P_1 + S_{1,1,3}}$
- *(vii) $H_1 = $ diamond *and* $H_2 \in \{P_1 + P_2 + P_3, P_1 + P_5\}$
- *(viii) $H_1 = 2P_1 + P_2$ *and* $H_2 \in \{\overline{P_1 + P_2 + P_3}, \overline{P_1 + P_5}\}$
- *(ix) $H_1 = $ gem *and* $H_2 = P_2 + P_3$, *or*
- *(x) $H_1 = P_1 + P_4$ *and* $H_2 = \overline{P_2 + P_3}$.

In particular, we ask if boundedness of clique-width of $(2P_2, \overline{P_2 + P_3})$-free atoms can be extended to $(P_5, \overline{P_2 + P_3})$-free atoms. Could this explain why COLOURING is polynomial-time solvable on $(P_5, \overline{P_2 + P_3})$-free graphs [49]? Is boundedness of clique-width the underlying reason? Brandstädt and Hoàng [9] showed that $(P_5, \overline{P_2 + P_3})$-free atoms with no dominating vertices and no vertex pairs $\{x, y\}$ with $N(x) \subseteq N(y)$ are either isomorphic to some specific graph G^* or all their induced C_5s are dominating. Recently, Huang and Karthick [43] proved a more refined decomposition. However, it is not clear how to use these results to prove boundedness of clique-width of $(P_5, \overline{P_2 + P_3})$-free atoms, and additional insights seem to be needed.

References

1. Alekseev, V.E.: On easy and hard hereditary classes of graphs with respect to the independent set problem. Discrete Appl. Math. **132**(1–3), 17–26 (2003). https://doi.org/10.1016/S0166-218X(03)00387-1
2. Belmonte, R., Vatshelle, M.: Graph classes with structured neighborhoods and algorithmic applications. Theor. Comput. Sci. **511**, 54–65 (2013). https://doi.org/10.1016/j.tcs.2013.01.011

3. Blanché, A., Dabrowski, K.K., Johnson, M., Lozin, V.V., Paulusma, D., Zamaraev, V.: Clique-width for graph classes closed under complementation. SIAM J. Discrete Math. **34**(2), 1107–1147 (2020). https://doi.org/10.1137/18M1235016

4. Bodlaender, H.L.: A linear-time algorithm for finding tree-decompositions of small treewidth. SIAM J. Comput. **25**(6), 1305–1317 (1996). https://doi.org/10.1137/S0097539793251219

5. Bodlaender, H.L., Brettell, N., Johnson, M., Paesani, G., Paulusma, D., van Leeuwen, E.J.: Steiner trees for hereditary graph classes. Proc. LATIN 2020, LNCS (2020, to appear)

6. Boliac, R., Lozin, V.: On the clique-width of graphs in hereditary classes. In: Bose, P., Morin, P. (eds.) ISAAC 2002. LNCS, vol. 2518, pp. 44–54. Springer, Heidelberg (2002). https://doi.org/10.1007/3-540-36136-7_5

7. Brandstädt, A., Dabrowski, K.K., Huang, S., Paulusma, D.: Bounding the clique-width of H-free split graphs. Discrete Appl. Math. **211**, 30–39 (2016). https://doi.org/10.1016/j.dam.2016.04.003

8. Brandstädt, A., Dabrowski, K.K., Huang, S., Paulusma, D.: Bounding the clique-width of H-free chordal graphs. J. Graph Theory **86**(1), 42–77 (2017). https://doi.org/10.1002/jgt.22111

9. Brandstädt, A., Hoàng, C.T.: On clique separators, nearly chordal graphs, and the maximum weight stable set problem. Theor. Comput. Sci. **389**(1–2), 295–306 (2007). https://doi.org/10.1016/j.tcs.2007.09.031

10. Brandstädt, A., Klembt, T., Mahfud, S.: P_6- and triangle-free graphs revisited: structure and bounded clique-width. Discrete Math. Theor. Comput. Sci. **8**(1), 173–188 (2006). https://dmtcs.episciences.org/372

11. Brandstädt, A., Le, H.O., Mosca, R.: Gem- and co-gem-free graphs have bounded clique-width. Int. J. Found. Comput. Sci. **15**(1), 163–185 (2004). https://doi.org/10.1142/S0129054104002364

12. Brandstädt, A., Le, H.O., Mosca, R.: Chordal co-gem-free and (P_5,gem)-free graphs have bounded clique-width. Discrete Appl. Math. **145**(2), 232–241 (2005). https://doi.org/10.1016/j.dam.2004.01.014

13. Brandstädt, A., Mahfud, S.: Maximum weight stable set on graphs without claw and co-claw (and similar graph classes) can be solved in linear time. Inf. Process. Lett. **84**(5), 251–259 (2002). https://doi.org/10.1016/S0020-0190(02)00291-0

14. Brandstädt, A., Mosca, R.: On distance-3 matchings and induced matchings. Discrete Appl. Math. **159**(7), 509–520 (2011). https://doi.org/10.1016/j.dam.2010.05.022

15. Brettell, N., Horsfield, J., Munaro, A., Paesani, G., Paulusma, D.: Bounding the mim-width of hereditary graph classes. In: Cao, Y., Pilipczuk, M. (eds.) IPEC 2020. LIPIcs vol. 180, pp. 6:1–6:19 (2020). https://doi.org/10.4230/LIPIcs.IPEC.2020.6

16. Bui-Xuan, B., Telle, J.A., Vatshelle, M.: Fast dynamic programming for locally checkable vertex subset and vertex partitioning problems. Theor. Comput. Sci. **511**, 66–76 (2013). https://doi.org/10.1016/j.tcs.2013.01.009

17. Cameron, K., da Silva, M.V.G., Huang, S., Vušković, K.: Structure and algorithms for (cap, even hole)-free graphs. Discrete Math. **341**(2), 463–473 (2018). https://doi.org/10.1016/j.disc.2017.09.013

18. Chudnovsky, M., Seymour, P.D.: The structure of claw-free graphs. London Math. Soc. Lecture Note Series **327**, 153–171 (2005). https://doi.org/10.1017/CBO9780511734885.008

19. Corneil, D.G., Rotics, U.: On the relationship between clique-width and treewidth. SIAM J. Comput. **34**, 825–847 (2005). https://doi.org/10.1137/S0097539701385351

20. Courcelle, B.: The monadic second-order logic of graphs. I. Recognizable sets of finite graphs. Inf. Comput. **85**(1), 12–75 (1990). https://doi.org/10.1016/0890-5401(90)90043-H

21. Courcelle, B., Engelfriet, J.: Graph Structure and Monadic Second-Order Logic: A Language-Theoretic Approach, Encyclopedia of Mathematics and its Applications, vol. 138. Cambridge University Press (2012). https://doi.org/10.1017/CBO9780511977619

22. Courcelle, B., Makowsky, J.A., Rotics, U.: Linear time solvable optimization problems on graphs of bounded clique-width. Theory Comput. Syst. **33**(2), 125–150 (2000). https://doi.org/10.1007/s002249910009

23. Courcelle, B., Olariu, S.: Upper bounds to the clique width of graphs. Discrete Appl. Math. **101**(1–3), 77–114 (2000). https://doi.org/10.1016/S0166-218X(99)00184-5

24. Dabrowski, K.K., Dross, F., Paulusma, D.: Colouring diamond-free graphs. J. Comput. Syst. Sci. **89**, 410–431 (2017). https://doi.org/10.1016/j.jcss.2017.06.005

25. Dabrowski, K.K., Huang, S., Paulusma, D.: Bounding clique-width via perfect graphs. J. Comput. Syst. Sci. **104**, 202–215 (2019). https://doi.org/10.1016/j.jcss.2016.06.007

26. Dabrowski, K.K., Johnson, M., Paulusma, D.: Clique-width for hereditary graph classes. London Math. Soc. Lecture Note Series **456**, 1–56 (2019). https://doi.org/10.1017/9781108649094.002

27. Dabrowski, K.K., Lozin, V.V., Paulusma, D.: Clique-width and well-quasi-ordering of triangle-free graph classes. J. Comput. Syst. Sci. **108**, 64–91 (2020). https://doi.org/10.1016/j.jcss.2019.09.001

28. Dabrowski, K.K., Lozin, V.V., Raman, R., Ries, B.: Colouring vertices of triangle-free graphs without forests. Discrete Math. **312**(7), 1372–1385 (2012). https://doi.org/10.1016/j.disc.2011.12.012

29. Dabrowski, K.K., Masařík, T., Novotná, J., Paulusma, D., Rzążewski, P.: Clique-width: Harnessing the power of atoms. CoRR abs/2006.03578 (2020). https://arxiv.org/abs/2006.03578

30. Dabrowski, K.K., Paulusma, D.: Classifying the clique-width of H-free bipartite graphs. Discrete Appl. Math. **200**, 43–51 (2016). https://doi.org/10.1016/j.dam.2015.06.030

31. Dabrowski, K.K., Paulusma, D.: Clique-width of graph classes defined by two forbidden induced subgraphs. Comput. J. **59**(5), 650–666 (2016). https://doi.org/10.1093/comjnl/bxv096

32. Espelage, W., Gurski, F., Wanke, E.: How to solve NP-hard graph problems on clique-width bounded graphs in polynomial time. In: Brandstädt, A., Le, V.B. (eds.) WG 2001. LNCS, vol. 2204, pp. 117–128. Springer, Heidelberg (2001). https://doi.org/10.1007/3-540-45477-2_12

33. Foley, A.M., Fraser, D.J., Hoàng, C.T., Holmes, K., LaMantia, T.P.: The intersection of two vertex coloring problems. Graphs Comb. **36**(1), 125–138 (2020). https://doi.org/10.1007/s00373-019-02123-1

34. Fraser, D.J., Hamel, A.M., Hoàng, C.T., Holmes, K., LaMantia, T.P.: Characterizations of $(4K_1, C_4, C_5)$-free graphs. Discrete Appl. Math. **231**, 166–174 (2017). https://doi.org/10.1016/j.dam.2016.08.016

35. Gaspers, S., Huang, S., Paulusma, D.: Colouring square-free graphs without long induced paths. J. Comput. Syst. Sci. **106**, 60–79 (2019). https://doi.org/10.1016/j.jcss.2019.06.002
36. Gerber, M.U., Kobler, D.: Algorithms for vertex-partitioning problems on graphs with fixed clique-width. Theor. Comput. Sci. **299**(1), 719–734 (2003). https://doi.org/10.1016/S0304-3975(02)00725-9
37. Golovach, P.A., Johnson, M., Paulusma, D., Song, J.: A survey on the computational complexity of colouring graphs with forbidden subgraphs. J. Graph Theory **84**(4), 331–363 (2017). https://doi.org/10.1002/jgt.22028
38. Grötschel, M., Lovász, L., Schrijver, A.: Polynomial algorithms for perfect graphs. Ann. Discrete Math. **21**, 325–356 (1984). https://doi.org/10.1016/S0304-0208(08)72943-8
39. Gurski, F.: The behavior of clique-width under graph operations and graph transformations. Theory Comput. Syst. **60**(2), 346–376 (2017). https://doi.org/10.1007/s00224-016-9685-1
40. Hermelin, D., Mnich, M., van Leeuwen, E.J., Woeginger, G.J.: Domination when the stars are out. ACM Trans. Algorithms **15**(2), 25:1–25:90 (2019). https://doi.org/10.1145/3301445
41. Hliněný, P., Oum, S., Seese, D., Gottlob, G.: Width parameters beyond tree-width and their applications. Comput. J. **51**(3), 326–362 (2008). https://doi.org/10.1093/comjnl/bxm052
42. Hoàng, C.T., Lazzarato, D.A.: Polynomial-time algorithms for minimum weighted colorings of $(P_5, \overline{P_5})$-free graphs and similar graph classes. Discrete Appl. Math. **186**, 106–111 (2015). https://doi.org/10.1016/j.dam.2015.01.022
43. Huang, S., Karthick, T.: On graphs with no induced five-vertex path or paraglider. CoRR abs/1903.11268 (2019). https://arxiv.org/abs/1903.11268
44. Jansen, K., Scheffler, P.: Generalized coloring for tree-like graphs. Discrete Appl. Math. **75**(2), 135–155 (1997). https://doi.org/10.1016/S0166-218X(96)00085-6
45. Kamiński, M., Lozin, V.V., Milanič, M.: Recent developments on graphs of bounded clique-width. Discrete Appl. Math. **157**(12), 2747–2761 (2009). https://doi.org/10.1016/j.dam.2008.08.022
46. Kobler, D., Rotics, U.: Edge dominating set and colorings on graphs with fixed clique-width. Discrete Appl. Math. **126**(2–3), 197–221 (2003). https://doi.org/10.1016/S0166-218X(02)00198-1
47. Lozin, V.V., Rautenbach, D.: On the band-, tree-, and clique-width of graphs with bounded vertex degree. SIAM J. Discrete Math. **18**(1), 195–206 (2004). https://doi.org/10.1137/S0895480102419755
48. Makowsky, J.A., Rotics, U.: On the clique-width of graphs with few P_4's. Int. J. Found. Comput. Sci. **10**(03), 329–348 (1999). https://doi.org/10.1142/S0129054199000241
49. Malyshev, D.S., Lobanova, O.O.: Two complexity results for the vertex coloring problem. Discrete Appl. Math. **219**, 158–166 (2017). https://doi.org/10.1016/j.dam.2016.10.025
50. Oum, S., Seymour, P.D.: Approximating clique-width and branch-width. J. Comb. Theory Ser. B **96**(4), 514–528 (2006). https://doi.org/10.1016/j.jctb.2005.10.006
51. Rao, M.: MSOL partitioning problems on graphs of bounded treewidth and clique-width. Theor. Comput. Sci. **377**(1–3), 260–267 (2007). https://doi.org/10.1016/j.tcs.2007.03.043

52. Sæther, S.H., Vatshelle, M.: Hardness of computing width parameters based on branch decompositions over the vertex set. Theor. Comput. Sci. **615**, 120–125 (2016). https://doi.org/10.1016/j.tcs.2015.11.039
53. Tarjan, R.E.: Decomposition by clique separators. Discrete Math. **55**(2), 221–232 (1985). https://doi.org/10.1016/0012-365X(85)90051-2
54. Vatshelle, M.: New Width Parameters of Graphs. Ph.D. thesis, University of Bergen (2012)

Edge Elimination and Weighted Graph Classes

Jesse Beisegel[1], Nina Chiarelli[2,3], Ekkehard Köhler[1], Matjaž Krnc[2],
Martin Milanič[2,3], Nevena Pivač[2,3], Robert Scheffler[1(✉)], and Martin Strehler[1]

[1] Brandenburg University of Technology, Cottbus, Germany
{jesse.beisegel,ekkehard.koehler,robert.scheffler,martin.strehler}@b-tu.de
[2] FAMNIT, University of Primorska, Koper, Slovenia
nina.chiarelli@famnit.upr.si, {matjaz.krnc,martin.milanic}@upr.si
[3] IAM, University of Primorska, Koper, Slovenia
nevena.pivac@iam.upr.si

Abstract. Edge-weighted graphs play an important role in the theory of
Robinsonian matrices and similarity theory, particularly via the concept
of level graphs, that is, graphs obtained from an edge-weighted graph by
removing all sufficiently light edges. This naturally leads to a generalization of the concept of a graph class to the weighted case by requiring that
all level graphs belong to the class. We examine some types of monotonicity of graph classes, such as sandwich monotonicity, to construct edge
elimination schemes of edge-weighted graphs. This leads to linear-time
recognition algorithms of weighted graphs for which all level graphs are
split, threshold, or chain graphs.

Keywords: Edge elimination · Weighted graphs · Split graphs ·
Threshold graphs · Chain graphs · Linear-time recognition algorithm

1 Introduction

Background and Motivation. Vertex and edge elimination orderings are well
established concepts in graph theory (see Chap. 5 in [6]). For example, chordal
graphs can be characterized as the graphs with a perfect vertex elimination ordering [13,34]. In 2017 Laurent and Tanigawa [27] extended the classical notion of
perfect (vertex) elimination ordering for graphs to edge-weighted graphs, giving a framework capturing common vertex elimination orderings of families of
chordal graphs, Robinsonian matrices, and ultrametrics. They showed that an
edge-weighted graph G has a perfect elimination ordering if and only if it has a
vertex ordering that is a simultaneous perfect elimination ordering of all its level
graphs. Here, the *i-th level graph* of G is the graph obtained from G by removing
all edges with weights smaller than i. In particular, this latter condition implies
that all the level graphs must be chordal.

Similarly, edge elimination orderings can be used to characterize graph
classes. Adding an edge between a two-pair of a weakly chordal graph always

© Springer Nature Switzerland AG 2020
I. Adler and H. Müller (Eds.): WG 2020, LNCS 12301, pp. 134–147, 2020.
https://doi.org/10.1007/978-3-030-60440-0_11

maintains weak chordality [37]. Since the class of weakly chordal graphs is self-complementary, a graph is weakly chordal if and only if it has an edge elimination ordering where every edge is a two-pair in the complement of the current graph. On a bipartite graph, an edge elimination ordering is said to be perfect (also called perfect edge-without-vertex-elimination ordering) if every edge is bisimplicial at the time of elimination, that is, the closed neighborhood of both endpoints of the edge induces a complete bipartite subgraph. A bipartite graph is chordal bipartite if and only if it admits a perfect edge elimination ordering (see, e.g., [24]).

What the above concepts have in common is that a certain property of the graph is maintained even after a certain vertex or edge has been deleted. Some graph properties achieve this in a trivial way. A graph class $\mathcal{G}$ is said to be *monotone* if every subgraph of a graph in $\mathcal{G}$ is also in $\mathcal{G}$. For example, planar graphs and bipartite graphs are monotone graph classes. In particular, an arbitrary edge can be deleted without leaving the monotone graph class. This definition can be relaxed as follows. A graph class $\mathcal{G}$ is said to be *weakly monotone* if every graph $G = (V, E)$ in $\mathcal{G}$ is either edgeless or has a $\mathcal{G}$-safe edge, that is, an edge $e \in E$ such that $G - e$ is in $\mathcal{G}$. Obviously, a monotone graph class is also weakly monotone. Note that we do not delete vertices here.

Instead of deleting edges one at a time, one can also consider the deletion of sequences of pairwise disjoint sets of edges. Every set is deleted at once, and the requirement is that all the intermediate graphs belong to a fixed graph class $\mathcal{G}$. Here, the level graphs of an edge-weighted graph, as used by Laurent and Tanigawa [27], are equivalent to such edge set eliminations. Another example is given by the class of threshold graphs (see [7]), where edges are deleted when the threshold is raised.

Such more general edge elimination sequences are naturally related to the following concept also studied in the literature. A graph class $\mathcal{G}$ is said to be *sandwich monotone* if for any two graphs G and G' in $\mathcal{G}$ such that G is a spanning subgraph of G', graph G can be obtained from G' by a sequence of edge deletions such that all intermediate graphs are in $\mathcal{G}$. This property was studied in 1976 by Rose et al. [35] who showed that chordal graphs have this property. In 2007 Heggernes and Papadopoulos [18] (see also [19]) introduced the term *sandwich monotonicity* for this property and showed that the classes of threshold graphs and chain graphs are sandwich monotone. The same was shown for split graphs by Heggernes and Mancini [16] as well as for both strongly chordal graphs and chordal bipartite graphs by Heggernes et al. [17].

Our Contributions.   Applying edge elimination orderings on edge-weighted graphs, we generalize the definition of a graph class as follows. Given a graph class $\mathcal{G}$, we say that an edge-weighted graph is *level-$\mathcal{G}$* if all its level graphs are in $\mathcal{G}$. A particularly nice situation occurs when $\mathcal{G}$ is sandwich monotone. In this case, all the edges of an edge-weighted level-$\mathcal{G}$ graph can be eliminated one at a time, from lightest to heaviest, so that all the intermediate graphs are in $\mathcal{G}$, which yields an edge elimination ordering with increasing edge weights. Such an edge elimination ordering is called a *sorted $\mathcal{G}$-safe edge elimination ordering of G*.

In Sect. 3 we discuss relations between various types of monotonicity of graph classes and identify several examples of weakly monotone graph classes.

In Sect. 4 we consider weighted analogs of split, threshold, and chain graphs. Furthermore, we introduce the concept of a degree-minimal edge in a given set of edges and show that for the classes of split, threshold, and chain graphs, every degree-minimal member of a safe set of edges is a safe edge. Additionally, we develop our key technical contribution, a linear-time algorithm for computing a degree-minimal edge elimination scheme of an arbitrary weighted graph. Combining this algorithm with known results yields linear-time recognition algorithms of level-split, level-threshold and level-chain weighted graphs. This is a significant improvement over the naive $\mathcal{O}(m^2)$-algorithm which checks for each level graph whether it belongs to the respective class. Due to lack of space, most proofs are omitted and will appear in the full version of this work.

Related Work. Our results related to elimination schemes of weighted graphs can be seen as part of a more general research framework aimed at generalizing theoretical and algorithmic aspects of graphs to (edge-)weighted graphs, or, equivalently, from binary to real-valued symmetric matrices. For example, Robinsonian similarities are weighted analogues of unit interval graphs [29,33], Robinsonian dissimilarities (see [32]) are weighted analogues of co-comparability graphs [12], and Similarity-first search is a weighted graph analogue of Lexicographic breadth-first search [25]. Other concepts that were generalized to the weighted case include perfect elimination orderings [27] and asteroidal triples [26].

All these works, including ours, share a common feature that is often applicable to weighted problems: instead of the exact numerical values of the input, only the structure of these values, that is, the ordinal aspects of the distances or weights, matter. This is a common situation for problems arising in social network analysis (see, e.g., [11]), combinatorial data analysis (see, e.g., [28]), in phylogenetics (see, e.g., [20,22]), as well as in greedy algorithms for some combinatorial optimization problems such as Kruskal's or Prim's algorithms for the minimum spanning tree problem, or the greedy algorithm for the problem of finding a minimum-weight basis of a matroid [8].

Unsurprisingly, concepts similar to that of an edge-weighted graph and its level graphs appeared in the literature in different contexts and under different names. For example, Berry et al. were interested in weighted graphs derived from an experimentally obtained dissimilarity matrix, motivated by questions related to phylogeny reconstruction [4]. The level graphs of a weighted graph can also be seen as a special case of a *temporal graph*, a dynamically changing graph in which each edge can appear and disappear over a certain time period (see, e.g., [31]). In the terminology of Fluschnik et al. [9], the level graphs of a weighted graph form a "1-monotone temporal graph" (see also [23]). Furthermore, the special case of weighted graphs when all edges have different weights corresponds to an *edge ordered graph*, a concept of interest in extremal graph theory (see, e.g., [39]).

2   Preliminaries

All graphs considered in this paper are finite, simple, and undirected. A graph is *nontrivial* if it has at least two vertices and *edgeless* if it contains no edge. For definitions of standard terms like neighborhood, degree, and induced subgraphs we refer to [40]. For definitions of graph classes such as weakly chordal graphs, interval graphs, and comparability graphs see [6].

In the following we present the definitions of the three main graph classes considered in this paper. A *split graph* G is a graph whose vertex set can be partitioned into sets C and I such that C is a clique and I is an independent set in G (see [10]). We call (C, I) a *split partition* of G.

A graph $G = (V, E)$ is said to be *threshold* if there exists a labeling $\ell : V \to \mathbb{N}_0$ and a threshold value $t \in \mathbb{N}_0$ such that a set $X \subseteq V$ is independent in G if and only if $\sum_{x \in X} \ell(x) \leq t$ (see [7]).

A bipartite graph $G = (V, E)$ is a *chain graph* if its vertex set can be partitioned into two independent sets X and Y such that vertices in X can be ordered linearly with respect to set inclusion of their neighborhoods (see [41]). We will refer to such a pair (X, Y) as a *chain bipartition* of G.

An *ordering of the vertices* in G is a bijection $\sigma : V(G) \to \{1, 2, \ldots, n\}$. For an arbitrary ordering σ of the vertices in G we denote by $\sigma(v)$ the position of vertex $v \in V(G)$. Given two vertices u and v in G, we say that u is *to the left* (resp. *to the right*) of v if $\sigma(u) < \sigma(v)$ (resp. $\sigma(u) > \sigma(v)$) and we denote this by $u \prec_\sigma v$ (resp. $u \succ_\sigma v$). Analogously, we define an *ordering of the edges* of G as a bijection $\tau : E(G) \to \{1, 2, \ldots, m\}$. Given an edge ordering $\tau = (e_1, \ldots, e_m)$ of G we denote by G_τ^i its spanning subgraph $G - \{e_1, \ldots, e_i\}$.

3   Monotonicity Properties of Graph Classes

A graph class is said to be *hereditary* if every induced subgraph of every graph in that class belongs to the class. This can also be defined in a different way by saying that we can delete arbitrary vertices from the graph and remain in the same class, a concept that is also known as *vertex monotonicity*. *Edge monotonicity* is defined in the same way: we can remove arbitrary edges from a graph and remain in the class. If a class is both edge and vertex monotone, we simply call it *monotone*. Edge monotonicity is a rather restrictive property and many well-studied graph classes are not edge monotone. In order to include more graph classes we also consider a relaxation of this property called *weak (edge) monotonicity*. A graph class $\mathcal{G}$ is called *weakly (edge) monotone* if every member G of this class is either edgeless, or there is an edge e in G such that $G - e \in \mathcal{G}$. We say that such an edge is $\mathcal{G}$-*safe* for G. More generally, a set F of edges of a graph $G \in \mathcal{G}$ is said to be $\mathcal{G}$-*safe* if $G - F$ is a member of $\mathcal{G}$. Note that weak (edge) monotonicity was already mentioned by Heggernes and Papadopoulos [19] as "edge monotonicity".

For a graph class $\mathcal{G}$ and a graph $G \in \mathcal{G}$, a $\mathcal{G}$-*safe edge elimination scheme* of G is defined as an ordering $\tau = (e_1, \ldots, e_m)$ of the edges of G such that for each $i \in \{1, \ldots, m\}$ the spanning subgraph G_τ^i is in $\mathcal{G}$. Such elimination schemes always exist precisely for weakly monotone graph classes.

Theorem 1. *Let $\mathcal{G}$ be a graph class. Then, every graph in $\mathcal{G}$ has a $\mathcal{G}$-safe edge elimination scheme if and only if $\mathcal{G}$ is weakly monotone.*

In [19] Heggernes and Papadopoulos introduce the concept of *sandwich monotonicity*. A graph class $\mathcal{G}$ is sandwich monotone if for each graph $G \in \mathcal{G}$ and each non-empty $\mathcal{G}$-safe set $F \subseteq E(G)$ there is a $\mathcal{G}$-safe edge in F. Equivalently, one can say that a graph class is sandwich monotone if between two of its members $G = (V, E)$ and $G' = (V, E \cup F)$ with $E \cap F = \emptyset$ there is sequence of graphs $(G = G_0, G_1, \ldots, G_{|F|} = G')$ such that for all $i \in \{1, \ldots, |F|\}$ we have $G_i = G_{i-1} + e$ with $e \in F$ and each graph G_i is a member of $\mathcal{G}$. Therefore, it makes no difference whether we say that we can delete edges one by one from the larger graph until we get the smaller one such that all intermediate graphs are in $\mathcal{G}$, or we consider in a similar way the reverse process of adding edges to the smaller graph until we obtain the larger one. This also leads to the observation already stated in [19] that for a sandwich monotone graph class $\mathcal{G}$ the complementary graph class co-$\mathcal{G}$ is also sandwich monotone.

Note that this property does not hold for weak monotonicity, i.e., there are weakly monotone graph classes whose complementary graph class is not weakly monotone. An example is the class of bipartite graphs, which is monotone and, therefore, also weakly monotone. However, the class of co-bipartite graphs is not weakly monotone. Counterexamples are the complements of complete bipartite graphs with at least two vertices in one part.

Many known graph classes are weakly monotone. Obviously, monotone graph classes fulfill this condition. But not every weakly monotone graph class is monotone. For example, the class of graphs with at most one nontrivial component is weakly monotone but not monotone. Furthermore, monotone graph classes are also sandwich monotone.

On the other hand, there are graph classes which are sandwich monotone but not weakly monotone. An example is the class of connected graphs. However, for many graph classes it can be shown that weak monotonicity is a generalization of sandwich monotonicity. A graph class is called *grounded* if it fulfills the following condition for every positive integer n: If there is a graph $G \in \mathcal{G}$ with n vertices, then the edgeless graph with n vertices is also in $\mathcal{G}$. Obviously, every weakly monotone graph class is grounded. Furthermore, we have the following proposition.

Proposition 2. *Let $\mathcal{G}$ be a graph class. If $\mathcal{G}$ is grounded and sandwich monotone, then $\mathcal{G}$ is weakly monotone.*

The above connection between sandwich monotone and weakly monotone graph classes was already mentioned by Heggernes and Papadopoulos [19]. However, there they did not emphasize the fact that the implication only holds for grounded graph classes. We summarize the implications between the mentioned properties of graph classes in Fig. 1.

The following graph classes have been shown to be sandwich monotone: chordal graphs [2, 35], split graphs [16], threshold and chain graphs [19], as well as strongly chordal and chordal bipartite graphs [17]. Since all these graph classes

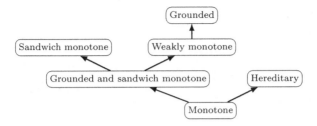

Fig. 1. Relationships between various monotonicity properties of graph classes. Arrows represent implications, e.g., every monotone graph class is also hereditary.

are grounded, by Proposition 2 they are also weakly monotone. The following proposition gives further examples of weakly monotone graph classes.

Proposition 3. *The classes of interval graphs, unit interval graphs, comparability graphs, co-comparability graphs, weakly chordal graphs, and permutation graphs are weakly monotone.*

With the possible exception of weakly chordal graphs, none of these graph classes are sandwich monotone (see [19] for counterexamples). For weakly chordal graphs it is an open question whether they are sandwich monotone.

On the other hand, many graph classes are not weakly monotone. We have already seen that graph classes that are not grounded cannot be weakly monotone. There are also grounded graph classes that are not weakly monotone. An example of this are the perfect graphs, as verified by the line graph of $K_{3,3}$. Moreover, as shown in [5], there exists an infinite family of perfect graphs that become non-perfect upon either deletion or addition of any single edge.

4   Degree-Minimal Edge Elimination Schemes and Recognition of Level-$\mathcal{G}$ Weighted Graphs

Before moving on to the main results of this article, we state a formal definition of the notion of weighted graphs already mentioned in the introduction. Given a positive integer k, a *k-weighted graph* is a pair (G, ω) where G is a graph and $\omega : E(G) \to \{1, \ldots, k\}$ is a surjective *weight function*. We will often denote a k-weighted graph (G, ω) simply by G and call it *weighted*. In many applications a graph is equipped with a weight function of arbitrary real values. In this case, sorting the edges by weight yields the required surjective function $\omega : E(G) \to \{1, \ldots, k\}$, where k is the number of distinct weights.

Definition 4. *The i-th level graph of a weighted graph (G, ω) is the graph obtained from G by removing all edges e with $\omega(e) < i$. Given a graph class $\mathcal{G}$, we say that a weighted graph (G, ω) is level-$\mathcal{G}$ if all level graphs of G are in $\mathcal{G}$. A $\mathcal{G}$-safe edge elimination scheme $(e_1, \ldots, e_m)$ of a weighted graph $G \in \mathcal{G}$ is called sorted if for any pair of edges e_i and e_j with $i < j$ it holds that $\omega(e_i) \leq \omega(e_j)$.*

Recall that by Theorem 1 the graph classes in which $\mathcal{G}$-safe edge elimination schemes always exist are exactly the weakly monotone graph classes. Analogously, the existence of sorted $\mathcal{G}$-safe edge elimination schemes of weighted graphs is guaranteed exactly by the sandwich monotonicity property.

Theorem 5. *Let $\mathcal{G}$ be a graph class. Then, every level-$\mathcal{G}$ graph has a sorted $\mathcal{G}$-safe edge elimination scheme if and only if $\mathcal{G}$ is sandwich monotone.*

In the remainder of this section, we will discuss how to compute such a sorted $\mathcal{G}$-safe edge elimination scheme and how this scheme can be used to recognize level-split, level-threshold, and level-chain weighted graphs.

4.1   Degree-Minimal Edge Elimination Schemes of Weighted Graphs

The first step in the computation of a sorted $\mathcal{G}$-safe edge elimination scheme is to efficiently identify a safe edge among all edges with minimal weight. In the following we will show that an edge with "minimal degree" is in fact $\mathcal{G}$-safe for the considered graph classes.

Definition 6. *Let G be a graph and let F be a set of edges of G. A degree-minimal edge in F is an edge $xy \in F$ such that:*

1. *Vertex x has the smallest degree in G among all vertices incident to an edge in F, and*
2. *The degree of y in G is the smallest among all neighbors of x that are adjacent to x via an edge in F.*

In other words, a degree-minimal edge in F is a lexicographically smallest edge $e = xy \in F$ with respect to the pair $(d_1(e), d_2(e))$, where $d_1(e) = \min\{d_G(x), d_G(y)\}$ and $d_2(e) = \max\{d_G(x), d_G(y)\}$.

For split, threshold and chain graphs degree minimality can be used to find a safe edge, giving a very simple certificate for this property which, however, is only sufficient and not necessary.

Theorem 7. *Let $\mathcal{G}$ be one of the following graph classes: split graphs, threshold graphs, chain graphs. Then, for every two graphs $G = (V, E)$ and $G' = (V, E \cup F)$ in $\mathcal{G}$, where $E \cap F = \emptyset$, every degree-minimal edge in F is $\mathcal{G}$-safe.*

We now introduce a particular edge elimination scheme of weighted graphs, where each edge is degree-minimal among all edges with minimal weight in the remaining graph, and devise a linear-time algorithm that constructs such a scheme for arbitrary graph. Furthermore, we show that for special graph classes $\mathcal{G}$ these orderings are the $\mathcal{G}$-safe edge elimination schemes. In particular, Theorem 7 implies that this holds for the classes of split, threshold, and chain graphs. The proposed elimination scheme is used for a linear-time recognition for level-split, level-threshold, and level-chain graphs in the subsequent subsection.

Recall that given an edge ordering $\tau = (e_1, \ldots, e_m)$ of G and an integer $i \in \{1, \ldots, m\}$, we denote by G_τ^i its spanning subgraph $G - \{e_1, \ldots, e_i\}$.

Definition 8. *Let (G, ω) be a weighted graph. A linear ordering $\tau = (e_1, \ldots, e_m)$ of edges in G is said to be a degree-minimal edge elimination scheme of (G, ω) if for every $i \in \{1, \ldots, m\}$ the edge e_i is a degree-minimal edge in the set of all minimum-weight edges in the graph G_τ^{i-1}.*

The next result connects the concepts of degree-minimal and sorted $\mathcal{G}$-safe edge elimination schemes for sandwich monotone graph classes satisfying an additional condition.

Theorem 9. *Let $\mathcal{G}$ be a sandwich monotone graph class such that for every graph $G \in \mathcal{G}$ and every $\mathcal{G}$-safe set $F \subseteq E(G)$, each degree-minimal edge in F is $\mathcal{G}$-safe. Then, for any level-$\mathcal{G}$ weighted graph (G, ω) every degree-minimal edge elimination scheme is also a sorted $\mathcal{G}$-safe edge elimination scheme.*

Theorems 7 and 9 have the following consequence.

Corollary 10. *Let $\mathcal{G}$ be one of the following graph classes: split graphs, threshold graphs, chain graphs. Then, for any level-$\mathcal{G}$ weighted graph (G, ω) every degree-minimal edge elimination scheme is also a sorted $\mathcal{G}$-safe edge elimination scheme.*

We now present a linear-time algorithm that computes a degree-minimal edge elimination scheme for an arbitrary weighted graph.

Theorem 11. *Given a k-weighted graph $(G = (V, E), \omega)$, we can compute a degree-minimal edge elimination scheme of G in time $\mathcal{O}(|V| + |E|)$.*

Proof. We describe and analyze an algorithm with the above properties. We start with the description of the main ideas of the algorithm. For every vertex v and every weight i appearing on an edge incident with v we create a copy of v named v_i. Then, we order the vertex copies non-decreasingly with respect to their indices. The vertex copies with the same index are ordered such that the resulting linear order σ of all the vertex copies satisfies the following condition: For every copy v_i vertex v is a vertex of smallest degree in the graph obtained from G by deleting from it all edges $e = xy$ with $x_{\omega(e)} \prec_\sigma v_i$ or $y_{\omega(e)} \prec_\sigma v_i$. For every vertex copy v_i we define the v_i-star as the edge set $\Phi(v_i) = \{vw \mid \omega(vw) = i$ and $v_i \prec_\sigma w_i\}$. We create an ordered partition of the edge set based on the order σ of the vertex copies by replacing each vertex copy v_i with its respective v_i-star $\Phi(v_i)$. Any ordering $\rho = (e_1, \ldots, e_m)$ of the edges of G respecting this partition is sorted with respect to the edge weights and satisfies the following condition: For every edge e_i one of the two incident vertices has minimal degree in G_ρ^{i-1} among all vertices that are incident to an edge with weight $\omega(e_i)$. Finally, we reorder the edges within the sets $\Phi(v_i)$ such that also Condition 2 of Definition 6 holds for every edge.

Phase 1: Slicing the Input Graph. For all $i \in \{1, \ldots, k\}$ we compute the set V_i defined as $V_i = \{v_i \mid v \in V$ is a vertex of G incident to an edge with weight $i\}$. We will refer to $v_i \in V_i$ as the i-th *copy* of v. We denote by Ξ the set $\bigcup_{i=1}^k V_i$ and we will call the sets V_i the *slices* of Ξ. Note that $|\Xi| \leq 2|E|$, since each edge

$e = xy \in E$ can generate at most two vertex copies, namely $x_{\omega(e)}$ and $y_{\omega(e)}$. For all $i \in \{1, \dots, k\}$ and all $v_i \in V_i$, we compute the value of $d_i(v)$, where $d_i(v)$ denotes the degree of v in the i-th level graph of (G, ω).

Phase 2: Ordering of Ξ. We construct an ordering σ of Ξ which respects the fact that degrees of the vertices change during the transition from one level graph to the next, while the edges are eliminated one by one. The ordering σ has to fulfill the following properties. First, if $i < j$, then $v_i \prec_\sigma w_j$ for any two vertices $v, w \in V$. Secondly, for vertex copies v_i and w_i it holds that $v_i \prec_\sigma w_i$ if the degree of v is smaller or equal to the degree of w in the graph $G - \bigcup_{x_j \prec_\sigma v_i} \Phi(x_j)$. This is achieved by processing V_i one element at a time. Suppose that we have already appended some (possibly none, but not all) elements of V_i to σ. Let V_i' denote the set of remaining elements of V_i and let Z be the set of vertices of V for which there still exist copies in V_i'. Furthermore, let F_i be the set of edges vw with weight i such that $\{v_i, w_i\} \subseteq V_i'$. For all $j > i$ let F_j be the set of edges vw with weight j and let $F = \bigcup_{j \geq i} F_j$. We choose a vertex v in Z with smallest degree in the graph (V, F) and append vertex v_i to σ.

Phase 3: Ordering the Edges. The linear order σ induces an ordered partition of the edges of G by replacing each vertex copy v_i with its respective v_i-star $\Phi(v_i)$ in σ. Any ordering $\rho = (e_1, \dots, e_m)$ of the edges of G respecting this partition is sorted with respect to the edge weights and satisfies the following condition: For every edge e_i one of the two incident vertices has minimal degree in G_ρ^{i-1} among all vertices that are incident to an edge with weight $\omega(e_i)$. However, such an order ρ does not necessarily satisfy Condition 2 of Definition 6. This phase of the algorithm computes the final edge order τ of the edges by sorting the elements of the v_i-stars. For every $i \in \{1, \dots, k\}$ and every copy v_i we order the edges vw of the v_i-star non-decreasingly with respect to the degree of w in the graph (V, F) where F is the union of all w_j-stars where $w_j = v_i$ or $v_i \prec_\sigma w_j$.

Implementation Details. We will now describe how we can achieve a linear running time. First we sort the edges in E non-decreasingly according to their weights, which can be done in time $\mathcal{O}(|E|)$ using counting sort. To create Ξ we store for every vertex the copy created last. We traverse the edges according to their order. If for one of the vertices incident to the current edge e there is no copy for the weight of e, we create it. All edges are assigned a pointer to their corresponding vertex copies. This process can be done in linear time.

The values $d_i(v)$ can be computed in linear time by traversing the edges according to their order and updating the degrees. We assign the value $d_i(v)$ to v_i. Since the degrees lie between 0 and $|V| - 1$, we can order all vertex copies in linear time with counting sort with regard to the values d_i and then place them in their corresponding sets V_i in the order of their degrees.

In Phase 2 we use a data structure introduced by Ibarra [21] to dynamically recognize split graphs. It contains a list of vertices ordered by their degree, can be constructed in linear time and can be updated in constant time when an edge is deleted. We use such a data structure for every slice of Ξ separately, where the copies correspond to the vertices and as degrees the values d_i are used. Since the slices are disjoint and have an overall size in $\mathcal{O}(|E|)$, we can create these data structures in linear time. Furthermore, the vertex copy v_i with minimal degree can be found in constant time and updating the degrees of the other copies when the v_i-star is deleted only costs linear time overall.

In Phase 3 we compute for every edge $e = vw$ with $v_{\omega(e)} \prec_\sigma w_{\omega(e)}$ the degree of w in the graph obtained by deleting all x_i-stars with $x_i \prec_\sigma v_{\omega(e)}$. As in Phase 1, this can be done in linear time by traversing the x_i-stars with respect to σ. Afterwards we sort all edges with respect to the computed degrees in linear time with counting sort and reinsert them into their v_i-stars according to the computed degree ordering leading to the desired edge ordering τ.   □

4.2   Linear-Time Recognition of Level-Split, Level-Threshold, and Level-Chain Weighted Graphs

Combining the results of the previous sections we present linear-time algorithms that decide whether a given weighted graph (G, ω) is a level-split, level-threshold, or level-chain graph, respectively. The idea is the following. First we check whether G is in $\mathcal{G}$, which can be done in linear time for all three graph classes [14,15,30]. If this is not the case, then (G, ω) is not a level-$\mathcal{G}$ graph. Next, we compute a degree-minimal edge elimination scheme τ of G in linear time, using Theorem 11. By Corollary 10, we see that (G, ω) is a level-$\mathcal{G}$ weighted graph if and only if τ is a sorted $\mathcal{G}$-safe edge elimination scheme of (G, ω), or, equivalently if and only if τ is a $\mathcal{G}$-safe edge elimination scheme of G. To check this we use different approaches for the three graph classes. For chain graphs we introduce a characterizing vertex partition.

Definition 12. Let $G = (V, E)$ be a chain graph. A chain partition of G is an ordered partition $(A_1, B_1, \ldots, A_k, B_k, I)$ of V where I is the (possibly empty) set of isolated vertices in G, sets A_i and B_i are non-empty for all $1 \leq i \leq k$, and $xy \in E$ if and only if $x \in A_i$ and $y \in B_j$ or vice versa with $i \leq j$.

Note that Heggernes and Papadopoulos give a different but equivalent definition in [19]. Chain partitions are a very efficient way of storing chain graphs and can be computed in linear time.

Lemma 13. A graph is a chain graph if and only if it has a chain partition. For a given chain graph $G = (V, E)$ we can compute a chain partition in time $\mathcal{O}(|V| + |E|)$.

Using this notion of chain partitions, we can characterize chain-safe edges.

Lemma 14. Let $G = (V, E)$ be a chain graph with chain partition $(A_1, B_1, \ldots, A_k, B_k, I)$. Then, an edge $xy \in E$ is chain-safe if and only if $x \in A_i$ and $y \in B_i$ or vice versa.

This property can be checked in constant time and the chain partition can also be updated in constant time. For split graph we use the algorithm presented by Ibarra [21] which constructs a data structure in linear time that allows to check whether a graph is still split after the deletion of an edge in constant time. For threshold graphs we use a similar algorithm presented by Shamir and Sharan [36].

Therefore, we can check whether a computed degree-minimal edge elimination scheme is also a $\mathcal{G}$-safe edge elimination scheme for each of the three classes in linear time. Combining these two algorithms leads to our main theorem.

Theorem 15. *Let $\mathcal{G}$ be one of the following graph classes: split graphs, threshold graphs, chain graphs. Then, for a given weighted graph $(G = (V, E), \omega)$ we can decide whether G is a level-$\mathcal{G}$ graph in time $\mathcal{O}(|V| + |E|)$.*

5   Conclusion

By combining the concepts of level graphs – an important tool for the theory of Robinsonian matrices – and edge elimination schemes, we give a sufficient condition for split-, threshold-, and chain-safe edges in order to generate so-called sorted safe-edge elimination schemes. This yields linear-time recognition algorithms for level-split, level-threshold, and level-chain weighted graphs. Furthermore, we study the notion of weak edge-monotonicity – an analog to weak vertex-monotonicity studied in [1]. We show that, among others, the classes of permutation graphs, comparability graphs, and co-comparability graphs are weakly edge-monotone.

The above-mentioned contributions raise some interesting questions. As the classes of chordal, chordal bipartite, and strongly chordal graphs are all sandwich monotone, it is natural to ask whether the weighted analogs of these classes can be recognized faster than checking every level graph separately. Also, it would be interesting to find similar results for graph classes which are not sandwich monotone, for example comparability graphs or interval graphs. Furthermore, it remains open whether weakly chordal graphs are sandwich monotone, a question raised already in [3, 17, 38].

Finally, let us mention some natural extensions of the concepts discussed in this article that seem worthy of future investigations. One could define and study the concepts of weakly k-monotone and sandwich k-monotone graph classes for a positive integer k, by replacing the condition requiring the existence of a $\mathcal{G}$-safe edge in a particular set with the existence of a non-empty $\mathcal{G}$-safe subset of edges of cardinality at most k. For graph classes that are not hereditary (for example, the connected graphs), one could examine their vertex-weighted analogs in which the level graphs are defined by deleting all sufficiently light vertices.

Acknowledgements. The authors would like to thank Ulrik Brandes, Caroline Brosse, Christophe Crespelle, and Petr Golovach for their valuable discussions.

This research was funded in part by German Academic Exchange Service and the Slovenian Research Agency (BI-DE/17-19-18 and BI-DE/19-20-007), and by the Slovenian Research Agency (I0-0035, research programs P1-0285, P1-0383, P1-0404, research projects J1-1692, J1-9110, J1-9187, N1-0102, and N1-0160, and a Young Researchers Grant).

References

1. Andrade, D.V., Boros, E., Gurvich, V.: Not complementary connected and not CIS d-graphs form weakly monotone families. Discrete Math. **310**(5), 1089–1096 (2010). https://doi.org/10.1016/j.disc.2009.11.006
2. Bakonyi, M., Constantinescu, T.: Inheritance principles for chordal graphs. Linear Algebra Appl. **148**, 125–143 (1991). https://doi.org/10.1016/0024-3795(91)90090-J
3. Bakonyi, M., Bono, A.: Several results on chordal bipartite graphs. Czechoslov. Math. J. **47**(4), 577–583 (1997). https://doi.org/10.1023/A:1022806215452
4. Berry, A., Sigayret, A., Sinoquet, C.: Maximal sub-triangulation in pre-processing phylogenetic data. Soft Comput. **10**(5), 461–468 (2006). https://doi.org/10.1007/s00500-005-0507-7
5. Boros, E., Gurvich, V.: Vertex-and edge-minimal and locally minimal graphs. Discrete Math. **309**(12), 3853–3865 (2009). https://doi.org/10.1016/j.disc.2008.10.020
6. Brandstädt, A., Le, V.B., Spinrad, J.P.: Graph Classes: A Survey. SIAM, Philadelphia (1999)
7. Chvátal, V., Hammer, P.L.: Aggregation of inequalities in integer programming. In: Hammer, P.L., Johnson, E.L., Korte, B.H., Nemhauser, G.L. (eds.) Studies in Integer Programming, Annals of Discrete Mathematics, vol. 1, pp. 145–162. Elsevier (1977). https://doi.org/10.1016/S0167-5060(08)70731-3
8. Edmonds, J.: Matroids and the greedy algorithm. Math. Programming **1**, 127–136 (1971). https://doi.org/10.1007/BF01584082
9. Fluschnik, T., Molter, H., Niedermeier, R., Renken, M., Zschoche, P.: Temporal graph classes: a view through temporal separators. Theor. Comput. Sci. **806**, 197–218 (2020). https://doi.org/10.1016/j.tcs.2019.03.031
10. Foldes, S., Hammer, P.L.: Split graphs. In: Proceedings of the Eighth Southeastern Conference on Combinatorics, Graph Theory and Computing. Louisiana State University, Baton Rouge, La., 1977, pp. 311–315. Congressus Numerantium, no. XIX (1977)
11. Forsyth, E., Katz, L.: A matrix approach to the analysis of sociometric data: preliminary report. Sociometry **9**(4), 340–347 (1946). https://doi.org/10.2307/2785498
12. Fortin, D.: Robinsonian matrices: recognition challenges. J. Classif. **34**(2), 191–222 (2017). https://doi.org/10.1007/s00357-017-9230-1
13. Fulkerson, D.R., Gross, O.: Incidence matrices and interval graphs. Pac. J. Math. **15**(3), 835–855 (1965). https://doi.org/10.2140/pjm.1965.15.835
14. Hammer, P.L., Simeone, B.: The splittance of a graph. Combinatorica **1**(3), 275–284 (1981). https://doi.org/10.1007/BF02579333
15. Heggernes, P., Kratsch, D.: Linear-time certifying recognition algorithms and forbidden induced subgraphs. Nord. J. Comput. **14**(1–2), 87–108 (2007)
16. Heggernes, P., Mancini, F.: Minimal split completions. Discrete Appl. Math. **157**(12), 2659–2669 (2009). https://doi.org/10.1016/j.dam.2008.08.010
17. Heggernes, P., Mancini, F., Papadopoulos, C., Sritharan, R.: Strongly chordal and chordal bipartite graphs are sandwich monotone. J. Comb. Optim. **22**(3), 438–456 (2011). https://doi.org/10.1007/s10878-010-9322-x
18. Heggernes, P., Papadopoulos, C.: Single-Edge Monotonic Sequences of Graphs and Linear-Time Algorithms for Minimal Completions and Deletions. In: Lin, G. (ed.) COCOON 2007. LNCS, vol. 4598, pp. 406–416. Springer, Heidelberg (2007). https://doi.org/10.1007/978-3-540-73545-8_40

19. Heggernes, P., Papadopoulos, C.: Single-edge monotonic sequences of graphs and linear-time algorithms for minimal completions and deletions. Theor. Comput. Sci. **410**(1), 1–15 (2009). https://doi.org/10.1016/j.tcs.2008.07.020

20. Huson, D.H., Nettles, S., Warnow, T.J.: Obtaining highly accurate topology estimates of evolutionary trees from very short sequences. In: Istrail, S., Pevzner, P.A., Waterman, M.S. (eds.) Proceedings of the Third Annual International Conference on Research in Computational Molecular Biology, RECOMB 1999, Lyon, France, April 11–14, 1999, pp. 198–207. ACM (1999). https://doi.org/10.1145/299432.299484

21. Ibarra, L.: Fully dynamic algorithms for chordal graphs and split graphs. ACM Trans. Algorithms Art. 40 4(4), 20 (2008). https://doi.org/10.1145/1383369.1383371

22. Kearney, P.E., Hayward, R.B., Meijer, H.: Inferring evolutionary trees from ordinal data. In: Proceedings of the Eighth Annual ACM-SIAM Symposium on Discrete Algorithms, New Orleans, LA, 1997, pp. 418–426. ACM, New York (1997)

23. Khodaverdian, A., Weitz, B., Wu, J., Yosef, N.: Steiner network problems on temporal graphs. arXiv preprint arXiv:1609.04918 (2016)

24. Kloks, T., Kratsch, D.: Treewidth of chordal bipartite graphs. J. Algorithms **19**(2), 266–281 (1995). https://doi.org/10.1006/jagm.1995.1037

25. Laurent, M., Seminaroti, M.: Similarity-first search: a new algorithm with application to Robinsonian matrix recognition. SIAM J. Discrete Math. **31**(3), 1765–1800 (2017). https://doi.org/10.1137/16M1056791

26. Laurent, M., Seminaroti, M., Tanigawa, S.i.: A structural characterization for certifying Robinsonian matrices. Electron. J. Combin. Paper 2.21 **24**(2), 22 (2017). https://doi.org/10.37236/6701

27. Laurent, M., Tanigawa, S.: Perfect elimination orderings for symmetric matrices. Optimization Letters **14**(2), 339–353 (2017). https://doi.org/10.1007/s11590-017-1213-y

28. Liiv, I.: Seriation and matrix reordering methods: an historical overview. Stat. Anal. Data Min. ASA Data Sci. J. **3**(2), 70–91 (2010). https://doi.org/10.1002/sam.10071

29. Looges, P.J., Olariu, S.: Optimal greedy algorithms for indifference graphs. Comput. Math. Appl. **25**(7), 15–25 (1993). https://doi.org/10.1016/0898-1221(93)90308-I

30. Mahadev, N.V.R., Peled, U.N.: Threshold Graphs and Related Topics. Annals Discrete Mathematics, vol. 56. North-Holland Publishing Co., Amsterdam (1995)

31. Mertzios, G.B., Michail, O., Chatzigiannakis, I., Spirakis, P.G.: Temporal network optimization subject to connectivity constraints. In: Fomin, F.V., Freivalds, R., Kwiatkowska, M., Peleg, D. (eds.) ICALP 2013. LNCS, vol. 7966, pp. 657–668. Springer, Heidelberg (2013). https://doi.org/10.1007/978-3-642-39212-2_57

32. Préa, P., Fortin, D.: An optimal algorithm to recognize robinsonian dissimilarities. J. Classif. **31**(3), 351–385 (2014). https://doi.org/10.1007/s00357-014-9150-2

33. Roberts, F.S.: Indifference graphs. In: Proof Techniques in Graph Theory: Proceedings of the Second Ann Arbor Graph Theory Conference, pp. 139–146. Academic Press (1969)

34. Rose, D.J.: Triangulated graphs and the elimination process. J. Math. Anal. Appl. **32**(3), 597–609 (1970). https://doi.org/10.1016/0022-247X(70)90282-9

35. Rose, D.J., Tarjan, R.E., Lueker, G.S.: Algorithmic aspects of vertex elimination on graphs. SIAM J. Comput. **5**(2), 266–283 (1976). https://doi.org/10.1137/0205021

36. Shamir, R., Sharan, R.: A fully dynamic algorithm for modular decomposition and recognition of cographs. Discrete Appl. Math. **136**(2–3), 329–340 (2004). https://doi.org/10.1016/S0166-218X(03)00448-7

37. Spinrad, J., Sritharan, R.: Algorithms for weakly triangulated graphs. Discrete Appl. Math. **59**(2), 181–191 (1995). https://doi.org/10.1016/0166-218X(93)E0161-Q

38. Sritharan, R.: Graph modification problem for some classes of graphs. J. Discrete Algorithms **38**(41), 32–37 (2016). https://doi.org/10.1016/j.jda.2016.06.003

39. Tardos, G.: Extremal theory of vertex or edge ordered graphs. In: Surveys in Combinatorics 2019. London Mathematical Society Lecture Note Series, vol. 456, pp. 221–236. Cambridge University Press, Cambridge (2019). https://doi.org/10.1017/9781108649094.008

40. West, D.B.: Introduction to Graph Theory, vol. 2. Prentice Hall, Upper Saddle River (2001)

41. Yannakakis, M.: Computing the minimum fill-in is NP-complete. SIAM J. Algebraic Discrete Methods **2**(1), 77–79 (1981). https://doi.org/10.1137/0602010

Well-Partitioned Chordal Graphs: Obstruction Set and Disjoint Paths

Jungho Ahn[1,2] ⓘ, Lars Jaffke[3(✉)] ⓘ, O-joung Kwon[2,4] ⓘ, and Paloma T. Lima[3]

[1] Department of Mathematical Sciences, KAIST, Daejeon, South Korea
junghoahn@kaist.co.kr
[2] Discrete Mathematics Group, IBS, Daejeon, South Korea
[3] Department of Informatics, University of Bergen, Bergen, Norway
{lars.jaffke,paloma.lima}@uib.no
[4] Department of Mathematics, Incheon National University, Incheon, South Korea
ojoungkwon@gmail.com

Abstract. We introduce a new subclass of chordal graphs that generalizes split graphs, which we call well-partitioned chordal graphs. Split graphs are graphs that admit a partition of the vertex set into cliques that can be arranged in a star structure, the leaves of which are of size one. Well-partitioned chordal graphs are a generalization of this concept in the following two ways. First, the cliques in the partition can be arranged in a tree structure, and second, each clique is of arbitrary size. We provide a characterization of well-partitioned chordal graphs by forbidden induced subgraphs, and give a polynomial-time algorithm that given any graph, either finds an obstruction, or outputs a partition of its vertex set that asserts that the graph is well-partitioned chordal. We demonstrate the algorithmic use of this graph class by showing that two variants of the problem of finding pairwise disjoint paths between k given pairs of vertices is in FPT parameterized by k on well-partitioned chordal graphs, while on chordal graphs, these problems are only known to be in XP. From the other end, we observe that there are problems that are polynomial-time solvable on split graphs, but become NP-complete on well-partitioned chordal graphs.

Keywords: Well-partitioned chordal graph · Chordal graph · Split graph · Disjoint paths · Forbidden induced subgraphs

1 Introduction

A central methodology in the study of the complexity of computationally hard graph problems is to impose additional structure on the input graphs,

J. Ahn and O. Kwon are supported by the Institute for Basic Science (IBS-R029-C1). O. Kwon is also supported by the National Research Foundation of Korea (NRF) grant funded by the Ministry of Education (No. NRF-2018R1D1A1B07050294). L. Jaffke is supported by the Trond Mohn Foundation (TMS). P. T. Lima is supported by the Research Council of Norway via the project "CLASSIS".

I. Adler and H. Müller (Eds.): WG 2020, LNCS 12301, pp. 148–160, 2020.
https://doi.org/10.1007/978-3-030-60440-0_12

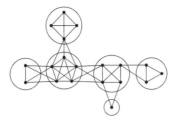

Fig. 1. A well-partitioned chordal graph.

and determine if this can be exploited in the design of an efficient algorithm. Typically, one restricts the input to be contained in a *graph class*, which is a set of graphs that share a common structural property. Following the establishment of the theory of NP-hardness, numerous problems were investigated in specific classes of graphs; either providing a polynomial-time algorithm for a problem Π on a specific graph class, while Π is NP-hard in a more general setting, or showing that Π remains NP-hard on a graph class. A key question in this field is to find for a given problem Π that is hard on a graph class $\mathcal{A}$, a subclass $\mathcal{B} \subsetneq \mathcal{A}$ such that Π is efficiently solvable on $\mathcal{B}$. Naturally, the goal is to narrow down the gap $\mathcal{A} \setminus \mathcal{B}$ as much as possible, and several notions of hardness/efficiency can be applied. For instance, we can require our target problem to be NP-hard on $\mathcal{A}$ and polynomial-time solvable on $\mathcal{B}$; or, from the viewpoint of parameterized complexity [6,7], we require a target parameterized problem Π to be W[1]-hard on $\mathcal{A}$, while Π is in FPT on $\mathcal{B}$, or a separation in the kernelization complexity [8] of Π between $\mathcal{A}$ and $\mathcal{B}$.

Chordal graphs are arguably one of the main characters in the algorithmic study of graph classes. They find applications for instance in computational biology [21] and sparse matrix computations [10]. Split graphs are an important subclass of chordal graphs. The complexities of computational problems on chordal and split graphs often coincide, however, this is not always the case. For instance, several variants of graph (vertex) coloring problems are polynomial-time solvable on split graphs and NP-hard on chordal graphs, see the works of Havet et al. [12], and of Silva [22]. Also, the SPARSEST k-SUBGRAPH [24] and DENSEST k-SUBGRAPH [5] problems are polynomial-time solvable on split graphs and NP-hard on chordal graphs. Other problems, for instance the TREE 3-SPANNER problem [3], are easy on split graphs, while their complexity on chordal graphs is still unresolved.

In this work, we introduce the class of *well-partitioned chordal graphs*, a subclass of chordal graphs that generalizes split graphs, which can be used as a tool for narrowing down complexity gaps for problems that are hard on chordal graphs, and easy on split graphs. The definition of well-partitioned chordal graphs is mainly motivated by a property of split graphs: the vertex set of a split graph can be partitioned into sets that can be viewed as a central clique of arbitrary size and cliques of size one that have neighbors only in the central clique. Thus, this partition has the structure of a star. Well-partitioned chordal

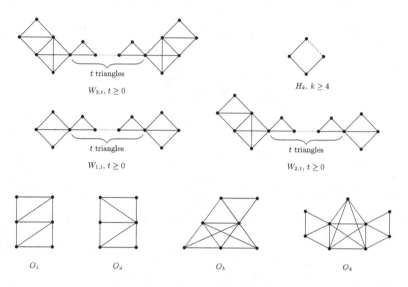

Fig. 2. The set of obstructions $\mathbb{O}$ for well-partitioned chordal graphs.

graphs relax these ideas in two ways: by allowing the parts of the partition to be arranged in a tree structure instead of a star, and by allowing the cliques in each part to have arbitrary size. The interaction between adjacent parts P and Q remains simple: it induces a complete bipartite graph between a subset of P, and a subset of Q. Such a tree structure is called a partition tree, and we give an example of a well-partitioned chordal graph in Fig. 1. Now, it is not difficult to observe that the graphs constructed in the NP-hardness proofs in the works [12, 22] are in fact well-partitioned chordal graphs.

The main structural contribution of this work is a characterization of well-partitioned chordal graphs by forbidden induced subgraphs (see Fig. 2).

Theorem 1. *A graph is a well-partitioned chordal graph if and only if it has no induced subgraph isomorphic to a graph in $\mathbb{O}$. Furthermore, there is a polynomial-time algorithm that given a graph G, outputs either an induced subgraph of G isomorphic to a graph in $\mathbb{O}$, or a partition tree for each connected component which confirms that G is a well-partitioned chordal graph.*

Before we proceed with the discussion of the algorithmic results of this paper, we would like to briefly touch on the relationship of well-partitioned chordal graphs and width parameters. Each split graph is a well-partitioned chordal graph, and there are split graphs of whose *maximum induced matching width* (mim-width) depends linearly on the number of vertices [17]. This rules out the applicability of any algorithmic meta-theorem based on one of the common width parameters such as tree-width or clique-width, to the class of well-partitioned chordal graphs. It is known that mim-width is a lower bound for them [23].

Besides narrowing the complexity gap between the classes of chordal and split graphs, the class of well-partitioned chordal graphs can also be useful as

Table 1. Complexity of the DISJOINT PATHS and SET-RESTRICTED DISJOINT PATHS problems parameterized by the number k of terminal pairs. Size bounds for kernels are in terms of the number of *vertices* of the kernelized instances. *Given a partition tree.

Graph Class	DISJOINT PATHS	SET-RESTRICTED DISJOINT PATHS
Chordal	linear FPT [14]	XP [1]
Well-partitioned chordal	$\mathcal{O}(k^3)$ kernel [T. 5]	linear* FPT [T. 2, 3]
Split	$\mathcal{O}(k^2)$ kernel [13]	$\mathcal{O}(k^2)$ kernel [C. 1]

a step towards determining the yet unresolved complexity of a problem Π on chordal graphs when it is known that Π is easy on split graphs. This is the case in our current work. Specifically, we study the DISJOINT PATHS problem, formally defined as follows, and generalizations thereof. Two paths P_1 and P_2 are called *internally vertex-disjoint*, if for $i \in [2]$, no internal vertex of P_i is contained in P_{3-i}. (Note that this excludes the possibility that an endpoint of one path is used as an internal vertex in the other path.)

The classical DISJOINT PATHS problem takes as input a graph G and a set $\mathcal{X} = \{(s_1, t_1), \ldots, (s_k, t_k)\}$ of k pairs of vertices of G, called *terminals*, and asks whether G contain k pairwise internally vertex-disjoint paths $P_1, \ldots, P_k$ such that for all $i \in [k]$, P_i is an (s_i, t_i)-path. This problem has already been shown by Karp to be NP-complete [15], and as a cornerstone result in the early days of fixed-parameter tractability theory, Robertson and Seymour showed that DISJOINT PATHS parameterized by k is in FPT [16,19]. From the viewpoint of kernelization complexity, Bodlaender et al. showed that DISJOINT PATHS does not admit a polynomial kernel unless NP $\subseteq$ coNP/poly [2].

Restricting the problem to chordal and split graphs, Heggernes et al. showed that DISJOINT PATHS remains NP-complete on split graphs, and that it admits a polynomial kernel parameterized by k [13], and Kammer and Tholey showed that it has an FPT-algorithm with linear dependence on the size of the input chordal graph [14]. The question whether DISJOINT PATHS has a polynomial kernel on chordal graphs remains open. We go one step towards such a polynomial kernel, by showing that DISJOINT PATHS has a polynomial kernel on well-partitioned chordal graphs; generalizing the polynomial kernel on split graphs [13].

We also study a generalization of the DISJOINT PATHS problem, where in a solution, each path P_i can only use a restricted set of vertices U_i, which is specified for each terminal pair at the input. This problem was recently introduced by Belmonte et al. and given the name SET-RESTRICTED DISJOINT PATHS [1]. Since this problem contains DISJOINT PATHS as a special case (setting all domains equal to the whole vertex set), it is NP-complete. Belmonte et al. showed that SET-RESTRICTED DISJOINT PATHS parameterized by k is in XP on chordal graphs, and leave as an open question whether it is in FPT or W[1]-hard on chordal graphs. Towards showing the former, we give an FPT-algorithm on well-partitioned chordal graphs. While we do not settle the kernelization complexity of SET-RESTRICTED DISJOINT PATHS on well-partitioned chordal graphs, we observe that our FPT-algorithm implies a polynomial kernel on split graphs. We summarize these results in Table 1.

Finally, we also consider the SET-RESTRICTED DISJOINT CONNECTED SUB-GRAPHS problem where we are given k terminal *sets* instead of pairs, and k domains, and the question is whether there are k pairwise disjoint connected subgraphs, each one connecting one of the terminal sets, using only vertices from the specified domain. This problem was also introduced in [1] and shown to be in XP on chordal graphs, when the parameter is the total number of vertices in all terminal sets. Extending our ideas of the above mentioned algorithms, we show that this problem is in fact FPT on well-partitioned chordal graphs.

Throughout, proofs of statements marked '♣' are deferred to the full version.

2   Preliminaries

For a positive integer n, we let $[n] := \{1, 2, \ldots, n\}$. All graphs considered here are simple and finite. For a graph G we denote by $V(G)$ and $E(G)$ the vertex set and edge set of G, respectively. Given $uv \in E(G)$, we call u and v its *endpoints*. Let G and H be two graphs. For a vertex v of a graph G, $N_G(v) := \{w \in V(G) \mid vw \in E(G)\}$ is the set of *neighbors* of v in G. The *degree* of v is $\deg_G(v) := |N_G(v)|$. The *subgraph induced by* X, denoted by $G[X]$, is the graph $(X, \{uv \in E(G) \mid u, v \in X\})$. We denote by $G - X$ the graph $G[V(G) \setminus X]$, and for a single vertex $x \in V(G)$, we use the shorthand '$G - x$' for '$G - \{x\}$'. For two sets $X, Y \subseteq V(G)$, we denote by $G[X, Y]$ the graph $(X \cup Y, \{xy \in E(G) \mid x \in X, y \in Y\})$. We say that X is *complete to* Y if $X \cap Y = \emptyset$ and each vertex in X is adjacent to every vertex in Y. Let G be a graph. We say that G is *complete* if $uv \in E(G)$ for every $u, v \in V(G)$. A set $X \subseteq V(G)$ is a *clique* if $G[X]$ is complete. A graph G is *connected* if for each 2-partition (X, Y) of $V(G)$ with $X \neq \emptyset$ and $Y \neq \emptyset$, there is a pair $x \in X$, $y \in Y$ such that $xy \in E(G)$. A tree with at most one vertex of degree at least two is a *star*.

A *hole* in a graph G is an induced cycle of G of length at least 4. A graph is *chordal* if it has no hole as an induced subgraph. A vertex is *simplicial* if $N_G(v)$ is a clique. We say that a graph G has a *perfect elimination ordering* $v_1, \ldots, v_n$ if v_i is simplicial in $G[\{v_i, v_{i+1}, \ldots, v_n\}]$ for each $i \in [n-1]$. It is known that a graph is chordal if and only if it has a perfect elimination ordering [9]. A graph G is a *split graph* if there is a 2-partition (C, I) of $V(G)$ such that C is a clique and I is an independent set. For a family $\mathcal{S}$ of subsets of some set, the *intersection graph of* $\mathcal{S}$ is the graph on vertex set $\mathcal{S}$ and edge set $\{ST \mid S, T \in \mathcal{S}$ and $S \cap T \neq \emptyset\}$.

3   Well-Partitioned Chordal Graphs

A connected graph G is a *well-partitioned chordal graph* if there exist a partition $\mathcal{P}$ of $V(G)$ and a tree $\mathcal{T}$ having $\mathcal{P}$ as a vertex set such that the following hold.

(i) Each part $X \in \mathcal{P}$ is a clique in G.
(ii) For each edge $XY \in E(\mathcal{T})$, there are subsets $X' \subseteq X$ and $Y' \subseteq Y$ such that $E(G[X, Y]) = \{xy \mid x \in X', y \in Y'\}$.
(iii) For each pair of distinct $X, Y \in V(\mathcal{T})$ with $XY \notin E(\mathcal{T})$, $E(G[X, Y]) = \emptyset$.

The tree $\mathcal{T}$ is called a *partition tree of G*, and the elements of $\mathcal{P}$ are called its *bags*. A graph is a well-partitioned chordal graph if all of its connected components are well-partitioned chordal graphs. We remark that a well-partitioned chordal graph can have more than one partition tree. Also, observe that well-partitioned chordal graphs are closed under taking induced subgraphs.

A useful concept when considering partition trees of well-partitioned chordal graphs is that of a *boundary of a bag*. Let $\mathcal{T}$ be a partition tree of a well-partitioned chordal graph G and let $X, Y \in V(\mathcal{T})$ be two bags that are adjacent in $\mathcal{T}$. The *boundary of X with respect to Y*, denoted by $\mathrm{bd}(X, Y)$, is the set of vertices of X that have a neighbor in Y, i.e. $\mathrm{bd}(X, Y) := \{x \in X \mid N_G(x) \cap Y \neq \emptyset\}$. By item (ii) of the definition of the class, $\mathrm{bd}(X, Y)$ is complete to $\mathrm{bd}(Y, X)$.

We now consider the relation between well-partitioned chordal graphs and other well-studied classes of graphs. It is easy to see that every well-partitioned chordal graph G is a chordal graph because every leaf of the partition tree of a component of G contains a simplicial vertex of G, and after removing this vertex, the remaining graph is still a well-partitioned chordal graph. Thus, we may construct a perfect elimination ordering. We show that, in fact, well-partitioned chordal graphs constitute a subclass of substar graphs. A graph is a *substar graph* [4] if it is an intersection graph of substars of a tree.

Proposition 1 (♣). *Every well-partitioned chordal graph is a substar graph.*

From the definition of well-partitioned chordal graphs, one can also see that every split graph is a well-partitioned chordal graph. Indeed, if G is a split graph with clique K and independent set S, the partition tree of G is a star, with the clique K as its central bag and each vertex of S contained in a different leaf bag. We show that, in fact, every starlike graph is a well-partitioned chordal graph. A *starlike graph* [11] is an intersection graph of substars of a star.

Proposition 2 (♣). *Every starlike graph is a well-partitioned chordal graph.*

We show that the graph O_1 in Fig. 2 is not a well-partitioned chordal graph. On the other hand, O_1 is a substar graph. Also a path graph on 5 vertices is a well-partitioned chordal graph but not a starlike graph. These observations with Propositions 1 and 2 show that we have the following hierarchy:

$$\begin{matrix} \text{split} \\ \text{graphs} \end{matrix} \underset{\neq}{\subseteq} \begin{matrix} \text{starlike} \\ \text{graphs} \end{matrix} \underset{\neq}{\subseteq} \begin{matrix} \text{well-partitioned} \\ \text{chordal graphs} \end{matrix} \underset{\neq}{\subseteq} \begin{matrix} \text{substar} \\ \text{graphs} \end{matrix} \underset{\neq}{\subseteq} \begin{matrix} \text{chordal} \\ \text{graphs} \end{matrix}$$

4   Characterization by Forbidden Induced Subgraphs

This section is entirely devoted to the proof of Theorem 1. That is, we show that the set $\mathbb{O}$ of graphs depicted in Fig. 2 is the set of all forbidden induced subgraphs for well-partitioned chordal graphs, and give a polynomial-time recognition algorithm for this graph class. For convenience, we say that an induced subgraph of a graph that is isomorphic to a graph in $\mathbb{O}$ is an *obstruction* for well-partitioned chordal graphs, or simply an *obstruction*.

Proposition 3 (♣). *The graphs in $\mathbb{O}$ are not well-partitioned chordal graphs.*

In the rest, we outline the implementation of the algorithm, which also proves that the set $\mathbb{O}$ is a complete set of forbidden induced subgraphs of well-partitioned chordal graphs.

Proposition 4. *Given a graph G, one can in polynomial time output either an obstruction in G or a partition tree of each connected component of G confirming that G is a well-partitioned chordal graph.*

We introduce the main concept in the algorithm, called a boundary-crossing path. Let G be a connected well-partitioned chordal graph with a partition tree $\mathcal{T}$. For a bag X of $\mathcal{T}$ and $B \subseteq X$, a vertex $z \in V(G) \setminus X$ is said to *cross B in X*, if it has a neighbor both in B and in $X \setminus B$. In this case, we also say that B has a crossing vertex. In the following definitions, a path $X_1 X_2 \ldots X_\ell$ in $\mathcal{T}$ is considered to be ordered from X_1 to X_ℓ. Let $\ell \geq 3$ be an integer. A path $X_1 X_2 \ldots X_\ell$ in $\mathcal{T}$ is called a *boundary-crossing path* if for each $1 \leq i \leq \ell - 2$, there is a vertex in X_i that crosses $\mathrm{bd}(X_{i+1}, X_{i+2})$. If for each $1 \leq i \leq \ell - 2$, there is no bag $Y \in V(\mathcal{T}) \setminus \{X_i\}$ containing a vertex that crosses $\mathrm{bd}(X_{i+1}, X_{i+2})$, then we say the path is *exclusive*. If for each $1 \leq i \leq \ell - 2$, $\mathrm{bd}(X_i, X_{i+1})$ is complete to X_{i+1}, then we say the path is *complete*. If a boundary-crossing path is both complete and exclusive, then we call it *good*. For convenience, we say that any path in $\mathcal{T}$ with at most two bags is a boundary-crossing path.

The outline of the recognition algorithm is as follows. First we may assume that a given graph G is chordal, otherwise we find a hole in polynomial time [18]. We may also assume that G is connected. So, it has a simplicial vertex v, and by an inductive argument, we can assume that $G - v$ is a well-partitioned chordal graph. As v is simplicial, $G - v$ is also connected, and thus it admits a partition tree $\mathcal{T}$. If v has neighbors in one bag of $\mathcal{T}$, then we can simply put v as a new bag adjacent to that bag. Thus, we may assume that v has neighbors in two distinct bags, say C_1 and C_2. Then our algorithm is divided into three parts:

1. We find a maximal good boundary-crossing path ending in $C_2 C_1$ (or $C_1 C_2$). To do this, given a good boundary-crossing path $C_i C_{i-1} \ldots C_2 C_1$, find a bag C_{i+1} containing a vertex crossing $\mathrm{bd}(C_i, C_{i-1})$. If there is no such bag, then this path is maximal. Otherwise, we argue that in polynomial time either we can find an obstruction, or verify that $C_{i+1} C_i \ldots C_2 C_1$ is good.

2. Assume that $C_k C_{k-1} \ldots C_2 C_1$ is the obtained maximal good boundary-crossing path. Then we can in polynomial time modify $\mathcal{T}$ so that no vertex crosses $\mathrm{bd}(C_2, C_1)$.
3. We show that if no vertex crosses $\mathrm{bd}(C_2, C_1)$ and no vertex crosses $\mathrm{bd}(C_1, C_2)$, then we can extend $\mathcal{T}$ to a partition tree of G.

Steps 2 and 3 can be handled immediately. Step 1 is the most technically involved one. We first prove a handful of auxiliary lemmas that we can use to find pieces of obstructions in boundary-crossing paths that are not good, and puzzle them together. We separately deal with the following three cases, and in each case, we show that either one can in polynomial time find an obstruction or output a partition tree of G.

- (Lemma A) $C_1 \subseteq N_G(v)$.
- (Lemma B) $\mathrm{bd}(C_1, C_2) \setminus N_G(v) \neq \emptyset$ and $C_2 \setminus N_G(v) \neq \emptyset$.
- (Lemma C) $C_1 \setminus N_G(v) \neq \emptyset$, $C_2 \setminus N_G(v) \neq \emptyset$ and $N_G(v) = \mathrm{bd}(C_1, C_2) \cup \mathrm{bd}(C_2, C_1)$.

In the proofs of these lemmas, we crucially use the aforementioned auxiliary lemmas that came out of our line of attack at Step 1 above. We sketch the idea of the proof of Lemma A.

Proof (Sketch of the proof of Lemma A). Since v is a simplicial vertex, we have that $\mathrm{bd}(C_1, C_2) = C_1$. If $N_G(v) \cap C_2 = \mathrm{bd}(C_2, C_1)$, then we can obtain a partition tree for G by adding v to C_1. Thus, we may assume that $N_G(v) \cap C_2 \neq \mathrm{bd}(C_2, C_1)$. Assume that $C_2 = \mathrm{bd}(C_2, C_1)$. Since $\mathrm{bd}(C_2, C_1)$ is complete to C_1, we have that $C_1 \cup C_2$ is a clique. Hence, we can obtain a partition tree $\mathcal{T}'$ for G from $\mathcal{T}$ by removing C_1 and C_2, adding a new bag $C^* = C_1 \cup C_2$, making all neighbors of C_1 and C_2 in $\mathcal{T}$ adjacent to C^*, and adding a new bag $C_v := \{v\}$ and making C_v adjacent to C^*. Thus, we may assume that $C_2 \setminus \mathrm{bd}(C_2, C_1) \neq \emptyset$.

Since $C_1 = \mathrm{bd}(C_1, C_2)$, no vertex of $G - v$ crosses $\mathrm{bd}(C_1, C_2)$. If no vertex of $G - v$ crosses $\mathrm{bd}(C_2, C_1)$, then using Step 3, we can obtain a partition tree for G in polynomial time. Thus, we may assume that there is a bag C_3 having a vertex that crosses $\mathrm{bd}(C_2, C_1)$. So, $C_3 C_2 C_1$ is a boundary-crossing path.

We find either an obstruction or a maximal good boundary-crossing path ending in $C_3 C_2 C_1$. First check whether $\mathrm{bd}(C_3, C_2)$ is complete to C_2. Otherwise, choose a vertex $p \in \mathrm{bd}(C_3, C_2)$, and a non-neighbor q of p in C_2. As p crosses $\mathrm{bd}(C_2, C_1)$, p has a neighbor a in $C_2 \setminus \mathrm{bd}(C_2, C_1)$ and a neighbor b in $\mathrm{bd}(C_2, C_1)$. There are three possibilities; q is contained in one of $N_G(v) \cap C_2$, $\mathrm{bd}(C_2, C_1) \setminus N_G(v)$, or $C_2 \setminus \mathrm{bd}(C_2, C_1)$. In each case, we can find an obstruction. So, we may assume that $\mathrm{bd}(C_3, C_2)$ is complete to C_2. Next, we check if there exists another neighbor bag $D \neq C_3$ of C_2 having a vertex q that crosses $\mathrm{bd}(C_2, C_1)$. In this case, we can find O_3. Otherwise, $C_3 C_2 C_1$ is a good boundary-crossing path.

We now extend a given good boundary-crossing path $C_i C_{i-1} \cdots C_2 C_1$ by recursively finding a bag C_{i+1} having a vertex crossing $\mathrm{bd}(C_i, C_{i-1})$, and if the new sequence is not good, then we output an obstruction. This recursive step stops at some point, and we end up with a maximal good boundary-crossing path $C_k C_{k-1} \cdots C_1$. Now, it is not difficult to see that replacing the sequence $C_k, C_{k-1}, \ldots, C_1$ with $C_k \setminus \mathrm{bd}(C_k, C_{k-1})$, $\mathrm{bd}(C_k, C_{k-1}) \cup (C_{k-1} \setminus \mathrm{bd}(C_{k-1}, C_{k-2}))$, $\ldots$, $\mathrm{bd}(C_2, C_1) \cup C_1$ makes a new tree partition where no vertex crosses $\mathrm{bd}(C_2', C_1')$ where C_2' and C_1' are the new last two bags. Then we can apply Step 3 to obtain a partition tree for the entire graph G in polynomial time. □

Proof (of Proposition 4). We use the polynomial-time algorithm of [18] to find a hole[1] in G if one exists. We may assume that G is chordal. Since a graph is a well-partitioned chordal graph if and only if its connected components are well-partitioned chordal graphs, it is sufficient to show it for each connected component. From now on, we assume that G is connected. We can find a perfect elimination ordering $(v_1, v_2, \ldots, v_n)$ of G in polynomial time [20].

For each $i \in \{1, 2, \ldots, n\}$, let $G_i := G[\{v_i, v_{i+1}, \ldots, v_n\}]$. Observe that since G is connected and v_i is simplicial in G_i for all $1 \le i \le n-1$, each G_i is connected. From $i = n$ to 1, we recursively find either an obstruction or a partition tree of G_i. Clearly, G_n admits a partition tree. Let $1 \le i \le n - 1$, and assume that we obtained a partition tree $\mathcal{T}$ of G_{i+1}. Recall that v_i is simplicial in G_i.

Since v_i is simplicial in G_i, $N_{G_i}(v_i)$ is a clique. This implies that there are at most two bags in $V(\mathcal{T})$ that have a non-empty intersection with $N_{G_i}(v_i)$. If there is only one such bag in $V(\mathcal{T})$, say C, we can construct a partition tree for G_i by simply adding a bag consisting of v_i and making it adjacent to C.

Hence, from now on, we can assume that there are precisely two distinct adjacent bags $C_1, C_2 \in V(\mathcal{T})$ that have a non-empty intersection with $N_{G_i}(v_i)$. As $N_{G_i}(v_i)$ is a clique, we can observe that $N_{G_i}(v_i) \subseteq \mathrm{bd}(C_1, C_2) \cup \mathrm{bd}(C_2, C_1)$.

If $C_1 \subseteq N_{G_i}(v_i)$ or $C_2 \subseteq N_{G_i}(v_i)$, then by Lemma A, we can in polynomial time either output an obstruction or output a partition tree of G_i. Thus, we may assume that $C_1 \setminus N_{G_i}(v_i) \ne \emptyset$ and $C_2 \setminus N_{G_i}(v_i) \ne \emptyset$. If $\mathrm{bd}(C_1, C_2) \setminus N_{G_i}(v_i) \ne \emptyset$ or $\mathrm{bd}(C_2, C_1) \setminus N_{G_i}(v_i) \ne \emptyset$, then by Lemma B, we can in polynomial time either output an obstruction or output a partition tree of G_i. Thus, we may further assume that $\mathrm{bd}(C_1, C_2) \setminus N_{G_i}(v_i) = \emptyset$ and $\mathrm{bd}(C_2, C_1) \setminus N_{G_i}(v_i) = \emptyset$. Then by Lemma C, we can in polynomial time either output an obstruction or output a partition tree of G_i, and this concludes the proposition. □

5   Algorithmic Applications

In this section, we give several FPT-algorithms and kernels for problems on well-partitioned chordal graphs. Specifically, we consider variants of the DIS-JOINT PATHS problem, called the SET-RESTRICTED DISJOINT PATHS and SET-RESTRICTED TOTALLY DISJOINT PATHS problems, where each path additionally has to be from a predefined domain. Recall that P_1 and P_2 are internally vertex-disjoint, if for $i \in [2]$, $(V(P_i) \setminus \{s_i, t_i\}) \cap V(P_{3-i}) = \emptyset$. Given a graph G, a set

---

[1] Note that holes in the sense of [18] are chordless cycles on at least five vertices; we can check for C_4 separately by brute force. While there are algorithms that verify chordality more directly, we use this procedure to fulfil the promise that we can always output an obstruction if there is one.

$\mathcal{X} = \{(s_1, t_1), \ldots, (s_k, t_k)\}$ of k pairs of vertices of G, called *terminals*, and a set $\mathcal{U} = \{U_1, \ldots, U_k\}$ of k vertex subsets of G, called *domains*, the SET-RESTRICTED DISJOINT PATHS problem asks if G contains k pairwise internally vertex-disjoint paths $P_1, \ldots, P_k$ such that for $i \in [k]$, P_i is an (s_i, t_i)-path with $V(P_i) \subseteq U_i$.

First, we can remove any adjacent terminal pair from the input, since we can always use the corresponding edge as the path in a solution. Next, we observe that finding pairwise internally vertex-disjoint paths is equivalent to finding pairwise internally vertex-disjoint *induced* paths. We call such a solution a *minimal* solution. We then use the following marking procedure. For each $i \in [k]$, we consider the path in $\mathcal{T}$ that connects that bag containing s_i with the bag containing t_i. For each edge $C_1 C_2$ on the path, we mark a maximal subset of $U_i \cap \mathrm{bd}(C_1, C_2)$ of size at most $2k$, and a maximal subset of $U_i \cap \mathrm{bd}(C_2, C_1)$ of size at most $2k$. We show that if our instance is a YES-instance, then it has some minimal solution that only uses marked vertices. We can therefore guess the intersection of such a solution with each bag, and we only have to consider its marked vertices. Formally, this is captured by the following notion.

Definition 1 (*I*-Feasible Bag). *Let $I \subseteq [k]$. Let $B \in V(\mathcal{T})$ be a bag and $M_i \subseteq V(G)$, $i \in I$, be sets of vertices. Then, we say that B is I-feasible w.r.t. $\{M_i \mid i \in I\}$, if there is a set $X \subseteq B$ and a labeling $\lambda: X \to [k]$ such that the following hold. For each $i \in I$ such that B lies on the path from the bag containing s_i to the bag containing t_i in $\mathcal{T}$, and each neighbor C of B on that path, either $\{s_i, t_i\} \cap \mathrm{bd}(B, C) \neq \emptyset$, or there is a vertex $x_i \in X \cap M_i \cap \mathrm{bd}(B, C)$ such that $\lambda(x_i) = i$. We use the shorthand 'feasible' for '$[k]$-feasible'.*

The algorithm works as follows. We apply the above marking procedure to obtain the marked sets $M_1, \ldots, M_k$. Note that for each bag B, $|B \cap \bigcup_{i \in [k]} M_i| = \mathcal{O}(k^2)$: for each $i \in [k]$ we marked at most $4k$ vertices in B, and only if B lies on the path from the bag containing s_i to the bag containing t_i in $\mathcal{T}$. Then, for each bag $B \in V(\mathcal{T})$, we verify whether B is feasible w.r.t. $M_1, \ldots, M_k$. If this is the case for all bags, then we conclude that we are dealing with a YES-instance, and otherwise, that we are dealing with a NO-instance.

Theorem 2 (♣). *There is an algorithm that solves each instance $(G, k, \mathcal{X}, \mathcal{U})$ of SET-RESTRICTED DISJOINT PATHS where G is a well-partitioned chordal graph given along with a partition tree $\mathcal{T}$, in time $2^{\mathcal{O}(k \log k)} \cdot n$.*

In the SET-RESTRICTED TOTALLY DISJOINT PATHS problem, we additionally require the paths in a solution to be pairwise distinct, i.e. if there is an edge xy in the graph and $\{s_i, t_i\} = \{s_j, t_j\} = \{x, y\}$, then only one of the paths P_i and P_j may consist of the edge xy. We call edges xy such that for some $w \geq 2$, $\{x, y\} = \{s_{i_1}, t_{i_1}\} = \ldots = \{s_{i_w}, t_{i_w}\}$ a *heavy edge of weight w*. We call the indices $i_1, \ldots, i_w$ *heavy indices*. Instead of looking for minimal solutions, we look for *minimum* solutions, meaning that no other solution has fewer edges. In such a solution, in any chordal graph, the paths corresponding to a heavy edge of weight w are $w - 1$ paths of length two, and one path consisting only of the edge itself. For each such index i_j, either s_{i_j} and t_{i_j} are in a common bag, or they are contained in the union of the boundaries of adjacent bags. In the former case, the middle vertex of a length two path may be from the bag itself, or from one of its neighbors, if both terminals are in the boundary to that neighbor. In the latter case, the middle vertex has to be in the union of the boundaries.

These observations allow for an adaption of the marking procedure to take into account heavy indices; the remaining indices can be treated as before. The algorithm works as follows. First, we apply the adapted marking procedure to obtain $M_1, \ldots, M_k$. Then, we guess the part of the solution corresponding to heavy indices, again among the marked vertices. Let I be the indices that are not heavy. It then suffices to check whether for one of these guesses, all bags are I-feasible w.r.t. $\{M_i \mid i \in I\}$. If we have a successful guess, then we conclude that we have a YES-instance, and otherwise, that we have a NO-instance.

Theorem 3 (♣). *There is an algorithm that solves each instance $(G, k, \mathcal{X}, \mathcal{U})$ of* SET-RESTRICTED TOTALLY DISJOINT PATHS *where G is a well-partitioned chordal graph given along with a partition tree $\mathcal{T}$, in time $2^{\mathcal{O}(k \log k)} \cdot n$.*

We then observe that the techniques used in the previous algorithms can solve the more general SET-RESTRICTED DISJOINT CONNECTED SUBGRAPHS problem on well-partitioned chordal graphs as well. Here, the parameter s is the sum of the sizes of all terminal sets.

Theorem 4 (♣). *There is an algorithm that solves each instance $(G, k, \mathcal{X}, \mathcal{U})$ of* SET-RESTRICTED DISJOINT CONNECTED SUBGRAPHS *where G is a well-partitioned chordal graph given with a partition tree $\mathcal{T}$, in time $2^{\mathcal{O}(s \log s)} \cdot n$.*

As a consequence of the marking procedures, we have the following polynomial kernels on split graphs.

Corollary 1 (♣). SET-RESTRICTED DISJOINT PATHS *and* SET-RESTRICTED TOTALLY DISJOINT PATHS *on split graphs admit kernels on $\mathcal{O}(k^2)$ vertices.*

Moreover, with two more reduction rules, we obtain polynomial kernels on well-partitioned chordal graphs. This can be seen as follows. The subgraph of the partition tree that only has bags with marked vertices has at most $2k$ degree one bags, and therefore $\mathcal{O}(k)$ bags of degree at least three. In the DISJOINT PATHS and TOTALLY DISJOINT PATHS problems, where we do not need to consider the domains of the paths, we can get rid of degree two bags that do not contain terminals as follows. If in such a degree two bags, the boundaries are large enough, then we can always bypass that bag in any solution. If one of the boundaries is too small, then no solution can pass through the bag.

Theorem 5 (♣). DISJOINT PATHS *and* TOTALLY DISJOINT PATHS *on well-partitioned chordal graphs parameterized by k admit kernels on $\mathcal{O}(k^3)$ vertices.*

6   Conclusions

In this paper, we introduced the class of *well-partitioned chordal graphs*, a subclass of chordal graphs that generalizes split graphs. We provided a characterization by a set of forbidden induced subgraphs which also gave a polynomial-time recognition algorithm, together with algorithmic applications in variants of the DISJOINT PATHS problem. Another typical characterization of (subclasses of) chordal graphs is via vertex orderings. For instance, chordal graphs are famously characterized as the graphs admitting perfect elimination orderings [9]. It would be interesting to see if well-partitioned chordal graphs admit a concise characterization in terms of vertex orderings as well. While the degree of the polynomial in the runtime of our recognition algorithm is moderate, our algorithm does not run in linear time. We therefore ask if it is possible to recognize well-partitioned chordal graphs in linear time; and note that a characterization in terms of vertex orderings can be a promising step in this direction.

References

1. Belmonte, R., Golovach, P.A., Heggernes, P., Hof, P.V., Kamiński, M., Paulusma, D.: Detecting fixed patterns in chordal graphs in polynomial time. Algorithmica **69**(3), 501–521 (2014)
2. Bodlaender, H.L., Thomassé, S., Yeo, A.: Kernel bounds for disjoint cycles and disjoint paths. Theoret. Comput. Sci. **412**(35), 4570–4578 (2011)
3. Brandstädt, A., Dragan, F.F., Le, H.O., Le, V.B.: Tree spanners on chordal graphs: complexity and algorithms. Theoret. Comput. Sci. **310**(1–3), 329–354 (2004)
4. Chang, Y.W., Jacobson, M.S., Monma, C.L., West, D.B.: Subtree and substar intersection numbers. Discret. Appl. Math. **44**(1–3), 205–220 (1993)
5. Corneil, D.G., Perl, Y.: Clustering and domination in perfect graphs. Discret. Appl. Math. **9**(1), 27–39 (1984)
6. Cygan, M.: Parameterized Algorithms. Springer, Cham (2015). https://doi.org/10.1007/978-3-319-21275-3
7. Downey, R.G., Fellows, M.R.: Fundamentals of Parameterized Complexity. TCS. Springer, London (2013). https://doi.org/10.1007/978-1-4471-5559-1
8. Fomin, F.V., Lokshtanov, D., Saurabh, S., Zehavi, M.: Kernelization. Cambridge University Press, Cambridge (2019)
9. Fulkerson, D., Gross, O.: Incidence matrices and interval graphs. Pac. J. Math. **15**(3), 835–855 (1965)
10. George, A., Gilbert, J.R., Liu, J.W.: Graph Theory and Sparse Matrix Computation. IMA, vol. 56. Springer, New York (2012). https://doi.org/10.1007/978-1-4613-8369-7
11. Gustedt, J.: On the pathwidth of chordal graphs. Discret. Appl. Math. **45**(3), 233–248 (1993)
12. Havet, F., Sales, C.L., Sampaio, L.: b-coloring of tight graphs. Discret. Appl. Math. **160**(18), 2709–2715 (2012)
13. Heggernes, P., van't Hof, P.V., van Leeuwen, E.J., Saei, R.: Finding disjoint paths in split graphs. Theor. Comput. Syst. **57**(1), 140–159 (2015)
14. Kammer, F., Tholey, T.: The k-disjoint paths problem on chordal graphs. In: Paul, C., Habib, M. (eds.) WG 2009. LNCS, vol. 5911, pp. 190–201. Springer, Heidelberg (2010). https://doi.org/10.1007/978-3-642-11409-0_17
15. Karp, R.M.: On the computational complexity of combinatorial problems. Networks **5**(1), 45–68 (1975)
16. Kawarabayashi, K.I., Kobayashi, Y., Reed, B.: The disjoint paths problem in quadratic time. J. Combin. Theor. Ser. B **102**(2), 424–435 (2012)

17. Mengel, S.: Lower bounds on the mim-width of some graph classes. Discret. Appl. Math. **248**, 28–32 (2018)
18. Nikolopoulos, S.D., Palios, L.: Detecting holes and antiholes in graphs. Algorithmica **47**(2), 119–138 (2007)
19. Robertson, N., Seymour, P.D.: Graph minors. XIII. The disjoint paths problem. J. Combin. Theor. Ser. B **63**(1), 65–110 (1995)
20. Rose, D.J., Tarjan, R.E., Lueker, G.S.: Algorithmic aspects of vertex elimination on graphs. SIAM J. Comput. **5**(2), 266–283 (1976)
21. Semple, C., Steel, M.: Phylogenetics. Oxford Lecture Series in Mathematics and its Applications, vol. 24. Oxford University Press, Oxford (2003)
22. Silva, A.: Graphs with small fall-spectrum. Discret. Appl. Math. **254**, 183–188 (2019)
23. Vatshelle, M.: New Width Parameters of Graphs. Ph.D. thesis, University of Bergen, Norway (2012)
24. Watrigant, R., Bougeret, M., Giroudeau, R.: Approximating the Sparsest k-subgraph in chordal graphs. Theor. Comput. Syst. **58**(1), 111–132 (2016)

Plattenbauten: Touching Rectangles in Space

Stefan Felsner[1], Kolja Knauer[2,3], and Torsten Ueckerdt[4(✉)]

[1] Institute of Mathematics, Technische Universität Berlin (TUB), Berlin, Germany
[2] Departament de Matemàtiques i Informàtica, Universitat de Barcelona (UB),
Barcelona, Spain
[3] Aix-Marseille Univ, Université de Toulon, CNRS, LIS, Marseille, France
[4] Institute for Theoretical Informatics, Karlsruhe Institute of Technology (KIT),
Karlsruhe, Germany
torsten.ueckerdt@kit.edu

Abstract. Planar bipartite graphs can be represented as touching graphs of horizontal and vertical segments in $\mathbb{R}^2$. We study a generalization in space, namely, touching graphs of axis-aligned rectangles in $\mathbb{R}^3$. We prove that planar 3-colorable graphs can be represented as touching graphs of axis-aligned rectangles in $\mathbb{R}^3$. The result implies a characterization of corner polytopes previously obtained by Eppstein and Mumford. A by-product of our proof is a distributive lattice structure on the set of orthogonal surfaces with given skeleton.

Moreover, we study the subclass of strong representations, i.e., families of axis-aligned rectangles in $\mathbb{R}^3$ in general position such that all regions bounded by the rectangles are boxes. We show that the resulting graphs correspond to octahedrations of an octahedron. This generalizes the correspondence between planar quadrangulations and families of horizontal and vertical segments in $\mathbb{R}^2$ with the property that all regions are rectangles.

Keywords: Touching graphs · Contact graphs · Boxicity · Planar graphs

1 Introduction

The importance of contact and intersection representations of graphs stems not only from their numerous applications including information visualization, chip design, bio informatics and robot motion planning (see for example the references in [2,9]), but also from the structural and algorithmic insights accompanying the investigation of these intriguing geometric arrangements. From a structural point of view, the certainly most fruitful contact representations (besides the "Kissing Coins" of Koebe, Andrew, and Thurston [1,16,23]) are axis-aligned segment

Omitted proofs and more figures can be found in the full version [11].

© Springer Nature Switzerland AG 2020
I. Adler and H. Müller (Eds.): WG 2020, LNCS 12301, pp. 161–173, 2020.
https://doi.org/10.1007/978-3-030-60440-0_13

contact representations: families of interior-disjoint horizontal and vertical segments in $\mathbb{R}^2$ where the intersection of any two segments is either empty or an endpoint of at least one of the segments. The corresponding touching graph[1] has the segments as its vertices and the pairs of segments as its edges for which an endpoint of one segment is an interior point of the other segment, see the left of Fig. 1. It has been discovered several times [15, 21] that any such touching graph is bipartite and planar, and that these two obviously necessary conditions are in fact already sufficient: Every planar bipartite graph is the touching graph of interior-disjoint axis-aligned segments in $\mathbb{R}^2$. In fact, edge-maximal segment contact representations endow their associated plane graphs with many useful combinatorial structures such as 2-orientations [9], separating decompositions [4], bipolar orientations [22, 24], transversal structures [12], and Schnyder woods [26].

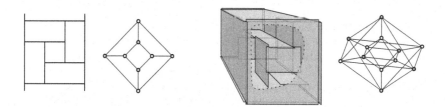

Fig. 1. An axis-aligned segment contact representation (left) and a Plattenbau (right) together with the respective touching graphs.

In this paper we extend axis-aligned segment contact representations in $\mathbb{R}^2$ to axis-aligned rectangle contact representations in $\mathbb{R}^3$. That is, we consider families $\mathcal{R}$ of axis-aligned closed and bounded rectangles in $\mathbb{R}^3$ with the property that for all $R, R' \in \mathcal{R}$ the intersection $R \cap R'$ is a subset of the boundary of at least one of them, i.e., the rectangles are interiorly disjoint. We call such a family a *Plattenbau*[2]. Given a Plattenbau $\mathcal{R}$ one can consider its *intersection graph* $I_\mathcal{R}$. However, for us the more important concept is a certain subgraph of $I_\mathcal{R}$, called the *touching graph* $G_\mathcal{R}$ of $\mathcal{R}$. There is one vertex in $G_\mathcal{R}$ for each rectangle in $\mathcal{R}$ and two vertices are adjacent if the corresponding rectangles *touch*, i.e., their intersection is non-empty and contains interior points of one and only one of the rectangles. We say that G is a *Plattenbau graph* if there is a Plattenbau $\mathcal{R}$ such that $G \cong G_\mathcal{R}$. In this case we call $\mathcal{R}$ a *Plattenbau representation* of G.

Plattenbauten are a natural generalization of axis-aligned segment contact representations in $\mathbb{R}^2$ and thus Plattenbau graphs are a natural generalization of

---

[1] We use the term touching graphs rather than the more standard contact graph to underline the fact that segments with coinciding endpoints (e.g., two horizontal segments touching a vertical segment in the same point but from different sides, but also non-parallel segments with coinciding endpoint) do not form an edge.

[2] Plattenbau (plural Plattenbauten) is a German word describing a building (*Bau*) made of prefabricated concrete panels (*Platte*).

planar bipartite graphs. While clearly all Plattenbau graphs are tripartite (properly vertex 3-colorable), it is an interesting challenge to determine the exact topological properties in $\mathbb{R}^3$ that hold for all Plattenbau graphs, thus generalizing the concept of planarity from 2 to 3 dimensions (for tripartite graphs). We present results towards a characterization of Plattenbau graphs in three directions.

Our Results and Organization of the Paper. In Sect. 2 we provide some simple examples of Plattenbau graphs and give some simple necessary conditions for all Plattenbau graphs. We observe that unlike touching graphs of segments, general Plattenbau graphs are not closed under taking subgraphs. We circumvent this issue by introducing the subclass of proper Plattenbauten where for any two touching rectangles R, R' the intersection $R \cap R'$ must be a boundary edge of one of R, R'. Moreover, we introduce boxed Plattenbauten where every bounded region of $\mathbb{R}^3$ is a box, and discuss questions of augmentability.

In Sect. 3 we show that within planar graphs the necessary condition of 3-colorability is also sufficient for Plattenbau graphs. Thus, the topological characterization of Plattenbau graphs must fully contain planarity (which is not obvious as we consider 3-colorable graphs and not only bipartite graphs).

Theorem 1. *Every 3-colorable planar graph is the touching graph of a proper Plattenbau.*

Along the proof of Theorem 1, we obtain a characterization of skeletons of orthogonal surfaces which is implicit already in work of Eppstein and Mumford [6]. Another proof of Theorem 1 can be obtained from Gonçalves' recent proof that 3-colorable planar graphs admit segment intersection representations with segments of 3 slopes [13]. A consequence of our approach is a natural partial order - namely a distributive lattice - on the set of orthogonal surfaces with a given skeleton.

In Sect. 4 we consider proper and boxed Plattenbau graphs as the 3-dimensional correspondence to the edge-maximal planar bipartite graphs, the quadrangulations. We give a complete characterization of these graphs.

Theorem 2. *A graph G is the touching graph of a proper boxed Plattenbau $\mathcal{R}$ if and only if there are six outer vertices in G such that each of the following holds:*

(P1) G is connected and the outer vertices of G induce an octahedron.
(P2) The edges of G admit an orientation such that
  – the bidirected edges are exactly the outer edges,
  – each vertex has exactly 4 outgoing edges.
(P3) The neighborhood $N(v)$ of each vertex v induces a spherical quadrangulation $SQ(v)$ in which the out-neighbors of v induce a 4-cycle.
  – If v is an outer vertex, this 4-cycle bounds a face of $SQ(v)$.
(P4) For every edge uv of G with common neighborhood $C = N(u) \cap N(v)$, the cyclic ordering of C around u in $SQ(v)$ is the reverse of the cyclic ordering of C around v in $SQ(u)$.

A *spherical quadrangulation* is a graph embedded on the 2-dimensional sphere without crossings with all faces bounded by a 4-cycle. Spherical quadrangulations are 2-connected, planar, and bipartite. We remark that Theorem 2 does not give a complete characterization of proper Plattenbau graphs since some proper Plattenbau graphs are not contained in any proper boxed Plattenbau graph as discussed in Sect. 2.

Omitted figures and omitted proofs can be found in the full version [11].

2   Types of Plattenbauten and Questions of Augmentation

Let us observe some properties of Plattenbau graphs. Clearly, the class of all Plattenbau graphs is closed under taking induced subgraphs. Examples of Plattenbau graphs are $K_{2,2,n}$ and the class of grid intersection graphs, i.e., bipartite intersection graphs of axis-aligned segments in the plane [15]. For the latter take the segment intersection representation of a graph, embed it into the xy-plane in $\mathbb{R}^3$ and thicken all horizontal segments a small amount into y-direction and all vertical segments a bit into z-direction outwards the xy-plane. In particular, $K_{m,n}$ is a Plattenbau graph. In order to exclude some graphs, we observe some necessary properties of all Plattenbau graphs.

Observation 1. *If G is a Plattenbau graph, then*

1. *the chromatic number of G is at most 3,*
2. *the neighborhood of any vertex of G is planar.*
3. *the boxicity of G, i.e., the smallest dimension d such that G is the intersection graph of boxes in $\mathbb{R}^d$, is at most 3.*

Proof. Item 1: Each orientation class is an independent set.

Item 2: Let v be a vertex of G represented by $R \in \mathcal{R}$. Let H be the supporting hyperplane of R and H^+, H^- the corresponding closed halfspaces. The neighborhood $N(v)$ consists of rectangles R^+ intersecting H^+ and those R^- intersecting H^-. The rectangles in each of these sets have a plane touching graph, since it corresponds to the touching graph of the axis-aligned segments given by their intersections with R. The neighboring rectangles in $R^+ \cap R^-$ are on the outer face in both graphs in opposite order, so identifying them gives a planar drawing of the graph induced by $N(v)$.

Item 3: A Plattenbau $\mathcal{R}$ can be transformed into a set $\mathcal{B}$ of boxes such that the touching graph of $\mathcal{R}$ is the intersection graph of $\mathcal{B}$ as follows: First, shrink each rectangle orthogonal to the i-axis by a small enough $\varepsilon > 0$ in both dimensions different from i. As a result, we obtain a set of pairwise disjoint rectangles. Then, expand each such rectangle by ε in dimension i. The obtained set $\mathcal{B}$ of boxes are again interiorly disjoint and all intersections are touchings.   □

Note that for Items 2 and 3 of Observation 1 it is crucial that G is the touching graph and not the intersection graph. Moreover, Observation 1 allows to reject some graphs as Plattenbau graphs:

- K_4 is not a Plattenbau graph (by Item 1 of Observation 1),
- $K_{1,3,3}$ is not a Plattenbau graph (by Item 2 of Observation 1).
- The full subdivision of $K_{2^{2^5}+1}$ is not a Plattenbau graph (by Item 3 of Observation 1 and [3]).

In particular, some bipartite graphs are not Plattenbau graphs. Together with $K_{m,n}$ being a Plattenbau graph, this shows that the class of Plattenbau graphs is not closed under taking subgraphs; an unusual situation for touching graphs, which prevents us from solely focusing on edge-maximal Plattenbau graphs. To overcome this issue, we say that a Plattenbau $\mathcal{R}$ is *proper* if for any two touching rectangles an edge of one is contained in the other rectangle. (Note the ambiguity of the term "edge" here, which refers to a maximal line segment in the boundary of a rectangle as well as to a pair of adjacent vertices in a graph.) If $\mathcal{R}$ is proper, then each edge of the touching graph $G_{\mathcal{R}}$ can be removed by shortening one of the participating rectangles slightly. That is, the class of graphs with proper Plattenbau representations is closed under subgraphs.

We furthermore say that a Plattenbau $\mathcal{R}$ is *boxed* if six *outer rectangles* constitute the sides of a box that contains all other rectangles and all regions inside this box are also boxes. (A box is an axis-aligned full-dimensional cuboid, i.e., the Cartesian product of three bounded intervals of non-zero length.) For boxed Plattenbauten we use the additional convention that the edge-to-edge intersections of outer rectangles yield edges in the touching graph, even though these intersections contain no interior points. In particular, the outer rectangles of a proper boxed Plattenbau induce an octahedron in the touching graph.

Observation 2. *The touching graph $G_{\mathcal{R}}$ of a proper Plattenbau $\mathcal{R}$ with $n \geq 6$ vertices has at most $4n - 12$ edges. Equality holds if and only if $\mathcal{R}$ is boxed.*

Proof. For a proper Plattenbau $\mathcal{R}$ with touching graph $G_{\mathcal{R}}$ there is an injection from the edges of $G_{\mathcal{R}}$ to the edges of rectangles in $\mathcal{R}$: For each edge uv in $G_{\mathcal{R}}$ with corresponding rectangles $R_u, R_v \in \mathcal{R}$, take the edge of R_u or R_v that forms their intersection $R_u \cap R_v$. This way, each of the four edges of each of the n rectangles in $\mathcal{R}$ corresponds to at most one edge in $G_{\mathcal{R}}$.

Moreover, if $\mathcal{R}$ contains at least two rectangles of each orientation, the bounding box of $\mathcal{R}$ contains at least 12 edges of rectangles in its boundary, none of which corresponds to an edge in $G_{\mathcal{R}}$. Thus, in this case $G_{\mathcal{R}}$ has at most $4n - 12$ edges. Otherwise, for one of the three orientations, $\mathcal{R}$ contains at most one rectangle in that orientation. In this case, $G_{\mathcal{R}}$ is a planar bipartite graph plus possibly one additional vertex. In particular, $G_{\mathcal{R}}$ has at most $2(n-1) - 4 + (n-1) < 4n - 12$ edges, as long as $n \geq 6$.

Finally, in order to have exactly $4n - 12$ edges, the above analysis must be tight. This implies that $\mathcal{R}$ has at least two rectangles of each orientation and its bounding box contains exactly 12 edges of rectangles, i.e., $\mathcal{R}$ is boxed.   □

An immediate consequence of Observation 2 is that $K_{5,6}$ is a Plattenbau graph which has no proper Plattenbau representation. Contrary to the case of axis-aligned segments in $\mathbb{R}^2$, not every proper Plattenbau in $\mathbb{R}^3$ can be completed to a boxed Plattenbau. This example can also easily be extended to give a graph G that is the touching graph of a proper Plattenbau, but that is not a subgraph of any Plattenbau graph with a proper and boxed Plattenbau representation.

3   Planar 3-Colorable Graphs

Let us recall what shall be the main result of this section:

Theorem 1. *Every 3-colorable planar graph is the touching graph of a proper Plattenbau.*

The proof of this theorem is in several steps. First we introduce orthogonal surfaces and show that the dual graph of the skeleton of an orthogonal surface is a Plattenbau graph (Proposition 1). In the second step we characterize triangulations whose dual is the skeleton of an orthogonal surface (Proposition 2). One consequence of this is a natural very well-behaved partial order, namely a distributive lattice, on the set of orthogonal surfaces with given skeleton. We then show that a Plattenbau representation of a 3-colorable triangulation can be obtained by patching orthogonal surfaces in corners of orthogonal surfaces.

We begin with an easy observation.

Observation 3. *Every 3-colorable planar graph G is an induced subgraph of a 3-colorable planar triangulation.*

Proof (Sketch). Consider G with a plane embedding. It is easy to find a 2-connected 3-colorable G' which has G as an induced subgraph.

Fix a 3-coloring of G'. Let f be a face of G' of size at least four and c be a color such that at least three vertices of f are not colored c. Stack a vertex v inside f and connect it to the vertices on f that are not colored c. The new vertex v is colored c and the sizes of the new faces within f are 3 or 4. After stacking in a 4-face, the face is either triangulated or there is a color which is not used on the newly created 4-face. A second stack triangulates it.   □

A plane triangulation T is 3-colorable if and only if it is Eulerian. Hence, the dual graph T^* of T apart from being 3-connected, cubic, and planar is also bipartite. The idea of the proof is to find an orthogonal surface $\mathfrak{S}$ such that T^* is the skeleton of $\mathfrak{S}$. This is not always possible but with a technique of patching one orthogonal surface in an appropriate corner of a Plattenbau representation obtained from another orthogonal surface, we shall get to a proof of the theorem.

Consider $\mathbb{R}^3$ with the dominance order, i.e., $x \leq y$ if and only if $x_i \leq y_i$ for $i = 1, 2, 3$. The join and meet of this distributive lattice are the componentwise max and min. Let $\mathcal{V} \subseteq \mathbb{R}^3$ be a finite *antichain*, i.e., a set of mutually incomparable points. The *filter* of $\mathcal{V}$ is the set $\mathcal{V}^\uparrow := \{x \in \mathbb{R}^3 \mid \exists v \in \mathcal{V} : v \leq x\}$ and the boundary $\mathfrak{S}_\mathcal{V}$ of $\mathcal{V}^\uparrow$ is the *orthogonal surface* generated by $\mathcal{V}$.

Orthogonal surfaces have been studied by Scarf [14] in the context of test sets for integer programs. They later became of interest in commutative algebra, cf. the monograph of Miller and Sturmfels [20]. Miller [19] observed the connections between orthogonal surfaces, Schnyder woods and the Brightwell-Trotter Theorem about the order dimension of polytopes, see also [7].

A maximal set of points of an orthogonal surface which is constant in one of the coordinates is called a *flat*. A non-empty intersection of two flats is an *edge*. A point contained in three flats is called a *vertex*. An edge incident to only one vertex is a *ray*. We will only consider orthogonal surfaces obeying the following non-degeneracy conditions: (1) The boundary of every bounded flat is a simple closed curve. (2) There are exactly three rays. Note that from (1) it can be deduced that every vertex is contained in exactly three flats.

The skeleton graph $G_\mathfrak{S}$ of an orthogonal surface consists of the vertices and edges of the surface, in addition there is a vertex v_∞ which serves as second vertex of each ray. The skeleton graph is planar, cubic, and bipartite. The bipartition consists of the maxima and minima of the surface in one class and of the saddle vertices in the other class. The vertex v_∞ is a saddle vertex. The dual of $G_\mathfrak{S}$ is a triangulation with a designated outer face, the dual of v_∞.

The boundary of a bounded flat consists of two zig-zag paths sharing the two *extreme points* of the flat. The minima of the *lower* zig-zag are elements of the generating set $\mathcal{V}$, they are minimal elements of the orthogonal surface $\mathfrak{S}$. The maxima of the *upper* zig-zag are maximal elements of $\mathfrak{S}$, they can be considered to be *dual generators*.

With the following proposition we establish a first connection between orthogonal surfaces and Plattenbau graphs.

Proposition 1. *The dual triangulation of the skeleton of an orthogonal surface $\mathfrak{S}$ is a Plattenbau graph and admits a proper Plattenbau representation.*

Proof. Choose a point not on $\mathcal{V}$ on each of the three rays of $\mathfrak{S}$ and call these points the extreme points of their incident unbounded flats.

The two extreme points a_f, b_f of a flat f of $\mathfrak{S}$ span a rectangle $R(f)$. Note that the other two corners of $R(f)$ are $\max(a_f, b_f)$ and $\min(a_f, b_f)$. We claim that the collection of rectangles $R(f)$ is a *weak* rectangle contact representation of the dual triangulation T of the skeleton of $\mathfrak{S}$. Here weak means that the contacts of pairs of rectangles of different orientation can be an edge to edge contact. If f and f' share an edge e of the skeleton, then since one of the ends of e is a saddle point of $\mathfrak{S}$ and thus extreme in two of its incident flats, it is extreme for at least one of f and f'. This shows that e is contained in the boundary of at least one of the rectangles $R(f), R(f')$, i.e., the intersection of the open interiors of the rectangles is empty.

Let f and f' be two flats which share no edge. Let H_f and $H_{f'}$ be the supporting planes. If f is contained in an open halfspace O defined by $H_{f'}$, then $\max(a_f, b_f)$ and $\min(a_f, b_f)$, the other two corners of $R(f)$, are also in O, hence $R(f) \subset O$ and $R(f) \cap R(f') = \emptyset$. If f intersects $H_{f'}$ and f' intersects H_f, then consider the line $\ell = H_{f'} \cap H_f$. This line is parallel to one of the axes, hence it intersects $\mathfrak{S}$ in a closed interval $I_\mathfrak{S}$. If I_f and $I_{f'}$ are the intervals obtained by intersecting ℓ with f and f' respectively, then one of them equals $I_\mathfrak{S}$ and the other is an edge of the skeleton of $\mathfrak{S}$, i.e., (f, f') is an edge of T.

It remains to expand some of the rectangles to change weak contacts into true contacts. Let $e = f \cap f'$ be an edge such that the contact of $R(f)$ and $R(f')$ is weak. Select one of f and f', say f. Now expand the rectangle $R(f)$ with a small parallel shift of the boundary segment containing e. This makes the contact of $R(f)$ and $R(f')$ a true contact. The expansion can be taken small enough as to avoid that new contacts or intersections are introduced. Iterating over the edges we eventually get rid of all weak contacts.   □

Recall that we aim at realizing T^*, the dual of the 3-colorable triangulation T as the skeleton of an orthogonal surface. Since T is Eulerian its dual T^* is bipartite. Let U (black) and U' (white) be the bipartition of the vertices of T^* such that the dual v_∞ of the outer face of T is in U. The critical task is to assign two extreme vertices to each face of T^* which does not contain v_∞. This has to be done so that each vertex in U (except v_∞) is extremal for exactly two of the faces.

To solve the assignment problem we will work with an auxiliary graph H_T. The faces of T^* which do not contain v_∞ are the interior vertices of T, we denote this set with V°. As the vertices of T^* are the facial triangles of T, we think of U as representing the black triangles of T. We also let $U^\circ = U - v_\infty$, this is the set of bounded black triangles of T. The vertices of H_T are $V^\circ \cup U^\circ$ the edges of H_T correspond to the incidence relation in T^* and T respectively, i.e., v, u with $v \in V^\circ$ and $u \in U^\circ$ is an edge if vertex v is a corner of the black triangle u. A valid assignment of extreme vertices is equivalent to an orientation of H_T such that each vertex $v \in V^\circ$ has outdegree two and each vertex $u \in U^\circ$ has indegree two, i.e., the outdegrees of the vertices are prescribed by the function α with $\alpha(v) = 2$ for $v \in V^\circ$ and $\alpha(u) = \deg(u) - 2$ for $u \in U^\circ$. Since $|V^\circ| = |U^\circ| = n - 3$ it is readily seen that the sum of the α-values of all vertices equals the number of edges of H_T.

Orientations of graphs with prescribed out-degrees have been studied e.g. in [8], there it is shown that the following necessary condition is also sufficient for the existence of an α-orientation. For all $W \subset V^\circ$ and $S \subset U^\circ$ and $X = W \cup S$

$$\sum_{x \in X} \alpha(x) \le |E[X]| + |E[X, \overline{X}]|. \qquad (\alpha)$$

Here $E[X]$ and $E[X, \overline{X}]$ denote the set of edges induced by X, and the set of edges in the cut defined by X, respectively.

Inequality (α) does not hold for all triangulations T and all X. We next identify specific sets X violating the inequality, they are associated to certain badly behaving triangles, we call them *babets*. In Proposition 2 we then show that babets are the only obstructions for the validity of (α).

Let Δ be a separating triangle of T such that the faces of T bounding Δ from the outside are white. Let W be the set of vertices inside Δ and let S be the collection of black triangles of T which have all vertices in W. We claim that $X = W \cup S$ is violating (α). If $|W| = k$ and $|S| = s$, then $\sum_{x \in X} \alpha(x) = 2|W| + |S| = 2k + s$. The triangulation whose outer boundary is Δ has $2(k + 3) - 4$ triangles, half of them, i.e., $k + 1$, are black and interior. The right side of (α) is counting the number of incidences between W and black triangles. Black triangles in Δ have $3(k+1)$ incidences in total. There are $k+1-s$ black triangles have an incidence with a corner of Δ and 3 of them have incidences with two corners of Δ. Hence the value on the right side is $3(k+1) - (k+1-s) - 3 = 2k+s-1$. This shows that the inequality is violated. A separating triangle Δ of T with white touching triangles on the outside is a *babet*.

Proposition 2. *If T has no babet, then there is an orientation of H_T whose outdegrees are as prescribed by α.*

First, we construct a Plattenbau representation in the babet-free case based on an auxiliary graph G arising from the bipartition of T^*, and a Schnyder wood S for G. We then find an orthogonal surface $\mathfrak{S}$ based on the Schnyder wood S and show that the skeleton $G_{\mathfrak{S}}$ of $\mathfrak{S}$ is T^*, which together with Proposition 1 gives a Plattenbau for T. Then, in case T contains some babets, we cut the triangulation T along an innermost babet, find orthogonal surfaces for the orthogonal surfaces for the inside and outside, and patch the former into a saddle point of the latter.

The set of α-orientations of a fixed planar graph carries the structure of a distributive lattice [8]. With Proposition 2 we can establish a correspondence of such a set with the orthogonal surfaces having a given skeleton. We obtain that this set carries such a structure.

4   Proper Boxed Plattenbauten and Octahedrations

In this section we characterize the touching graphs of proper boxed Plattenbauten, that is, we prove Theorem 2.

First, in any proper Plattenbau any two touching rectangles R, R' have a proper contact, i.e., a boundary edge of one rectangle, say R, is completely contained in the other rectangle R'. We denote this as $R \rightarrow R'$ and remark that this orientation has already been used in the proof of Observation 2.

Secondly, in any proper boxed Plattenbau $\mathcal{R}$ there are six rectangles that are incident to the unbounded region. We refer to them as *outer rectangles* and to the six corresponding vertices in the touching graph G for $\mathcal{R}$ as the *outer vertices*. The corners incident to three outer rectangles are the *outer corners*, and the inner regions/cells of $\mathcal{R}$ will be called *rooms*.

Whenever we have specified some vertices of a graph to be outer vertices, this defines inner vertices, outer edges, and inner edges as follows: The *inner vertices* are exactly the vertices that are not outer vertices; the *outer edges* are those between two outer vertices; the *inner edges* are those with at least one inner vertex as endpoint. We shall use these notions for a Plattenbau graph, as well as for some planar quadrangulations we encounter along the way.

Let us start with the necessity of Items (P1) to (P4) in Theorem 2.

Proposition 3. *Every touching graph of a proper boxed Plattenbau satisfies Items (P1) to (P4) in Theorem 2.*

Next, we prove the sufficiency in Theorem 2, i.e., for every graph G satisfying Items (P1) to (P4) we find a proper boxed Plattenbau with touching graph G.

Fix a graph $G = (V, E)$ with six outer vertices and edge orientation fulfilling Items (P1) to (P4). For each vertex $v \in V$ denote by $SQ(v)$ the spherical quadrangulation induced by $N(v)$ given in Item (P3). By Item (P3), the out-neighbors of vertex v induce a 4-cycle in $SQ(v)$, which we call the *equator* O_v of $SQ(v)$. The equator O_v splits the spherical quadrangulation $SQ(v)$ into two *hemispheres*, each being a plane embedded quadrangulation with outer face O_v with the property that each vertex of $SQ(v) - O_v$ is contained in exactly one hemisphere. The vertices of O_v are the outer vertices of either hemisphere. Note that one hemisphere (or even both) may be trivial, namely when the equator bounds a face of $SQ(v)$.

We proceed with a number of claims.

Claim 1. *In each hemisphere, each inner vertex has exactly two outgoing edges and no outer vertex has an outgoing inner edge.*

Each equator edge of $SQ(v)$ induces together with v a triangle in G, and we call these four triangles the *equator triangles* of v.

Claim 2. *Every triangle in G is an equator triangle.*

Clearly, a vertex w forms a triangle with two vertices u and v if and only if uv is an edge and w is a common neighbor of u and v. Equivalently, w is adjacent to v in $SQ(u)$, which in turn is equivalent to w being adjacent to u in $SQ(v)$. Hence, the set $N(u) \cap N(v)$ of all common neighbors (and thus also the set of all triangles sharing edge uv) is endowed with the clockwise cyclic ordering around v in $SQ(u)$, as well as with the clockwise cyclic ordering around u in $SQ(v)$. By Item (P4), these two cyclic orderings are reversals of each other.

Let us define for a triangle Δ in G with vertices u, v, w the two *sides* of Δ as the two cyclic permutations of u, v, w, which we denote by $[u, v, w]$ and $[u, w, v]$. So triangle Δ has the two sides $[u, v, w] = [v, w, u] = [w, u, v]$ and $[u, w, v] = [w, v, u] = [v, u, w]$. We define a binary relation $\sim$ on the set of all sides of triangles in G as follows.

$$[u, v, a] \sim [v, u, b] \text{ if } \begin{cases} a \text{ comes immediately before } b \\ \text{in the clockwise cyclic ordering} \\ \text{of } N(u) \cap N(v) \text{ around } v \text{ in } SQ(u) \end{cases} \tag{1}$$

Note that by (P4) a comes immediately before b in the clockwise ordering around v if and only if b comes immediately before a in the clockwise ordering around u. Thus $[u, v, a] \sim [v, u, b]$ also implies $[v, u, b] \sim [u, v, a]$, i.e., $\sim$ is a symmetric relation and as such encodes an undirected graph H on the sides of triangles.

Claim 3. *Each connected component of H is a cube. The corresponding subgraph in G is an octahedron.*

With Claim 3 we have identified a family of octahedra in G such that each side of each triangle in G is contained in exactly one octahedron. We call these octahedra the *cells* of G, as these correspond in the 2-dimensional case to the 4-cycles bounding faces of the quadrangulation. As Eq. (1) puts two triangle sides $[u, v, a]$ and $[v, u, b]$ into a common cell if and only if the edges va and vb bound the same facial 4-cycle in $SQ(u)$, we obtain the following correspondence between the cells of G and the faces in the spherical quadrangulations.

Claim 4. *If $O \subseteq G$ is a cell of G and C is an induced 4-cycle in O, then C bounds a face of $SQ(v)$ and a face of $SQ(u)$ for the two vertices $u, v \in O - C$. Conversely, if C is a 4-cycle bounding a face of $SQ(v)$, then there is a cell O of G containing $\{v\} \cup V(C)$.*

Having identified the cells, we can now construct a proper boxed Plattenbau for G by identifying two opposite vertices in a particular cell, calling induction, and then splitting the rectangle corresponding to the identification vertex into two. The cells of G will then correspond to the rooms in $\mathcal{R}$, except that one cell in G will correspond to the unbounded region of $\mathcal{R}$ (which is not a room). To this end, we prove the following stronger statement:

Lemma 1. *Let G be a graph satisfying Items (P1) to (P4) and let A, B, C be three outer vertices forming a triangle in G. Then there exists a proper boxed Plattenbau $\mathcal{R}$ whose touching graph is G such that each of the following holds.*

(I1) *The six outer vertices of G correspond to the outer rectangles of $\mathcal{R}$.*
(I2) *The cells of G correspond to the rooms of $\mathcal{R}$, except for one cell that is formed by all six outer vertices.*
(I3) *For any two vertices u, v with corresponding rectangles R_u, R_v we have $u \to v$ in the orientation of G if and only if $R_u \cap R_v$ contains an edge of R_u.*
(I4) *For each vertex v corresponding to rectangle R_v, the rectangles touching R_v come in the same spherical order as their corresponding vertices in $SQ(v)$.*

Lemma 1 shows the sufficiency of Items (P1) to (P4) . The necessity is given in Proposition 3. Together this proves Theorem 2 and concludes this section.

5   Conclusions

Touching graphs of proper boxed Plattenbauten are natural generalizations of quadrangulations from the plane to space. As we have shown these graphs are octahedrations of 3-space. This can be seen as a novel way of going beyond planarity. A question in this spirit was asked by Jean Cardinal at the Order & Geometry Workshop at Gułtowy Palace in 2016: What is the 3-dimensional analogue of Baxter permutations? This is based on Baxter permutations being in bijection with boxed arrangements of axis-parallel segments in $\mathbb{R}^2$ [10].

A continuation of this project to higher dimensions would be to consider proper boxed Plattenbauten in $\mathbb{R}^d$ and study the resulting touching graphs as generalizations of plane quadrangulations to arbitrary dimensions.

Considering the intersection graph $I_\mathcal{R}$ instead of the touching graph of a Plattenbau, yields a very different graph class. Its plane analogue is known as B_0-CPG graphs, see [5]. For example, every 4-connected triangulation has a rectangle contact representation in $\mathbb{R}^2$, see [17,18,22,25,27]. Also K_{12} is the intersection graph of the Plattenbau $\mathcal{R}$ consisting of the twelve axis-parallel unit squares in $\mathbb{R}^3$ that have a corner on the origin.

References

1. Bowers, P.L.: Circle packing: a personal reminiscence. In: The Best Writing on Mathematics, pp. 330–345 (2010)
2. Buchsbaum, A.L., Gansner, E.R., Procopiuc, C.M., Venkatasubramanian, S.: Rectangular layouts and contact graphs. ACM Trans. Algorithms (TALG) 4(1), 1–28 (2008)
3. Chandran, L.S., Mathew, R., Sivadasan, N.: Boxicity of line graphs. Discr. Math. 311(21), 2359–2367 (2011)
4. De Fraysseix, H., de Mendez, P.O.: On topological aspects of orientations. Discr. Math. 229(1–3), 57–72 (2001)
5. Deniz, Z., Galby, E., Munaro, A., Ries, B.: On Contact Graphs of Paths on a Grid. In: Biedl, T., Kerren, A. (eds.) GD 2018. LNCS, vol. 11282, pp. 317–330. Springer, Cham (2018). https://doi.org/10.1007/978-3-030-04414-5_22
6. Eppstein, D., Mumford, E.: Steinitz theorems for simple orthogonal polyhedra. J. Comput. Geom. 5(1), 179–244 (2014)
7. Felsner, S.: Geometric Graphs and Arrangements. Advanced Lectures in Mathematics. Vieweg+Teubner Verlag, Wiesbaden (2004). https://doi.org/10.1007/978-3-322-80303-0
8. Felsner, S.: Lattice structures from planar graphs. Electr. J. Comb. 11(R15), 24p (2004)
9. Felsner, S.: Rectangle and square representations of planar graphs. In: Pach, J. (ed.) Thirty Essays on Geometric Graph Theory. Mathematics and StatisticsMathematics and Statistics, pp. 213–248. Springer, New York (2013). https://doi.org/10.1007/978-1-4614-0110-0_12
10. Felsner, S., Fusy, É., Noy, M., Orden, D.: Bijections for Baxter families and related objects. J. Comb. Theor. Ser. A 118(3), 993–1020 (2011)
11. Felsner, S., Knauer, K., Ueckerdt, T.: Plattenbauten: touching rectangles in space. ArXiv preprints (2020)
12. Fusy, É.: Combinatoire des cartes planaires et applications algorithmiques. Ph.D., thesis (2007)
13. Gonçalves, D.: 3-Colorable Planar Graphs Have an Intersection Segment Representation Using 3 Slopes. In: Sau, I., Thilikos, D.M. (eds.) WG 2019. LNCS, vol. 11789, pp. 351–363. Springer, Cham (2019). https://doi.org/10.1007/978-3-030-30786-8_27
14. Hansen, T., Scarf, H.: The Computation of Economic Equilibria. Cowles Foundation Monograph, vol. 24. Yale University Press, London (1973)
15. Hartman, I.B.-A., Newman, I., Ziv, R.: On grid intersection graphs. Discr. Math. 87(1), 41–52 (1991)
16. Koebe, P.: Kontaktprobleme der konformen Abbildung. Hirzel (1936)
17. Koźmiński, K., Kinnen, E.: Rectangular duals of planar graphs. Networks 15(2), 145–157 (1985)

18. Leinwand, S.M., Lai, Y.-T.: An algorithm for building rectangular floor-plans. In: 21st Design Automation Conference Proceedings, pp. 663–664. IEEE (1984)
19. Miller, E.: Planar graphs as minimal resolutions of trivariate monomial ideals. Doc. Math. **7**, 43–90 (2002)
20. Miller, E., Sturmfels, B.: Combinatorial Commutative Algebra. Graduate Texts in Mathematics. Springer, New York (2004)
21. Pach, J., de Fraysseix, H., de Mendez, P.: Representation of planar graphs by segments. North-Holland, Technical report (1994)
22. Rosenstiehl, P., Tarjan, R.E.: Rectilinear planar layouts and bipolar orientations of planar graphs. Discrete & Computational Geometry **1**(4), 343–353 (1986). https://doi.org/10.1007/BF02187706
23. Stephenson, K.: Introduction to Circle Packing: The Theory of Discrete Analytic Functions. Cambridge University Press, Cambridge (2005)
24. Tamassia, R., Tollis, I.G.: A unified approach to visibility representations of planar graphs. Discrete & Computational Geometry **1**(4), 321–341 (1986). https://doi.org/10.1007/BF02187705
25. Thomassen, C.: Interval representations of planar graphs. J. Comb. Theor. Ser. B **40**(1), 9–20 (1986)
26. Ueckerdt, T.: Geometric representations of graphs with low polygonal complexity. Doctoral thesis, Technische Universität Berlin, Fakultät II - Mathematik und Naturwissenschaften, Berlin (2012). https://doi.org/10.14279/depositonce-3190
27. Ungar, P.: On diagrams representing graphs. J. London Math. Soc. **28**, 336–342 (1953)

Universal Geometric Graphs

Fabrizio Frati[1]([✉]), Michael Hoffmann[2], and Csaba D. Tóth[3,4]

[1] Roma Tre University, Rome, Italy
frati@dia.uniroma3.it
[2] ETH Zürich, Zürich, Switzerland
hoffmann@inf.ethz.ch
[3] California State University Northridge, Los Angeles, CA, USA
csaba.toth@csun.edu
[4] Tufts University, Medford, MA, USA

Abstract. We introduce and study the problem of constructing geometric graphs that have few vertices and edges and that are universal for planar graphs or for some sub-class of planar graphs; a geometric graph is *universal* for a class $\mathcal{H}$ of planar graphs if it contains an embedding, i.e., a crossing-free drawing, of every graph in $\mathcal{H}$.

Our main result is that there exists a geometric graph with n vertices and $O(n \log n)$ edges that is universal for n-vertex forests; this extends to the geometric setting a well-known graph-theoretic result by Chung and Graham, which states that there exists an n-vertex graph with $O(n \log n)$ edges that contains every n-vertex forest as a subgraph. Our $O(n \log n)$ bound on the number of edges is asymptotically optimal.

We also prove that, for every $h > 0$, every n-vertex convex geometric graph that is universal for the class of the n-vertex outerplanar graphs has $\Omega_h(n^{2-1/h})$ edges; this almost matches the trivial $O(n^2)$ upper bound given by the n-vertex complete convex geometric graph.

Finally, we prove that there is an n-vertex convex geometric graph with n vertices and $O(n \log n)$ edges that is universal for n-vertex caterpillars.

1 Introduction

A graph G is *universal* for a class $\mathcal{H}$ of graphs if G contains every graph in $\mathcal{H}$ as a subgraph. The study of universal graphs was initiated by Rado [20] in the 1960s. Obviously, the complete graph K_n is universal for any family $\mathcal{H}$ of n-vertex graphs. Research focused on finding the minimum size (i.e., number of edges) of universal graphs for various families of sparse graphs on n vertices. Babai et al. [3] proved that if $\mathcal{H}$ is the family of all graphs with m edges, then the size of a universal graph for $\mathcal{H}$ is in $\Omega(m^2 / \log^2 m)$ and $O(m^2 \log \log m / \log m)$. Alon et al. [1,2] constructed a universal graph of optimal $\Theta(n^{2-2/k})$ size for n-vertex graphs with maximum degree k.

Partially supported by the MSCA-RISE project "CONNECT" No 734922, the MIUR Project "AHeAD" under PRIN 20174LF3T8, the NSF award DMS-1800734, and the SNSF Project "Arrangements and Drawings" No 200021E-171681.

© Springer Nature Switzerland AG 2020
I. Adler and H. Müller (Eds.): WG 2020, LNCS 12301, pp. 174–186, 2020.
https://doi.org/10.1007/978-3-030-60440-0_14

Significantly better bounds exist for minor-closed families. Babai et al. [3] proved that there exists a universal graph with $O(n^{3/2})$ edges for n-vertex planar graphs. For bounded-degree planar graphs, Capalbo [10] constructed universal graphs of linear size, improving an earlier bound by Bhatt et al. [5], which extends to other families with bounded bisection width. Böttcher et al. [7,8] proved that every n-vertex graph with minimum degree $\Omega(n)$ is universal for n-vertex bounded-degree planar graphs. For n-vertex trees, Chung and Graham [12,13] constructed a universal graph of size $O(n \log n)$, and showed that this bound is asymptotically optimal apart from constant factors.

In this paper, we extend the concept of universality to geometric graphs. A *geometric graph* is a graph together with a straight-line drawing in the plane in which the vertices are distinct points and the edges are straight-line segments not containing any vertex in their interiors. We investigate the problem of constructing, for a given class $\mathcal{H}$ of planar graphs, a geometric graph with few vertices and edges that is *universal for* $\mathcal{H}$, that is, it contains an *embedding* of every graph in $\mathcal{H}$. For an (abstract) graph G_1 and a geometric graph G_2, an *embedding of* G_1 *onto* G_2 is an injective graph homomorphism $\varphi : V(G_1) \to V(G_2)$ such that (i) every edge $uv \in E(G_1)$ is mapped to a line segment $\varphi(u)\varphi(v) \in E(G_2)$; and (ii) every pair of edges $u_1 v_2, u_2 v_2 \in E(G_1)$ is mapped to a pair of noncrossing line segments $\varphi(u_1)\varphi(v_1)$ and $\varphi(u_2)\varphi(v_2)$ in the plane.

Previous research in the geometric setting was limited to finding the smallest *complete* geometric graph that is universal for the planar graphs on n vertices. The intersection pattern of the edges of a geometric graph is determined by the location of its vertices; hence universal complete geometric graphs are commonly referred to as n-*universal point sets*. De Fraysseix et al. [16] proved that the $2n \times n$ section of the integer lattice is an n-universal point set. Over the last 30 years, the upper bound on the size of an n-universal point set has been improved from $2n^2$ to $n^2/4 + O(n)$ [4]; the current best lower bound is $(1.293 - o(1))n$ [21] (based on stacked triangulations, i.e., maximal planar graphs of treewidth three; see also [11,19]). It is known that every set of n points in general position is universal for n-vertex outerplanar graphs [6,18]. An $O(n^{3/2} \log n)$ upper bound is known for n-vertex stacked triangulations [17].

Our Results. The results on universal point sets yield an upper bound of $O(n^4)$ for the size of a geometric graph that is universal for n-vertex planar graphs and $O(n^2)$ for n-vertex outerplanar graphs, including trees. We improve the upper bound for n-vertex trees to an optimal $O(n \log n)$, and show that the quadratic upper bound for outerplanar graphs is essentially tight for convex geometric graphs. More precisely, we prove the following results:

– For every $n \in \mathbb{N}$, there is a geometric graph G with n vertices and $O(n \log n)$ edges that is universal for forests with n vertices (Theorem 1 in Sect. 2). The $O(n \log n)$ bound is asymptotically optimal, even in the abstract setting, for caterpillars (a *caterpillar* is a tree such that the removal of its leaves results in a path, called *spine*), and if the universal graph is allowed to have more than n vertices [12, Theorem 1]. The proof of universality is constructive and yields a polynomial-time algorithm that embeds any n-vertex forest onto G.

– For every $h \in \mathbb{N}$ and $n \geq 3h^2$, every n-vertex convex geometric graph that is universal for the family of n-vertex cycles with h disjoint chords has $\Omega_h(n^{2-1/h})$ edges (Theorem 2 in Sect. 3); this almost matches the trivial $O(n^2)$ bound, which hence cannot be improved by polynomial factors even for n-vertex outerplanar graphs of maximum degree three. For n-vertex cycles with 2 disjoint chords, there is an n-vertex convex geometric graph with $O(n^{3/2})$ edges (Theorem 3 in Sect. 3), which matches the lower bound above.

– For every $n \in \mathbb{N}$, a convex geometric graph with n vertices and $O(n \log n)$ edges exists that is universal for n-vertex caterpillars (Theorem 4 in Sect. 3).

A full version of the paper can be found in [14].

2 Universal Geometric Graphs for Forests

In this section, we prove the following theorem.

Theorem 1. *For every $n \in \mathbb{N}$, there exists a geometric graph G with n vertices and $O(n \log n)$ edges that is universal for forests with n vertices.*

Construction. We adapt a construction of Chung and Graham [13] to the geometric setting. For $n \in \mathbb{N}$, they construct an n-vertex graph G with $O(n \log n)$ edges that contains every n-vertex forest as a subgraph. We present this construction. For simplicity assume $n = 2^h - 1$ with $h \geq 2$. Let B be an n-vertex complete rooted ordered binary tree. A *level* is a set of vertices at the same distance from the root. The levels are labeled $1, \ldots, h$, from the one of the root to the one of the leaves. A preorder traversal of B (visiting first the root, then recursively the vertices in its left subtree, and then recursively the vertices in its right subtree) determines a total order on the vertices, which also induces a total order on the vertices in each level. On each level, we call two consecutive elements in this order *level-neighbors*; in particular, siblings are level-neighbors. We denote by $B(v)$ the subtree of B rooted at a vertex v. The graph G contains B and three additional groups of edges (see Fig. 1): (E1) Every vertex v is adjacent to all vertices in $B(v)$; (E2) every vertex v with a level-neighbor u in B is adjacent to all vertices in $B(u)$; and (E3) every vertex v whose parent has a left level-neighbor p is adjacent to all vertices in $B(p)$.

Number of Edges. The tree B has 2^{i-1} vertices on level i, for $i = 1, \ldots, h$. A vertex v on level i has $2^{h-i+1} - 1$ descendants (including itself), and its at most two level-neighbors have the same number of descendants. In addition, the left level-neighbor of the parent of v (if present) has $2 \cdot (2^{h-i+1} - 1)$ descendants (excluding itself). Altogether v is adjacent to less than $5 \cdot 2^{h-i+1}$ vertices at the same or at lower levels of B. Hence, the number of edges in G is less than $5 \cdot \sum_{i=1}^{h} 2^{i-1} \cdot 2^{h-i+1} = 5 \cdot 2^h \cdot h = 5(n+1) \cdot \log_2(n+1) \in O(n \log n)$.

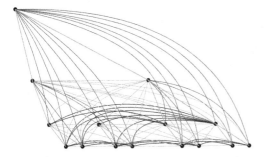 

Fig. 1. A schematic drawing of the 15-vertex universal graph (left) and a geometric drawing of the 7-vertex universal graph. The edges of B are black; the edges of the groups (E1), (E2), and (E3) are red, orange, and blue, respectively. Edges in several groups have the color of the first group they belong to. (Color figure online)

Chung and Graham [13] showed that G is universal for n-vertex forests.[1]

Geometric Representation. We next describe how to embed the vertices of G into $\mathbb{R}^2$; see Fig. 2(left) for an illustration. First, the x-coordinates of the vertices are assigned in the order determined by a preorder traversal of B. For simplicity, let us take these x-coordinates to be $0, \ldots, n-1$, so that the root of B is placed on the y-axis. The vertex of G with x-coordinate i is denoted by v_i.

The y-coordinates of the vertices are determined by a BFS traversal of B starting from the root, in which at every vertex the right sibling is visited before the left sibling. If a vertex u is visited before a vertex v by this traversal, then u gets a larger y-coordinate than v. The gap between two consecutive y-coordinates is chosen so that every vertex is above every line through two vertices with smaller y-coordinate; this implies that, for any vertex v, all vertices with larger y-coordinate than v, if any, see the vertices below v in the same circular order as v. The vertices of G are in general position, that is, no three are collinear.

Our figures display the vertices of B in the correct x- and y-order, but—with the exception of Fig. 1(right)—they are not to scale. The y-coordinates in our construction are rapidly increasing (similarly to [9,17]). For this reason, in our figures we draw the edges in B as straight-line segments and all other edges as Jordan arcs. We have the following property (refer to Fig. 2(right)).

Observation 1. *If $ab, cd \in E(G)$ are such that (1) a has larger y-coordinate than b, c, and d, and (2) b has smaller or larger x-coordinate than both c and d, then ab and cd do not cross.*

Intervals. A geometric graph is *plane* if it contains no crossings. For every interval $[i,j] \subseteq [0, n-1]$ we define $G[i,j]$ as the subgraph of G induced by the

---

[1] The construction by Chung and Graham uses fewer edges: in the edge groups (E2) and (E3), they use siblings instead of level-neighbors. But we were unable to verify their proof with the smaller edge set, namely we do not see why the graph G_2 in [13, Fig. 7] is admissible. However, their proof works with the edge set we define here.

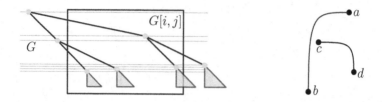

Fig. 2. Illustration for the assignment of x- and y-coordinates to the vertices of G, and for the definition of interval (left). Illustration for Observation 1 (right).

vertices with x-coordinate in $[i, j]$. Then $G[i, j]$ is an *interval* of G. The *length* of $G[i, j]$ is defined as $|G[i, j]| = j - i + 1$, which is the number of vertices in $G[i, j]$. If I is an interval of integers, then we denote by $G(I)$ the corresponding interval of G. For a subset $U \subset V(G)$, we denote by $G[U]$ the subgraph of G induced by U. We will show eventually that every tree on h vertices admits an embedding onto every interval of length h of G. We now present the following lemma.

Lemma 1. *Every interval $G[i, j]$ of G on at least two vertices contains two plane spanning stars, centered at the highest vertex v_k and at the second highest vertex v_s of $G[i, j]$. If $k < j$, then $G[i, j]$ contains a plane spanning star centered at the highest vertex of $G[k + 1, j]$ (which may or may not be v_s).*

Our upcoming recursive algorithm sometimes embeds a subtree of T onto an induced subgraph of G that is "almost" an interval, in the sense that it can be obtained from an interval of G by deleting its highest vertex or by replacing its highest vertex with a vertex that does not belong to the interval.

We first prove that the "structure" of an interval without its highest vertex is similar to that of an interval. Let U and W be two subsets of $V(G)$ with $h = |U| = |W|$. Let $u_1, \ldots, u_h$ and $w_1, \ldots, w_h$ be the vertices of U and W, respectively, ordered by increasing x-coordinates. We say that $G[U]$ and $G[W]$ are *crossing-isomorphic* if: (C1) for any $p, q \in \{1, \ldots, h\}$, the edge $u_p u_q$ belongs to $G[U]$ if and only if the edge $w_p w_q$ belongs to $G[W]$; (C2) for any $p, q, r, s \in \{1, \ldots, h\}$ such that the edges $u_p u_q$ and $u_r u_s$ belong to $G[U]$, the edges $u_p u_q$ and $u_r u_s$ cross if and only if the edges $w_p w_q$ and $w_r w_s$ cross; and (C3) if u_i is the highest vertex of $G[U]$, for some $i \in \{1, \ldots, h\}$, then w_i is the highest vertex of $G[W]$. The graph isomorphism given by $\lambda(u_i) = w_i$, for all $i = 1, \ldots, n$, is a *crossing-isomorphism*. Clearly, the inverse of a crossing-isomorphism is also a crossing-isomorphism. We have the following.

Lemma 2. *Let v_k be the highest vertex in an interval $G[i, j]$, and assume that $G[i, j]$ contains neither the right child of v_k nor any descendant of the left child of its left sibling (if it exists). Then $G[i, j] - v_k$ is crossing-isomorphic to some interval $G(I)$ of G; the interval I can be computed in $O(1)$ time.*

We now present our tools for embedding trees onto "almost" intervals.

Lemma 3. *Let v_k be the highest vertex in an interval $G[i,j]$ with $h+1$ vertices. Suppose that there is a crossing-isomorphism λ from $G[i,j]-v_k$ to some interval $G(I)$ of G with h vertices. Further, suppose that a tree T with h vertices admits an embedding φ onto $G(I)$. Then $\varphi' = \lambda^{-1} \circ \varphi$ is an embedding of T onto $G[i,j] - v_k$, and if a is the vertex of T such that $\varphi(a)$ is the highest vertex of $G(I)$, then $\varphi'(a)$ is the highest vertex of $G[i,j] - v_k$.*

Lemma 4. *Let $G[i,j]$ be an interval of G with h vertices and let v_k be its highest vertex. Let v_x be a vertex of G that is higher than all vertices in $G[i,j] - v_k$ and that does not belong to $G[i,j]$. Suppose that a tree T with h vertices admits an embedding φ onto $G[i,j]$. Let a be the vertex of T such that $\varphi(a) = v_k$; further, let $\varphi'(a) = v_x$ and $\varphi'(b) = \varphi(b)$ for every vertex b of T other than a. Then φ' is an embedding of T onto $G[i,j] - v_k + v_x$.*

The following lemma is a variant of the (unique) lemma in [13].

Lemma 5. *Given a rooted tree T on $m \geq 2$ vertices and an integer s, with $1 \leq s \leq m$, there is a vertex c of T such that $|V(T(c))| \geq s$ but $|V(T(d))| \leq s-1$, for all children d of c. Such a vertex c can be computed in time $O(m)$.*

Proof Strategy. Given a tree T on h vertices and an interval $G[i,j]$ of length h, we describe a recursive algorithm that constructs an embedding φ of T onto $G[i,j]$. For a subtree T' of T, we denote by $\varphi(T')$ the image of φ restricted to T'. A step of the algorithm explicitly embeds some vertices; the remaining vertices form subtrees that are recursively embedded onto pairwise disjoint subintervals of $G[i,j]$. We insist that in every subtree at most two vertices, called *portals*, are adjacent to vertices not in the subtree. We also ensure that whenever a subtree is embedded onto a subinterval, the vertices not in the subtree that connect to the portals of that subtree are embedded above the subinterval.

For a point p, we denote $Q^+(p) = \{q \in \mathbb{R}^2 : x(p) < x(q) \text{ and } y(p) < y(q)\}$ and $Q^-(p) = \{q \in \mathbb{R}^2 : x(q) < x(p) \text{ and } y(p) < y(q)\}$.

We inductively prove the following lemma, which immediately implies Theorem 1 with $G[i,j] = G[0, n-1]$ and a portal a chosen arbitrarily.

Lemma 6. *We are given a tree T on h vertices, an interval $G[i,j]$ of length h, and either (1) a single portal a in T, or (2) two distinct portals a and b in T. Then there exists an embedding φ of T onto $G[i,j]$ with the following properties:*

1. *If only one portal is given, then*
   (a) *$\varphi(a)$ is the highest vertex in $G[i,j]$; and*
   (b) *if $\deg_T(a) = 1$ and a' is the unique neighbor of a in T, then $Q^-(\varphi(a'))$ does not intersect any vertex or edge of the embedding $\varphi(T(a'))$.*
2. *If two distinct portals are given, then*
   (a) *$\varphi(a)$ is to the left of $\varphi(b)$;*
   (b) *$Q^-(\varphi(a))$ does not intersect any edge or vertex of $\varphi(T)$; and*
   (c) *$Q^+(\varphi(b))$ does not intersect any edge or vertex of $\varphi(T)$.*

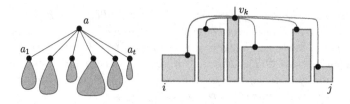

Fig. 3. Case 1.1: Tree T (left) and its embedding onto $G[i, j]$ (right).

Proof sketch: We proceed by induction on h. In the base case $h = 1$, hence T has one vertex, which must be the portal a, and the map $\varphi(a) = v_i$ maps a to the highest vertex of $G[i, i]$. For the induction step we assume that $h \geq 2$.

Case 1: There is only one portal a. Let v_k denote the highest vertex in $G[i, j]$. We need to find an embedding of T onto $G[i, j]$ where $\varphi(a) = v_k$. Consider T to be rooted at a. We distinguish two cases depending on the degree of a in T.

Case 1.1: $\deg_T(a) \geq 2$. Assume that a has t children $a_1, \ldots, a_t$. Refer to Fig. 3. Partition the integers $[i, j] \setminus \{k\}$ into t contiguous subsets $I_1, \ldots, I_t$ such that $|I_x| = |V(T(a_x))|$, for $x = 1, \ldots, t$. W.l.o.g. assume that I_q contains $k - 1$ or $k + 1$, and so $I_q \cup \{k\}$ is an interval of integers.

By induction, there is an embedding φ_x of $T(a_x)$ onto $G(I_x)$ such that $\varphi_x(a_x)$ is the highest vertex of $G(I_x)$, for all $x \neq q$, and there is an embedding φ_q of $T - \bigcup_{x \neq q} T(a_x)$ onto $G(I_q \cup \{k\})$ such that $\varphi_q(a) = v_k$. Then the combination of these embeddings is an embedding φ of T onto $G[i, j]$ satisfying Properties 1(a) and 1(b). In particular, the edges $\varphi(a)\varphi(a_x)$ are in $G[i, j]$ by Lemma 1, and an edge $\varphi(a)\varphi(a_x)$ does not cross $\varphi(T(a_y))$, where $y \neq x$, by Observation 1. ◀

Case 1.2: $\deg_T(a) = 1$. Let a' be the neighbor of a in T and let $T' = T(a')$.

Case 1.2.1: $k = j$. Set $\varphi(a) = v_k$ and recursively embed T' onto $G[i, k - 1]$ with a single portal a', which is mapped to the highest vertex in $G[i, k - 1]$. Clearly, φ is an embedding of T onto $G[i, j]$ satisfying Properties 1(a) and 1(b), since the edge $\varphi(a)\varphi(a')$ is above, and hence does not cross, $\varphi(T')$. ◀

Case 1.2.2: $k = i$. This case is symmetric to Case 1.2.1. ◀

Case 1.2.3: $i < k < j$ and the left sibling v_ℓ of v_k exists and is in $G[i, j]$. It follows that $\ell = i$, as if $\ell > i$, then $v_{\ell-1}$, which is the parent of v_ℓ and v_k, would be a vertex in $G[i, j]$ higher than v_k. By construction, v_i is the second highest vertex in $G[i, j]$. Recursively construct an embedding ψ of T' onto $G[i + 1, j]$ with a single portal a'. By Property 1(a), we have $\psi(a') = v_k$. By Lemma 4, there exists an embedding φ of T' onto $G[i + 1, j] - v_k + v_i = G[i, j] - v_k$ in which $\varphi(a') = v_i$ (hence φ satisfies Property 1(b)). Finally, set $\varphi(a) = v_k$ (hence φ satisfies Property 1(a)). As in Case 1.2.1, the edge $\varphi(a)\varphi(a') = v_k v_i$ does not cross $\varphi(T')$, hence φ is an embedding of T onto $G[i, j]$. ◀

Case 1.2.4: $i < k < j$, the left sibling of v_k does not exist or is not in $G[i, j]$, and the right child of v_k is not in $G[i, j]$. Refer to Fig. 4. By construction, the left child of v_k is v_{k+1}, which is in $G[i, j]$. By the assumptions of this case, v_{k+1} is the second highest vertex in $G[i, j]$.

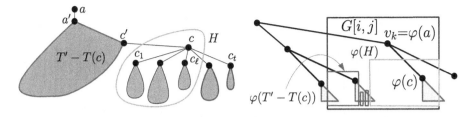

Fig. 4. Case 1.2.4. Tree T (left) and its embedding onto $G[i,j]$ (right).

Set $s = j - k + 1$; then $s < h$, given that $k > i$. By Lemma 5, there is a vertex c in T' such that $|V(T'(c))| \geq s$ but $|V(T'(d))| \leq s - 1$ for all children d of c. Label the children of c as $c_1, \ldots, c_t$ in an arbitrary order and let $\ell \in [1, t]$ be the smallest index such that $1 + \sum_{x=1}^{\ell} |V(T(c_i))| \geq s$. Since $|V(T(c_\ell))| \leq s - 1$, we have $s \leq 1 + \sum_{x=1}^{\ell} |V(T(c_i))| \leq 2s - 2$.

Let c' be the parent of c in T. Let H denote the subtree of T induced by c and by $V(T(c_1)), \ldots, V(T(c_\ell))$, and let $m = |V(H)|$. By the above inequalities, we have $s \leq m \leq 2s - 2$. On the one hand, $j - k + 1 \leq m$ implies that the subinterval $G[j - m, j]$ contains v_k, and so v_k is the highest vertex in $G[j - m, j]$. On the other hand, the interval $G[j - m, k - 1]$ contains $k - 1 - j + m + 1 = m - s + 1 \leq s - 1$ vertices, given that $m \leq 2s - 2$; however, since the right child of v_k is not in $G[i, j]$, we know that the size of a subtree of B rooted at any vertex at the level below v_k is larger than or equal to $s - 1$. It follows that $G[j - m, j]$ does not contain any descendants of the left child of the left sibling of v_k (if it exists). By Lemma 2, $G[j - m, j] - v_k$ is crossing-isomorphic to an interval $G(I)$ of size m.

Recursively embed H onto $G(I)$ with one portal c. By Lemma 3, there exists an embedding φ of H onto $G[j - m, j] - v_k$ such that $\varphi(c) = v_{k+1}$. Set $\varphi(a) = v_k$. If c has more than ℓ children, then embed the subtrees $T(c_{\ell+1}), \ldots, T(c_t)$ onto subintervals to the left of $G[j - m, j]$, with single portals $c_{\ell+1}, \ldots, c_t$, respectively. Finally, by induction, we can embed $T' - T(c)$ onto the remaining subinterval of $G[i, j]$ with two portals a' and c'. Then φ is an embedding of T onto $G[i, j]$ satisfying Properties 1(a) and 1(b). In particular, the edge $\varphi(c)\varphi(c')$ does not cross $\varphi(T' - T(c))$, since this satisfies Property 2(c) (note that $\varphi(c)$ is in $Q^+(\varphi(c'))$). ◄

Case 1.2.5: $i < k < j$, the left sibling of v_k does not exist or is not in $G[i, j]$, and the right child v_r of v_k is in $G[i, j]$. By assumption, we have $k + 1 < r \leq j$; further, the second highest vertex in $G[i, j]$ is v_r. Set $s = j - r + 1$. Lemma 5 yields a vertex c in T' such that $|V(T(c))| \geq s$ but $|V(T(d))| \leq s - 1$ for all children d of c. Let $T(c)$ be the subtree of T rooted at c, set $m = |V(T(c))|$, and label the children of c by $c_1, \ldots, c_t$ in an arbitrary order. Let c' be the parent of c and denote by $T_c(c')$ the subtree of T induced by c' and $V(T(c))$.

Case 1.2.5.1: $m \leq j - k - 1$. Then the interval $[j - m, j]$ contains r but does not contain k, hence v_r is the highest vertex in $G[j - m, m]$. By induction, there is an embedding ψ_1 of $T' - T(c)$ onto $G[i, j - m - 1]$ with two portals a' and c'. By Lemma 4, there is an embedding φ of $T' - T(c)$ onto $G[i, j - m - 1] - v_k + v_r$. Set $\varphi(a) = v_k$. Again by induction, there is an embedding ψ_2 of $T_c(c')$ onto

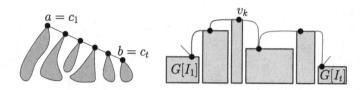

Fig. 5. Case 2. Tree T (left) and its embedding onto $G[i,j]$ (right).

$G[j-m,j]$ with a single portal c'. Let $\varphi(T(c)) = \psi_2(T(c))$ and note that $\varphi(c')$ may be different from $\psi_2(c') = v_r$. This completes the definition of $\varphi(T)$, which is an embedding of T onto $G[i,j]$ satisfying Properties 1(a) and 1(b). In particular, we argue that the edge $\varphi(c)\varphi(c')$ is in $G[i,j]$. Let $v_p = \varphi(c')$ and $v_q = \varphi(c)$, and note that $p < q$ or $p = r$. In the latter case, $\varphi(c)\varphi(c')$ exists as $\psi_2(c') = v_r$ and the edge cc' belongs to $T_c(c')$; hence, assume that $p < q$. By Property 2(c) of ψ_1, we have that $Q^+(\psi_1(c'))$ does not intersect $\psi_1(T'-T(c))$, hence $k < p$, as otherwise v_k would be in $Q^+(\psi_1(c'))$; hence, v_p is the highest vertex in $G[p, j-m-1]$. Further, by Property 1(b) of ψ_2, we have that $Q^-(\psi_2(c))$ does not intersect $\psi_2(T(c))$; hence, v_q is either the highest or the second highest vertex in $G[j-m,q]$ (as v_r might belong to such an interval). Overall, one of v_p or v_q is the highest or the second highest vertex in $G[p, q]$. By Lemma 1, $G[p, q]$ contains a star centered at v_p or v_q, and so it contains the edge $v_p v_q = \varphi(c')\varphi(c)$.

Case 1.2.5.2: $j-k-1 < m$. Then $[j-m,j]$ contains both k and r. Partition $[j-m,j] \setminus \{k,r\}$ into t contiguous subsets $I_1, \ldots, I_t$ such that $|I_x| = |V(T(c_x))|$, for $x = 1, \ldots, t$. W.l.o.g. assume that I_q contains $r-1$.

Let $\mathcal{I}(c)$ be the collection of the sets $I_q \cup \{v_r\}$ and I_x, for $x \in [1,t] \setminus \{q\}$. At least $t-1$ of these sets are intervals, and at most one of them, say I_p, is an interval minus its highest element. Since every tree $T(c_i)$ has at most $s-1$ vertices, I_p has at most s elements. Since $s = j-r-1$, we have $|I_p| \le |V(B(v_r))|$, hence I_p contains neither the right child of v_k nor any descendant of its left sibling. By Lemma 2, $G(I_p)$ is crossing-isomorphic to an interval. By Lemma 3, we can embed $T(c_p)$ onto $G(I_p)$. We also recursively embed $T(c_x)$ onto $G[I_x]$ for all $x \in [1,t] \setminus \{p,q\}$ and $T(c) - \bigcup_{x \ne q} T_x$ onto $G(I_q \cup \{u\})$. Embed a at v_k. Finally, embed $T' - T(c)$ onto $G[i, j-m-1]$ with portals a' and c'. The combination of these embeddings is an embedding φ of T onto $G[i,j]$ satisfying Properties 1(a) and 1(b). ◀

Case 2: Two portals a and b; refer to Fig. 5. Let $P = (a = c_1, \ldots, c_t = b)$ be the path between a and b in T, where $t \ge 2$. The deletion of the edges in P splits T into t trees rooted at $c_1, \ldots, c_t$. Partition $[i,j]$ into t subintervals $I_1, \ldots, I_t$ such that $|I_x| = |V(T(c_x))|$, for $x = 1, \ldots, t$.

For $x = 1, \ldots, t$, recursively embed $T(c_x)$ onto $G(I_x)$ with one portal c_x. The combination of these embeddings is an embedding φ of T onto $G[i,j]$ satisfying Properties 2(a), 2(b), and 2(c). □

3   Convex Geometric Graphs

Every graph embedded onto a *convex* geometric graph is outerplanar. Clearly, an n-vertex complete convex geometric graph has $O(n^2)$ edges and is universal for the n-vertex outerplanar graphs. We show that this trivial bound is almost tight. For $h \geq 0$ and $n \geq 2h + 2$, let $\mathcal{O}_h(n)$ be the family of n-vertex outerplanar graphs consisting of a spanning cycle plus h pairwise disjoint chords.

Theorem 2. *For every positive integer h and $n \geq 3h^2$, every convex geometric graph C on n vertices that is universal for $\mathcal{O}_h(n)$ has $\Omega_h(n^{2-1/h})$ edges.*

Proof. Denote by ∂C the outer (spanning) cycle of C. The *length* of a chord uv of ∂C is the length of a shortest path between u and v along ∂C. For $k \geq 2$, denote by E_k the set of length-k chords in C, and let $m \in \{2, \ldots, \lfloor n/(3h) \rfloor\}$ be an integer such that $|E_m| = \min\{|E_2|, \ldots, |E_{\lfloor n/(3h) \rfloor}|\}$.

Let $\mathcal{L}$ be the set of *labeled* n-vertex outerplanar graphs that consist of a spanning cycle $(v_0, \ldots, v_{n-1})$ plus h pairwise-disjoint chords of length m such that one chord is $v_0 v_m$ and all h chords have both vertices on the path $P = (v_0, \ldots, v_{\lfloor n/3 \rfloor + hm - 1})$. Every graph $G \in \mathcal{L}$ has a unique spanning cycle H, which is embedded onto ∂C. Since they all have the same length, the h chords of H have a well-defined cyclic order along H. A *gap* of G is a path between two consecutive chords along H. Note that G has h gaps. The length of P is $\lfloor n/3 \rfloor + hm - 1 \leq \lfloor n/3 \rfloor + h \cdot \lfloor n/(3h) \rfloor - 1 < 2\lfloor n/3 \rfloor$, hence the length of the gap between the last and the first chords is more than $n - 2\lfloor n/3 \rfloor = \lceil n/3 \rceil$. This is the longest gap, as the lengths of the other gaps sum up to at most $\lfloor n/3 \rfloor$.

Let $\mathcal{U}$ denote the subset of *unlabeled* graphs in $\mathcal{O}_h(n)$ that correspond to some labeled graph in $\mathcal{L}$. Each graph in $\mathcal{L}$ is determined by the lengths of its $h - 1$ shortest gaps. The sum of these lengths is an integer between $h - 1$ and $(\lfloor n/3 \rfloor + hm - 1) - hm < \lfloor n/3 \rfloor$. The number of compositions of $\lfloor n/3 \rfloor$ into h positive integers (i.e., $h - 1$ lengths and a remainder) is $\binom{\lfloor n/3 \rfloor}{h-1} \in \Theta_h(n^{h-1})$. Each unlabeled graph in $\mathcal{U}$ corresponds to at most two labeled graphs in $\mathcal{L}$, since any graph automorphism setwise fixes the unique spanning cycle as well as the longest gap. Hence, $|\mathcal{U}| \in \Theta(|\mathcal{L}|) \subseteq \Theta_h(n^{h-1})$.

Since C is universal for $\mathcal{O}_h(n)$ and $\mathcal{U} \subset \mathcal{O}_h(n)$, every graph G in $\mathcal{U}$ embeds onto C. Since every embedding of G maps the spanning cycle of G onto ∂C and the h chords of G into a subset of E_m, we have that C contains at most $\binom{|E_m|}{h} \leq |E_m|^h$ graphs in $\mathcal{U}$. The combination of the lower and upper bounds for $|\mathcal{U}|$ yields $|E_m|^h \in \Omega_h(n^{h-1})$, hence $|E_m| \in \Omega_h(n^{1-1/h})$. Overall, the number of edges in C is at least $\sum_{i=1}^{\lfloor n/(3h) \rfloor} |E_i| \geq \lfloor n/(3h) \rfloor \cdot |E_m| \in \Omega_h(n^{2-1/h})$.   □

For the case $h = 2$, the lower bound of Theorem 2 is the best possible.

Theorem 3. *For every $n \in \mathbb{N}$, there exists a convex geometric graph C with n vertices and $O(n^{3/2})$ edges that is universal for $\mathcal{O}_2(n)$.*

Proof. The vertices $v_0, \ldots, v_{n-1}$ of C form a convex n-gon and the edges of this spanning cycle are in C. Let $S = \{0, \ldots, \lfloor \sqrt{n} \rfloor - 1\} \cup \{i \lfloor \sqrt{n} \rfloor : 1 \leq i \leq \lfloor \sqrt{n} \rfloor\}$

and add a star centered at v_s, for every $s \in S$, to C. Clearly, C contains $O(n^{3/2})$ edges. Moreover, for every $d \in \{1, \ldots, \lfloor n/2 \rfloor\}$ there exist $a, b \in S$ so that $b - a = d$. For any $G \in \mathcal{O}_2(n)$, let $a, b \in S$ so that the distance along the outer cycle between the two closest vertices of the two chords of G is $b - a$. As C contains stars centered at both v_a and v_b, the graph G embeds onto C. □

We next construct a convex geometric graph G with n vertices and $O(n \log n)$ edges that is universal for n-vertex caterpillars; this bound is asymptotically optimal [12]. In order to construct G, we define a sequence π_n of n integers. Let $\pi_1 = (1)$. For every integer m of the form $m = 2^h - 1$, where $h \geq 2$, let $\pi_m = \pi_{(m-1)/2}(m)\pi_{(m-1)/2}$. For any $n \in \mathbb{N}$, the sequence π_n consists of the first n integers in π_m, where $m \geq n$ and $m = 2^h - 1$, for some $h \geq 1$. For example, $\pi_{10} = (1, 3, 1, 7, 1, 3, 1, 15, 1, 3)$. Let $\pi_n(i)$ be the ith term of π_n.

Property 1 [15]. For every $n \in \mathbb{N}$ and for every x with $1 \leq x \leq n$, the maximum of any x consecutive elements in π_n is at least x.

The graph G has vertices $v_1, \ldots, v_n$, placed in counterclockwise order along a circle c. Further, for $i = 1, \ldots, n$, we have that G contains edges connecting v_i to the $\pi_n(i)$ vertices preceding v_i and to the $\pi_n(i)$ vertices following v_i along c.

Theorem 4. *For every $n \in \mathbb{N}$, there exists a convex geometric graph G with n vertices and $O(n \log n)$ edges that is universal for n-vertex caterpillars.*

Proof sketch: The number of edges of G is at most twice the sum of the integers in π_n; the latter is less than or equal to the sum of the integers in π_m, where $m < 2n$ and $m = 2^h - 1$, for some integer $h \geq 1$. Further, π_m is easily shown to be equal to $(h - 1) \cdot 2^h + 1 \in O(n \log n)$.

Let C be an n-vertex caterpillar and let $(u_1, \ldots, u_s)$ be the spine of C. For $i = 1, \ldots, s$, let S_i be the star composed of u_i and its adjacent leaves; let $n_i = |V(S_i)|$. Let $m_1 = 0$; for $i = 2, \ldots, s$, let $m_i = \sum_{j=1}^{i-1} n_j$. For $i = 1, \ldots, s$, we embed S_i onto the subgraph G_i of G induced by the vertices $v_{m_i+1}, v_{m_i+2}, \ldots, v_{m_i+n_i}$: This is done by embedding u_i at the vertex v_{x_i} of G_i whose degree (in G) is maximum, and by embedding the leaves of S_i at the remaining vertices of G_i. By Property 1, we have that v_{x_i} is adjacent in G to the n_i vertices preceding it and to the n_i vertices following it along c, hence it is adjacent to all other vertices of G_i; thus, the above embedding of S_i onto G_i is valid. The arguments showing that the edge $v_{x_i} v_{x_{i+1}}$ belongs to G for all $i = 1, \ldots, s-1$ are analogous. The proof is concluded by observing that the edges of the spine $(u_1, \ldots, u_s)$ do not cross each other, since the vertices $u_1, \ldots, u_s$ appear in this order along c. □

4   Conclusions and Open Problems

In this paper we introduced and studied the problem of constructing geometric graphs with few vertices and edges that are universal for families of planar graphs. Our research raises several challenging problems.

What is the minimum number of edges of an n-vertex convex geometric graph that is universal for n-vertex trees? We proved that the answer is in $O(n \log n)$ if convexity is not required, or if caterpillars, rather than trees, are considered, while it is close to $\Omega(n^2)$ if outerplanar graphs, rather than trees, are considered. What is the minimum number of edges in a geometric graph that is universal for all n-vertex planar graphs? For abstract graphs, Babai et al. [3] constructed a universal graph with $O(n^{3/2})$ edges based on separators. Can such a construction be adapted to a geometric setting? The current best lower bound is $\Omega(n \log n)$, same as for trees [13], while the best upper bound is only $O(n^4)$.

Finally, the problems we studied in this paper can be posed for *topological* (multi-)graphs, as well, in which edges are represented by Jordan arcs.

Theorem 5. *For every $n \in \mathbb{N}$, there is a topological multigraph with n vertices and $O(n^3)$ edges that contains a planar drawing of every n-vertex planar graph.*

Theorem 6. *For every $n \in \mathbb{N}$, there is a topological multigraph with n vertices and $O(n^2)$ edges that contains a planar drawing of every n-vertex subhamiltonian planar graph.*

References

1. Alon, N., Capalbo, M.R.: Optimal universal graphs with deterministic embedding. In: Proceedings of the 19th ACM-SIAM Symposium on Discrete Algorithms (SODA), pp. 373–378 (2008). https://dl.acm.org/doi/abs/10.5555/1347082. 1347123
2. Alon, N., Capalbo, M.R., Kohayakawa, Y., Rödl, V., Rucinski, A., Szemerédi, E.: Universality and tolerance. In: Proceedings of the 41st IEEE Symposium on Foundations of Computer Science (FOCS), pp. 14–21 (2000). https://doi.org/10. 1109/SFCS.2000.892007
3. Babai, L., Chung, F.R.K., Erdős, P., Graham, R.L., Spencer, J.H.: On graphs which contain all sparse graphs. Ann. Discrete Math. **12**, 21–26 (1982) https:// doi.org/10.1016/S0304-0208(08)73486-8
4. Bannister, M.J., Cheng, Z., Devanny, W.E., Eppstein, D.: Super patterns and universal point sets. J. Graph Algorithms Appl. **18**(2), 177–209 (2014). https:// doi.org/10.7155/jgaa.00318
5. Bhatt, S.N., Chung, F.R.K., Leighton, F.T., Rosenberg, A.L.: Universal graphs for bounded-degree trees and planar graphs. SIAM J. Discrete Math. **2**(2), 145–155 (1989). https://doi.org/10.1137/0402014
6. Bose, P.: On embedding an outer-planar graph in a point set. Comput. Geom. **23**(3), 303–312 (2002). https://doi.org/10.1016/S0925-7721(01)00069-4
7. Böttcher, J., Pruessmann, K.P., Taraz, A., Würfl, A.: Bandwidth, expansion, treewidth, separators and universality for bounded-degree graphs. Eur. J. Comb. **31**(5), 1217–1227 (2010). https://doi.org/10.1016/j.ejc.2009.10.010
8. Böttcher, J., Schacht, M., Taraz, A.: Proof of the bandwidth conjecture of Bollobás and Komlós. Math. Ann. **343**(1), 175–205 (2009). https://doi.org/10.1007/s00208-008-0268-6
9. Bukh, B., Matoušek, J., Nivasch, G.: Lower bounds for weak epsilon-nets and stairconvexity. Israel J. Math. **182**, 199–228 (2011). https://doi.org/10.1007/s11856-011-0029-1

10. Capalbo, M.R.: Small universal graphs for bounded-degree planar graphs. Combinatorica **22**(3), 345–359 (2002). https://doi.org/10.1007/s004930200017
11. Cardinal, J., Hoffmann, M., Kusters, V.: On universal point sets for planar graphs. J. Graph Algorithms Appl. **19**(1), 529–547 (2015). https://doi.org/10.7155/jgaa.00374
12. Chung, F.R.K., Graham, R.L.: On graphs which contain all small trees. J. Combin. Theory Ser. B **24**(1), 14–23 (1978). https://doi.org/10.1016/0095-8956(78)90072-2
13. Chung, F.R.K., Graham, R.L.: On universal graphs for spanning trees. J. London Math. Soc. **27**(2), 203–211 (1983). https://doi.org/10.1112/jlms/s2-27.2.203
14. Frati, F., Hoffmann, M., Tóth, C.D.: Universal geometric graphs. CoRR abs/2006.11262 (2020). https://arxiv.org/abs/2006.11262
15. Frati, F., Patrignani, M., Roselli, V.: LR-drawings of ordered rooted binary trees and near-linear area drawings of outerplanar graphs. J. Comput. Syst. Sci. **107**, 28–53 (2020). https://doi.org/10.1016/j.jcss.2019.08.001
16. de Fraysseix, H., Pach, J., Pollack, R.: How to draw a planar graph on a grid. Combinatorica **10**(1), 41–51 (1990). https://doi.org/10.1007/BF02122694
17. Fulek, R., Tóth, C.D.: Universal point sets for planar three-trees. J. Discrete Algorithms **30**, 101–112 (2015). https://doi.org/10.1016/j.jda.2014.12.005
18. Gritzmann, P., Mohar, B., Pach, J., Pollack, R.: Embedding a planar triangulation with vertices at specified points. Amer. Math. Monthly **98**(2), 165–166 (1991)
19. Kurowski, M.: A 1.235 lower bound on the number of points needed to draw all n-vertex planar graphs. Inf. Process. Lett. **92**(2), 95–98 (2004). https://doi.org/10.1016/j.ipl.2004.06.009
20. Rado, R.: Universal graphs and universal functions. Acta Arithmetica **9**(4), 331–340 (1964). https://doi.org/10.4064/aa-9-4-331-340
21. Scheucher, M., Schrezenmaier, H., Steiner, R.: A note on universal point sets for planar graphs. J. Graph Algorithms Appl. **24**(3), 247–267 (2020). https://doi.org/10.7155/jgaa.00529

Computing Subset Transversals
in H-Free Graphs

Nick Brettell$^{(\boxtimes)}$, Matthew Johnson, Giacomo Paesani, and Daniël Paulusma

Department of Computer Science, Durham University, Durham, UK
{nicholas.j.brettell,matthew.johnson2,giacomo.paesani,
daniel.paulusma}@durham.ac.uk

Abstract. We study the computational complexity of two well-known graph transversal problems, namely SUBSET FEEDBACK VERTEX SET and SUBSET ODD CYCLE TRANSVERSAL, by restricting the input to H-free graphs, that is, to graphs that do not contain some fixed graph H as an induced subgraph. By combining known and new results, we determine the computational complexity of both problems on H-free graphs for every graph H except when $H = sP_1 + P_4$ for some $s \geq 1$. As part of our approach, we introduce the SUBSET VERTEX COVER problem and prove that it is polynomial-time solvable for $(sP_1 + P_4)$-free graphs for every $s \geq 1$.

1 Introduction

The central question in Graph Modification is whether or not a graph G can be modified into a graph from a prescribed class $\mathcal{G}$ via at most k graph operations from a prescribed set S of permitted operations such as vertex or edge deletion. The *transversal* problems VERTEX COVER, FEEDBACK VERTEX SET and ODD CYCLE TRANSVERSAL are classical problems of this kind. For example, the VERTEX COVER problem is equivalent to asking if one can delete at most k vertices to turn G into a member of the class of edgeless graphs. The problems FEEDBACK VERTEX SET and ODD CYCLE TRANSVERSAL ask if a graph G can be turned into, respectively, a forest or a bipartite graph by deleting vertices.

We can relax the condition on belonging to a prescribed class to obtain some related *subset transversal* problems. We state these formally after some definitions. For a graph $G = (V, E)$ and a set $T \subseteq V$, an *(odd) T-cycle* is a cycle of G (with an odd number of vertices) that intersects T. A set $S_T \subseteq V$ is a T-*vertex cover*, a T-*feedback vertex set* or an *odd T-cycle transversal* of G if S_T has at least one vertex of, respectively, every edge incident to a vertex of T, every T-cycle, or every odd T-cycle. For example, let G be a star with centre vertex c, whose leaves form the set T. Then, both $\{c\} = V \setminus T$ and T are T-vertex covers of G but the first is considerably smaller than the second. See Fig. 1 for some more examples.

This paper received support from the Leverhulme Trust (RPG-2016-258).

I. Adler and H. Müller (Eds.): WG 2020, LNCS 12301, pp. 187–199, 2020.
https://doi.org/10.1007/978-3-030-60440-0_15

SUBSET VERTEX COVER
  Instance: a graph $G = (V, E)$, a subset $T \subseteq V$ and a positive integer k.
  Question: does G have a T-vertex cover S_T with $|S_T| \leq k$?

SUBSET FEEDBACK VERTEX SET
  Instance: a graph $G = (V, E)$, a subset $T \subseteq V$ and a positive integer k.
  Question: does G have a T-feedback vertex set S_T with $|S_T| \leq k$?

SUBSET ODD CYCLE TRANSVERSAL
  Instance: a graph $G = (V, E)$, a subset $T \subseteq V$ and a positive integer k.
  Question: does G have an odd T-cycle transversal S_T with $|S_T| \leq k$?

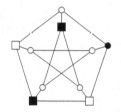 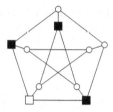

Fig. 1. In both examples, the square vertices of the Petersen graph form a set T and the black vertices form an odd T-cycle transversal S_T, which is also a T-feedback vertex set. In the left example, $S_T \setminus T \neq \emptyset$, and in the right example, $S_T \subseteq T$.

The SUBSET FEEDBACK VERTEX SET and SUBSET ODD CYCLE TRANSVERSAL problems are well known. The SUBSET VERTEX COVER problem is introduced in this paper, and we are not aware of past work on this problem. On general graphs, SUBSET VERTEX COVER is polynomially equivalent to VERTEX COVER: to solve SUBSET VERTEX COVER remove edges in the input graph that are not incident to any vertex of T to yield an equivalent instance of VERTEX COVER. However, this equivalence no longer holds for graph classes that are *not* closed under edge deletion.

As the three problems are NP-complete, we consider the restriction of the input to special graph classes in order to better understand which graph properties cause the computational hardness. Instead of classes closed under edge deletion, we focus on classes of graphs closed under vertex deletion. Such classes are called *hereditary*. The reasons for this choice are threefold. First, hereditary graph classes capture many well-studied graph classes. Second, every hereditary graph class $\mathcal{G}$ can be characterized by a (possibly infinite) set $\mathcal{F}_\mathcal{G}$ of forbidden induced subgraphs. This enables us to initiate a *systematic* study, starting from the case where $|\mathcal{F}_\mathcal{G}| = 1$. Third, we aim to extend and strengthen existing complexity results (that are for hereditary graph classes). If $\mathcal{F}_\mathcal{G} = \{H\}$ for some graph H, then $\mathcal{G}$ is *monogenic*, and every $G \in \mathcal{G}$ is H-free. Our research question is: *How does the structure of a graph H influence the computational complexity of a subset transversal problem for input graphs that are H-free?*

As a general strategy one might first try to prove that the restriction to H-free graphs is NP-complete if H contains a cycle or an induced claw (the 4-vertex star). This is usually done by showing, respectively, that the problem is NP-complete on graphs of arbitrarily large girth (the length of a shortest cycle) and on line graphs, which form a subclass of claw-free graphs. If this is the case, then it remains to consider the case where H has no cycle, and has no claw either. So H is a *linear forest*, that is, the disjoint union of one or more paths.

Existing Results. As NP-completeness results for transversal problems carry over to subset transversal problems, we first discuss results on FEEDBACK VERTEX SET and ODD CYCLE TRANSVERSAL for H-free graphs. By Poljak's construction [28], FEEDBACK VERTEX SET is NP-complete for graphs of girth at least g for every integer $g \geq 3$. The same holds for ODD CYCLE TRANSVERSAL [6]. Moreover, FEEDBACK VERTEX SET [30] and ODD CYCLE TRANSVERSAL [6] are NP-complete for line graphs and thus for claw-free graphs. Hence, both problems are NP-complete for H-free graphs if H has a cycle or claw. Both problems are polynomial-time solvable for P_4-free graphs [4], for sP_2-free graphs for every $s \geq 1$ [6] and for $(sP_1 + P_3)$-free graphs for every $s \geq 1$ [9]. In addition, ODD CYCLE TRANSVERSAL is NP-complete for (P_2+P_5, P_6)-free graphs [9]. Very recently, Abrishami et al. showed that FEEDBACK VERTEX SET is polynomial-time solvable for P_5-free graphs [1]. We summarize as follows ($F \subseteq_i G$ means that F is an induced subgraph of G; see Sect. 2 for the other notation used).

Theorem 1. *For a graph H, FEEDBACK VERTEX SET on H-free graphs is polynomial-time solvable if $H \subseteq_i P_5$, $H \subseteq_i sP_1 + P_3$ or $H \subseteq_i sP_2$ for some $s \geq 1$, and NP-complete if $H \supseteq_i C_r$ for some $r \geq 3$ or $H \supseteq_i K_{1,3}$.*

Theorem 2. *For a graph H, ODD CYCLE TRANSVERSAL on H-free graphs is polynomial-time solvable if $H = P_4$, $H \subseteq_i sP_1 + P_3$ or $H \subseteq_i sP_2$ for some $s \geq 1$, and NP-complete if $H \supseteq_i C_r$ for some $r \geq 3$, $H \supseteq_i K_{1,3}$, $H \supseteq_i P_6$ or $H \supseteq_i P_2 + P_5$.*

We note that no integer r is known such that FEEDBACK VERTEX SET is NP-complete for P_r-free graphs. This situation changes for SUBSET FEEDBACK VERTEX SET which is, unlike FEEDBACK VERTEX SET, NP-complete for split graphs (that is, $(2P_2, C_4, C_5)$-free graphs), as shown by Fomin et al. [12]. Papadopoulos and Tzimas [26,27] proved that SUBSET FEEDBACK VERTEX SET is polynomial-time solvable for sP_1-free graphs for any $s \geq 1$, co-bipartite graphs, interval graphs and permutation graphs, and thus P_4-free graphs. Some of these results were generalized by Bergougnoux et al. [2], who solved an open problem of Jaffke et al. [17] by giving an $n^{O(w^2)}$-time algorithm for SUBSET FEEDBACK VERTEX SET given a graph and a decomposition of this graph of mim-width w. This does not lead to new results for H-free graphs: a class of H-free graphs has bounded mim-width if and only if $H \subseteq_i P_4$ [5].

We are not aware of any results on SUBSET ODD CYCLE TRANSVERSAL for H-free graphs, but note that this problem generalizes ODD MULTIWAY CUT, just as SUBSET FEEDBACK VERTEX SET generalizes NODE MULTIWAY CUT, another well-studied problem. We refer to [7,8,12,13,15,16,19–22] for further details, in particular for parameterized and exact algorithms for SUBSET FEEDBACK VERTEX SET and SUBSET ODD CYCLE TRANSVERSAL. These algorithms are beyond the scope of this paper.

Our Results. By a significant extension of the known results for the two problems on H-free graphs we obtain two almost-complete dichotomies:

Theorem 3. *Let H be a graph with $H \neq sP_1 + P_4$ for all $s \geq 1$. Then* SUBSET FEEDBACK VERTEX SET *on H-free graphs is polynomial-time solvable if $H = P_4$ or $H \subseteq_i sP_1 + P_3$ for some $s \geq 1$ and* NP-*complete otherwise.*

Theorem 4. *Let H be a graph with $H \neq sP_1 + P_4$ for all $s \geq 1$. Then* SUBSET ODD CYCLE TRANSVERSAL *on H-free graphs is polynomial-time solvable if $H = P_4$ or $H \subseteq_i sP_1 + P_3$ for some $s \geq 1$ and* NP-*complete otherwise.*

Though the proved complexities of SUBSET FEEDBACK VERTEX SET and SUBSET ODD CYCLE TRANSVERSAL are the same on H-free graphs, the algorithm that we present for SUBSET ODD CYCLE TRANSVERSAL on $(sP_1 + P_3)$-free graphs is more technical compared to the algorithm for SUBSET FEEDBACK VERTEX SET, and considerably generalizes the transversal algorithms for $(sP_1 + P_3)$-free graphs of [9]. There is further evidence that SUBSET ODD CYCLE TRANSVERSAL is a more challenging problem than SUBSET FEEDBACK VERTEX SET. For example, the best-known parameterized algorithm for SUBSET FEEDBACK VERTEX SET runs in $O^*(4^k)$ time [16], but the best-known run-time for SUBSET ODD CYCLE TRANSVERSAL is $O^*(2^{O(k^3 \log k)})$ [22]. Moreover, it is not known if there is an XP algorithm for SUBSET ODD CYCLE TRANSVERSAL in terms of mim-width in contrast to the known XP algorithm for SUBSET FEEDBACK VERTEX SET [2].

In Sect. 2 we introduce our terminology. In Sect. 3 we present some results for SUBSET VERTEX COVER: the first result shows that SUBSET VERTEX COVER is polynomial-time solvable for $(sP_1 + P_4)$-free graphs for every $s \geq 1$, and we later use this as a subroutine to obtain a polynomial-time algorithm for SUBSET ODD CYCLE TRANSVERSAL on P_4-free graphs. We present our results on SUBSET FEEDBACK VERTEX SET and SUBSET ODD CYCLE TRANSVERSAL in Sects. 4 and 5, respectively. In Sect. 6 on future work we discuss SUBSET VERTEX COVER in more detail.

2   Preliminaries

We consider undirected, finite graphs with no self-loops and no multiple edges. Let $G = (V, E)$ be a graph, and let $S \subseteq V$. The graph $G[S]$ is the subgraph of G induced by S. We write $G - S$ to denote the graph $G[V \setminus S]$. Recall that for a graph F, we write $F \subseteq_i G$ if F is an induced subgraph of G. The cycle and path on r vertices are denoted C_r and P_r, respectively. We say that S is *independent* if $G[S]$ is edgeless, and that S is a *clique* if $G[S]$ is *complete*, that is, contains every possible edge between two vertices. We let K_r denote the complete graph on r vertices, and sP_1 denote the graph whose vertices form an independent set of size s. A *(connected) component* of G is a maximal connected subgraph of G. The graph $\overline{G} = (V, \{uv \mid uv \notin E \text{ and } u \neq v\})$ is the *complement* of G. The *neighbourhood* of a vertex $u \in V$ is the set $N_G(u) = \{v \mid uv \in E\}$. For $U \subseteq V$, we let $N_G(U) = \bigcup_{u \in U} N(u) \setminus U$. The *closed* neighbourhoods of u and U are denoted by $N_G[u] = N_G(u) \cup \{u\}$ and $N_G[U] = N_G(U) \cup U$, respectively. We omit subscripts when there is no ambiguity.

Let $T \subseteq V$ be such that $S \cap T = \emptyset$. Then S is *complete* to T if every vertex of S is adjacent to every vertex of T, and S is *anti-complete* to T if there are no edges between S and T. In the first case, S is also said to be *complete* to $G[T]$, and in the second case we say it is *anti-complete* to $G[T]$.

We say that G is a *forest* if it has no cycles, and, furthermore, that G is a *linear forest* if it is the disjoint union of one or more paths. The graph G is *bipartite* if V can be partitioned into at most two independent sets. A graph is *complete bipartite* if its vertex set can be partitioned into two independent sets X and Y such that X is complete to Y. We denote such a graph by $K_{|X|,|Y|}$. If X or Y has size 1, the complete bipartite graph is a *star*; recall that $K_{1,3}$ is also called a *claw*. A graph G is a *split graph* if it has a bipartition (V_1, V_2) such that $G[V_1]$ is a clique and $G[V_2]$ is an independent set. A graph is split if and only if it is $(C_4, C_5, 2P_2)$-free [11].

Let G_1 and G_2 be two vertex-disjoint graphs. The *union* operation $+$ creates the disjoint union $G_1 + G_2$ of G_1 and G_2 (recall that $G_1 + G_2$ is the graph with vertex set $V(G_1) \cup V(G_2)$ and edge set $E(G_1) \cup E(G_2)$).

We also consider optimization versions of subset transversal problems, in which case we have instances (G, T) (instead of instances (G, T, k)). We say that a set $S \subseteq V(G)$ is a *solution* for an instance (G, T) if S is a T-transversal (of whichever kind we are concerned with). A solution S is *smaller* than a solution S' if $|S| < |S'|$, and a solution S is *minimum* if (G, T) does not have a solution smaller than S, and it is *maximum* if there is no larger solution. We will use the following general lemma, which was implicitly used in [27].

Lemma 1. *Let S be a minimum solution for an instance (G, T) of a subset transversal problem. Then $|S \setminus T| \leq |T \setminus S|$.*

Let $T \subseteq V$ be a vertex subset of a graph $G = (V, E)$. Recall that a cycle is a T-cycle if it contains a vertex of T. A subgraph of G is a T-*forest* if it has no T-cycles. Recall also that a cycle is odd if it contains an odd number of edges. A subgraph of G is T-*bipartite* if it has no odd T-cycles. Recall that a set $S_T \subseteq V$ is a T-*vertex cover*, a T-*feedback vertex set* or an *odd T-cycle transversal* of G if S_T has at least one vertex of, respectively every edge incident to a vertex of T, every T-cycle, or every odd T-cycle. Note that S_T is a T-feedback vertex set if and only if $G[V \setminus S_T]$ is a T-forest, and S_T is an odd T-cycle transversal if and only if $G[V \setminus S_T]$ is T-bipartite. A T-*path* is a path that contains a vertex of T. A T-path is *odd* (or *even*) if the number of edges in the path is odd (or even, respectively).

We will use the following easy lemma, which proves that T-forests and T-bipartite graphs can be recognized in polynomial time. It combines results claimed but not proved in [22,27].

Lemma 2. *Let $G = (V, E)$ be a graph and $T \subseteq V$. Then deciding whether or not G is a T-forest or T-bipartite takes $O(n + m)$ time.*

3   Subset Vertex Cover

We need the following two results on SUBSET VERTEX COVER (proofs omitted).

Lemma 3. SUBSET VERTEX COVER *can be solved in polynomial time for P_4-free graphs.*

Lemma 4. *Let H be a graph. If* SUBSET VERTEX COVER *is polynomial-time solvable for H-free graphs, then it is for $(P_1 + H)$-free graphs as well.*

Lemma 3, combined with s applications of Lemma 4, yields the following result.

Theorem 5. *For every integer $s \geq 1$,* SUBSET VERTEX COVER *can be solved in polynomial time for $(sP_1 + P_4)$-free graphs.*

4   Subset Feedback Vertex Set

To prove Theorem 3. We require two lemmas. In the first lemma (whose proof we omit), the bound of $4s - 2$ is not necessarily tight, but suffices for our needs.

Lemma 5. *Let s be a non-negative integer, and let R be an $(sP_1 + P_3)$-free tree. Then either*

(i) $|V(R)| \leq \max\{7, 4s - 2\}$, *or*
(ii) *R has precisely one vertex r of degree more than 2 and at most $s - 1$ vertices of degree 2, each adjacent to r. Moreover, r has at least $3s - 1$ neighbours.*

We can extend "partial" solutions to full solutions in polynomial time as follows.

Lemma 6. *Let $G = (V, E)$ be a graph with a set $T \subseteq V$. Let $V' \subseteq V$ and $S'_T \subseteq V'$ such that S'_T is a T-feedback vertex set of $G[V']$, and let $Z = V \setminus V'$. Suppose that $G[Z]$ is P_3-free, and $|N_{G-S'_T}(Z)| \leq 1$. Then there is a polynomial-time algorithm that finds a minimum T-feedback vertex set S_T of G such that $S'_T \subseteq S_T$ and $V' \setminus S'_T \subseteq V \setminus S_T$.*

Proof. Since $G[Z]$ is P_3-free, it is a disjoint union of complete graphs. Let $G' = G - S'_T$, and consider a T-cycle C in G'. Then C contains at least one vertex of Z. If $N_{G'}(Z) = \emptyset$, then C is contained in a component of $G[Z]$. On the other hand, if $N_{G'}(Z) = \{y\}$, say, then y is a cut-vertex of G', so there exists a component $G[U]$ of $G[Z]$ such that C is contained in $G[U \cup \{y\}]$. Hence, we can consider each component of $G[Z]$ independently: for each component $G[U]$ it suffices to find the maximum subset U' of U such that $G[U' \cup N_{G'}(U)]$ contains no T-cycles. Then $U' \subseteq F_T$ and $U \setminus U' \subseteq S_T$, where $F_T = V \setminus S_T$.

Let $U \subseteq Z$ such that $G[U]$ is a component of $G[Z]$. Either $N_{G'}(U) \cap T = \emptyset$, or $N_{G'}(U) = \{y\}$ for some $y \in T$. First, consider the case where $N_{G'}(U) \cap T = \emptyset$. We find a set U' that is a maximum subset of U such that $G[U' \cup N_{G'}(U)]$ has no T-cycles. Clearly if $|U| = 1$, then we can set $U' = U$. If $|U'| \geq 3$, then, since U' is a clique, $U' \subseteq V \setminus T$. Thus, if $|U \setminus T| \geq 2$, then we set $U' = U \setminus T$. So it remains to consider when $|U| \geq 2$ but $|U \setminus T| \leq 1$. If there is some $u \in U$ that is anti-complete to $N_{G'}(U)$, then we can set U' to be any 2-element subset of U containing u. Otherwise $N_{G'}(U) = \{y\}$ and y is complete to U. In this case, for any $u \in U$, we set $U' = \{u\}$.

Now we may assume that $N_{G'}(U) = \{y\}$ and $y \in T$. Again, we find a set U' that is a maximum subset of U such that $G[U' \cup \{y\}]$ has no T-cycles. Partition U into $\{U_0, U_1\}$ where $u \in U_1$ if and only if u is a neighbour of y. Since $y \in V' \setminus S_T'$, observe that U' contains at most one vertex of U_1, otherwise $G[U' \cup \{y\}]$ has a T-cycle. Since U' is a clique, if $|U'| \geq 3$ then $U' \subseteq U \setminus T$. So if $|U_0 \setminus T| \geq 2$ and there is an element $u \in U_1 \setminus T$, then we can set $U' = \{u\} \cup (U_0 \setminus T)$. If $|U_0 \setminus T| \geq 2$ but $U_1 \setminus T = \emptyset$, then we can set $U' = U_0 \setminus T$. So we may now assume that $|U_0 \setminus T| \leq 1$. If $U_0 \neq \emptyset$ and $|U| \geq 2$, then we set U' to any 2-element subset of U containing some $u \in U_0$. Clearly if $|U| = 1$, then we can set $U' = U$. So it remains to consider when $U_0 = \emptyset$ and $|U_1| \geq 2$. In this case, we set $U' = \{u\}$ for an arbitrary $u \in U_1$. $\square$

We now prove the main result of this section.

Theorem 6. *For every integer $s \geq 0$,* SUBSET FEEDBACK VERTEX SET *can be solved in polynomial time for $(sP_1 + P_3)$-free graphs.*

Proof. Let $G = (V, E)$ be an $(sP_1 + P_3)$-free graph for some $s \geq 0$, and let $T \subseteq V$. We describe a polynomial-time algorithm for the optimization version of the problem on input (G, T). Let $S_T \subseteq V$ such that S_T is a minimum T-feedback vertex set of G, and let $F_T = V \setminus S_T$, so $G[F_T]$ is a maximum T-forest. Note that $G[F_T \cap T]$ is a forest. We consider three cases: either

1. $G[F_T \cap T]$ has at least $2s$ components;
2. $G[F_T \cap T]$ has fewer than $2s$ components, and each of these components consists of at most $\max\{7, 4s - 2\}$ vertices; or
3. $G[F_T \cap T]$ has fewer than $2s$ components, one of which consists of at least $\max\{8, 4s - 1\}$ vertices.

We describe polynomial-time subroutines that find a set F_T such that $G[F_T]$ is a maximum T-forest in each of these three cases, giving a minimum solution $S_T = V \setminus F_T$ in each case. We obtain an optimal solution by running each of these subroutines in turn: of the (at most) three potential solutions, we output the one with minimum size.

Case 1: $G[F_T \cap T]$ has at least $2s$ components.

We begin by proving a sequence of claims that describe properties of a maximum T-forest F_T, when in Case 1. Since G is $(sP_1 + P_3)$-free, $F_T \cap T$ induces a P_3-free forest, so $G[F_T \cap T]$ is a disjoint union of graphs isomorphic to P_1 or P_2. Let $A \subseteq F_T \cap T$ such that $G[A]$ consists of precisely $2s$ components. Note that $|A| \leq 4s$. We also let $Y = N(A) \cap F_T$, and partition Y into $\{Y_1, Y_2\}$ where $y \in Y_1$ if y has only one neighbour in A, whereas $y \in Y_2$ if y has at least two neighbours in A.

Claim 1. $|Y_2| \leq 1$.

Proof of Claim 1. Let $v \in Y_2$. Then v has neighbours in at least $s + 1$ of the components of $G[A]$, otherwise $G[A \cup \{v\}]$ contains an induced $sP_1 + P_3$. Note also that v has at most one neighbour in each component of $G[A]$, otherwise $G[F_T]$ has a T-cycle. Now suppose that Y_2 contains distinct vertices v_1 and v_2. Then, of the $2s$ components of $G[A]$, the vertices v_1 and v_2 each have some neighbour in $s + 1$ of these components. So there are at least two components of $G[A]$ containing both a vertex adjacent to v_1, and a vertex adjacent to v_2. Let A' and A'' be the vertex sets of two such components. Then $A' \cup A'' \cup \{v_1, v_2\} \subseteq F_T$, but $G[A' \cup A'' \cup \{v_1, v_2\}]$ has a T-cycle; a contradiction.   ◇

Claim 2. $|Y| \leq 2s + 1$.

Proof of Claim 2. By Claim 1, it suffices to prove that $|Y_1| \leq 2s$. We argue that each component of $G[A]$ has at most one neighbour in Y_1, implying that $|Y_1| \leq 2s$. Indeed, suppose that there is a component C_A of $G[A]$ having two neighbours in Y_1, say u_1 and u_2. Then $G[V(C_A) \cup \{u_1, u_2\}]$ contains an induced P_3 that is anti-complete to $A \setminus V(C_A)$, contradicting that G is $(sP_1 + P_3)$-free.   ◇

Claim 3. Y_1 is independent, and no component of $G[A]$ of size 2 has a neighbour in Y_1.

Proof of Claim 3. Suppose that there are adjacent vertices u_1 and u_2 in Y_1. Let a_i be the unique neighbour of u_i in A for $i \in \{1, 2\}$. Note that $a_1 \neq a_2$, for otherwise $G[F_T]$ has a T-cycle. Then $\{a_1, u_1, u_2\}$ induces a P_3, so $G[\{u_1, u_2\} \cup A]$ contains an induced $sP_1 + P_3$, which is a contradiction. We deduce that Y_1 is independent.

Now let $\{a_1, a_2\} \subseteq A$ such that $G[\{a_1, a_2\}]$ is a component of $G[A]$, and suppose that $u_1 \in Y_1$ is adjacent to a_1. Then a_1 is the unique neighbour of u_1 in A, so $G[\{u_1, a_1, a_2\}] \cong P_3$. Thus $G[\{u_1\} \cup A]$ contains an induced $sP_1 + P_3$, which is a contradiction.   ◇

Claim 4. Let $Z = V \setminus N[A]$. Then $N(Z) \cap F_T \subseteq Y_2$.

Proof of Claim 4. Suppose that there exists $y \in Y_1$ that is adjacent to a vertex $c \in Z$. Let a be the unique neighbour of y in A. Then $G[\{c, y\} \cup A]$ contains an induced $sP_1 + P_3$, which is a contradiction. So Y_1 is anti-complete to Z. Now, if $c \in Z$ is adjacent to a vertex in $N[A] \cap F_T$, then c is adjacent to y_2 where $Y_2 = \{y_2\}$.   ◇

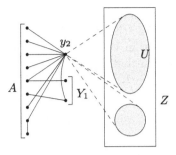

Fig. 2. An example of the structure obtained in Case 1 when $Y_2 = \{y_2\}$.

We now describe the subroutine that finds an optimal solution in Case 1. In this case, for any maximum forest F_T, there exists some set $A \subseteq T$ of size at most $4s$ such that $A \subseteq F_T$, and $G[A]$ consists of exactly $2s$ components, each isomorphic to either P_1 or P_2. Moreover, there is such an A for which $N(A) \cap T \subseteq S_T$. Thus we guess a set $A' \subseteq T$ in $O(n^{4s})$ time, discarding those sets that do not induce a forest with exactly $2s$ components, and those that induce a component consisting of more than two vertices.

For any such F_T and A', the set $N(A') \cap F_T$ has size at most $2s + 1$, by Claim 2. Thus, in $O(n^{2s+1})$ time, we guess $Y' \subseteq N(A')$ with $|Y'| \leq 2s + 1$, and assume that $Y' \subset F_T$ whereas $N(A') \setminus Y' \subset S_T$. Let Y_2' be the subset of Y' that contains vertices that have at least two neighbours in A'. We discard any sets Y' that do not satisfy Claims 1 or 3, or those sets for which $G[A' \cup Y']$ has a T-cycle on three vertices, one of which is the unique vertex of Y_2'.

Let $Z = V \setminus N[A']$ (for example, see Fig. 2). Since $G[A']$ contains an induced sP_1, the subgraph $G[Z]$ is P_3-free. Now $N(Z) \cap F_T \subseteq Y_2'$ by Claim 4, where $|Y_2'| \leq 1$ by Claim 1. Thus, by Lemma 6, we can extend a partial solution $S_T' = N[A'] \setminus (A' \cup Y')$ of $G[N[A']]$ to a solution S_T of G, in polynomial time.

Case 2: $G[F_T \cap T]$ has at most $2s - 1$ components, each of size at most $\max\{7, 4s - 2\}$.

We guess sets $F \subseteq T$ and $S \subseteq V \setminus T$ such that $F_T \cap T = F$ and $S_T \setminus T = S$. Since F has size at most $(2s-1)\max\{7, 4s-2\}$ vertices, there are $O(n^{\max\{14s-7, 8s^2-8s+2\}})$ possibilities for F. By Lemma 1, we may assume that $|S_T \cap T| \leq |F|$. So for each guessed F, there are at most $O(n^{\max\{14s-7, 8s^2-8s+2\}})$ possibilities for S. For each S and F, we set $S_T = (T \setminus F) \cup S$ and check, in $O(n + m)$-time by Lemma 2, if $G - S_T$ is a T-forest. In this way we exhaustively find all solutions satisfying Case 2, in $O(n^{\max^2\{14s-7, 8s^2-8s+2\}})$ time; we output the one of minimum size.

Case 3: $G[F_T \cap T]$ has at most $2s - 1$ components, one of which has size at least $\max\{8, 4s - 1\}$.

By Lemma 5, there is some subset $B_T \subseteq F_T \cap T$ such that $|B| \geq \max\{8, 4s - 1\}$, and $G[B]$ is a component of $G[F_T \cap T]$ that is a tree satisfying Lemma 5(ii). In particular, there is a unique vertex $r \in B$ such that r has degree more than 2 in

$G[B]$. Moreover, $G[F_T]$ has a component $G[D]$ that contains B, where $G[D]$ is a tree that also satisfies Lemma 5(ii). Note that there are at most $s-1$ vertices in $N_{G[B]}(r)$ having a neighbour in $V \setminus T$.

We guess a set $B' \subseteq T$ such that $|B'| = \max\{8, 4s-1\}$. We also guess a set $L' \subseteq V \setminus T$ such that $|L'| \le s-1$. Let $D' = B' \cup L'$. We check that $G[D']$ has the following properties:

- $G[D']$ is a tree,
- $G[D']$ has a unique vertex r' of degree more than 2, with $r' \in B'$,
- $G[D']$ has at most $s-1$ vertices with distance 2 from r', and each of these vertices has degree 1, and
- each vertex $v \in L'$ has degree 1 in $G[D']$, and distance 2 from r'.

We assume that D' induces a subtree of the large component $G[D]$, where $r = r'$, and D' contains r, all neighbours of r with degree 2 in $G[D]$, and all vertices at distance 2 from r. In other words, $G[D']$ can be obtained from $G[D]$ by deleting some subset of the leaves of $G[D]$ that are adjacent to r. In particular, $D' \subseteq F_T$. We also assume that L' is the set of all vertices of $V(D) \setminus T$ that have distance 2 from r.

It follows from these assumptions that $N(D' \setminus \{r\}) \setminus \{r\} \subseteq S_T$. Let $Z = V \setminus N[D' \setminus \{r\}]$, and observe that each $z \in Z$ has at most one neighbour in D' (if it has such a neighbour, this neighbour is r). So $N(Z) \cap F_T \subseteq \{r\}$.

Towards an application of Lemma 6, we claim that $G[Z]$ is P_3-free. Let $B_1 = B' \cap N(r)$. As r has at least $3s-1$ neighbours in $G[B']$, by Lemma 5, $G[B_1]$ contains an induced sP_1. Moreover, $N(B_1) \cap F_T \subseteq D'$. Since G is $(sP_1 + P_3)$-free, $G[Z]$ is P_3-free. We now apply Lemma 6, which completes the proof. $\quad\square$

We are now ready to prove Theorem 3.

Theorem 3 (restated). *Let H be a graph with $H \ne sP_1 + P_4$ for all $s \ge 1$. Then* SUBSET FEEDBACK VERTEX SET *on H-free graphs is polynomial-time solvable if $H = P_4$ or $H \subseteq_i sP_1 + P_3$ for some $s \ge 1$ and is* NP*-complete otherwise.*

Proof. If H has a cycle or claw, we use Theorem 1. The cases $H = P_4$ and $H = 2P_2$ follow from the corresponding results for permutation graphs [26] and split graphs [12]. The remaining case $H \subseteq_i sP_1 + P_3$ follows from Theorem 6.
$\quad\square$

5   Subset Odd Cycle Transversal

At the end of this section we prove Theorem 4. We show three new results (proofs omitted). Our first result uses the reduction of [26] which proved the analogous result for SUBSET FEEDBACK VERTEX SET. Our third result is the main result of this section. Its proof uses the same approach as the proof of Theorem 6 but we need more advanced arguments for distinguishing cycles according to parity.

Theorem 7. SUBSET ODD CYCLE TRANSVERSAL *is* NP*-complete for the class of split graphs (or equivalently, $(C_4, C_5, 2P_2)$-free graphs).*

Theorem 8. SUBSET ODD CYCLE TRANSVERSAL *can be solved in polynomial time for P_4-free graphs.*

Theorem 9. *For every integer $s \geq 0$,* SUBSET ODD CYCLE TRANSVERSAL *can be solved in polynomial time for $(sP_1 + P_3)$-free graphs.*

We are now ready to prove our almost-complete classification.

Theorem 4 (restated). *Let H be a graph with $H \neq sP_1 + P_4$ for all $s \geq 1$. Then* SUBSET ODD CYCLE TRANSVERSAL *on H-free graphs is polynomial-time solvable if $H = P_4$ or $H \subseteq_i sP_1 + P_3$ for some $s \geq 1$ and* NP-*complete otherwise.*

Proof. If H has a cycle or claw, we use Theorem 2. The cases $H = P_4$ and $H = 2P_2$ follow from Theorems 7 and 8, respectively. The remaining case, where $H \subseteq_i sP_1 + P_3$, follows from Theorem 9. □

6   Conclusions

We gave almost-complete classifications of the complexity of SUBSET FEEDBACK VERTEX SET and SUBSET ODD CYCLE TRANSVERSAL for H-free graphs. The only open case in each classification is when $H = sP_1 + P_4$ for some $s \geq 1$, which is also open for FEEDBACK VERTEX SET and ODD CYCLE TRANSVERSAL for H-free graphs. Our proof techniques for $H = sP_1 + P_3$ do not carry over and new structural insights are needed in order to solve the missing cases where $H = sP_1 + P_4$ for $s \geq 1$.

We also introduced the SUBSET VERTEX COVER problem and showed that this problem is polynomial-time solvable on $(sP_1 + P_4)$-free graphs for every $s \geq 0$. Lokshtanov et al. [23] proved that VERTEX COVER is polynomial-time solvable for P_5-free graphs. Grzesik et al. [14] extended this result to P_6-free graphs. What is the complexity of SUBSET VERTEX COVER for P_5-free graphs? Does there exist an integer $r \geq 5$ such that SUBSET VERTEX COVER is NP-complete for P_r-free graphs. By Poljak's construction [28], VERTEX COVER is NP-complete for H-free graphs if H has a cycle. However, VERTEX COVER becomes polynomial-time solvable on $K_{1,3}$-free graphs [24,29]. We did not research the complexity of SUBSET VERTEX COVER on $K_{1,3}$-free graphs and also leave this as an open problem for future work.

Finally, several related transversal problems have been studied but not yet for H-free graphs. For example, the parameterized complexity of EVEN CYCLE TRANSVERSAL and SUBSET EVEN CYCLE TRANSVERSAL has been addressed in [25] and [19], respectively. Moreover, several other transversal problems have been studied for H-free graphs, but not the subset version: for example, CONNECTED VERTEX COVER, CONNECTED FEEDBACK VERTEX SET and CONNECTED ODD CYCLE TRANSVERSAL, and also for INDEPENDENT FEEDBACK VERTEX SET and INDEPENDENT ODD CYCLE TRANSVERSAL; see [3,6,10,18] for a number of recent results. It would be interesting to solve the subset versions of these transversal problems for H-free graphs and to determine the connections amongst all these problems in a more general framework.

References

1. Abrishami, T., Chudnovsky, M., Pilipczuk, M., Rzążewski, P., Seymour, P.: Induced subgraphs of bounded treewidth and the container method. CoRR, abs/2003.05185 (2020)
2. Bergougnoux, B., Papadopoulos, C., Telle, J.A.: Node multiway cut and subset feedback vertex set on graphs of bounded mim-width. In: Adler, I., Müller, H. (eds.) WG 2020. LNCS, vol. 12301, pp. 388–400. Springer, Cham (2020)
3. Bonamy, M., Dabrowski, K.K., Feghali, C., Johnson, M., Paulusma, D.: Independent feedback vertex set for P_5-free graphs. Algorithmica 81, 1342–1369 (2019)
4. Brandstädt, A., Kratsch, D.: On the restriction of some NP-complete graph problems to permutation graphs. In: Budach, L. (ed.) FCT 1985. LNCS, vol. 199, pp. 53–62. Springer, Heidelberg (1985). https://doi.org/10.1007/BFb0028791
5. Brettell, N., Horsfield, J., Munaro, A., Paesani, G., Paulusma, D.: Bounding the mim-width of hereditary graph classes. CoRR, abs/2004.05018 (2020)
6. Chiarelli, N., Hartinger, T.R., Johnson, M., Milanič, M., Paulusma, D.: Minimum connected transversals in graphs: new hardness results and tractable cases using the price of connectivity. Theoret. Comput. Sci. 705, 75–83 (2018)
7. Chitnis, R., Fomin, F.V., Lokshtanov, D., Misra, P., Ramanujan, M.S., Saurabh, S.: Faster exact algorithms for some terminal set problems. J. Comput. Syst. Sci. 88, 195–207 (2017)
8. Cygan, M., Pilipczuk, M., Pilipczuk, M., Wojtaszczyk, J.O.: Subset feedback vertex set is fixed-parameter tractable. SIAM J. Discrete Math. 27, 290–309 (2013)
9. Dabrowski, K.K., Feghali, C., Johnson, M., Paesani, G., Paulusma, D., Rzążewski, P.: On cycle transversals and their connected variants in the absence of a small linear forest. Algorithmica. To appear
10. Dabrowski, K.K., Johnson, M., Paesani, G., Paulusma, D., Zamaraev, V.: On the price of independence for vertex cover, feedback vertex set and odd cycle transversal. In: Proceedings of the MFCS 2018. LIPIcs, vol. 117, pp. 63:1–63:15 (2018)
11. Földes, S., Hammer, P.L.: Split graphs. Congressus Numerantium 19, 311–315 (1977)
12. Fomin, F.V., Heggernes, P., Kratsch, D., Papadopoulos, C., Villanger, Y.: Enumerating minimal subset feedback vertex sets. Algorithmica 69, 216–231 (2014)
13. Golovach, P.A., Heggernes, P., Kratsch, D., Saei, R.: Subset feedback vertex sets in chordal graphs. J. Discrete Algorithms 26, 7–15 (2014)
14. Grzesik, A., Klimošová, T., Pilipczuk, M., Pilipczuk, M.: Polynomial-time algorithm for maximum weight independent set on P_6-free graphs. In: Proceedings of the SODA 2019, pp. 1257–1271 (2019)
15. Hols, E.C., Kratsch, S.: A randomized polynomial kernel for subset feedback vertex set. Theory Comput. Syst. 62(1), 63–92 (2018)
16. Iwata, Y., Wahlström, M., Yoshida, Y.: Half-integrality, LP-branching, and FPT algorithms. SIAM J. Comput. 45(4), 1377–1411 (2016)
17. Jaffke, L., Kwon, O., Telle, J.A.: Mim-width II. The feedback vertex set problem. Algorithmica 82(1), 118–145 (2020)
18. Johnson, M., Paesani, G., Paulusma, D.: Connected vertex cover for $(sP_1 + P_5)$-free graphs. Algorithmica 82, 20–40 (2020)
19. Kakimura, N., Kawarabayashi, K., Kobayashi, Y.: Erdős-Pósa property and its algorithmic applications: parity constraints, subset feedback set, and subset packing. In: Proceedings of the SODA 2012, pp. 1726–1736 (2012)

20. Kawarabayashi, K., Kobayashi, Y.: Fixed-parameter tractability for the subset feedback set problem and the S-cycle packing problem. J. Comb. Theory Ser. B **102**(4), 1020–1034 (2012)
21. Kratsch, S., Wahlström, M.: Representative sets and irrelevant vertices: new tools for kernelization. In: Proceedings of the FOCS 2012, pp. 450–459 (2012)
22. Lokshtanov, D., Misra, P., Ramanujan, M.S., Saurabh, S.: Hitting selected (odd) cycles. SIAM J. Discrete Math. **31**(3), 1581–1615 (2017)
23. Lokshtanov, D., Vatshelle, M., Villanger, Y.: Independent set in P_5-free graphs in polynomial time. In: Proceedings of the SODA 2014, pp. 570–581 (2014)
24. Minty, G.J.: On maximal independent sets of vertices in claw-free graphs. J. Comb. Theory Ser. B **28**(3), 284–304 (1980)
25. Misra, P., Raman, V., Ramanujan, M.S., Saurabh, S.: Parameterized algorithms for Even Cycle Transversal. In: Golumbic, M.C., Stern, M., Levy, A., Morgenstern, G. (eds.) WG 2012. LNCS, vol. 7551, pp. 172–183. Springer, Heidelberg (2012). https://doi.org/10.1007/978-3-642-34611-8_19
26. Papadopoulos, C., Tzimas, S.: Polynomial-time algorithms for the subset feedback vertex set problem on interval graphs and permutation graphs. Discrete Appl. Math. **258**, 204–221 (2019)
27. Papadopoulos, C., Tzimas, S.: Subset feedback vertex set on graphs of bounded independent set size. Theoret. Comput. Sci. **814**, 177–188 (2020)
28. Poljak, S.: A note on stable sets and colorings of graphs. Commentationes Math. Univ. Carol. **15**, 307–309 (1974)
29. Sbihi, N.: Algorithme de recherche d'un stable de cardinalité maximum dans un graphe sans étoile. Discrete Math. **29**(1), 53–76 (1980)
30. Speckenmeyer, E.: Untersuchungen zum Feedback Vertex Set Problem in ungerichteten Graphen. Ph.D. thesis, Universität Paderborn (1983)

Feedback Edge Sets in Temporal Graphs

Roman Haag, Hendrik Molter[iD], Rolf Niedermeier[iD], and Malte Renken[✉][iD]

Algorithmics and Computational Complexity, Faculty IV, TU Berlin,
Berlin, Germany
{h.molter,rolf.niedermeier,m.renken}@tu-berlin.de

Abstract. The classical, linear-time solvable FEEDBACK EDGE SET problem is concerned with finding a minimum number of edges intersecting all cycles in a (static, unweighted) graph. We provide a first study of this problem in the setting of *temporal graphs*, where edges are present only at certain points in time. We find that there are four natural generalizations of FEEDBACK EDGE SET, all of which turn out to be NP-hard. We also study the tractability of these problems with respect to several parameters (solution size, lifetime, and number of graph vertices, among others) and obtain some parameterized hardness but also fixed-parameter tractability results.

1 Introduction

A temporal graph $\mathcal{G} = (V, \mathcal{E}, \tau)$ has a fixed vertex set V and each time-edge in $\mathcal{E}$ has a discrete time-label $t \in \{1, 2, \ldots, \tau\}$, where τ denotes the *lifetime* of the temporal graph $\mathcal{G}$. A *temporal cycle* in a temporal graph is a cycle of time-edges with increasing time-labels. We study the computational complexity of searching for small *feedback edge sets*, i.e., edge sets whose removal from the temporal graph destroys all temporal cycles. We distinguish between the following two variants of feedback edge set problems.

1. *Temporal feedback edge sets,* which consist of time-edges, that is, connections between two specific vertices at a specific point in time.
2. *Temporal feedback connection sets,* which consist of vertex pairs $\{v, w\}$ causing that all time-edges between v and w will be removed.

Defining feedback edge set problems in temporal graphs is not straightforward because for temporal graphs the notions of paths and cycles are more involved than for static graphs. First, we consider two different, established models of temporal paths. Temporal paths are time-respecting paths in a temporal graph. *Strict* temporal paths have strictly increasing time-labels on consecutive time-edges. *Non-strict* temporal paths have non-decreasing time-labels on consecutive time-edges. Non-strictness can be used whenever the traversal time per edge is very short compared to the scale of the time dimension.

Supported by the DFG, project MATE (NI 369/17).

I. Adler and H. Müller (Eds.): WG 2020, LNCS 12301, pp. 200–212, 2020.
https://doi.org/10.1007/978-3-030-60440-0_16

We focus on finding temporal feedback edge sets and temporal feedback connection sets (formalized in Sect. 2) of small cardinality in unweighted temporal graphs, each time using both the strict and non-strict temporal cycle model. We call the corresponding problems (STRICT) TEMPORAL FEEDBACK EDGE SET and (STRICT) TEMPORAL FEEDBACK CONNECTION SET, respectively.

(STRICT) TEMPORAL FEEDBACK EDGE SET ((S)TFES)
Input: A temporal graph $\mathcal{G} = (V, \mathcal{E}, \tau)$ and $k \in \mathbb{N}$.
Question: Is there a (strict) temporal feedback edge set $\mathcal{E}' \subseteq \mathcal{E}$ of $\mathcal{G}$ with $|\mathcal{E}'| \leq k$?

(STRICT) TEMPORAL FEEDBACK CONNECTION SET ((S)TFCS)
Input: A temporal graph $\mathcal{G} = (V, \mathcal{E}, \tau)$ and $k \in \mathbb{N}$.
Question: Is there a (strict) temporal feedback connection set C' of $\mathcal{G}$ with $|C'| \leq k$?

Related Work. In static connected graphs, removing a minimum-cardinality feedback edge set results in a spanning tree. This can be done in linear time via depth-first or breadth-first search. Thus, it is natural to compare temporal feedback edge sets to the temporal analogue of a spanning tree. This analogue is known as the *minimum temporally connected (sub)graph*, which is a graph containing a time-respecting path from each vertex to every other vertex. The concept was first introduced by Kempe et al. [16], and Axiotis and Fotakis [4] showed that in an n-vertex graph such a minimum temporally connected subgraph can have $\Omega(n^2)$ edges while Casteigts et al. [7] showed that complete temporal graphs admit sparse temporally connected subgraphs. Additionally, Akrida et al. [2] and Axiotis and Fotakis [4] proved that computing a minimum temporally connected subgraph is APX-hard. Considering *weighted* temporal graphs, there is also (partially empirical) work on computing minimum spanning trees, mostly focusing on polynomial-time approximability [15].

While feedback edge sets in temporal graphs seemingly have not been studied before, Agrawal et al. [1] investigated the related problem α−SIMULTANEOUS FEEDBACK EDGE SET, where the edge set of a graph is partitioned into α color classes and one wants to find a set of at most k edges intersecting all monochromatic cycles. They show that this is NP-hard for $\alpha \geq 3$ colors and give a $2^{O(k\alpha)}\text{poly}(n)$-time algorithm.

Another related problem is finding s-t-separators in temporal graphs; this was studied by Berman [5], Kempe et al. [16], and Zschoche et al. [23]. Already here some differences were found between the strict and the non-strict setting, a distinction that also matters for our results.

Our Contributions. Based on a polynomial-time many-one reduction from 3-SAT, we show NP-hardness for all four problem variants. The properties of the corresponding construction yield more insights concerning special cases. More specifically, the constructed graph uses $\tau = 8$ distinct time-labels for the strict variants and $\tau = 3$ labels for the non-strict variants. Similarly, we observe that our constructed graph has at most one time-edge between any pair of vertices (i.e., is *simple*), implying that the problems remain NP-hard when restricted to simple temporal graphs. Assuming the Exponential Time Hypothesis, we can additionally prove that there is no subexponential-time algorithm solving (S)TFES or (S)TFCS. Moreover, we show that all four problem variants are W[1]-hard when parameterized by the solution size k, using a parameterized reduction from the W[1]-hard problem MULTICUT IN DAGs [17].

Table 1. Overview of our results for (STRICT) TEMPORAL FEEDBACK EDGE SET (marked with *) and (STRICT) TEMPORAL FEEDBACK CONNECTION SET (marked with **). Unmarked results apply to both variants. The parameter k denotes the solution size, τ the lifetime of the temporal graph, L the maximum length of a minimal temporal cycle, and $tw_\downarrow$ resp. $td_\downarrow$ the treewidth resp. treedepth of the underlying graph.

Param.	Complexity	
	Strict variant	Non-strict variant
None	NP-hard [Theorem 10*/Corollary 11**]	NP-hard [Corollary 12]
k	W[1]-hard [Theorem 14]	W[1]-hard [Theorem 14]
τ	$\tau \geq 8$: NP-h. [Theorem 10*/Corollary 11**]	$\tau \geq 3$: NP-h. [Corollary 12]
$k + L$	$O(L^k \cdot |\mathcal{E}|^2)$ [Observation 6]	$O(L^k \cdot |\mathcal{E}|^2 \log |\mathcal{E}|)$ [Observation 6]
$k + \tau$	$O(\tau^k \cdot |\mathcal{E}|^2)$ [Corollary 7]	Open
$k + td_\downarrow$	$2^{O(td_\downarrow \cdot k)} \cdot |\mathcal{E}|^2$ [Corollary 8]	$2^{O(td_\downarrow \cdot k)} \cdot |\mathcal{E}|^2 \log |\mathcal{E}|$ [Corollary 8]
$|V|$	$O(2^{2|V|^2} \cdot |V|^3 \cdot \tau)$* [Theorem 15*] $O(2^{\frac{1}{2}(|V|^2 - |V|)} \cdot |\mathcal{E}|^2)$** [Observation 9**]	$O(2^{3|V|^2} \cdot |V|^2 \cdot \tau)$* [Theorem 15*] $O(2^{\frac{1}{2}(|V|^2 - |V|)} \cdot |\mathcal{E}|^2 \log |\mathcal{E}|)$** [Observation 9**]
$tw_\downarrow + \tau$	FPT [Theorem 21]	FPT [Theorem 21]

On the positive side, based on a simple search tree, we first observe that all problem variants are fixed-parameter tractable with respect to the combined parameter $k + L$, where L is the maximum length of a *minimal* temporal cycle. For the strict problem variants, this also implies fixed-parameter tractability for the combined parameter $\tau + k$. Our main algorithmic result is to prove fixed-parameter tractability for (S)TFES with respect to the number of vertices $|V|$. (For (S)TFCS, the corresponding result is straightforward as there are $\frac{1}{2}(|V|^2 - |V|)$ vertex pairs to consider.) Finally, studying the combined parameter τ plus treewidth of the underlying graph, we show fixed-parameter tractability based on an MSO formulation.

Our results are summarized in Table 1. Notable distinctions between the different settings include the combined parameter $k + \tau$ where the non-strict case remains open, and the parameter $|V|$ where the proof for (S)TFES is much more involved than for (S)TFCS.

Due to space constraints, several details (marked with $(\star)$) are omitted and can be found in a long version [12].

2   Preliminaries and Basic Observations

We assume familiarity with standard notion from graph theory and from (param-eterized) complexity theory. We denote the set of positive integers with $\mathbb{N}$. For $a \in \mathbb{N}$, we set $[a] := \{1, \ldots, a\}$.

We use the following definition of temporal graphs in which the vertex set does not change and each time-edge has a discrete time-label [13,14,19].

Definition 1 (Temporal Graph, Underlying Graph). *An (undirected) temporal graph* $\mathcal{G} = (V, \mathcal{E}, \tau)$ *is an ordered triple consisting of a set V of vertices, a set $\mathcal{E} \subseteq \binom{V}{2} \times [\tau]$ of (undirected)* time-edges, *and a lifetime $\tau \in \mathbb{N}$.*

The underlying graph $G_{\downarrow}$ *is the static graph obtained by removing all time-labels from $\mathcal{G}$ and keeping only one edge from every multi-edge. We call a temporal graph* simple *if each vertex pair is connected by at most one time-edge.*

Let $\mathcal{G} = (V, \mathcal{E}, \tau)$ be a temporal graph. For $i \in [\tau]$, let $E_i(\mathcal{G}) := \{\{v, w\} \mid (\{v, w\}, i) \in \mathcal{E}\}$ be the set of edges with time-label i. We call the static graph $G_i(\mathcal{G}) = (V, E_i(\mathcal{G}))$ *layer i* of $\mathcal{G}$. For $t \in [\tau]$, we denote the temporal subgraph consisting of the first t layers of $\mathcal{G}$ by $G_{[t]}(\mathcal{G}) := (V, \{(e, i) \mid i \in [t] \wedge e \in E_i(\mathcal{G})\}, t)$. We omit the function parameter $\mathcal{G}$ if it is clear from the context. For some $\mathcal{E}' \subseteq \binom{V}{2} \times [\tau]$, we denote $\mathcal{G} - \mathcal{E}' := (V, \mathcal{E} \setminus \mathcal{E}', \tau)$.

Definition 2 (Temporal Path, Temporal Cycle). *Given a temporal graph* $\mathcal{G} = (V, \mathcal{E}, \tau)$, *a temporal path of length ℓ in $\mathcal{G}$ is a sequence $P = (e_1, e_2, \ldots, e_\ell)$ of time-edges $e_i = (\{v_i, v_{i+1}\}, t_i) \in \mathcal{E}$ where $v_i \neq v_j$ for all $i, j \in [\ell]$ and $t_i \leq t_{i+1}$ for all $i \in [\ell - 1]$.*

A temporal cycle *is a temporal path of length at least three, except that the first and last vertex are identical.*

A temporal path or cycle is called strict *if $t_i < t_{i+1}$ for all $i \in [\ell - 1]$.*

The definitions of (STRICT) TEMPORAL FEEDBACK EDGE SET and (STRICT) TEMPORAL FEEDBACK CONNECTION SET (see Sect. 1) are based on the following two sets (problem and set names are identical).

Definition 3 ((Strict) Temporal Feedback Edge Set). *Let $\mathcal{G} = (V, \mathcal{E}, \tau)$ be a temporal graph. A time-edge set $\mathcal{E}' \subseteq \mathcal{E}$ is called a (strict) temporal feedback edge set of $\mathcal{G}$ if $\mathcal{G}' = (V, \mathcal{E} \setminus \mathcal{E}', \tau)$ does not contain a (strict) temporal cycle.*

Definition 4 ((Strict) Temporal Feedback Connection Set). *Let $\mathcal{G} = (V, \mathcal{E}, \tau)$ be a temporal graph with underlying graph $G_{\downarrow} = (V, E_{\downarrow})$. An edge set $C' \subseteq E_{\downarrow}$ is a (strict) temporal feedback connection set of $\mathcal{G}$ if $\mathcal{G}' = (V, \mathcal{E}', \tau)$ with $\mathcal{E}' = \{(\{v, u\}, t) \in \mathcal{E} \mid \{v, u\} \notin C'\}$ does not contain a (strict) temporal cycle.*

The elements in a feedback connection set are known as *underlying edges* (edges of $G_{\downarrow}$).

Simple Observations. We can compute shortest temporal paths from any given vertex to all other vertices in $O(|\mathcal{E}| \log |\mathcal{E}|)$ time [21], respectively $O(|\mathcal{E}|)$ time for strict temporal paths [23, Proposition 3.7]. Thus, by searching for each time-edge $(\{v, w\}, t)$ for a shortest temporal path from w to v which starts at time t and avoids the edge $\{v, w\}$, we can record the following observation.

Observation 5. *In $O(|\mathcal{E}|^2 \log |\mathcal{E}|)$ time, we can find a shortest temporal cycle or confirm that none exists. For the strict case $O(|\mathcal{E}|^2)$ time suffices.*

Given a shortest temporal cycle of length L, any temporal feedback edge or connection set must contain an edge or connection used by that cycle. By repeatedly searching for a shortest temporal cycle and then branching over all of its edges or connections, we obtain the following (again the log-factor is only required in the non-strict case).

Observation 6 ($\star$). *Let $\mathcal{G} = (V, \mathcal{E}, \tau)$ be a temporal graph where each temporal cycle has length at most $L \in \mathbb{N}$. Then, (S)TFES and (S)TFCS can be solved in $O(L^k \cdot |\mathcal{E}|^2 \log |\mathcal{E}|)$ time. For the strict cases $O(L^k \cdot |\mathcal{E}|^2)$ time suffices.*

Clearly, a strict temporal cycle cannot be longer than the lifetime τ. Thus, Observation 6 immediately gives the following result.

Corollary 7. *STFES and STFCS can be solved in $O(\tau^k \cdot |\mathcal{E}|^2)$ time.*

Alternatively, we can also upper-bound L in terms of the length of any cycle of the underlying graph $G_\downarrow$, which in turn can be upper-bounded by $2^{O(\mathrm{td}_\downarrow)}$ [20, Proposition 6.2], where $\mathrm{td}_\downarrow$ is the treedepth of the underlying graph.

Corollary 8. *Let $\mathcal{G}$ be a temporal graph and $\mathrm{td}_\downarrow$ be the treedepth of $G_\downarrow$. Then, (S)TFES and (S)TFCS can be solved in $2^{O(\mathrm{td}_\downarrow \cdot k)} \cdot |\mathcal{E}|^2 \log |\mathcal{E}|$ time. For the strict cases $2^{O(\mathrm{td}_\downarrow \cdot k)} \cdot |\mathcal{E}|^2$ time suffices.*

In contrast to static graphs, $|V|$ is to be considered as a useful parameter for temporal graphs because the maximum number of time-edges $|\mathcal{E}|$ can be arbitrarily much larger than $|V|$. However, the number of underlying edges is at most $\frac{1}{2}(|V|^2 - |V|)$ which yields the following fixed-parameter tractability result for (S)TFCS.

Observation 9 ($\star$). *TFCS can be solved in $\mathcal{O}(2^{\frac{1}{2}(|V|^2 - |V|)} \cdot |\mathcal{E}|^2 \log |\mathcal{E}|)$ time and STFCS can be solved in $\mathcal{O}(2^{\frac{1}{2}(|V|^2 - |V|)} \cdot |\mathcal{E}|^2)$ time.*

3   Computational Hardness Results

We now show that all four problem variants, (S)TFES and (S)TFCS, are NP-hard on simple temporal graphs with constant lifetime. The proofs work by polynomial-time many-one reduction from the classical 3-SAT problem.

Theorem 10 ($\star$). *STFES is NP-hard for simple temporal graphs with $\tau = 8$.*

The temporal graph constructed in the proof for Theorem 10 does not contain any pair of vertices which is connected by more than one time-edge. Hence, each underlying edge corresponds to a single time-edge and thus the reduction implies the following corollary. A similar reduction can also be used for the following.

Corollary 11. *STFCS is NP-hard even for simple temporal graphs with $\tau = 8$.*

Corollary 12 ($\star$). *TFES and TFCS are both NP-hard even for simple temporal graphs with $\tau \geq 3$.*

We can also observe that the strict problem variants are NP-hard even if all edges are present at all times. This problem is essentially equivalent to selecting a set of edges of the underlying graph that intersects all cycles of length at most τ, which is known to be NP-hard [22, Theorem 1].

Observation 13 ($\star$). *STFES and STFCS are NP-hard even on temporal graphs where all edges are present at all times, even with $\tau = 3$, planar underlying graph $G_\downarrow$, and $\Delta(G_\downarrow) = 7$.*

We next show that our problems are W[1]-hard when parameterized by the solution size k with a parameterized reduction from MULTICUT IN DAGs [17]. The idea here is that we can simulate a DAG D by an undirected temporal graph by first subdividing all edges of D and then assigning time-labels according to a topological ordering. This ensures that each path in D corresponds to a path in the resulting temporal graph and vice versa. By adding a reverse edge from t to s for each terminal pair (s, t) of the MULTICUT instance, an s-t-path in D produces a temporal cycle involving s, t and vice versa.

Theorem 14 ($\star$). *(S)TFES and (S)TFCS, parameterized by the solution size k, are W[1]-hard.*

4 Fixed-Parameter Tractability Results

After having shown computational hardness for the single parameters solution size k and lifetime τ in Sect. 3, we now consider larger and combined parameters, and present fixed-parameter tractability results.

4.1 Parameterization by Number of Vertices

As shown in Observation 9, (S)TFCS is trivially fixed-parameter tractable with respect to the number of vertices $|V|$. For (S)TFES, however, the same result is much more difficult to show as the size of the search space is only upper-bounded by $2^{\tau(|V|^2 - |V|)}$. Here, the dependence on τ prevents us from using the (brute-force) approach that worked for (S)TFCS.

Theorem 15. *STFES can be solved in $\mathcal{O}(2^{2|V|^2} \cdot |V|^3 \cdot \tau)$ time and TFES can be solved in $\mathcal{O}(2^{3|V|^2} \cdot |V|^2 \cdot \tau)$ time, both requiring $\mathcal{O}(2^{|V|^2})$ space.*

We prove Theorem 15 using a dynamic program which computes the minimum number of time-edges which have to be removed to achieve a specified connectivity at a specified point in time. The key idea is that we can efficiently determine which edges need to be removed from layer t if we know for evey pair of vertices v, w whether w should be unreachable from v until time $t-1$ and time t, respectively. To exclude temporal cycles, it then suffices that every vertex is unreachable from itself until time τ (except by a trivial path of course).

Before formally describing the dynamic program, we need to introduce some notations and intermediate results. Let $\mathcal{G} = (V, \mathcal{E}, \tau)$ be a temporal graph with $V = \{v_1, v_2, \ldots, v_n\}$. The first dimension of the dynamic programming table will be the connectivity between the vertices of $\mathcal{G}$. We will store this as a *connectivity matrix* $A \in \{0, ?\}^{n \times n}$ encoding the following connectivity relationships:

$$a_{ij} = 0 \quad \Rightarrow \quad \text{there is no temporal path from vertex } v_i \text{ to } v_j$$
$$\text{(resp. no temporal cycle if } i = j) \text{ and}$$
$$a_{ij} = ? \quad \Rightarrow \quad \text{there might be a temporal path (resp. cycle) from } v_i \text{ to } v_j.$$

Next, we define two functions, $\mathrm{srd}(G, B, A)$ (strict required deletions) and $\mathrm{nrd}(G, B, A)$ (non-strict required deletions), which return the solution to the following subproblem. Given connectivity B (before) at time $t - 1$, what is the minimum number of edge deletions required in G_t to ensure connectivity A (after) at time t? Figure 1 illustrates this problem for two vertices v_i and v_j. If $a_{ij} = 0$ and there is some vertex v_k which might be reachable from v_i (i.e., $b_{ik} = ?$ represented by the dotted path), then we must remove the edge between v_k and v_j. In order to guarantee correctness, we have to assume that every "?" in B represents an existing path. Additionally, if A and B encode incompatible connectivity, then the function value is defined as ∞.

Definition 16. *Let $G = (V, E)$ be a static graph with $|V| = n$ and let $A, B \in \{0, ?\}^{n \times n}$ be two connectivity matrices. Function $\mathrm{srd}(G, B, A)$ is as follows.*

If $\exists i, j \in [n] : b_{ij} = ? \wedge a_{ij} = 0$, then $\mathrm{srd}(G, B, A) := \infty$. Otherwise, $\mathrm{srd}(G, B, A) := |\{\{v_k, v_j\} \in E \mid \exists i \in [n] : a_{ij} = 0 \wedge (b_{ik} = ? \vee i = k)\}|$.

Note the clause $b_{ik} = ? \vee i = k$ in the formulation of Definition 16. This is due to the fact that a vertex v_k is always reachable from itself by a trivial temporal path, regardless of whether a temporal cycle at v_k exists.

Next, we show that $\mathrm{srd}(G, B, A)$ can be computed in polynomial time.

Lemma 17 ($\star$). *Algorithm 1 computes function $\mathrm{srd}(G, B, A)$ in $\mathcal{O}(|V|^3)$ time.*

Since, in the non-strict case, a temporal path can successively use multiple edges from G_t, it is not possible to consider each entry $a_{ij} = 0$ separately (a single edge might be part of multiple unwanted temporal walks). Instead, we have to find an optimal edge-cut disconnecting all "problematic" pairs (v_k, v_j) in G_t where $\exists i \in [n] : a_{ij} = 0 \wedge (b_{ik} = ? \vee i = k)\}$. This problem is known as the MULTICUT problem. We will use MULTICUT to define the second function, $\mathrm{nrd}(G, B, A)$.

vertices

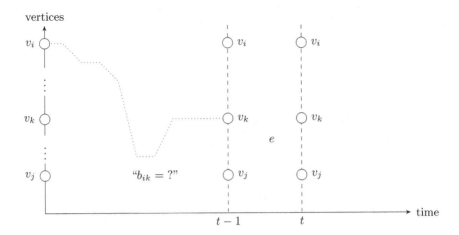

Fig. 1. Illustration of the subproblem solved by $\mathrm{nrd}(G, B, A)$ resp. $\mathrm{srd}(G, B, A)$. If we want to make sure that no temporal path from v_i to v_j exists at time t (that is, we have $a_{ij} = 0$), then the time-edge e has to be removed from the graph because there might be a temporal path from v_i to v_k ($b_{ik} = ?$, dotted line).

MULTICUT (OPTIMIZATION VARIANT)
Input: An undirected, static graph $G = (V, E)$ and a set of r terminal pairs $\mathcal{T} = \{(s_i, t_i) \mid i \in [r] \text{ and } s_i, t_i \in V\}$.
Output: A minimum-cardinality edge set $E' \subseteq E$ whose removal disconnects all terminal pairs in $\mathcal{T}$.

Definition 18. *Let $G = (V, E)$ be a static graph with $|V| = n$ and let $A, B \in \{0, ?\}^{n \times n}$ be two connectivity matrices. Function $\mathrm{nrd}(G, B, A)$ is as follows.*
    If $\exists i, j \in [n] : b_{ij} = ? \wedge a_{ij} = 0$, then $\mathrm{nrd}(G, B, A) := \infty$.
    Otherwise, let E' be a solution to MULTICUT $(G, \mathcal{T})$ with $\mathcal{T} = \{(v_k, v_j) \mid \exists i \in [n] : a_{ij} = 0 \wedge (b_{ik} = ? \vee i = k)\}$. Then, $\mathrm{nrd}(G, B, A) := |E'|$.

In order to compute $\mathrm{nrd}(G, B, A)$, we have to solve MULTICUT which was shown to be APX-hard [9, Theorem 5 and Sect. 5]. While there exist FPT algorithms [6,18] for the parameter solution size, our best upper bound for the solution size is $|V|^2 - |V|$ and thus using these algorithms would result in a worse running time than the brute-force approach we will use to prove the next lemma.

Lemma 19 ($\star$). *Function $\mathrm{nrd}(G, B, A)$ can be computed in $\mathcal{O}(2^{|V|^2} \cdot |V|^2)$ time.*

We can now define the dynamic program which we will use to prove Theorem 15. Let $A \in \{0, ?\}^{n \times n}$ be a connectivity matrix. The table entry $T(A, t) \in \mathbb{N}$ contains the minimum number of time-edges which have to be removed from $G_{[t]}$ in order to achieve the connectivity specified by A. We define T as follows.

$$T(A, 0) := 0 \quad \forall A \in \{0, ?\}^{n \times n} \tag{1}$$

Algorithm 1. Algorithm computing $\mathrm{srd}(G, B, A)$ (for strict temporal paths)

Parameters
  G: static graph
  A, B: connectivity matrices
Output
  $\mathrm{srd}(G, B, A)$

1: **function** STRICTREQUIREDDELETIONS(G, B, A)
2:    $E' \leftarrow \{\}$
3:    **for** $i, j \in [n]$ **do**
4:        **if** $b_{ij} = ? \wedge a_{ij} = 0$ **then**
5:            **return** ∞
6:        **else if** $b_{ij} = 0 \wedge a_{ij} = 0$ **then**
7:            **for all** $e := \{v_k, v_j\} \in E(G)$ with $b_{ik} = ? \vee i = k$ **do**
8:                $E' \leftarrow E' \cup \{e\}$
9:            **end for**
10:        **end if**
11:    **end for**
12:    **return** $|E'|$
13: **end function**

$$\text{strict paths: } T(A, t) := \min_{B \in \{0,?\}^{n \times n}} T(B, t-1) + \mathrm{srd}(G_t, B, A) \tag{2a}$$

$$\text{non-strict paths: } T(A, t) := \min_{B \in \{0,?\}^{n \times n}} T(B, t-1) + \mathrm{nrd}(G_t, B, A) \tag{2b}$$

Lemma 20. *Let $\mathcal{G} = (V, \mathcal{E}, \tau)$ be a temporal graph with $|V| = n$ and let $A \in \{0,?\}^{n \times n}$ be a connectivity matrix. Then, $T(A, t)$ is the minimum cardinality of a set $\mathcal{E}' \subseteq \mathcal{E}$ for which $(\mathcal{G} - \mathcal{E}')_{[t]}$ possesses the connectivity specified by A.*

Proof. We prove the lemma for the strict case via induction over t. The non-strict case works analogously. Recall that the connectivity matrix A can only encode that certain temporal paths must not exist. Thus, the correctness of the initialization $T(A, 0) = 0$ is easy to see since no temporal paths exist at time $t = 0$. For the correctness of the update step (Equation (2a)), we note that, by minimizing over all possible $B \in \{0,?\}^{n \times n}$, we always find the optimal state B for the time $t - 1$. With some B fixed, it remains to show that $T(B, t - 1) + \mathrm{srd}(G_t, B, A)$ is minimal for achieving both connectivity B at time $t - 1$ and connectivity A at time t. By induction hypothesis, we know that $T(B, t - 1)$ is minimal. To show correctness and minimality of $\mathrm{srd}(G_t, B, A)$, we analyze how the time-edges of layer G_t influence the possible temporal paths up to time t and which changes (i.e., time-edge deletions) are required to achieve connectivity A. For any two vertices $v_i, v_j \in V$, we compare the connectivity for time $t - 1$ given by b_{ij} to the target connectivity given by a_{ij} and identify the following four cases.

(Case 1) $b_{ij} = ? \land a_{ij} = ?$: Here, we do not care if v_j is reachable from v_i and, thus, we do not need to remove any edges.

(Case 2) $b_{ij} = ? \land a_{ij} = 0$: We cannot disconnect a temporal path that already exists at time $t - 1$ by removing edges in layer t and, therefore, cannot guarantee $a_{ij} = 0$. In this case, there is no solution for the input parameters. By Definition 16, srd(G_t, B, A) returns ∞ in this case.

(Case 3) $b_{ij} = 0 \land a_{ij} = ?$: Identical to Case 1.

(Case 4) $b_{ij} = 0 \land a_{ij} = 0$: We have to ensure that v_j is not reachable from v_i in $G_{[t]}$. The definition of srd(G_t, B, A) mirrors the following argument. If there is a vertex v_k which was reachable from v_i in the past, then we cannot keep an edge (strict case) or any path (non-strict case) connecting v_k and v_j in layer t. Assuming that Case 2 is already excluded for all vertex pairs, it can be easily verified that Definition 16 uses exactly the set of such edges $\{v_k, v_j\}$.

Finally, we must show that assuming "$b_{ij} = ? \Rightarrow$ there is a temporal path from v_i to v_j" at time $t - 1$ for the function srd(G_t, B, A) did not result in unnecessary time-edge deletions. To this end, assume $b_{ij} = ?$ and that there is no temporal path from v_i to v_j at time $t - 1$. Let B' be a connectivity matrix identical to B except for $b'_{ij} = 0$. As the path from v_i to v_j does not exist, we have $T(B, t - 1) = T(B', t - 1)$. If the entry $b_{ij} = ?$ resulted in an unnecessary time-edge deletion, i.e., srd$(G_t, B, A) > $ srd(G_t, B', A), then the minimum function in Equation (2a) will not choose the value computed using B.   □

Since we are only interested in specifying that certain paths (cycles) must not exist, we choose to use "?"-entries in the connectivity matrices to represent entries we do not care about. The advantage is evident in Cases 1 and 3 of the previous proof. We now have all required ingredients to prove Theorem 15.

Proof. (Theorem 15). Let $(\mathcal{G}, k)$ be an instance of (S)TFES. Further, let A^* be an $n \times n$ connectivity matrix with $a^*_{ij} = 0$ if $i = j$ and $a^*_{ij} = ?$ otherwise. As "?"-entries cannot require more time-edge deletions than "0"-entries, A^* is the cheapest connectivity specification that does not allow any temporal cycle to exist. Thus, it follows from Lemma 20 that $(\mathcal{G}, k)$ is a yes-instance if $T(A^*, \tau) \leq k$, and a no-instance otherwise.

For the running time, we first note that a connectivity matrix has size $|V|^2$ with two possible choices for each entry resulting in $2^{|V|^2}$ possible connectivity matrices. Thus, the table size of the dynamic program is $2^{|V|^2} \cdot \tau$. To compute each table entry, we have to compute srd(G_t, B, A) (resp. nrd(G_t, B, A)) for each of the $2^{|V|^2}$ possible choices for B. Together with Lemmas 17 and 19, we obtain the running times stated in the theorem. The computation requires $\mathcal{O}(2^{|V|^2})$ space as we only need the table entries for time $t - 1$ in order to compute the entries for time t. Thus, it is not necessary to store more than two columns of the table, each of size $2^{|V|^2}$.   □

We note that our dynamic program indeed solves the optimization variant of (S)TFES. That is, given a temporal graph $\mathcal{G}$, it finds the smallest k for which $(\mathcal{G}, k)$ is a yes-instance of the decision variant stated defined in Sect. 1. As shown in the previous proof, we can easily use the result to solve any instance $(\mathcal{G}, k')$ of the decision variant by comparing k' to k.

For the ease of presentation, we did not store the actual solution, that is, the feedback edge set of size $T(A, t)$. However, the functions $\mathrm{srd}(G_t, B, A)$ and $\mathrm{nrd}(G_t, B, A)$ can easily be changed to return the solution edge sets for each layer t. Using linked lists, which can be concatenated in constant time, it is possible to include the solutions sets in the dynamic programming table without changing the asymptotic running time.

4.2   Parameterization by Treewidth and Lifetime

In this last part, we show that all our problem variants are fixed-parameter tractable when parameterized by the combination of the treewidth of the underlying graph and the lifetime. To this end we employ an optimization variant of Courcelle's famous theorem on graph properties expressible in monadic second-order (MSO) logic [3,8] and apply it in the temporal setting [10].

Theorem 21 (★). (S)TFES *and* (S)TFCS *are fixed-parameter tractable when parameterized by the combination of the treewidth of the underlying graph and the lifetime.*

5   Conclusion

We conclude with some challenges for future research. For the parameter lifetime τ, it remains open whether there exists a polynomial-time algorithm for instances with $3 \leq \tau \leq 7$ in the strict case and $\tau = 2$ in the non-strict case. We believe that, for the strict case, our 3-SAT reduction can be modified to use only seven time-labels. Similarly to the work of Zschoche et al. [23] in the context of temporal separators, we could not resolve the question whether the non-strict variants are fixed-parameter tractable for the combined parameter $\tau + k$, whereas for the strict case, this is almost trivial.

We further leave as a future research challenge to investigate whether our fixed-parameter tractability result for the parameter "number of vertices" can be improved: On the hand we would like to improve the running time of the algorithm or show some conditional running time lower bound to show that it likely cannot be improved significantly. On the other hand, we leave open whether it is possible to obtain a polynomial-size problem kernel for the number of vertices as a parameter.

Additionally, it seems natural to study (S)TFES and (S)TFCS variants restricted to specific temporal graph classes (e.g., see Fluschnik et al. [11]). In particular, we could not settle the parameterized complexity of our problem variants when parameterized by (solely) the treewidth of the underlying graph.

References

1. Agrawal, A., Panolan, F., Saurabh, S., Zehavi, M.: Simultaneous feedback edge set: a parameterized perspective. In: Proceedings of the 27th International Symposium on Algorithms and Computation (ISAAC 2016). LIPIcs, vol. 64, pp. 5:1–5:13. Schloss Dagstuhl - Leibniz-Zentrum fuer Informatik (2016). https://doi.org/10.4230/LIPIcs.ISAAC.2016.5

2. Akrida, E.C., Gąsieniec, L., Mertzios, G.B., Spirakis, P.G.: On temporally connected graphs of small cost. In: Sanitá, L., Skutella, M. (eds.) WAOA 2015. LNCS, vol. 9499, pp. 84–96. Springer, Cham (2015). https://doi.org/10.1007/978-3-319-28684-6_8

3. Arnborg, S., Lagergren, J., Seese, D.: Easy problems for tree-decomposable graphs. J. Algorithm **12**(2), 308–340 (1991). https://doi.org/10.1016/0196-6774(91)90006-K

4. Axiotis, K., Fotakis, D.: On the size and the approximability of minimum temporally connected subgraphs. In: Proceedings of the 43rd International Colloquium on Automata, Languages, and Programming (ICALP 2016). LIPIcs, vol. 55, pp. 149:1–149:14. Schloss Dagstuhl - Leibniz-Zentrum für Informatik (2016). https://doi.org/10.4230/LIPIcs.ICALP.2016.149

5. Berman, K.A.: Vulnerability of scheduled networks and a generalization of Menger's theorem. Networks **28**(3), 125–134 (1996). https://doi.org/bvfwgx

6. Bousquet, N., Daligault, J., Thomassé, S.: Multicut is FPT. SIAM J. Comput. **47**(1), 166–207 (2018). https://doi.org/10.1137/140961808

7. Casteigts, A., Peters, J.G., Schoeters, J.: Temporal cliques admit sparse spanners. In: Proceedings of the 46th International Colloquium on Automata, Languages and Programming (ICALP 2019). LIPIcs, vol. 132, pp. 134:1–134:14. Schloss Dagstuhl - Leibniz-Zentrum für Informatik (2019). https://doi.org/10.4230/LIPIcs.ICALP.2019.134

8. Courcelle, B., Engelfriet, J.: Graph Structure and Monadic Second-order Logic: A Language-theoretic Approach. Cambridge University Press, Cambridge (2012)

9. Dahlhaus, E., Johnson, D.S., Papadimitriou, C.H., Seymour, P.D., Yannakakis, M.: The complexity of multiterminal cuts. SIAM J. Comput. **23**(4), 864–894 (1994). https://doi.org/10.1137/S0097539792225297

10. Fluschnik, T., Molter, H., Niedermeier, R., Renken, M., Zschoche, P.: As time goes by: reflections on treewidth for temporal graphs. In: Fomin, F.V., Kratsch, S., van Leeuwen, E.J. (eds.) Treewidth, Kernels, and Algorithms. LNCS, vol. 12160, pp. 49–77. Springer, Cham (2020). https://doi.org/10.1007/978-3-030-42071-0_6

11. Fluschnik, T., Molter, H., Niedermeier, R., Renken, M., Zschoche, P.: Temporal graph classes: a view through temporal separators. Theor. Comput. Sci. **806**, 197–218 (2020). https://doi.org/10.1016/j.tcs.2019.03.031

12. Haag, R., Molter, H., Niedermeier, R., Renken, M.: Feedback edge sets in temporal graphs. CoRR 2003.13641 (2020). https://arxiv.org/abs/2003.13641

13. Holme, P., Saramäki, J. (eds.) Temporal Networks. Springer, Berlin, Heidelberg (2013). https://doi.org/10.1007/978-3-642-36461-7

14. Holme, P., Saramäki, J. (eds.): Temporal Network Theory. CSS. Springer, Cham (2019). https://doi.org/10.1007/978-3-030-23495-9

15. Huang, S., Fu, A.W., Liu, R.: Minimum spanning trees in temporal graphs. In: Proceedings of the 2015 ACM SIGMOD International Conference on Management of Data, pp. 419–430 (2015). https://doi.org/10.1145/2723372.2723717

16. Kempe, D., Kleinberg, J.M., Kumar, A.: Connectivity and inference problems for temporal networks. J. Comput. Syst. Sci. **64**(4), 820–842 (2002). https://doi.org/10.1006/jcss.2002.1829

17. Kratsch, S., Pilipczuk, M., Pilipczuk, M., Wahlström, M.: Fixed-parameter tractability of multicut in directed acyclic graphs. SIAM J. Disc. Math. **29**(1), 122–144 (2015). https://doi.org/10.1137/120904202

18. Marx, D., Razgon, I.: Fixed-parameter tractability of multicut parameterized by the size of the cutset. SIAM J. Comput. **43**(2), 355–388 (2014). https://doi.org/10.1137/110855247

19. Michail, O.: An introduction to temporal graphs: an algorithmic perspective. Internet Math. **12**(4), 239–280 (2016). https://doi.org/10.1080/15427951.2016.1177801

20. Nešetřil, J., De Mendez, P.O.: Sparsity: Graphs, Structures, and Algorithms. AC, vol. 28. Springer, Heidelberg (2012). https://doi.org/10.1007/978-3-642-27875-4

21. Wu, H., et al.: Efficient algorithms for temporal path computation. IEEE Trans. Knowl. Data Eng. **28**(11), 2927–2942 (2016). https://doi.org/10.1109/TKDE.2016.2594065

22. Yannakakis, M.: Edge-deletion problems. SIAM J. Comput. **10**(2), 297–309 (1981). https://doi.org/10.1137/0210021

23. Zschoche, P., Fluschnik, T., Molter, H., Niedermeier, R.: The complexity of finding small separators in temporal graphs. J. Comput. Syst. Sci. **107**, 72–92 (2020). https://doi.org/10.1016/j.jcss.2019.07.006

On Flips in Planar Matchings

Marcel Milich[1], Torsten Mütze[2(✉)], and Martin Pergel[3]

[1] Institut für Mathematik, Technische Universität Berlin, Berlin, Germany
`marcel.milich@campus.tu-berlin.de`
[2] Department of Computer Science, University of Warwick, Coventry, UK
`torsten.mutze@warwick.ac.uk`
[3] Department of Software and Computer Science Education,
Charles University Prague, Prague, Czech Republic
`perm@kam.mff.cuni.cz`

Abstract. In this paper we investigate the structure of flip graphs on non-crossing perfect matchings in the plane. Consider all non-crossing straight-line perfect matchings on a set of $2n$ points that are placed equidistantly on the unit circle. The graph $\mathcal{H}_n$ has those matchings as vertices, and an edge between any two matchings that differ in replacing two matching edges that span an empty quadrilateral with the other two edges of the quadrilateral, provided that the quadrilateral contains the center of the unit circle. We show that the graph $\mathcal{H}_n$ is connected for odd n, but has exponentially many small connected components for even n, which we characterize and count via Catalan and generalized Narayana numbers. For odd n, we also prove that the diameter of $\mathcal{H}_n$ is linear in n. Furthermore, we determine the minimum and maximum degree of $\mathcal{H}_n$ for all n, and characterize and count the corresponding vertices. Our results imply the non-existence of certain rainbow cycles, and they answer several open questions and conjectures raised in a recent paper by Felsner, Kleist, Mütze, and Sering.

Keywords: Flip graph · Matching · Diameter · Cycle

1 Introduction

Flip graphs are a powerful tool to study different classes of basic combinatorial objects, such as binary trees, strings, permutations, partitions, triangulations, matchings, spanning trees etc. A flip graph has as vertex set all the combinatorial objects of interest, and an edge between any two objects that differ only by a small local change operation called a *flip*. It thus equips the underlying objects with a structure that reveals interesting properties about the objects, and that allows one to solve different fundamental algorithmic tasks for them.

Torsten Mütze is also affiliated with the Faculty of Mathematics and Physics, Charles University Prague, Czech Republic, and he was supported by Czech Science Foundation grant GA 19-08554S and by German Science Foundation grant 413902284.
Martin Pergel was also supported by Czech Science Foundation grant GA 19-08554S.

© Springer Nature Switzerland AG 2020
I. Adler and H. Müller (Eds.): WG 2020, LNCS 12301, pp. 213–225, 2020.
https://doi.org/10.1007/978-3-030-60440-0_17

A classical example is the flip graph of binary trees under rotations, which has as vertices all binary trees on n nodes, and an edge between any two trees that differ in a tree rotation. A problem that has received a lot of attention is to determine the diameter of this flip graph, i.e., how many rotations are always sufficient to transform two binary trees into each other. This question was first considered by Sleator, Tarjan, and Thurston [27] in the 1980s, and answered conclusively only recently by Pournin [22]. This problem has also been studied extensively from an algorithmic point of view, with the goal of computing a short sequence of rotations between two given binary trees [6,7,16,26]. It is a well-known open question whether computing a shortest flip sequence is NP-hard.

Another important property of flip graphs is whether they have a Hamilton path or cycle. The reason is that computing such a path corresponds to an algorithm that exhaustively generates the underlying combinatorial objects [14]. It is known that the flip graph of binary trees mentioned before has a Hamilton cycle [13], and that a Hamilton path in this graph can be computed efficiently [17].

Flip graphs also have deep connections to lattices and polytopes [3,21,24,25]. For instance, the aforementioned flip graph of binary trees under rotations arises as the cover graph of the well-known Tamari lattice, and can be realized as an $(n-1)$-dimensional polytope in several different ways [5,15]. Other properties of interest that have been investigated for the flip graph of binary trees are its automorphism group [15], the vertex-connectivity [13], the chromatic number [4, 8], its genus [20], and the eccentricities of vertices [23]. Similar results are known for flip graphs of several geometric configurations, such as matchings, spanning trees, partitions and dissections, etc.; see e.g. [1,2,11,12].

In this paper, we consider the flip graph of non-crossing perfect matchings in the plane. For any integer $n \geq 2$, we consider a set of $2n$ points placed equidistantly on a unit circle. We let $\mathcal{M}_n$ denote the set of all non-crossing straight-line perfect matchings with n edges on this point set. It is well-known that the cardinality of $\mathcal{M}_n$ is the n-th *Catalan number* $C_n = \frac{1}{n+1}\binom{2n}{n}$. For any matching $M \in \mathcal{M}_n$, consider two matching edges $e, f \in M$ that span an empty quadrilateral, i.e., the convex hull of these two edges does not contain any other edges of M; see Fig. 1. Replacing the two edges e and f by the other two edges of the quadrilateral yields another matching $M' \in \mathcal{M}_n$, and we say that M and M' differ in a *flip*. The flip graph $\mathcal{G}_n$ has $\mathcal{M}_n$ as its vertex set, and an undirected edge between any two matchings that differ in a flip; see Fig. 2. Hernando, Hurtado, and Noy [10] proved that the graph $\mathcal{G}_n$ has diameter $n-1$ and connectivity $n-1$, is bipartite for all n, has a Hamilton cycle for all even $n \geq 4$, and no Hamilton cycle or path for any odd $n \geq 5$.

We now distinguish two different kinds of flips. A flip is *centered* if and only if the quadrilaterial that determines the flip contains the center of the unit circle. For odd n, the circle center may lie on the boundary of the quadrilateral, which still counts as a centered flip. In Fig. 1, the flip between M and M' is centered, whereas the flip between M and M'' is not. In all our figures, the circle center is

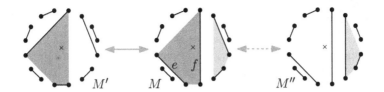

Fig. 1. A centered flip (left) and a non-centered flip (right).

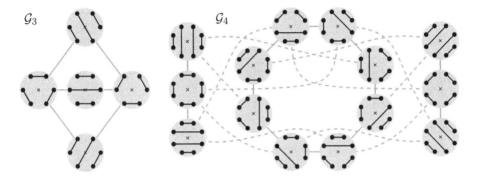

Fig. 2. Flip graphs $\mathcal{G}_3$ (left) and $\mathcal{G}_4$ (right). Solid edges correspond to centered flips and are present in the subgraphs $\mathcal{H}_3 \subseteq \mathcal{G}_3$ and $\mathcal{H}_4 \subseteq \mathcal{G}_4$, whereas the dashed edges correspond to non-centered flips and are not present in these subgraphs.

marked with a cross. We let $\mathcal{H}_n$ denote the spanning subgraph of $\mathcal{G}_n$ obtained by taking all edges that correspond to centered flips, omitting edges that correspond to non-centered flips; see Figs. 2 and 3. Clearly, both graphs $\mathcal{H}_n$ and $\mathcal{G}_n$ have the same vertex set.

The main motivation for considering centered flips comes from the study of rainbow cycles in flip graphs, a direction of research that was initiated in a recent paper by Felsner, Kleist, Mütze, and Sering [9]. Roughly speaking, along a rainbow cycle in $\mathcal{G}_n$ all possible lengths of quadrilateral edges that are involved in flip operations must appear equally often, which leads to non-centered flips becoming unusable, so we may restrict our attention to the subgraph $\mathcal{H}_n$ given by centered flips only. In other words, edges of a rainbow cycle in $\mathcal{G}_n$ must be edges of $\mathcal{H}_n$ also.

Let us address another potential concern right away: Our assumption that the $2n$ points of the point set are placed equidistantly on a unit circle is not necessary for expressing or proving any of our results. Our results and proofs are indeed robust under moving the $2n$ points to any configuration in convex position, by suitably replacing all geometric notions by purely combinatorial ones. In particular, centered flips can be defined without reference to the center of the unit circle (see Sect. 2). Nevertheless, in the rest of the paper we stick to the equidistancedness assumption, to be able to use both geometric and combinatorial arguments, whatever is more convenient.

1.1   Our Results

In this work we investigate the structure of the graph $\mathcal{H}_n \subseteq \mathcal{G}_n$. In particular, we solve some of the open questions and conjectures from the aforementioned paper [9] on rainbow cycles in flip graphs. For several graph parameter, we observe an intriguing dichotomy between the cases where n is odd or even, i.e., the graph $\mathcal{H}_n$ has an entirely different structure in those two cases. Table 1 contains a summary of our results, with references to the theorems where they are established.

Table 1. Summary of properties of the graph $\mathcal{H}_n$.

graph property	odd $n \geq 3$	even $n \geq 2$
max. degree ([19, Theorem 4])	n	$n/2$
# of max. deg. vertices	2	?
min. degree ([19, Theorem 5])	2	1
# of min. deg. vertices	$n \cdot (C_{(n-3)/2})^2$	$n \cdot (C_{(n-2)/2})^2$
diameter	between $n-1$ and $11n - 29$ (Theorem 4)	∞
# of components	1 (Theorem 2)	$\geq C_{n/2} + n - 3$ (Theorem 5+Corollary 7); $\binom{n}{n/2}$ vertices form trees of size $n/2 + 1$ each; comp. sizes bounded by Narayana numbers (Theorem 8 and Corollary 9)
r-rainbow cycles (Theorem 10)	none for any $r \geq 1$	none for large r
Hamilton path/cycle	none for $n \geq 4$	
colorability	bipartite ([10])	

Most importantly, the graph $\mathcal{H}_n$ is connected for odd n (Theorem 2), but has exponentially many connected components for even n (Theorem 5 and Corollary 7). For odd n, we show that the diameter of $\mathcal{H}_n$ is linear in n (Theorem 4). For even n, we provide a fine-grained picture of the component structure of the graph (Theorems 5 and 8, and Corollary 9). We also describe the degrees of vertices in $\mathcal{H}_n$ for all n, and we characterize and count the vertices of minimum and maximum degree (Theorems 3, 4 and 5 in [19]). Finally, we easily see that $\mathcal{H}_n$ does not admit a Hamilton cycle or path for any $n \geq 4$. This follows from the non-Hamiltonicity of $\mathcal{G}_n$ for odd $n \geq 5$ proved in [10], and for even $n \geq 4$ this is trivial as $\mathcal{H}_n$ has more than one component. Our results also imply the non-existence of certain rainbow cycles in the graph $\mathcal{G}_n$ (Theorem 10). Their exact definition will be provided later.

In all of these results, Catalan numbers and generalized Narayana numbers make their appearance, and in our proofs we encounter several new bijections between different combinatorial objects counted by these numbers.

1.2   Outline of This Paper

In Sect. 2 we discuss some preliminaries that will be used throughout the paper. In Sect. 3 we present the structural results when the number n of matching edges is odd. In Sect. 4 we discuss the properties of $\mathcal{H}_n$ when n is even. Finally, in Sect. 5 we discuss the implications of our results with regards to rainbow cycles in $\mathcal{G}_n$. We conclude with some open questions in Sect. 6. Results and proofs that are omitted due to space constraints in this extended abstract can be found in the preprint [19].

2   Preliminaries

We first explain the combinatorial characterization of centered flips. Given a matching $M \in \mathcal{M}_n$, and any of its edges $e \in M$, we let $\ell(e)$ denote the minimum number of other matching edges from M on either of the two sides of the edge e. We refer to $\ell(e)$ as the *length* of the edge e. We let $\mu = \mu(n)$ denote the maximum possible length of an edge, so the possible edge lengths are $0, 1, \ldots, \mu$. Clearly, we have $\mu = (n-1)/2$ if n is odd and $\mu = (n-2)/2$ if n is even. The following lemma can be verified easily.

Lemma 1. *A flip is centered if and only if the sum of the lengths of the four edges of the corresponding quadrilateral equals $n - 2$. On the other hand, the flip is non-centered if and only if this sum is strictly less than $n - 2$.*

We say that an edge $e \in M$ is *visible* from the circle center, if the rays from the center to both edge endpoints do not cross any other matching edges. If n is odd, there may be an edge through the circle center, and then we decide visibility of the other edges by ignoring this edge, and declare the edge itself to *not* be visible. Moreover, we say that a matching edge f is *hidden behind* another edge e, if the rays from the circle center to the endpoints of f cross e.

3   Connectedness and Diameter for Odd n

In this section, we assume that the number n of matching edges is odd. We show that the graph $\mathcal{H}_n$ is connected in this case (Theorem 2), and that its diameter is linear in n (Theorem 4).

Theorem 2. *For odd $n \geq 3$, the graph $\mathcal{H}_n$ is connected.*

For proving Theorem 2, we consider two special matchings, namely those that have only edges of length 0, and we denote them by M_0 and M_0'. Roughly, the proof proceeds in two steps: We first argue that in the graph $\mathcal{H}_n$, there is a path from any matching $M \in \mathcal{M}_n$ to either M_0 or M_0'. We then show that there is also a path between M_0 and M_0', and this will establish the theorem. The first step of the proof is based on the following key lemma.

Lemma 3. *Consider a matching $M \in \mathcal{M}_n$ that has no edge through the circle center and that is different from M_0 and M_0', i.e., M has an edge of length strictly more than 0. There is a sequence of at most 4 centered flips from M to another matching that has no edge through the circle center and that has at least one more visible edge than M.*

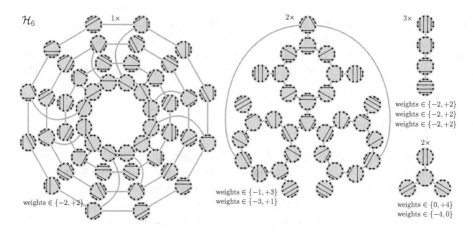

Fig. 3. The graph $\mathcal{H}_6$ with the weights of all of its components. Among the components, isomorphic copies are omitted, and the multiplicities are shown above the components. The isomorpic copies differ only by rotation of the matchings. For instance, there are two copies of the component shown at the bottom right.

Clearly, repeatedly applying this claim shows that there is a path from M to either M_0 or M_0', as these are the only two matchings with the maximum number of visible edges. For a proof of this lemma, see [19].

Recall that the *diameter* of a graph is the maximum length of all shortest paths between any two vertices of the graph. In the flip graph $\mathcal{H}_n$, the diameter measures how many centered flips are needed in the worst case to transform two matchings into each other. With computer help, we determined the diameter of $\mathcal{H}_n$ for $n = 3, 5, 7, 9, 11$ to be $2, 8, 14, 20, 26$, which equals $3n - 7$ for those values of n. In all those cases, this distance was attained for the two matchings with only length-0 edges (differing by a rotation of π/n). These are the extreme vertices on the left and right in Fig. 4 (cf. also the left hand side of Fig. 2). We conjecture that this is the correct value for all n. As a first step towards this conjecture, we can prove the following linear bounds.

Theorem 4. *For odd $n \geq 3$, the diameter of $\mathcal{H}_n$ is at least $n - 1$ and at most $11n - 29$.*

Proof. Hernando, Hurtado, and Noy [10] showed that the diameter of $\mathcal{G}_n$ is exactly $n-1$, and as $\mathcal{H}_n$ is a spanning subgraph of $\mathcal{G}_n$, its diameter is at least $n-1$.

It remains to prove the upper bound in the theorem. As before, we let M_0 and M_0' denote the two matchings that have only edges of length 0. We first argue that the distance between any matching $M \in \mathcal{M}_n$ and either M_0 or M_0' is at most $4n - 11$. Indeed, if M has no edge through the circle center, then it has at least 3 visible edges (2 visible edges are only possible when n is even). As a consequence of Lemma 3, we can reach M_0 or M_0', which have n visible edges each, from M with at most $4(n-3) = 4n - 12$ centered flips. On the other hand, if M has an edge through the circle center, then a single centered flip leads

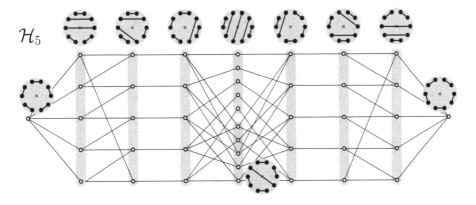

Fig. 4. The graph $\mathcal{H}_5$, drawn in a simplified way, where for every matching, we only show one representative of the equivalence class under rotation by $2\pi/5$.

from M to one of its neighbors that does not have an edge through the center, establishing the bound $4n - 11$.

In the remainder of the proof we show that the distance between M_0 and M_0' in $\mathcal{H}_n$ is at most $3n - 7$. With these two bounds, we can then bound the distance in $\mathcal{H}_n$ between any two matchings $M, M' \in \mathcal{M}_n$ as follows: We know that from both M and M' we can reach either M_0 or M_0' with at most $4n - 11$ centered flips each. If both of these flip sequences reach the same matching from $\{M_0, M_0'\}$, we have found a path in $\mathcal{H}_n$ of length at most $2(4n - 11)$ between M and M'. Otherwise we can connect M_0 and M_0' with a path of length $3n - 7$, yielding a path of length at most $2(4n - 11) + (3n - 7) = 11n - 29$ between M and M', proving the upper bound in the theorem.

We prove this claim by induction on all odd values of $n \geq 3$; see Fig. 5. For $n = 3$ the distance between M_0 and M_0' is $3n - 7 = 2$, as can be verified from the left hand side of Fig. 2. For the induction step, suppose that $n \geq 5$ is odd and that the claim holds for $n - 2$. Consider the flip sequence shown at the top part of Fig. 5, consisting of 3 centered flips, leading from the matching M_0 to a matching M_1 that contains $n - 1$ edges of length 0 and one edge of length 1. Consider the two edges a and a' in M_1 that lie on opposite sides of the circle, where the edge a is hidden behind the unique length-1 edge. We can ignore the edges a and a' from the configuration, and obtain a matching with $n - 2$ edges. As the ignored edges are antipodal on the circle, every centered flip operating on the remaining $n - 2$ edges in $\mathcal{H}_{n-2}$ is also a centered flip in $\mathcal{H}_n$. Also observe that ignoring those two edges from M_1 leaves us with a matching with only length-0 edges in $\mathcal{H}_{n-2}$. Consequently, by induction we have a flip sequence of length $3(n - 2) - 7$ from M_1 to a matching M_1' that has $n - 1$ edges of length 0 and one edge of length 1, that still contains the edges a and a', but now the edge a' is hidden behind the unique length-1 edge. By symmetry, we can reach M_0' from M_1' with at most 3 centered flips. Overall, the length of the flip sequence from M_0 to M_0' obtained in this way is $3(n-2)-7+2\cdot 3 = 3n-7$. This completes the inductive proof and thus the proof of the theorem. ⊓⊔

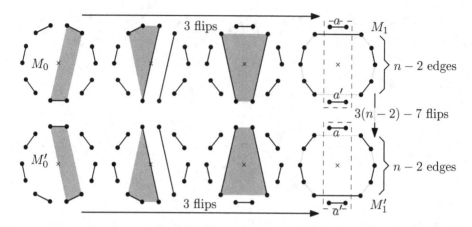

Fig. 5. Illustration of the inductive proof that the distance between M_0 and M_0' in $\mathcal{H}_n$ is at most $3n - 7$.

4 Component Structure for Even n

In this section, we assume that the number n of matching edges is even. It was proved in [9] that in this case the graph $\mathcal{H}_n$ has at least $n - 1$ components. We improve upon this considerably, by showing that $\mathcal{H}_n$ has *exponentially* many components, and we also provide a fine-grained picture of the component structure of the graph $\mathcal{H}_n$ (Theorem 5 and Corollary 7). We also prove explicit formulas for the number of matchings with certain weights, a parameter that is closely related to the component sizes of the graph $\mathcal{H}_n$, proving a conjecture raised in [9] (Theorem 8 and Corollary 9).

4.1 Point-Symmetric Matchings

We now consider the set of all point-symmetric matchings, i.e., matchings that are point-symmetric with respect to the circle center.

Theorem 5. *For even $n \geq 2$, there are $\binom{n}{n/2}$ point-symmetric matchings, and all those matchings form components in $\mathcal{H}_n$ that are trees. There are $C_{n/2}$ such components, and each of them contains exactly $n/2+1$ matchings. All matchings that are not point-symmetric form components that are not trees.*

The properties of the graph $\mathcal{H}_n$ stated in Theorem 5 can be seen nicely in Fig. 3 for the case $n = 6$. The proof of Theorem 5 can be found in [19].

4.2 Weights of Matchings

For our further investigations, we assign an integer weight to each matching. For this we give the points on the unit circle a fixed labelling by $1, 2, \ldots, 2n$ in clockwise direction. Consider a matching $M \in \mathcal{M}_n$ and one of its edges $e \in M$, and let i and j be the endpoints of e so that the circle center lies to the right of the ray from i to j. We define the *sign of the edge* e as $\text{sgn}(e) := 1$ if i is odd, and $\text{sgn}(e) := -1$ if i is even. We call an edge e *positive* if $\text{sgn}(e) = +1$, and we call it *negative* if $\text{sgn}(e) = -1$. Moreover, we define the *weight* of the matching M as $w(M) := \sum_{e \in M} \text{sgn}(e) \cdot \ell(e)$; see Fig. 6. Note that rotating a matching by π/n, in either direction, changes the weight by a factor of -1.

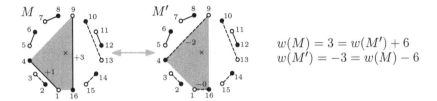

$$w(M) = 3 = w(M') + 6$$
$$w(M') = -3 = w(M) - 6$$

Fig. 6. The weight of two matchings with $n = 8$ edges. Odd points are drawn as white bullets, even points as black bullets. Positive edges are drawn solid, negative edges are drawn dashed. The centered flip changes the weight by $\pm(n - 2) = \pm 6$.

Observe also that a quadrilateral corresponding to a centered flip has two positive edges and two negative edges, and the edges with the same sign are opposite to each other. From this and Lemma 1 we see that any centered flip changes the weight of a matching by $\pm(n - 2)$. Moreover, it was shown in [9] that all possible weights are in a particular integer range; see Fig. 6.

Lemma 6 (Lemmas 11+12 in *[9]*). *Let $n \geq 2$ be even. Applying a centered flip to any matching from $\mathcal{M}_n$ changes its weight by $-(n-2)$ if the two negative edges appear in this flip, or by $+(n - 2)$ if the two positive edges appear in this flip, and flips of these two kinds must alternate along any sequence of centered flips. Moreover, for any matching $M \in \mathcal{M}_n$ we have $w(M) \in [-(n-2), n-2] := \bigl\{ -(n - 2), -(n - 2) + 1, \ldots, n - 3, n - 2 \bigr\}$, and each of these weight values is attained for some matching in $\mathcal{M}_n$.*

Our next result is an immediate consequence of this lemma.

Corollary 7. *For even $n \geq 4$, the graph $\mathcal{H}_n$ has at least $C_{n/2} + n - 3$ components.*

Proof. By Theorem 5, the graph $\mathcal{H}_n$ has exactly $C_{n/2}$ components that contain all point-symmetric matchings. Moreover, for $c = 1, \ldots, n - 3$, we can easily construct a matching that is not point-symmetric and has weight c. Indeed, for $c = 1, \ldots, \mu$, we take a matching that has a single edge of length c, and all other edges of length 0. For $c = \mu + 1, \ldots, 2\mu - 1 = n - 3$, we take a matching with a single edge of length μ, another edge of length c, and all other edges of length 0. By Lemma 6, these $n - 3$ matchings all lie in distinct components of $\mathcal{H}_n$, and they must be different from the $C_{n/2}$ components containing the point-symmetric matchings. This implies the claimed lower bound.   □

Motivated by Lemma 6, we partition the set of all matchings $\mathcal{M}_n$ according to their weights. Specifically, for any non-zero integer $c \in [-(n-2)(n-2)]$, we let $\mathcal{W}_{n,c}$ be the set of all matchings from $\mathcal{M}_n$ with weight exactly c. For the special case $c = 0$ we define $\mathcal{W}_{n,0} := \{M_0\}$ and $\mathcal{W}_{n,0}^- := \{M_0^-\}$, where M_0 is the matching that has only length-0 edges and all of them positive, and M_0^- is the matching that has only length-0 edges and all of them negative. Clearly, we have $w(M_0) = w(M_0^-) = 0$. Moreover, for $c = 0, 1, \dots, n-2$ we define

$$\mathcal{M}_{n,c} := \begin{cases} \mathcal{W}_{n,c} \cup \mathcal{W}_{n,c-(n-2)} & \text{if } c \leq n-3, \\ \mathcal{W}_{n,n-2} \cup \mathcal{W}_{n,0}^- & \text{if } c = n-2, \end{cases} \tag{1}$$

i.e., the set $\mathcal{M}_{n,c}$ contains all matchings that have either the same weight or whose weights differ by $n-2$.

We now establish explicit formulas for the cardinalities of the sets $\mathcal{W}_{n,c}$ and $\mathcal{M}_{n,c}$, answering a conjecture raised in [9] that expresses these quantities via *generalized Narayana numbers* $N_r(n,k)$, defined as

$$N_r(n,k) = \tfrac{r+1}{n+1} \binom{n+1}{k} \binom{n-r-1}{k-1} \tag{2}$$

for any integers $n \geq 1$, $r \geq 0$, and $1 \leq k \leq n-r$.

Theorem 8. *For even $n \geq 2$ and any $c = 0, 1, \dots, n-2$, we have $|\mathcal{W}_{n,c}| = N_1(n, |c| + 1)/2$.*

The proof of Theorem 8 can be found in [19]. By Lemma 6, there are no centered flips between any matchings from distinct sets $\mathcal{M}_{n,c}$, $c = 0, 1, \dots, n-2$, i.e., the cardinalities $|\mathcal{M}_{n,c}|$ are an upper bound for the size of the components of the graph $\mathcal{H}_n$. The following corollary makes these bounds explicitly. It follows immediately from Theorem 8, using (1) and (2) (for details, see [19]).

Corollary 9. *For even $n \geq 2$, every component of $\mathcal{H}_n$ has at most $N_1(n, n/2)$ vertices. Asymptotically, this is a $2/\sqrt{\pi n}(1 + o(1))$-fraction of all vertices.*

5   Rainbow Cycles

We now turn back our attention to the flip graph $\mathcal{G}_n$ discussed in the beginning, which contains all possible flips, not just the centered ones. For any integer $r \geq 1$, an *r-rainbow cycle* in the graph $\mathcal{G}_n$ is a cycle with the property that every possible matching edge appears exactly r times in flips along this cycle. The notion of rainbow cycles was introduced in [9], and studied for several different flip graphs, including the graph $\mathcal{G}_n$. The authors showed that $\mathcal{G}_n$ has a 1-rainbow cycle for $n = 2, 4$, and a 2-rainbow cycle for $n = 6, 8$. It was also proved that $\mathcal{G}_n$ has no 1-rainbow cycle for any odd $n \geq 3$ and for $n = 6, 8, 10$. The last result was extended to the case $n = 12$ in [18]. We complement these results as follows:

Theorem 10. *For odd $n \geq 3$ and any $r \geq 1$, there is no r-rainbow cycle in $\mathcal{G}_n$. For even $n \geq 2$ and any $r > 2/n^2 \cdot N_1(n, n/2)$, there is no r-rainbow cycle in $\mathcal{G}_n$.*

Proof. For any $n \geq 2$, the number possible matching edges is n^2. As in every flip, two edges appear in the matching and two edges disappear, an r-rainbow cycle must have length $rn^2/2$.

Let $n \geq 3$ be odd. There are $2n$ distinct possible matching edges for each length $c = 0, \ldots, \mu - 1$, and only n distinct possible matching edges of length $c = \mu$. Therefore, the average length of all edges appearing or disappearing along an r-rainbow cycle is $\left(\sum_{c=0}^{\mu-1} c \cdot 2n + \mu \cdot n\right)/n^2 = (n-2)/4 + 1/(4n) > (n-2)/4$, where we used $\mu = (n-1)/2$. However, by Lemma 1, the average length of the four edges appearing or disappearing in a centered flip is only $(n-2)/4$, and even smaller for non-centered flips. Consequently, there can be no r-rainbow cycle.

Let $n \geq 2$ be even. It was proved in [9, Lemma 10] (with a similar averaging argument as given before for odd n) that every r-rainbow cycle in $\mathcal{G}_n$ may only contain centered flips, i.e., we may restrict our attention to the subgraph $\mathcal{H}_n \subseteq \mathcal{G}_n$ given by centered flips. By Corollary 9, all components of this graph contain at most $N_1(n, n/2)$ vertices. Consequently, if the length of the cycle exceeds this bound, then no such cycle can exist. This is the case if $rn^2/2 > N_1(n, n/2)$, or equivalently, if $r > 2/n^2 \cdot N_1(n, n/2)$. $\qquad\square$

6 Open Questions

- For odd $n \geq 3$, a natural task is to narrow down the bounds for the diameter of the graph $\mathcal{H}_n$ given by Theorem 4. We believe that the answer is $3n - 7$, which is the correct value for $n = 3, 5, 7, 9, 11$.
- For even $n \geq 4$, it would be very interesting to prove that the number of components of the graph $\mathcal{H}_n$ is exactly $C_{n/2} + n - 3$, which we established as a lower bound in Corollary 7, and which is tight for $n = 4, 6, 8, 10, 12, 14$.
- It is open whether r-rainbow cycles exist in the graph $\mathcal{G}_n$ for even $n \geq 14$ and any $1 \leq r \leq 2/n^2 \cdot N_1(n, n/2)$. As mentioned in the proof of Theorem 10, we may restrict our search to the subgraph $\mathcal{H}_n \subseteq \mathcal{G}_n$.

Acknowledgements. We thank the anonymous reviewers of the extended abstract of this paper, who provided many insightful comments. In particular, one referee's observation about our proof of Lemma 3 improved our previous upper bound on the diameter of $\mathcal{H}_n$ for odd n from $\mathcal{O}(n \log n)$ to $\mathcal{O}(n)$ (recall Theorem 4).

Figure 3 was obtained by slightly modifying Fig. 10 from [9], and the authors of this paper kindly provided us with the source code of their figure.

References

1. Aichholzer, O., Asinowski, A., Miltzow, T.: Disjoint compatibility graph of non-crossing matchings of points in convex position. Electron. J. Combin. **22**(1), 53 (2015). https://www.combinatorics.org/ojs/index.php/eljc/article/view/v22i1p65. Paper 1.65
2. Aichholzer, O., Aurenhammer, F., Huemer, C., Vogtenhuber, B.: Gray code enumeration of plane straight-line graphs. Graphs Combin. **23**(5), 467–479 (2007). https://doi.org/10.1007/s00373-007-0750-z

3. Aichholzer, O., et al.: Flip distances between graph orientations. In: Sau, I., Thilikos, D.M. (eds.) WG 2019. LNCS, vol. 11789, pp. 120–134. Springer, Cham (2019). https://doi.org/10.1007/978-3-030-30786-8_10
4. Berry, L.A., Reed, B., Scott, A., Wood, D.R.: A logarithmic bound for the chromatic number of the associahedron. arXiv:1811.08972 (2018)
5. Ceballos, C., Santos, F., Ziegler, G.M.: Many non-equivalent realizations of the associahedron. Combinatorica **35**(5), 513–551 (2014). https://doi.org/10.1007/s00493-014-2959-9
6. Cleary, S., St. John, K.: Rotation distance is fixed-parameter tractable. Inf. Process. Lett. **109**(16), 918–922 (2009). https://doi.org/10.1016/j.ipl.2009.04.023
7. Cleary, S., St. John, K.: A linear-time approximation for rotation distance. J. Graph Algorithms Appl. 14(2), 385–390 (2010). http://dx.doi.org/10.7155/jgaa.00212
8. Fabila-Monroy, R., et al.: On the chromatic number of some flip graphs. Discr. Math. Theor. Comput. Sci. **11**(2), 47–56 (2009). http://dmtcs.episciences.org/460
9. Felsner, S., Kleist, L., Mütze, T., Sering, L.: Rainbow cycles in flip graphs. SIAM J. Discr. Math. **34**(1), 1–39 (2020). https://doi.org/10.1137/18M1216456
10. Hernando, C., Hurtado, F., Noy, M.: Graphs of non-crossing perfect matchings. Graphs Combin. **18**(3), 517–532 (2002). https://doi.org/10.1007/s003730200038
11. Houle, M.E., Hurtado, F., Noy, M., Rivera-Campo, E.: Graphs of triangulations and perfect matchings. Graphs Combin. **21**(3), 325–331 (2005). https://doi.org/10.1007/s00373-005-0615-2
12. Huemer, C., Hurtado, F., Noy, M., Omaña-Pulido, E.: Gray codes for non-crossing partitions and dissections of a convex polygon. Discr. Appl. Math. **157**(7), 1509–1520 (2009). https://doi.org/10.1016/j.dam.2008.06.018
13. Hurtado, F., Noy, M.: Graph of triangulations of a convex polygon and tree of triangulations. Comput. Geom. **13**(3), 179–188 (1999). https://doi.org/10.1016/S0925-7721(99)00016-4
14. Knuth, D.E.: The Art of Computer Programming. Vol. 4A. Combinatorial Algorithms. Part 1. Addison-Wesley, Upper Saddle River, NJ (2011)
15. Lee, C.W.: The associahedron and triangulations of the n-gon. Eur. J. Combin. **10**(6), 551–560 (1989). https://doi.org/10.1016/S0195-6698(89)80072-1
16. Li, M., Zhang, L.: Better approximation of diagonal-flip transformation and rotation transformation. In: Hsu, W.-L., Kao, M.-Y. (eds.) COCOON 1998. LNCS, vol. 1449, pp. 85–94. Springer, Heidelberg (1998). https://doi.org/10.1007/3-540-68535-9_12
17. Lucas, J.M., van Baronaigien, D.R., Ruskey, F.: On rotations and the generation of binary trees. J. Algorithms **15**(3), 343–366 (1993). https://doi.org/10.1006/jagm.1993.1045
18. Milich, M.: Kreise für planare Matchings. Bachelor's thesis, TU Berlin, German (2018)
19. Milich, M., Mütze, T., Pergel, M.: On flips in planar matchings. arXiv:2002.02290. Preprint version of the present paper with full proofs (2020)
20. Parlier, H., Petri, B.: The genus of curve, pants and flip graphs. Discr. Comput. Geom. **59**(1), 1–30 (2018). https://doi.org/10.1007/s00454-017-9922-7
21. Pilaud, V., Santos, F.: Quotientopes. Bull. Lond. Math. Soc. **51**(3), 406–420 (2019). https://doi.org/10.1112/blms.12231
22. Pournin, L.: The diameter of associahedra. Adv. Math. **259**, 13–42 (2014). https://doi.org/10.1016/j.aim.2014.02.035
23. Pournin, L.: Eccentricities in the flip-graphs of convex polygons. J. Graph Theor. **92**(2), 111–129 (2019). https://doi.org/10.1002/jgt.22443

24. Reading, N.: From the Tamari lattice to Cambrian lattices and beyond. In: Müller-Hoissen, F., Pallo, J., Stasheff, J. (eds.) Associahedra, Tamari Lattices and Related Structures. Progress in Mathematics, vol. 299, pp. 293–322. Birkhäuser/Springer, Basel (2012). https://doi.org/10.1007/978-3-0348-0405-9_15

25. Reading, N.: Lattice theory of the poset of regions. In: Grätzer, G., Wehrung, F. (eds.) COCOON 1998. LNCS, vol. 1449, pp. 399–487. Springer, Cham (2016). https://doi.org/10.1007/978-3-319-44236-5_9

26. Rogers, R.O.: On finding shortest paths in the rotation graph of binary trees. In: Proceedings of the Thirtieth Southeastern International Conference on Combinatorics, Graph Theory, and Computing, Boca Raton, FL, 1999, vol. 137, pp. 77–95 (1999)

27. Sleator, D.D., Tarjan, R.E., Thurston, W.P.: Rotation distance, triangulations, and hyperbolic geometry. J. Amer. Math. Soc. 1(3), 647–681 (1988). http://www.jstor.org/stable/1990951

Degree Distribution
for Duplication-Divergence Graphs:
Large Deviations

Alan Frieze[1] , Krzysztof Turowski[2](✉) , and Wojciech Szpankowski[3]

[1] Department of Mathematical Sciences, Carnegie Mellon University,
Pittsburgh, PA, USA
alan@random.math.cmu.edu
[2] Department of Theoretical Computer Science, Jagiellonian University,
Krakow, Poland
krzysztof.szymon.turowski@gmail.com
[3] Center for Science of Information, Department of Computer Science,
Purdue University, West Lafayette, IN, USA
spa@cs.purdue.edu

Abstract. We present a rigorous and precise analysis of the degree distribution in a dynamic graph model introduced by Solé et al. in which nodes are added according to a duplication-divergence mechanism, i.e. by iteratively copying a node and then randomly inserting and deleting some edges for a copied node. This graph model finds many applications since it well captures the growth of some real-world processes e.g. biological or social networks. However, there are only a handful of rigorous results concerning this model. In this paper we present rigorous results concerning the degree distribution.

We focus on two related problems: the expected value and large deviation for the *degree of a fixed node* through the evolution of the graph and the expected value and large deviation of the *average degree* in the graph. We present exact and asymptotic results showing that both quantities may decrease or increase over time depending on the model parameters. Our findings are a step towards a better understanding of the overall graph behaviors, especially, degree distribution, symmetry, and compression, important open problems in this area.

Keywords: Dynamic graphs · Duplication-divergence graphs · Degree distribution · Large deviation

1 Introduction

It is widely accepted that we live in the age of data deluge. On a daily basis we observe the increasing availability of data collected and stored in various forms,

This work was supported by NSF Center for Science of Information (CSoI) Grant CCF-0939370, and in addition by NSF Grant CCF-1524312, and National Science Center, Poland, Grant 2018/31/B/ST6/01294. This work was also supported by NSF Grant DMS1661063.

I. Adler and H. Müller (Eds.): WG 2020, LNCS 12301, pp. 226–237, 2020.
https://doi.org/10.1007/978-3-030-60440-0_18

as sequences, expressions, interactions or structures. A large part of this data is given in a complex form which also conveys a "shape" of the structure, such as network data. As examples we have various biological networks, social networks, and Web graphs.

Given a representation of these networks as graphs, there arises a natural question: what are the rules governing the growth and evolution of such networks? In fact, finding such rules should enable us to model real networks arising in many diverse applications. For example, there is experimental evidence [16] that the evolution of some biological networks is driven by the duplication mechanism, in which new nodes appear as copies of some already existing ones in the network. This is supplemented by a certain amount of divergence due to random mutations that leads to some differences between patterns of interaction for the source and the duplicate elements.

Fundamental questions arise about the structural properties of these networks. For example, Faloutsos et al. [5] brought to the front the issue of "scale-free" power law behavior. First, there is the question as to whether or not the degree distributions of the real-world networks do indeed have a tail close to a power law. Second and most important, one needs to verify whether the underlying random graph models may indeed generate graphs that exhibit the desired behavior e.g. in expectation. This is directly related to the broader question of the degree distributions in graphs that may be generated from these random models. We would expect that a good model for real-world networks generates graphs that typically are not much different in terms of the number of vertices with given degrees to what is seen in practice. However, to answer this question we first need a good theoretical understanding of the degree distribution of our models and this is the subject of this paper.

Another important problem in this area is the question of *symmetry*. It may be formulated as follows: given a probability distribution over graphs of size n, what is the distribution of $\log|\mathrm{Aut}(G)|$, where G is a random graph drawn from this distribution and $\mathrm{Aut}(G)$ is its automorphism group (i.e., permutation preserving adjacency). Clearly, it is related to the degree distribution problem since for example the number of small symmetrical structures like cherries and diamonds (vertices of degree 1 and 2, respectively, having the same neighborhood) is a lower bound on the number of automorphisms for any graph. Interestingly enough, many real-world networks such as protein-protein and social networks, exhibit a lot of symmetry as shown in Table 1.

It turns out that the most popular random graph models do not exhibit much symmetry. For example, it was proved in [2] that the Erdős-Renyi random graph model generates asymmetric graphs (for not too small and too large edge probability), that is, $\log|\mathrm{Aut}(G)| = 0$, with high probability. In a similar vein, it was proved for the preferential attachment model, also known as the Barabási-Albert model, that for $m \geq 3$ (which is necessary if we want to obtain sufficiently dense graphs that resemble real-world networks) is asymmetric with high probability [9].

Table 1. Symmetries of the real-world networks [13,15].

Network	Nodes	Edges	$\log |\mathrm{Aut}(G)|$
Baker's yeast protein-protein interactions	6,152	531,400	546
Fission yeast protein-protein interactions	4,177	58,084	675
Mouse protein-protein interactions	6,849	18,380	305
Human protein-protein interactions	17,295	296,637	3026
ArXiv high energy physics citations	7,464	116,268	13
Simple English Wikipedia hyperlinks	10,000	169,894	1019
CollegeMsg online messages	1,899	59,835	232

Therefore, in order to study and understand the behavior of real-world networks we need to look at dynamic graph models that naturally generate internal graph symmetries. As was mentioned before, one promising route is to investigate models that evolve according to the duplication and mutation rules. So let us consider the most popular duplication-divergence model introduced by Solé et al. [12], referred to below as $\mathrm{DD}(t, p, r)$. It is defined as follows: starting from a given graph on t_0 vertices (labeled from 1 to t_0) we add subsequent vertices labeled $t_0 + 1$, $t_0 + 2, \ldots, t$ as copies of some existing vertices in the graph and then we introduce divergence by adding and removing some edges connected to the new vertex independently at random. Finally, we remove the labels and return the structure, i.e. the unlabeled graph.

It has been shown that for a certain set of parameters, graphs generated according to the duplication-divergence mechanism fit very well empirically with the structure of some real-world networks (e.g., protein-protein and citation networks) in terms of the degree distribution [3] and small subgraph (graphlets) counts [11]. However, at the moment there do not exist any rigorous general results regarding symmetries and hence degree distribution for such graphs. Experimentally, when generating multiple graphs according to this model with different parameters, we observe the pattern presented in Fig. 1: There is a set of parameters (i.e., p and r) for which the generated graphs are highly symmetric with a large automorphisms group. It was shown by Sreedharan et al. [13] that possible values of the parameters for real-world networks lie in the blue-violet area, indicating a lot of symmetry. All these remarks suggest that there is a certain merit to study a possible link between the duplication-divergence model and certain types of real-world networks. To accomplish this we need to study the average and large deviation of their degree sequence, which is the main topic of this conference paper.

There exist only a handful previous rigorous results on the $\mathrm{DD}(t, p, r)$ model. In view of these, it is imperative that we understand the degree distribution for the duplication-divergence networks. Turowski et al. showed in [14] that for the special case of $p = 1$, $r = 0$ the expected logarithm of the number of automorphisms for graphs on t vertices is asymptotically $\Theta(t \log t)$, which

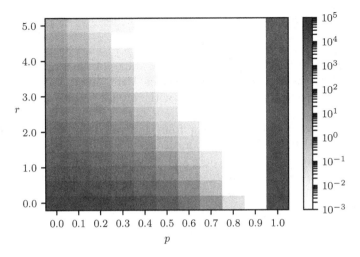

Fig. 1. Symmetry of graphs $(\log|\mathrm{Aut}(G)|)$ generated by the DD(t, p, r) model. (Color figure online)

indicates a lot of symmetry. This allows to construct asymptotically optimal compression algorithms for such graphs. However, the proposed approach used certain properties for this particular set of parameters that does not generalize to other sets of parameters.

For $r = 0$ and $p < 1$, it was recently proved by Hermann and Pfaffelhuber in [6] that depending on the value of p either there exists a limiting distribution of degree frequencies with almost all vertices isolated or there is no limiting distribution as $t \to \infty$. Moreover, it is shown in [8] that the number of vertices of degree one is $\Omega(\log t)$ but again the precise rate of growth of the number of vertices with any fixed degree $k > 0$ is currently unknown. Recently, also for $r = 0$, Jordan [7] showed that the *non-trivial connected component* has a degree distribution which conforms to a power-law behavior in size, but only for $p < e^{-1}$. In this case the exponent is equal to γ which is the solution of $3 = \gamma + p^{\gamma - 2}$.

In this paper we study the degree distribution from a different perspective. In particular, we present results concerning the degree of a given vertex s at time t (denoted by $\deg_t(s)$) and the average degree in the graph (denoted by $D(G_t)$). We show that the asymptotic values of the means $\mathbb{E}[\deg_t(s)]$ and $\mathbb{E}[D(G_t)]$ as $t \to \infty$ exhibit phase transitions over the parameter space as a function of p and r. We then present some results for the tails of the degree distribution for $D(G_t)$ and $\deg_t(s)$ for $s = O(1)$. It turns out that the deviation by a polylogarithmic factor under or over the respective means is sufficient to obtain a polynomial tail, that is to find an $O(t^{-A})$ tail probability. In this way we have proved that the distribution of $D(G_t)$ and $\deg_t(s)$ are in some sense concentrated around their means.

2   Main Results

In this section we first define formally the DD(t, p, r) model. Then, we present our main results, first concerning the expected values of the degree distribution, and then large deviations of their distributions. Here, due to the space limit we provide only the main lines of reasoning, with the full proofs to be included in the journal version.

We use standard graph notation, e.g. from [4]: $V(G)$ denotes the set of vertices of graph G, $\mathcal{N}_G(u)$ – the set of neighbors of vertex u in G, $\deg_G(u) = |\mathcal{N}_G(u)|$ – the degree of u in G. By G_t we denote a graph on t vertices. For brevity we use the abbreviations, e.g. $\deg_t(u)$ instead of $\deg_{G_t}(u)$ and $\mathcal{N}_t(u)$ instead of $\mathcal{N}_{G_t}(u)$. All graphs are simple. Let us also introduce the *average degree* $D(G)$ of G as

$$D(G) = \frac{1}{|V(G)|} \sum_{u \in V(G)} \deg_G(u).$$

It is worth noting that it is also known in the literature as the first moment of the degree distribution.

We formally define the model DD(t, p, r) as follows: let $0 \le p \le 1$ and $0 \le r \le t_0$ be the parameters of the model. Let also G_{t_0} be a graph on t_0 vertices, with $V(G_{t_0}) = \{1, \ldots, t_0\}$. Now, for every $t = t_0, t_0 + 1, \ldots$ we create G_{t+1} from G_t according to the following rules:

1. add a new vertex $t + 1$ to the graph,
2. pick vertex u from $V(G_t) = \{1, \ldots, t\}$ uniformly at random – and denote u as *parent*$(t + 1)$,
3. for every vertex $i \in V(G_t)$:
   (a) if $i \in \mathcal{N}_t(parent(t + 1))$, then add an edge between i and $t + 1$ with probability p,
   (b) if $i \notin \mathcal{N}_t(parent(t + 1))$, then add an edge between i and $t + 1$ with probability $\frac{r}{t}$.

2.1   Average Degree in the Graph

We start with the average degree in the graph $D(G_t)$. First, we find the following recurrence for the average degree of G_{t+1}:

$$\mathbb{E}[D(G_{t+1}) \mid G_t] = \frac{1}{t+1} \mathbb{E}\left[\sum_{i=1}^{t+1} \deg_{t+1}(i) \mid G_t\right]$$

$$= \frac{1}{t+1} \mathbb{E}\left[\sum_{i=1}^{t} \deg_t(i) + 2\deg_{t+1}(t+1) \mid G_t\right]$$

$$= \frac{1}{t+1} \left(\sum_{i=1}^{t} \deg_t(i) + 2\mathbb{E}\left[\deg_{t+1}(t+1) \mid G_t\right]\right)$$

$$= \frac{1}{t+1} \left(tD(G_t) + 2\mathbb{E}[\deg_{t+1}(t+1) \mid G_t]\right).$$

Next, we find the following relationship between the expected average degree $\mathbb{E}[D(G_t)]$ and the expected degree of the new vertex $\mathbb{E}[\deg_{t+1}(t+1)]$:

Lemma 1. *For any $t \geq t_0$ it holds that*

$$\mathbb{E}[\deg_{t+1}(t+1)] = \left(p - \frac{r}{t}\right)\mathbb{E}[D(G_t)] + r.$$

It is quite intuitive that the expected degree of a new vertex behaves as if we would choose a vertex with the average degree $\mathbb{E}[D(G_t)]$ as its parent, and then copy a p fraction of its edges, also adding r more edges (in expectation) to the other vertices in the graph.

From this lemma we find

$$\mathbb{E}[D(G_{t+1}) \mid G_t] = D(G_t)\left(1 + \frac{2p-1}{t+1} - \frac{2r}{t(t+1)}\right) + \frac{2r}{t+1}. \tag{1}$$

This recurrence falls under a general recurrence of the form

$$\mathbb{E}[f(G_{t+1}) \mid G_t] = f(G_t)g_1(t) + g_2(t) \tag{2}$$

where g_1 and g_2 are given functions, dependent on p and r. We will solve it exactly and asymptotically in the sequel. This allows us to find an asymptotic expression for the average degree.

In the sequel we present a series of lemmas that will be used to obtain the asymptotics of $\mathbb{E}[f(G_t)]$. These lemmas are based on martingale theory and they use various asymptotic properties of the Euler gamma function. For space reasons, proofs of Lemmas 2 and 5 are omitted.

Lemma 2. *Let $(G_n)_{n=n_0}^{\infty}$ be a Markov process for which $\mathbb{E}f(G_{n_0}) > 0$ and Eq. (2) holds with $g_1(n) > 0$, $g_2(n) \geq 0$ for all $n = n_0, n_0 + 1, \ldots$. Then for all $n \geq n_0$*

$$\mathbb{E}f(G_n) = f(G_{n_0}) \prod_{k=n_0}^{n-1} g_1(k) + \sum_{j=n_0}^{n-1} g_2(j) \prod_{k=j+1}^{n-1} g_1(k).$$

The above lemma shows that the solutions of recurrences of type Eq. (2) contain products and sum of products of g_1 and g_2. The next lemmas show how to handle such products. First, since in our case $g_1(n)$ and $g_2(n)$ are of form $\frac{W_1(n)}{W_2(n)}$ for certain polynomials $W_1(n)$, $W_2(n)$, we turn the products of polynomials into the products of Euler gamma functions.

Lemma 3. *Let $W_1(k)$, $W_2(k)$ be polynomials of degree d with respective (not necessarily distinct) roots a_i, b_i $(i = 1, \ldots, d)$, that is, $W_1(k) = \prod_{i=1}^{d}(k - a_i)$ and $W_2(k) = \prod_{j=1}^{d}(k - b_j)$. Then*

$$\prod_{k=n_0}^{n-1} \frac{W_1(k)}{W_2(k)} = \prod_{i=1}^{d} \frac{\Gamma(n - a_i)}{\Gamma(n - b_i)} \frac{\Gamma(n_0 - b_i)}{\Gamma(n_0 - a_i)}.$$

Proof. We have

$$\prod_{k=n_0}^{n-1} \frac{W_1(k)}{W_2(k)} = \prod_{k=n_0}^{n-1} \prod_{i=1}^{d} \frac{k-a_i}{k-b_i} = \prod_{i=1}^{d} \prod_{k=n_0}^{n-1} \frac{k-a_i}{k-b_i} = \prod_{i=1}^{d} \frac{\Gamma(n-a_i)}{\Gamma(n-b_i)} \frac{\Gamma(n_0-b_i)}{\Gamma(n_0-a_i)}$$

which completes the proof.

Next, for the sake of completeness we present a well-known asymptotic formula for the Euler gamma function, which helps us to deal with $\prod_{k=n_0}^{n-1} g_1(k)$:

Lemma 4 (Abramowitz and Stegun [1]). *For any $a, b \in \mathbb{R}$ if $n \to \infty$, then $\frac{\Gamma(n+a)}{\Gamma(n+b)} = \Theta(n^{a-b})$.*

Finally, we deal with the sum of products $\sum_{j=n_0}^{n-1} g_2(j) \prod_{k=j+1}^{n-1} g_1(k)$. In terms of Euler gamma functions, via Lemma 3, we are interested in the asymptotics of the following formulas

$$\sum_{j=n_0}^{n} \frac{\prod_{i=1}^{k} \Gamma(j+a_i)}{\prod_{i=1}^{k} \Gamma(j+b_i)}$$

with $a = \sum_{i=1}^{k} a_i$, $b = \sum_{i=1}^{k} b_i$.

Lemma 5. *Let $a_i, b_i \in \mathbb{R}$ ($k \in \mathbb{N}$) with $a = \sum_{i=1}^{k} a_i$, $b = \sum_{i=1}^{k} b_i$. Then it holds asymptotically for $n \to \infty$ that*

$$\sum_{j=n_0}^{n} \frac{\prod_{i=1}^{k} \Gamma(j+a_i)}{\prod_{i=1}^{k} \Gamma(j+b_i)} = \begin{cases} \Theta\left(n^{a-b+1}\right) & \text{if } a+1 > b, \\ \Theta\left(\log n\right) & \text{if } a+1 = b, \\ \Theta\left(1\right) & \text{if } a+1 < b. \end{cases}$$

With this background information, we are now in the position to solve recurrence (1) and present exact and asymptotic results for the average degree. From Lemma 2 with $g_1(t) = 1 + \frac{p}{t} - \frac{r}{t^2}$ and $g_2(t) = \frac{r}{t}$ we get that

$$\mathbb{E}[D(G_t)] = D(G_{t_0}) \prod_{k=t_0}^{t-1} \left(1 + \frac{2p-1}{k+1} - \frac{2r}{k(k+1)}\right)$$

$$+ \sum_{j=t_0}^{t-1} \frac{2r}{j+1} \prod_{k=j+1}^{t-1} \left(1 + \frac{2p-1}{k+1} - \frac{2r}{k(k+1)}\right).$$

By applying Lemma 3 we replace the products above by the products of Euler gamma functions and obtain

$$\mathbb{E}[D(G_t)] = D(G_{t_0}) \frac{\Gamma(t_0)\Gamma(t_0+1)}{\Gamma(t)\Gamma(t+1)} \frac{\Gamma(t+c_3)\Gamma(t+c_4)}{\Gamma(t_0+c_3)\Gamma(t_0+c_4)}$$

$$+ \sum_{j=t_0}^{t-1} \frac{2r\Gamma(j+1)}{\Gamma(j+2)} \frac{\Gamma(j+c_3)\Gamma(j+c_4)}{\Gamma(t_0+c_3)\Gamma(t_0+c_4)} \frac{\Gamma(t_0)\Gamma(t_0+1)}{\Gamma(j)\Gamma(j+1)}$$

for c_3, c_4 – the (possibly complex) solutions of the equation $k^2 + 2pk - 2r = 0$.

Next, we may use Lemmas 4 and 5 respectively for the first and second part of the equation above. Note that when $r = 0$, the second part vanishes. Using previous lemmas we can derive asymptotics for the average degree which we present next.

Theorem 1. *For $G_t \sim DD(t,p,r)$ asymptotically as $t \to \infty$ we have*

$$\mathbb{E}[D(G_t)] = \begin{cases} \Theta(1) & \text{if } p < \frac{1}{2} \text{ and } r > 0, \\ \Theta(\log t) & \text{if } p = \frac{1}{2} \text{ and } r > 0, \\ \Theta(t^{2p-1}) & \text{if } p > \frac{1}{2} \text{ or } r = 0. \end{cases}$$

The asymptotic behavior of $\mathbb{E}[D(G_t)]$ has a threefold characteristic: when $p < \frac{1}{2}$ and $r > 0$, the majority of the edges are not created by copying them from parents, but actually by attaching them according to the value of r. For $p = \frac{1}{2}$ and $r > 0$ we note the curious situation of a phase transition (still with non-copied edges dominating), and only if $p > \frac{1}{2}$ or $r = 0$ do the edges copied from the parents asymptotically contribute the major share of the edges.

The next question regarding the average degree $D(G_t)$ is how much it deviates from the expected value $\mathbb{E}[D(G_t)]$, in probability. It turns out that $D(G_t)$ is concentrated around $\mathbb{E}[D(G_t)]$ in such a way that with probability $1 - O(t^{-A})$ it falls within a polylogarithmic ratio from the mean. We observe that unlike the large deviations for say preferential attachment graphs, in the duplication-divergence model we need to consider three cases reflecting the different behavior of $\mathbb{E}[D(G_t)]$ for $p < 1/2$, $p = 1/2$ and $p > 1/2$.

Theorem 2. *Asymptotically for $G_t \sim DD(t,p,r)$ it holds that*

$$\Pr[D(G_t) \geq A\,C\,\log^2(t)] = O(t^{-A}) \qquad \qquad \text{for } p < \frac{1}{2},$$

$$\Pr[D(G_t) \geq A\,C\,\log^3(t)] = O(t^{-A}) \qquad \qquad \text{for } p = \frac{1}{2},$$

$$\Pr[D(G_t) \geq A\,C\,t^{2p-1}\log^2(t)] = O(t^{-A}) \qquad \qquad \text{for } p > \frac{1}{2}.$$

for some fixed constant $C > 0$ and any $A > 0$.

Here we outline the main steps of the proof. We begin by carefully bounding from above the moment generating function $\mathbb{E}[\exp(\lambda_t D(G_t)) | G_t]$. This way, we are able to show

$$\mathbb{E}\left[\exp(\lambda_{t+1} D(G_{t+1})) \mid G_t\right] \leq \exp\left(\lambda_{t+1} D(G_t) h_1(t) + \frac{\lambda_{t+1}}{t+1} h_2(t)\right)$$

for certain explicit functions h_1 and h_2. By defining $\lambda_t = \lambda_{t+1}h_1(t)$ we finally arrive at

$$\mathbb{E}[\exp\left(\lambda_t D(G_t)\right)] \leq \exp\left(\lambda_{t_0} D(G_{t_0})\right)\left(\frac{t}{t_0}\right)^{2r\varepsilon_{t+1}+C_1}$$

for some $\varepsilon_t = 1/\ln(t/t_0)$ and constant C_1. After choosing a suitable sequence[1] $(\lambda_i)_{i=t_0}^{t}$ and applying Chernoff's inequality we find the large deviation bound. In all three cases we need an extra $\log^2(t)$ factor over the mean, one logarithm coming from the choice of ε_t (which decays like $\frac{1}{\ln t}$), and one from our requirement to get $O(t^{-A}) = O\left(\exp(-A\ln(t))\right)$ tail.

The left tail behavior is similar, however the proof is slightly more complicated, as discussed briefly below.

Theorem 3. *For $G_t \sim DD(t,p,r)$ with $p > \frac{1}{2}$ asymptotically it holds that*

$$\Pr\left[D(G_t) \leq \frac{C}{A}t^{2p-1}\log^{-3-\varepsilon}(t)\right] = O(t^{-A}).$$

for some fixed constant $C > 0$ and any $\varepsilon, A > 0$.

Note that since $\mathbb{E}[D(G_t)] = O(\log t)$ for $p \leq \frac{1}{2}$, bounds of the above form are trivial in this range of p and therefore not interesting, since all smaller values are within polylogarithmic distance to the mean.

The whole proof may be sketched as following: first, we find a bound

$$\mathbb{E}[\exp\left(\lambda(D(G_{t+1}) - D(G_t))\right)|G_t, \neg\mathcal{B}_t]$$

for a certain event $\mathcal{B}_t$ that allows us to bound the right tail for the variable $D(G_{t+1}) - D(G_t)$. Then, we use an auxiliary variable $Y_k = D(G_{(k+1)t}) - D(G_{kt})$ for which we know both the value of $\mathbb{E}[Y_k]$ and that the right tail of Y_k is small in particular, it is $O(t^{-A})$ when we are only $\log^2(t)$ times over the mean. Therefore it may be shown that also the left tail of Y_k cannot be large. Finally, we use the result for Y_k to obtain a bound for $D(G_t)$.

2.2 Degree of a Given Vertex s

We focus now on the expected value of $\deg_t(s)$, that is, the degree of vertex s at time t. We start with a recurrence relation for $\mathbb{E}[\deg_t(s)]$. Observe that for any $t \geq s$ we know that vertex s may be connected to vertex $t + 1$ in one of the following two cases:

- either $s \in \mathcal{N}_t(parent(t+1))$ (which holds with probability $\frac{\deg_t(s)}{t}$) and we add an edge between s and $t+1$ (with probability p),
- or $s \notin \mathcal{N}_t(parent(t+1))$ (with probability $\frac{t-\deg_t(s)}{t}$) and we an add edge between s and $t+1$ (with probability $\frac{r}{t}$).

---

[1] We choose $\lambda_t = \varepsilon_t\left(\frac{t}{t_0}\right)^{-(2p-1)(1+O(\varepsilon_t))}$ so that $\lambda_{t_0} \leq \varepsilon_t$.

From the model description we directly obtain the following recurrence for $\mathbb{E}[\deg_t(s)]$:

$$\mathbb{E}[\deg_{t+1}(s) \mid G_t] = \left(\frac{\deg_t(s)}{t}p + \frac{t - \deg_t(s)}{t}\frac{r}{t}\right)(\deg_t(s) + 1)$$

$$+ \left(\frac{\deg_t(s)}{t}(1 - p) + \frac{t - \deg_t(s)}{t}\left(1 - \frac{r}{t}\right)\right)\deg_t(s)$$

$$= \deg_t(s)\left(1 + \frac{p}{t} - \frac{r}{t^2}\right) + \frac{r}{t}.$$

Again this recurrence falls under Eq. (2), so we may proceed in the same fashion as with $\mathbb{E}[D(G_t)]$. First, we apply Lemma 2 with $g_1(t) = 1 + \frac{p}{t} - \frac{r}{t^2}$ and $g_2(t) = \frac{r}{t}$ to obtain the equation for the exact behavior of the degree of a given node s at time t:

$$\mathbb{E}[\deg_t(s)] = \mathbb{E}[\deg_s(s)] \prod_{k=s}^{t-1}\left(1 + \frac{p}{k} - \frac{r}{k^2}\right) + \sum_{j=s}^{t-1}\frac{r}{j}\prod_{k=j+1}^{t-1}\left(1 + \frac{p}{k} - \frac{r}{k^2}\right).$$

Again we substitute the simple products by the products of Euler gamma functions using Lemma 3

$$\mathbb{E}[\deg_t(s)] = \mathbb{E}[\deg_s(s)]\frac{\Gamma(s)^2}{\Gamma(t)^2}\frac{\Gamma(t + c_1)\Gamma(t + c_2)}{\Gamma(s + c_1)\Gamma(s + c_2)}$$

$$+ \sum_{j=s}^{t-1}\frac{r\Gamma(j)}{\Gamma(j + 1)}\frac{\Gamma(j + c_1)\Gamma(j + c_2)}{\Gamma(s + c_1)\Gamma(s + c_2)}\frac{\Gamma(s)^2}{\Gamma(j)^2}$$

for c_1, c_2 – the (possibly complex) solutions of the equation $k^2 + pk - r = 0$.

We are finally in a position to state the asymptotic expressions for $\mathbb{E}[\deg_t(s)]$, using Lemmas 4 and 5.

Theorem 4. *For $G_t \sim DD(t, p, r)$ and $s = O(1)$ it holds asymptotically that*

$$\mathbb{E}[\deg_t(s)] = \begin{cases} \Theta(\log t) & \text{if } p = 0 \text{ and } r > 0 \\ \Theta(t^p) & \text{otherwise.} \end{cases}$$

Here we observe only two regimes. In the first, for the case when $p = 0$, when edges are added only due to the parameter r, we have logarithmic growth of $\mathbb{E}[\deg_t(s)]$. In the second one, edges attached to s accumulate mostly by choosing vertices adjacent to s as parents of the new vertices, and therefore the expected degree of s grows proportional to t^p.

If we assume that $s \to \infty$, that is, we consider asymptotics with respect to both s and t, we may combine Lemma 1 with Theorem 1 to obtain.

Theorem 5. *For $G_t \sim DD(t, p, r)$ asymptotically as $s, t \to \infty$ we have*

$$\mathbb{E}[\deg_t(s)] = \begin{cases} \Theta(\log\frac{t}{s}) & \text{if } p = 0 \text{ and } r > 0 \\ \Theta\left(\frac{t^p}{s^p}\right) & \text{if } 0 < p < \frac{1}{2} \text{ and } r > 0, \\ \Theta\left(\frac{t^p}{s^p}\log s\right) & \text{if } p = \frac{1}{2} \text{ and } r > 0, \\ \Theta\left(\frac{t^p}{s^p}s^{2p-1}\right) & \text{if } p > \frac{1}{2} \text{ or } r = 0. \end{cases}$$

Let us note that when $s = \Theta(t)$, the asymptotics of $\mathbb{E}[\deg_t(s)]$ are exactly like those for $\mathbb{E}[\deg_t(t)]$ and $\mathbb{E}[D(G_t)]$, only the leading coefficients are different for each case. For different ranges of p and r, we have rates of growth equal to $\Theta(1)$, $\Theta(\log t)$ or $\Theta(t^{2p-1})$, respectively.

Finally, we present bounds for the deviation of $\deg_t(s)$ from its mean when $s = O(1)$.

Theorem 6. *Asymptotically for $G_t \sim DD(t, p, r)$ and $s = O(1)$ it holds that*

$$\Pr[\deg_t(s) \geq A\,C\,t^p \log^2(t)] = O(t^{-A})$$

for some fixed constant $C > 0$ and any $A > 0$.

Theorem 7. *For $G_t \sim DD(t, p, r)$ with $p > 0$ and $s = O(1)$ it holds that*

$$\Pr\left[\deg_t(s) \leq \frac{C}{A} t^p \log^{-3-\varepsilon}(t)\right] = O(t^{-A})$$

for some fixed constant $C > 0$ and any $A > 0$.

The proofs are analogous to those for the tails of the distribution of $D(G_t)$, so we omit sketches.

3   Discussion

In this paper we have focused on rigorous and precise analyses of the average degree $D(G_t)$ and a fixed given node degree $\deg_t(s)$ in the divergence-duplication graph. We have derived asymptotic expressions for the expected values of these quantities and have also shown that with high probability they are only poly-logarithmic factors away from the means of their expected values.

It is worth pointing that it is the parameter p that drives the rate of growth of the expected value for these parameters. We note that exact analysis reveals the fact that the value of parameter r and the structure of the starting graph G_{t_0} impact only the leading constants and lower order terms.

We observe that there are several phase transitions of these quantities as a function of p and r. This distinguishes it from the preferential graph model [9]. However, as demonstrated in [13], it seems that all real-world networks fall within a range $\frac{1}{2} < p < 1$, $r > 0$ – and this case should probably be the main topic of further investigation.

Future work may go along the lines of investigating further properties of the degree distribution as a function of both degree and time t. For example we might investigate the number of nodes of (given) degree k or the maximum degree in the graph. The latter is clearly bounded from below by $\deg_t(1)$, so from Theorem 7 one concludes that it is a polylogarithmic factor below t^p – but the upper bound still remains an open question since it requires bounds on the right tail of $\deg_t(t)$. The problem is that $\deg_t(t)$ is depends on the whole degree distribution in G_{t-1}, and therefore it is very unlikely that for s closer to t we have similar tail bounds as for $\deg_t(s)$ when $s = O(1)$.

In order to study graph symmetry we need more information about the degree distribution. This will allow us to find the ranges of parameters (p, r) for which we obtain an asymmetric graph with high probability or the ranges where non-negligible symmetries occur. In other words, we could explain theoretically Fig. 1. This in turn will lead to finding efficient algorithms for graph compression extending [2, 9] to the duplication-divergence model. Moreover, the degree distribution may be useful in the problem of inferring node arrival order in networks [10].

References

1. Abramowitz, M., Stegun, I.: Handbook of Mathematical Functions: With Formulas, Graphs, and Mathematical Tables, vol. 55. Dover Publications, New York (1972)
2. Choi, Y., Szpankowski, W.: Compression of graphical structures: fundamental limits, algorithms, and experiments. IEEE Trans. Inf. Theory **58**(2), 620–638 (2012)
3. Colak, R., et al.: Dense graphlet statistics of protein interaction and random networks. In: Biocomputing 2009, pp. 178–189. World Scientific Publishing, Singapore (2009)
4. Diestel, R.: Graph Theory. Springer, Heidelberg (2005)
5. Faloutsos, M., Faloutsos, P., Faloutsos, C.: On power-law relationships of the internet topology. ACM SIGCOMM Comput. Commun. Rev. **29**(4), 251–262 (1999)
6. Hermann, F., Pfaffelhuber, P.: Large-scale behavior of the partial duplication random graph. ALEA **13**, 687–710 (2016)
7. Jordan, J.: The connected component of the partial duplication graph. ALEA - Lat. Am. J. Prob. Math. Stat. **15**, 1431–1445 (2018)
8. Li, S., Choi, K.P., Wu, T.: Degree distribution of large networks generated by the partial duplication model. Theoret. Comput. Sci. **476**, 94–108 (2013)
9. Luczak, T., Magner, A., Szpankowski, W.: Asymmetry and structural information in preferential attachment graphs. Random Struct. Algorithms **55**(3), 696–718 (2019)
10. Magner, A., Sreedharan, J., Grama, A., Szpankowski, W.: Inferring temporal information from a snapshot of a dynamic network. Nat. Sci. Rep. **9**, 3057–3062 (2019)
11. Shao, M., Yang, Y., Guan, J., Zhou, S.: Choosing appropriate models for protein-protein interaction networks: a comparison study. Brief. Bioinform. **15**(5), 823–838 (2013)
12. Solé, R., Pastor-Satorras, R., Smith, E., Kepler, T.: A model of large-scale proteome evolution. Adv. Complex Syst. **5**(01), 43–54 (2002)
13. Sreedharan, J., Turowski, K., Szpankowski, W.: Revisiting parameter estimation in biological networks: influence of symmetries (2019)
14. Turowski, K., Magner, A., Szpankowski, W.: Compression of dynamic graphs generated by a duplication model. In: 56th Annual Allerton Conference on Communication, Control, and Computing, pp. 1089–1096 (2018)
15. Turowski, K., Sreedharan, J., Szpankowski, W.: Temporal ordered clustering in dynamic networks (2019)
16. Zhang, J.: Evolution by gene duplication: an update. Trends Ecol. Evol. **18**(6), 292–298 (2003)

On Finding Balanced Bicliques
via Matchings

Parinya Chalermsook, Wanchote Po Jiamjitrak, and Ly Orgo[(⊠)]

Aalto University, Espoo, Finland
{parinya.chalermsook,wanchote.jiamjitrak,ly.orgo}@aalto.fi

Abstract. In the Maximum Balanced Biclique Problem (MBB), we are given an n-vertex graph $G = (V, E)$, and the goal is to find a balanced complete bipartite subgraph with q vertices on each side while maximizing q. The MBB problem is among the first known NP-hard problems, and has recently been shown to be NP-hard to approximate within a factor $n^{1-o(1)}$, assuming the *Small Set Expansion hypothesis* [Manurangsi, ICALP 2017]. An $O(n/\log n)$ approximation follows from a simple brute-force enumeration argument. In this paper, we provide the first approximation guarantees beyond brute-force: (1) an $O(n/\log^2 n)$ efficient approximation algorithm, and (2) a parameterized approximation that returns, for any $r \in \mathbb{N}$, an r-approximation algorithm in time $\exp\left(O(\frac{n}{r \log r})\right)$. To obtain these results, we translate the subgraph removal arguments of [Feige, SIDMA 2004] from the context of finding a clique into one of finding a balanced biclique. The key to our proof is the use of matching edges to guide the search for a balanced biclique.

1 Introduction

The Maximum Balanced Biclique (MBB) problem is among the oldest and most fundamental NP-hard graph problems. It was stated to be NP-hard (without proof) in Garey and Johnson's book [12]; a proof is provided, for instance, in [14]. In this problem, we are given an n-vertex graph G, and we are interested in finding a balanced complete bipartite subgraph with q vertices on each side while maximizing the value of q. Since the problem is NP-hard, the main theoretical interest so far has been on approximation algorithms [11,18], parameterized algorithms [16], and parameterized approximation [4]. All results so far have been on the negative side, suggesting that MBB is very highly intractable. First, in terms of approximation algorithms, Manurangsi [18] showed that the problem is NP-hard to approximate within a factor of $n^{1-o(1)}$ assuming the *Small Set Expansion (SSE)* hypothesis and that $\mathbf{NP} \not\subseteq \mathbf{BPP}$. The other hardness result, somewhat incomparable to Manurangsi's is shown by Khot [15], that MEB does not admit n^ϵ approximation, for some $\epsilon > 0$, unless $\mathbf{NP} \subseteq \mathbf{BPTIME}(2^{n^{\Omega(1)}})$. On the parameterized algorithm side, a recent remarkable result of Lin [16]

This manuscript is part of Ly Orgo's master thesis at Aalto University.

I. Adler and H. Müller (Eds.): WG 2020, LNCS 12301, pp. 238–247, 2020.
https://doi.org/10.1007/978-3-030-60440-0_19

has shown that MBB is $\mathbf{W}[1]$-hard, therefore not admiting an FPT algorithm unless $\mathbf{FPT} = \mathbf{W}[1]$. Finally, MBB does not even admit any $o(\mathcal{OPT})$ FPT approximation algorithm [4] assuming the *Gap Exponential-Time Hypothesis (Gap-ETH)*, where $\mathcal{OPT}$ is the number of vertices in the optimal biclique.

The focus of this paper is on approximation algorithms. The aforementioned negative results suggest that MBB is among the "highly intractable problems" (those that do not admit $n^{1-o(1)}$ approximation), likely to be in the same ballpark as the clique, independent set, induced matching, and graph coloring problem. All these problems admit an $O(n/\log n)$ approximation algorithm via brute-force enumeration techniques [6,9]. For maximum clique or graph coloring problems, a lot of attention in the approximation algorithms community has been on obtaining any algorithm that beats these brute-force algorithms. In the context of maximum clique and independent set problems, many LP/SDP (as well as combinatorial) approaches have been devised, that achieve guarantees beyond trivial algorithms (see for instance [2,10,13] and references therein). However, such results do not exist at all in the context of MBB.

1.1 Our Results and Techniques

In this paper, we provide the first set of approximation algorithms on MBB whose approximation guarantees are asymptotically better than brute-force. Our first result is an efficient algorithm that runs in polynomial time. Throughout the paper, we use *size* of a balanced biclique to denote the number of vertices on one side. In particular, the size of the complete bipartite graph with q vertices on each side ($K_{q,q}$) is q.

Theorem 1. *There is an $O(n/\log^2 n)$ polynomial time approximation algorithm for MBB.*

Our second result is a parameterized approximation that gives a tradeoff between approximation ratio and running time.

Theorem 2. *For any $r \in \mathbb{N}$, there exists an r-approximation algorithm running in time $\exp\left(O(\frac{n}{r\log r})\right)$.*

Now we give a high-level discussion that highlights our main technical ideas. Let $\mathcal{OPT}$ denote the maximum value of q such that $K_{q,q}$ exists in G. Notice that, when $\mathcal{OPT} < n/\log^2 n$, we are immediately done since we can return a single edge and it would be $n/\log^2 n$ approximation. Therefore, we may assume that $\mathcal{OPT} \geq n/\log^2 n$. Our main result shows that we can efficiently find a biclique containing roughly $\tilde{\Omega}(\log^2 n)$ vertices on each side[1], and this would also be an $O(n/\log^2 n)$ approximation.

---

[1] Here, we use the convention that $\tilde{\Omega}$ hides asymptotically smaller terms.

Enumeration by Vertices: By a standard brute-force enumeration argument, one can easily find a biclique of size $\Omega(\log n / \log \log n)$. We explain this algorithm as an intuitive starting point. Assume that the graph $G = (A \cup B, E)$ is bipartite[2] with $|A| = |B| = n$. Fix an optimal biclique Q in G. We partition A into $A_1 \cup A_2 \cup \ldots \cup A_\ell$ arbitrarily where each set A_j contains $|A_j| = \lceil \log^3 n \rceil$ vertices each (that is, we have $\ell = n / \lceil \log^3 n \rceil$ sets.) Recall that $\mathcal{OPT} \geq n / \log^2 n$. By averaging argument, there must be a "good" set A_j that contains $\frac{\log^3 n}{\log^2 n} = \log n$ vertices from the optimal biclique Q, that is, $|A_j \cap V(Q)| \geq \log n$. We then enumerate all subsets $X \subseteq A_j : |X| = (a \log n / \log \log n)$; for each X, let $C_X = \{ v \in B : N_G(v) \supseteq X \}$ be the neighbors that can be included to form a biclique with X (that is, $G[C \cup C_X]$ is a biclique). We return any pair of sets (X, C_X) such that $|C_X| \geq |X|$. The running time of this procedure is $\frac{n}{\lceil \log^3 n \rceil} \binom{\log^3 n}{(\log n / \log \log n)} \leq (\log^3 n)^{\log n / \log \log n} = \mathsf{poly}(n)$.

Feige's Subgraph Removal. The above enumeration trick has been used several times for many problems (including clique and biclique) in the literature. It provides immediately a search procedure for a clique/biclique of size $\Theta(\frac{\log n}{\log \log n})$ whenever $\mathcal{OPT} \geq n / (\log^{O(1)} n)$. For clique (as well as independent set), Feige [10] "augmented" a subgraph removal procedure on top of this vertex enumeration procedure, so that his algorithm returns a clique of size $\Omega((\log n / \log \log n)^2)$ instead. Unfortunately, the above-mentioned natural idea of vertex enumeration is not quite compatible with Feige's subgraph removal arguments.

New Idea: Enumeration by Matching Edges. This is where our new observation comes in handy. We perform a "matching-edge" enumeration instead of a vertex enumeration. We explain the intuition of our proof by describing another procedure that finds a biclique containing $\Omega(\log n / \log \log n)$ vertices on each side. The main benefit of matching-edge enumeration over vertex enumeration is its versatility, which allows us to use Feige's subgraph removal trick [10] to derive our desired result.

Again we fix an n-vertex m-edge bipartite graph $G = (A \cup B, E)$ and an optimal biclique Q in G. Since we focus on the case $\mathcal{OPT} \geq n / (\log^2 n)$, we have that $|E(Q)| = \mathcal{OPT}^2 \geq n^2 / (\log^4 n)$. We partition the edges in E into at most n matchings, that is, $E = E_1 \cup E_2 \cup \ldots \cup E_n$, where each set E_j is a matching. By dismissing all small sets, it is easy to see that there exists some larger set E_j such that $|E_j| \geq 16(\log^5 n)$ and $|E_j \cap E(Q)| \geq |E_j| / 2(\log^4 n)$. We again divide such set E_j into $E_{j,1} \cup E_{j,2} \cup \ldots \cup E_{j,s}$ such that $\lceil 2 \log^5 n \rceil \leq |E_{j,\alpha}| \leq \lceil 4 \log^5 n \rceil$ for all $\alpha = 1, 2, \ldots, s$. By averaging (using the fact that $|E_j \cap E(Q)| \geq |E_j| / (2 \log^4 n)$), there exists $E_{j,\alpha}$ such that $|E_{j,\alpha} \cap E(Q)| \geq \frac{2 \log^5 n}{2 \log^4 n} \geq \log n$. We can enumerate all size-$(\log n / \log \log n)$ subsets $M \subseteq E_{j,\alpha}$ in time $(\log^5 n)^{O(\log n / \log \log n)} = \mathsf{poly}(n)$. Each such subset M is a matching, and we can check whether it induces a biclique

---

[2] See Lemma 1 for a simple formal proof that it suffices to focus on the case of bipartite graphs.

in G. This concludes an algorithm that finds a biclique with $\Omega(\log n/\log\log n)$ vertices on each side, based on matching-edge enumeration.

In Sect. 3, we show how to implement Feige's subgraph removal procedure on top of the matching-edge enumeration procedure. The presentation there is self-contained and does not rely on the discussion in this section.

1.2 Further Related Results

Prior to Khot [15], Feige and Kogan showed that unless $\mathbf{NP} \subseteq \mathbf{DTIME}(2^{n^{\frac{3}{4}+o(1)}})$, the problem does not admit $2^{\Omega((\log n)^{\Omega(1)})}$ approximation. Due to interest on application sides, there have been several heuristics [20,21] proposed for MBB, but none of them give any theoretical guarantee on the approximation factor.

The Maximum Edge Biclique (MEB) problem is very similar to the MBB problem, except for the fact that MEB aims to maximize the number of edges in a (possibly not balanced) biclique. Similarly to MBB, MEB is known to be $n^{1-o(1)}$ hard to approximate under the same complexity-theoretic assumptions [18]. Assuming more standard complexity assumptions, the problem is known to be n^{ϵ} hard to approximate [1].

The covering variants of biclique problems (called *minimum biclique cover*) are better understood from approximation perspectives: There is a $n^{1-o(1)}$ hardness result assuming $\mathbf{P} \neq \mathbf{NP}$ [5] and some non-trivial algorithms exist in the contexts of both approximation and parameterized algorithms [7].

The trade-off between an approximation factor and the running time has recently received attention. See e.g. [2,3,8,19] and the references therein.

2   Preliminaries

This paper follows standard notation in graph theory. Given a graph $G = (V, E)$, denote by $N_G(v)$ the neighboring vertices of v in G (excluding v). For $q \in \mathbb{N}$, denote by $K_{q,q}$ the complete bipartite graph with q vertices on each side. For any subset $S \subseteq V$, denote by $G[S]$, the induced subgraph on S. We sometimes abuse notation and use, for each edge set $F \subseteq E$, $G[F]$ to represent $G[V(F)]$.

An s-edge coloring of graph $G = (V, E)$ is a partition of E into $E_1 \cup E_2 \cup \ldots \cup E_s$ where each E_i is a matching in G. We will use the following edge coloring theorem of König (see, for instance, [17]).

Theorem 3. *Given a bipartite graph $G = (A \cup B, E)$ where each node has a degree of at most Δ, there exists a Δ-edge coloring of G that can be computed efficiently.*

We show that we can focus only on designing approximation algorithms for bipartite graphs.

Lemma 1. *If there is an α-approximation algorithm for MBB in bipartite graphs, then there is an $O(\alpha)$ approximation for MBB in general graphs.*

Proof. We turn a general graph $G = (V_G, E_G)$ into a bipartite graph $H = (L \cup R, E_H)$ as follows: For each vertex $v \in V_G$, add the vertex into either L or R independently with probability $1/2$. Next, we keep only the edges between L and R, that is, $E_H = \{(u, v) : u \in L, v \in R\}$.

Let $A \cup B$ be an optimal complete bipartite subgraph in G where $|A| = |B| = \mathcal{OPT}$. Let M be a perfect matching in $G[A \cup B]$, so $|M| = \mathcal{OPT}$. We say that an edge $e = (u, v) \in M$ is good if $u \in L$ and $v \in R$. Let $M' \subseteq M$ be the set of good edges. Notice that, $\mathbb{E}[|M'|] = |M|/4$ and that the vertices of M' induce a biclique in H. Therefore, in expectation, the biclique $A \cup B$ appears in H as a biclique of size at least $\mathcal{OPT}/4$ on each side, so the presumed α-approximation would be able to return a biclique of size $\mathcal{OPT}/4\alpha$ in expectation.

To obtain a deterministic algorithm, notice that the above proof only relies on pairwise independence of the choice of random bits. □

3   Our Algorithms

3.1   Subgraph Removal Implies Approximation Algorithms

We prove an analogue of Feige's subgraph removal procedure in the context of MBB. By simple calculation, it implies both of our algorithmic results.

Theorem 4. *Given a graph $G = (V, E)$ with a maximum balanced biclique of size n/z, for each $t = O(\frac{n}{z^5})$, there exists an algorithm that runs in time $z^{O(t)} poly(n)$ and finds a balanced biclique of size $q = \Theta(t \log_z \frac{n}{t})$.*

Now we show that this theorem implies both Theorem 1 and Theorem 2. These proofs are standard (see [2,10]) and are only presented here for completeness of exposition.

Corollary 1. *For any $r \in \mathbb{N}$, there exists an r-approximation algorithm running in time $\exp\left(O(\frac{n}{r \log r})\right)$.*

Proof. We are given a bipartite graph $G = (A \cup B, E)$. If $\mathcal{OPT} \le n/(\log^2 r)$, we are done since we can enumerate all subsets $S \subseteq A$ of size $n/(r \log^2 r)$ (and this would be an r-approximation) in time

$$\binom{n}{n/(r \log^2 r)} \le (er \log^2 r)^{n/(r \log^2 r)} \le 2^{O(n/(r \log r))}$$

Otherwise, we have that $\mathcal{OPT} = n/z$ for $z \le \log^2 r$. Choose $t = \frac{n}{r \log r \log z}$ so that $z^{O(t)} = 2^{O(\frac{n}{r \log r})}$. It is easy to check that $t \le O(n/z^5)$ for sufficiently large n. Theorem 4 gives us a biclique with at least $\Omega(t \cdot \log_z(n/t)) = \Omega(\frac{n \log r}{r \log r \log^2 z}) = \Omega(\frac{n}{r \log^2 z})$ nodes on each side. The approximation factor obtained is:

$$\frac{n/z}{\Omega(\frac{n}{r \log^2 z})} = O(r)$$

□

Corollary 2. *There is a polynomial time $O(n/\log^2 n)$ approximation algorithm.*

Proof. If $\mathcal{OPT} \le n/(\log^2 n)$, we are immediately done. Otherwise, assume that $z = O(\log^2 n)$. Let $t = \log_z n$, we use Theorem 4 to find a balanced biclique $K_{q,q}$ where $q = \Theta(\log_z^2 n)$ (so this algorithm runs in polynomial time). Therefore, the approximation factor is $\mathcal{OPT}/q \le O(\frac{n\log^2 z}{z\log^2 n}) \le O(n/\log^2 n)$. $\square$

3.2   Proof of Theorem 4

We now describe our algorithm. It has two steps. In the first step, we perform a pre-processing by removing vertices whose degrees are too large compared to the size of the optimal biclique. This step will allow us to apply König's edge coloring theorem, decomposing the graph into a union of disjoint matchings.

Step 1: Degree Reduction. We are given an n-vertex bipartite graph $G' = (V', E')$ that contains an optimal biclique Q of size $p = n/z$; define a parameter k, such that $p^2 = m/k$ is the number of edges in the optimal biclique.

This step is summarized in the following lemma.

Lemma 2. *There is an efficient algorithm that produces a graph $G = (V, E)$ such that:*

- $|V(G)| \ge n/2$.
- *Each vertex in G has degree at most $2pk$.*
- *There exists a biclique containing $p/2$ vertices on each side of G.*

Proof. Whenever there is a vertex v whose degree is more than $2pk = 2\sqrt{mk}$, we remove v and all edges incident to v (but we keep vertices in $N_G(v)$.) Notice that we would remove at most $\frac{1}{2}\sqrt{m/k} = \frac{p}{2}$ vertices from the graph (since we have at most m edges), and therefore at least $p/2$ vertices remain on each side of Q after such removals. This completes the proof. $\square$

Step 2: An Analogue of Feige's Argument. Assume we are given a graph $G = (A \cup B, E)$ that satisfies the conditions in Lemma 2. Let Q be the optimal biclique in G containing at least $p/2$ vertices on each side (therefore containing at least $m/4k$ edges). This choice of optimal clique Q is fixed throughout the execution of the algorithm.

Definition 1. *We say that a subset of edges E'' is poor if $|E'' \cap E(Q)| < |E''|/8k$.*

Definition 2. *Let Q' be a balanced biclique. An edge e is said to be **consistent** with Q' if $Q' \cup \{e\}$ induces a biclique and the endpoints of e is are disjoint from $V(Q)$, that is, if Q' is $K_{r,r}$, we must have that $Q' \cup \{e\}$ induces a $K_{r+1,r+1}$. Denote by $\mathcal{CE}(Q')$ the set of edges that are consistent with biclique Q'.*

Now we present our algorithm. The algorithm has multiple phases, and each phase consists of many iterations that keep growing the set of matching edges F whose vertices induce a biclique.

In the beginning, $E' \leftarrow E$. In each phase, we start from edge set $E'' \leftarrow E'$ and $F \leftarrow \emptyset$. In each iteration, if we have that $|E''| \leq 10^6 pk^3 t$, then we say that the **phase terminates successfully and the algorithm terminates.** Otherwise, we partition E'' into $2kp$ subsets $E''_1 \cup E''_2 \cup \ldots \cup E''_{2kp}$ where each E''_j is a matching. We say that E''_j is large if $|E''_j| \geq 16kt$; otherwise, we say that E''_j is small. Consider each large set E''_j and partition it further (arbitrarily) into $E''_{j,1} \cup E''_{j,2} \cup \ldots \cup E''_{j,\ell(j)}$ where each set has size at least $16kt$ and at most $32kt$. A *good matching* is a subset $M \subseteq E''_{j,\alpha} : |M| = t$ that satisfies two conditions: (i) $G[M]$ is a biclique and (ii) the set of consistent edges $E_M = CE(G[M]) \cap E''$ is sufficiently large, that is, $|E_M| \geq |E''|/8k - (5pkt)$. If a good matching exists in some subset $E''_{j,\alpha}$, we update $F \leftarrow F \cup M$, $E'' \leftarrow E_M$, and then start the new iteration. Otherwise, if no good matching is found, we claim that E'' is a poor subset (proof provided below), the **phase terminates unsuccessfully,** and we start a new phase with $E' \leftarrow E' \setminus E''$.

Analysis of Running Time

A phase either ends with a poor subset of edges removed from the graph (when it is unsuccessful) or it ends with a collection of matching edges F (when it is successful). Each poor subset is a subset of size at least one, so there can be at most m unsuccessful phases.

Lemma 3. *The running time of each phase is at most $z^{O(t)} n^{O(1)}$.*

Proof. In each iteration, each set $E''_{j,\alpha}$ has size at most $32kt$ and we enumerate all subsets of size t inside it. There are at most n iterations in each phase. So, the total running time would be at most:

$$\binom{32kt}{t} n^{O(1)} \leq \left(\frac{32ekt}{t}\right)^t n^{O(1)} \leq k^{O(t)} n^{O(1)}.$$

Since $k = \frac{z^2 m}{n^2} < z^2$, then $k^{O(t)} = z^{O(t)}$.

The Size of Bicliques

Now we proceed to show that at some point, the algorithm would terminate successfully and return a relatively large biclique.

Lemma 4. *If no good matching is found in a phase, then E'' is a poor subset.*

Proof. Assume, by contrapositive, that E'' is not poor. Then $|E'' \cap E(Q)| \geq |E''|/8k$. Notice that the number of edges in the small sets E''_j is at most $16kt \cdot (2kp) \leq |E''|/16k$. Therefore, at least $|E''|/16k$ edges in Q appear in one of the large sets. Since $|E'' \cap E(Q)|/|E''| \geq 1/16k$, by averaging argument, we have that there exists a large set $E''_{j,\alpha}$ for which $|E''_{j,\alpha} \cap E(Q)|/|E''_{j,\alpha}| \geq 1/16k$ which implies that $|E''_{j,\alpha} \cap E(Q)| \geq |E''_{j,\alpha}|/16k \geq t$. Let M be an arbitrary size-t subset of $E''_{j,\alpha} \cap E(Q)$. Notice that M is good: Firstly, $G[M]$ is a biclique. Moreover, any edge e in $E'' \cap E(Q)$ that is not sharing a vertex with any edge in M, is consistent with $G[M]$. There are at least $|E''|/8k - 5pkt$ such edges, because the number of edges that have two vertices in M is less than t^2 and the number of edges that have one vertex in M is at most $\Delta(G) \cdot 2t \leq 4kpt$. Since $t \leq p$, then $t^2 + 4kpt < 5kpt$. □

So, the above lemma implies that whenever a subset of edges is removed from the graph, that subset must be poor. The following lemma says that, at some point, a poor subset would not exist anymore.

Lemma 5. *Let $\hat{E}_1, \hat{E}_2, \ldots, \hat{E}_\ell$ be a collection of poor subsets removed from the phases. Then, $|\bigcup_i \hat{E}_i \cap E(Q)| < m/8k$.*

Proof. This follows from the fact that $|\hat{E}_i \cap E(Q)| < |\hat{E}_i|/8k$. Summing over all i gives us the desired bound. □

Corollary 3. *At the beginning of each phase, we have that $|E' \cap E(Q)| \geq m/8k$.*

Proof. Since after the degree reduction we have at least $m/4k$ edges in $E' \cap E(Q)$, then by Lemma 5, we have at least $m/4k - m/8k = m/8k$ edges in $E(Q)$ left in each phase. □

With this, we know that the unsuccessful phase cannot remove too many edges from the optimal solution. Now, we argue that the result returned by a successful phase is a matching F that induces a biclique and it has the desired size.

Observation 5. *Let F_1 be a matching such that $G[F_1]$ is a biclique. Let $F_2 \subseteq CE(G[F_1])$ be another matching such that $G[F_2]$ is also a biclique. Then $G[F_1 \cup F_2]$ is a biclique containing $|F_1| + |F_2|$ vertices on each side.*

This observation implies that the result returned by the algorithm induces a biclique. Its size is equal to the product of t with the number of iterations in that phase. The following lemma will finish the proof.

Lemma 6. *A successful phase runs for at least $\Omega(\log_z \left(\frac{m}{t}\right))$ iterations.*

Proof. Notice that in the same phase, each iteration, that starts with E'', proceeds to the next iteration (starting with E_M) on the condition that the number of remaining edges is at least $|E_M| \geq |E''|/8k - 5pkt \geq |E''|/16k$. At the beginning of the phase, there are at least $m/8k$ edges in E', and the stopping condition is when $|E''| \leq 10^6 pk^3 t$. Therefore, we can proceed for at least $\log_{16k} \left(\frac{m}{10^7 k^4 pt}\right)$ iterations. Since $m = \Theta(p^2 k)$ and $k = \frac{z^2 m}{n^2} \leq z^2$, then the number of iterations is $\Omega(\log_k \left(\frac{m}{t}\right)) \geq \Omega(\log_z \left(\frac{m}{t}\right))$. □

4   Discussions and Open Problems

In this paper, we present approximation algorithms for MBB whose guarantee is better than that of brute-force. One obvious open question is to match the $O(n/\log^3 n)$ approximation of clique, which would put biclique and clique in the same ballpark. A truly interesting direction is to study the power of semi-definite programs for bicliques. While there are many such algorithms for cliques and coloring, we do not have them for any biclique problem (including maximum edge biclique or biclique covering problems).

Finally, as discussed in [10], an interesting aspect of the subgraph removal algorithm is its connection to algorithmic Ramsey theory. In particular, the poor subgraph detection algorithm can be seen as a constructive Ramsey-type argument (please refer to discussions in Feige's paper for more detail.) While standard Ramsey arguments (i.e. clique v.s. independent set) have found their applications in theoretical computer science, ours is perhaps the first algorithmic result of Ramsey-type theorem for balanced bicliques.

Acknowledgement. This project is supported by European Research Council (ERC) under the European Union's Horizon 2020 research and innovation programme (grant agreement No 759557) and by Academy of Finland Research Fellowship, under grant number 310415. Ly Orgo has been supported by European Research Council (ERC) under the European Union's Horizon 2020 research and innovation programme (grant agreement No 759557).

References

1. Ambühl, C., Mastrolilli, M., Svensson, O.: Inapproximability results for maximum edge biclique, minimum linear arrangement, and sparsest cut. SIAM J. Comput. **40**(2), 567–596 (2011)
2. Bansal, N., Chalermsook, P., Laekhanukit, B., Nanongkai, D., Nederlof, J.: New tools and connections for exponential-time approximation. Algorithmica **81**(10), 3993–4009 (2019)
3. Bonnet, É., Lampis, M., Paschos, V.Th.: Time-approximation trade-offs for inapproximable problems. J. Comput. Syst. Sci. **92**, 171–180 (2018)
4. Chalermsook, P., et al.: From Gap-ETH to FPT-inapproximability: clique, dominating set, and more. In: 2017 IEEE 58th Annual Symposium on Foundations of Computer Science (FOCS), pp. 743–754. IEEE (2017)
5. Chalermsook, P., Heydrich, S., Holm, E., Karrenbauer, A.: Nearly tight approximability results for minimum biclique cover and partition. In: Schulz, A.S., Wagner, D. (eds.) ESA 2014. LNCS, vol. 8737, pp. 235–246. Springer, Heidelberg (2014). https://doi.org/10.1007/978-3-662-44777-2_20
6. Chalermsook, P., Laekhanukit, B., Nanongkai, D.: Independent set, induced matching, and pricing: connections and tight (subexponential time) approximation hardnesses. In: 2013 IEEE 54th Annual Symposium on Foundations of Computer Science, pp. 370–379. IEEE (2013)
7. Chandran, S., Issac, D., Karrenbauer, A.: On the parameterized complexity of biclique cover and partition. In: 11th International Symposium on Parameterized and Exact Computation (IPEC 2016). Schloss Dagstuhl-Leibniz-Zentrum fuer Informatik (2017)

8. Cygan, M., Kowalik, L., Pilipczuk, M., Wykurz, M.: Exponential-time approximation of hard problems. arXiv preprint arXiv:0810.4934 (2008)
9. Cygan, M., Kowalik, Ł., Wykurz, M.: Exponential-time approximation of weighted set cover. Inf. Process. Lett. **109**(16), 957–961 (2009)
10. Feige, U.: Approximating maximum clique by removing subgraphs. SIAM J. Discrete Math. **18**(2), 219–225 (2004)
11. Feige, U., Kogan, S.: Hardness of approximation of the balanced complete bipartite subgraph problem. Dept. Comput. Sci. Appl. Math., Weizmann Inst. Sci., Rehovot, Israel, Technical report. MCS04-04 (2004)
12. Garey, M.R., Johnson, D.S.: Computers and Intractability, vol. 29. WH Freeman, New York (2002)
13. Halperin, E.: Improved approximation algorithms for the vertex cover problem in graphs and hypergraphs. SIAM J. Comput. **31**(5), 1608–1623 (2002)
14. Johnson, D.S.: The NP-completeness column: an ongoing guide. J. Alg. **6**(3), 434–451 (1985)
15. Khot, S.: Ruling out PTAS for graph min-bisection, dense k-subgraph, and bipartite clique. SIAM J. Comput. **36**(4), 1025–1071 (2006)
16. Lin, B.: The parameterized complexity of k-biclique. In: Proceedings of the Twenty-Sixth Annual ACM-SIAM Symposium on Discrete Algorithms, pp. 605–615. SIAM (2014)
17. Lovász, L., Plummer, M.D.: Matching Theory, vol. 367. American Mathematical Society, Providence (2009)
18. Manurangsi, P.: Inapproximability of maximum edge biclique, maximum balanced biclique and minimum k-cut from the small set expansion hypothesis. In: 44th International Colloquium on Automata, Languages, and Programming (ICALP 2017). Schloss Dagstuhl-Leibniz-Zentrum fuer Informatik (2017)
19. Manurangsi, P., Trevisan, L.: Mildly exponential time approximation algorithms for vertex cover, balanced separator and uniform sparsest cut. In: Approximation, Randomization, and Combinatorial Optimization. Algorithms and Techniques (APPROX/RANDOM 2018). Schloss Dagstuhl-Leibniz-Zentrum fuer Informatik (2018)
20. Wang, Y., Cai, S., Yin, M.: New heuristic approaches for maximum balanced biclique problem. Inf. Sci. **432**, 362–375 (2018)
21. Yuan, B., Li, B., Chen, H., Yao, X.: A new evolutionary algorithm with structure mutation for the maximum balanced biclique problem. IEEE Trans. Cybernet. **45**(5), 1054–1067 (2014)

Finding Large Matchings in 1-Planar Graphs of Minimum Degree 3

Therese Biedl[1] and Fabian Klute[2]([⊠])

[1] David R. Cheriton School of Computer Science, University of Waterloo,
Waterloo, ON N2L 1A2, Canada
biedl@uwaterloo.ca

[2] Utrecht University, Utrecht, The Netherlands
f.m.klute@uu.nl

Abstract. A *matching* is a set of edges without common endpoint. It was recently shown that every *1-planar graph* (i.e., a graph that can be drawn in the plane with at most one crossing per edge) that has minimum degree 3 has a matching of size at least $\frac{n+12}{7}$, and this is tight for some graphs. The proof did not come with an algorithm to find the matching more efficiently than a general-purpose maximum-matching algorithm. In this paper, we give such an algorithm. More generally, we show that any matching that has no augmenting paths of length 9 or less has size at least $\frac{n+12}{7}$ in a 1-planar graph with minimum degree 3.

1 Introduction

The *matching problem* (i.e., finding a large set of edges in a graph such that no two chosen edges have a common endpoint) is one of the oldest problem in graph theory and graph algorithms, see for example [3,19] for overviews.

To find a maximum matching in a graph $G = (V, E)$, the fastest algorithm is the one by Hopcroft and Karp if G is bipartite [16], and the one by Micali and Vazirani otherwise ([20], see also [25] for further clarifications). As pointed out in [25], for a graph with n vertices and m edges the run-time of the algorithm by Micali and Vazirani is $O(m\sqrt{n})$ in the RAM model and $O(m\sqrt{n}\alpha(m, n))$ in the pointer model, where $\alpha(\cdot)$ is the inverse Ackerman function. For *planar graphs* (graphs that can be drawn without crossing in the plane) there exists a linear-time approximation scheme for maximum matching [1], and it can easily be generalized to so-called H-minor-free graphs [10] and k-planar graphs [14].

For many graph classes, specialized results concerning matchings and matching algorithms have been found. To name just a few, every bipartite d-regular graph has a *perfect matching* (a matching of size $n/2$) [15] and it can be found in $O(m)$ time [9]. Every 3-regular biconnected graph has a perfect matching [22]

F. Klute—This work was conducted while FK was a member of the "Algorithms and Complexity Group" at TU Wien.

Research of TB supported by NSERC. Research initiated while FK was visiting the University of Waterloo.

I. Adler and H. Müller (Eds.): WG 2020, LNCS 12301, pp. 248–260, 2020.
https://doi.org/10.1007/978-3-030-60440-0_20

and it can be found in linear time for planar graphs and in near-linear time for arbitrary graphs [4]. Every graph with a Hamiltonian path has a *near-perfect matching* (of size $\lceil (n-1)/2 \rceil$); this includes for example the 4-connected planar graphs [24] for which the Hamiltonian path (and with it the near-perfect matching) can be found in linear time [8].

For graphs that do not have near-perfect matchings, one possible avenue of exploration is to ask for guarantees on the size of matchings. One of the first results in this direction is due to Nishizeki and Baybars [21], who showed that every planar graph with minimum degree 3^1 has a matching of size at least $\frac{n+4}{3}$. (This bound is tight for some planar graphs with minimum degree 3.) The proof relies on the Tutte-Berge theorem and does not give an algorithm to find such a matching (or at least, none faster than any maximum-matching algorithm). Over 30 years later, a linear-time algorithm to find a matching of this size in planar graphs of minimum degree 3 was finally developed by Franke, Rutter, and Wagner [13]. The latter paper was a major inspiration for our current work.

In recent years, there has been much interest in *near-planar* graphs, i.e., graphs that may be required to have crossings but that are "close" to planar graphs in some sense. We are interested here in *1-planar graphs*, which are those that can be drawn with at most one crossing per edge. (Detailed definitions can be found in Sect. 2.) See a recent annotated bibliography [18] for an overview of many results known for 1-planar graphs. The first author and Wittnebel [6] gave matching-bounds for 1-planar graphs of varying minimum degrees, and showed that any 1-planar graph with minimum degree 3 has a matching of size at least $\frac{n+12}{7}$. (This bound is again tight.)

The proof in [6] is again via the Tutte-Berge theorem and does not give rise to a fast algorithm to find a matching of this size. This is the topic of the current paper. We give an algorithm that finds, for any 1-planar graph with minimum degree 3, a matching of size at least $\frac{n+12}{7}$ in linear time in the RAM model and time $O(n\alpha(n))$ in the pointer-model. The algorithm consists simply of running the algorithm by Micali and Vazirani for a limited number of rounds (and in particular, does not require that a 1-planar drawing of the graph is given). The bulk of the work consists of the analysis, which states that if there are no augmenting paths of length 9 or less, then the matching has the desired size for graphs with minimum degree 3. Along the way, we prove some bounds obtained for graphs with higher minimum degree, though these are not tight.

The paper is structured as follows. After reviewing some background in Sect. 2, we state the algorithm in Sect. 3. The analysis proceeds in multiple steps in Sect. 4. We first delete short flowers from the graph (and account for free vertices in them directly). The remaining graph is basically bipartite, and we can use bounds known for independent sets in 1-planar graphs to obtain matching-bounds that are very close to the desired goal. Closing this gap requires non-trivial modifications; we give a sketch of the involved techniques in Sect. 5 and refer to the full paper for the technical details.

---

[1] In this paper, 'minimum degree k' stands for 'minimum degree at least k'; of course the bounds also hold if all degrees are higher.

2 Background

We assume familiarity with graphs and graph algorithms, see for example [11,23]. Throughout the paper, G is a simple graph with n vertices and m edges. A *matching* of G is a subset M of its edges without common endpoints; we say that $e = (x, y) \in M$ is *matched* and x and y are *matching-partners*. $V(M)$ denotes the endpoints of edges in M; we call $v \in V(M)$ *matched* and all other vertices *free*. An *alternating walk* of M in G is a walk that alternates between unmatched and matched edges. An *augmenting path* of M in G is an alternating walk that repeats no vertices and begins and ends at a free vertex; we use *k-augmenting path* for an augmenting path with at most k edges. If P is an augmenting path of M (and viewed as an edge-set), then $(M \setminus P) \cup (P \setminus M)$ is also a matching and has one edge more than M.

A *drawing* Γ of a graph consists of assigning points in $\mathbb{R}^2$ to vertices and simple curves to each edge such that curves of edges end at the points of its endpoints. We usually identify the graph-theoretic object (vertex, edge) with the geometric object (point, curve) that it has been assigned to. We only consider *good* drawings (see [23] for details) that avoid degeneracies such as an edge going through the point of a non-incident vertex or two edges intersecting in more than one point. The connected sets of $\mathbb{R}^2 \setminus \Gamma$ are called the *regions* of the drawing.

A *crossing* c of Γ is a pair of two edges (v, w) and (x, y) that have a point in their interior in common. A drawing Γ is called *k-planar* (or *planar* for $k = 0$) if every edge has at most k crossings. A graph is called *k-planar* if it has a k-planar drawing. While planarity can be tested in linear time [7,17], testing 1-planarity is NP-complete [14].

Fix a 1-planar drawing Γ and consider a crossing c between edges (v_0, v_2) and (v_1, v_3). Then we could draw edge (v_i, v_{i+1}) (for $i = 0, \ldots, 3$ and addition modulo 4) without crossing by walking "very close" to crossing c. We call the pair (v_i, v_{i+1}) a *potential kite-edge* and note that if we inserted (v_i, v_{i+1}) in the aforementioned manner, then it would be consecutive with the crossing edges in the cyclic orders of edges around v_i and v_{i+1} in Γ.

3 Finding the Matching

Our algorithm to find a large matching is a one-liner: repeatedly extend the matching via 9-augmenting paths (i.e., of length at most 9) until there are no more such paths. Note that the algorithm does not depend on the knowledge that the graph is 1-planar and does not require having a 1-planar drawing at hand. It could be executed on any graph; our contribution is to show (in the next section) that if it is executed on a 1-planar graph G with minimum degree 3 then the resulting matching M has size at least $\frac{n+12}{7}$.

Running Time. Finding a matching M in G such that there is no k-augmenting path can be done in time $O(k|E|)$ in the RAM model using the algorithm by Micali and Vazirani [20]. (We state all run-time bounds here in the RAM model;

for the pointer model add a factor of $\alpha(|E|, |V|)$.) This algorithm runs in phases, each of which has a running time of $O(|E|)$ and increases the length of the minimum-length augmenting path by at least two. See for example the paper by Bast et al. [2] for a more detailed explanation. Since for 1-planar graphs we have $|E| \in O(|V|)$ we get a linear time algorithm in the number of vertices of G to find a matching without 9-augmenting paths.

4   Analysis

Assume that M is a matching without augmenting paths of length at most 9, and let F be the free vertices; $|F| = n - 2|M|$. To analyze the size of M, we proceed in three stages. First we remove some vertices and matching-edges that belong to short flowers (defined below); these are "easy" to account for. Next we split the remaining vertices by their distance (measured along alternating paths) to free vertices. Since short flowers have been removed, no edges can exist between vertices of even small distance; they hence form an independent set. Using a crucial lemma from [6] on the size of independent sets in 1-planar graphs, this shows that $|M| \geq \frac{7}{50}(n + 12)$, which is very close to the desired bound of $\frac{n+12}{7}$. The last stage (which does the improvement from $\frac{7}{50}$ to $\frac{1}{7}$) will require non-trivial effort and is done mostly out of academic interest; a sketch is in Sect. 5 and details are in the full paper [5].

Flowers. A *flower*[2] is an alternating walk that begins and ends at the same free vertex; we write *k-flower* for a flower with at most k edges. We only consider 7-flowers; Fig. 1 illustrates all possible such flowers. Note that such short flowers split into a path (called *stem*) and an odd simple cycle (the *blossom*); we call a flower a *cycle-flower* if the stem is empty.

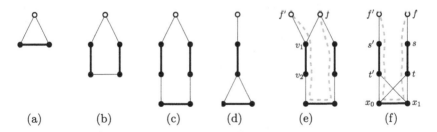

(a)   (b)   (c)   (d)   (e)   (f)

Fig. 1. (a–d) All possible 7-flowers. Free vertices are white, matched edges are thick. (e-f) Augmenting paths found in the proofs of (e) Claim 1 and (f) Claim 2.

Let V_C (the "C" reminds of "cycle") be all vertices that belong to some 7-cycle-flower, let F_C be all free vertices in V_C, and let M_C be all matching-edges within V_C, i.e., all edges with both endpoints in V_C.

---

[2] Our terminology follows the one in Edmonds' famous blossom-algorithm [12].

Claim 1. *Let M be a matching in a 1-planar graph G with minimum degree 3 such that there is no 9-augmenting path and let F_C and M_C be defined as above. Then $|F_C| \leq |M_C|$.*

Proof. For every $f \in F_C$ there exists some 7-cycle-flower f-v_1-v_2-...-v_k-f with $k \in \{2, 4, 6\}$. Assign f to edge (v_1, v_2). We claim that f is the only vertex in F_C assigned to (v_1, v_2), otherwise there would be an augmenting path of length less than 9. Since $(v_1, v_2) \in M_C$, this then proves the claim. So assume for contradiction that another vertex $f' \in F_C$ was also assigned to (v_1, v_2). Then f' is adjacent to one of v_1, v_2. If it is v_2, then f'-v_2-v_1-f is a 3-augmenting path. If it is v_1, then f'-v_1-...-v_k-f is a 7-augmenting path, see Fig. 1(e).   □

From now on we will only study the graph $G \backslash V_C$. Observe that M restricted to this graph is again a matching without augmenting paths up to length 9. All following definitions are only for vertices and edges in $G \setminus V_C$. Let F_B (the "B" reminds of "blossom") be all those free vertices f that are not in F_C and that belong to a 7-flower. By $f \notin F_C$ this flower has a non-empty stem, which is possible only if its length is exactly 7 and the stem has two edges f-s-t while the blossom is a 3-cycle t-x_0-x_1-t. Furthermore (s, t) and (x_0, x_1) are matching-edges. Let M_B be the set of such matching-edges (x_0, x_1) i.e., matching-edges that belong to the blossom of such a 7-flower. We do *not* include the matching-edge (s, t) in M_B (unless it belongs to a different 7-flower where it is in the blossom). Let T_B be the set of such vertices t, i.e., vertices that belong to a 7-flower and belong to both the stem and the blossom. Set $V_B = T_B \cup V(M_B)$ (see also Fig. 2).

Claim 2. *Let M be a matching in a 1-planar graph G with minimum degree 3 such that there is no 9-augmenting path and let T_B and M_B be defined as above. Then $|T_B| \leq |M_B|$.*

Proof. We argue similarly to the proof of Claim 1, i.e., assign each $t \in T_B$ to an edge in M_B and argue that no two vertices are assigned to the same edge unless there is a 9-augmenting path. Choose for each $t \in T_B$ a matching-edge $(x_0, x_1) \in M_B$ that is within the same blossom of some 7-flower of $G \setminus V_C$. Assume for contradiction that some other vertex $t' \in T_B$ is also assigned to (x_0, x_1). Let t-s-f and t'-s'-f' be the stems of the 7-flowers containing t and t', and note that $s \neq s'$ since they are matching-partners of $t \neq t'$. This gives an alternating path f-s-t-x_0-x_1-t'-s'-f', see Fig. 1(f). Depending on whether $f = f'$ this is a 7-augmenting path or 7-cycle-flower; the former contradicts the choice of M and the latter that $x_0, x_1 \in G \setminus V_C$.   □

The Auxiliary Graph H. For any vertex $v \in G \backslash V_C \backslash V_B$, let the *distance to a free vertex* be the number of edges in a shortest alternating path from a free vertex to v. Let D_k be the vertices of distance k to a free vertex. Since there are no 9-agumenting paths, one can easily see:

Observation 1. *In graph $G \setminus V_C \setminus V_B$, there are no matching-edges within D_k for $k = 1$ and $k = 3$, and no edges at all within D_k for $k = 0$ and $k = 2$.*

Proof. If there was such an edge (v, v'), then it, together with the alternating paths of length k that lead from free vertices to v, v', form a 7-augmenting path or a 7-flower.   □

From now on, we will only study the subgraph H induced by $D_0 \cup \cdots \cup D_3$, noting again that this does *not* include the vertices in $V_C \cup V_B$. For ease of referring to them, we rename the vertices of H as follows (see also Fig. 2):

- $F_H = F \setminus F_C = D_0$ are the free vertices in H.
- $S = D_1$ are the vertices in H that are adjacent to F_H.
- $T_H = D_2$ are the vertices in H that have matching-partners in S and are not in S.
- $U = D_3$ are the vertices in H that are adjacent to T_H and not in $F \cup S \cup T_H$.

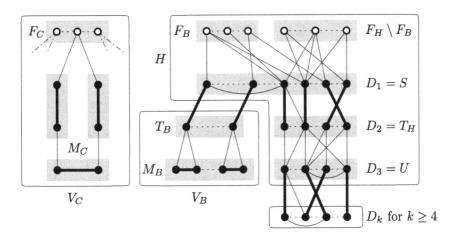

Fig. 2. Illustration of the partitioning of edges and vertices and graph H.

The following shortcuts will be convenient. For any vertex sets A, B, an A-*vertex* is a vertex in A, and an AB-*edge* is an edge between an A-vertex and a B-vertex. For any vertex v an A-*neighbour* is a neighbour of v in A. Using Observation 1 and the definition of V_C (which includes the *entire* flower) and V_B (which includes *both* ends of the matching-edge) one easily verifies the following:

Observation 2. – *There are no matching-edges within S or within U.*
- *There are no edges within F_H or within T_H.*
- *The matching-partner of an S-vertex is in $T_H \cup T_B$.*
- *The matching-partner of a U-vertex is not in H.*
- *All neighbours of an F_H-vertex belong to S or are not in H.*
- *All neighbours of a T_H-vertex belong to $S \cup U$ or are not in H.*

Let M_S be the set of matching-edges incident to S. Let M_U be the matching-edges incident to U. Since there are no matching-edges within S or U, we have $|S| = |M_S|$ and $|U| = |M_U|$.

We stated earlier that any neighbour of F_H is either in S or not in H. The latter is actually impossible (though this is non-trivial), and likewise for T_H.

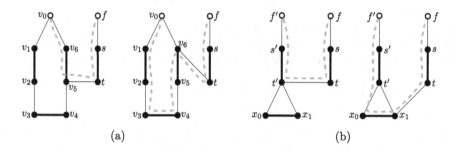

(a)   (b)

Fig. 3. Augmenting paths found in the proofs of (a) Lemma 1, $t \in T_H$ has a neighbour in V_C. (b) Lemma 1, $t \in T_H$ has a neighbour in V_B.

Lemma 1. *No vertex in $F_H \cup T_H$ has a neighbour in G that is outside H.*

Proof. First observe that no edge can connect a vertex in $F_H \cup T_H = D_0 \cup D_2$ with a vertex $z \in D_k$ for $k \geq 4$ since z would have been added to $D_1 = S$ or $D_3 = U$ instead. So we must only show that no vertex in $F_H \cup T_H$ has a neighbour in $V_C \cup V_B$. We show this only for $t \in T_H$; the proof is similar (and even easier) for $f \in F_H$ by replacing the path $t\text{-}s\text{-}f$ defined below with just f.

Consider Fig. 3(a). Fix some $t \in T_H$, let $s \in S$ be its matching-partner and let $f \in F_H$ be an arbitrary free vertex incident to s. Assume for contradiction that t has a neighbour v_i in V_C, so v_i belongs to some 7-cycle-flower $v_0\text{-}v_1\text{-}\ldots\text{-}v_k\text{-}v_0$ where $k \in \{2, 4, 6\}$ and $v_0 \in F$. Note that $v_0 \neq f$ since $v_0 \in F_C$ while $f \in F_H$. If i is odd then $f\text{-}s\text{-}t\text{-}v_i\text{-}\ldots\text{-}v_k\text{-}v_0$ is a 9-augmenting path, and if i is even then $f\text{-}s\text{-}t\text{-}v_i\text{-}v_{i-1}\text{-}\ldots\text{-}v_1\text{-}v_0$ is a 9-augmenting path; both are impossible.

Now consider some $(x_0, x_1) \in M_B$ that belongs to a 7-flower $f'\text{-}s'\text{-}t'\text{-}x_0\text{-}x_1\text{-}t'\text{-}s'\text{-}f'$ where (s', t') is a matching-edge and $t' \in T_B$. Note that $t' \neq t$ (hence $s' \neq s$) since $t' \in T_B$ while $t \in T_H$. If t and t' are adjacent, then $f\text{-}s\text{-}t\text{-}t'\text{-}s'\text{-}f'$ is a 5-augmenting path or a 5-cycle-flower. If t and x_i are adjacent for $i \in \{0, 1\}$, then $f\text{-}s\text{-}t\text{-}x_i\text{-}x_{1-i}\text{-}t'\text{-}s'\text{-}f'$ is a 7-augmenting path or 7-cycle-flower. See Fig. 3(b). Both are impossible since $t \notin T_C$.   □

In particular, if a vertex in $F_H \cup T_H$ had degree d in G, then it also has degree d in H; this will be important below.

Minimum Degree 3. With this, we can prove our first matching-bound. We need the following lemma by Biedl and Wittnebel, which is derived via (quite complicated) graph-augmentation and edge-counting:

Lemma 2 ([6]). *Let G be a simple 1-planar graph. Let A be a non-empty independent set of G where all vertices in A have degree 3 or more in G. Let A_d be the vertices of degree d in A. Then $2|A_3| + \sum_{d>3}(3d-6)|A_d| \leq 12|V \setminus A| - 24$.*

Lemma 3. *We have (i) $|F_H| \leq 6|S| - 12$ and (ii) $|F_H| + |T_H| \leq 6|S| + 6|U| - 12$.*

Proof. Consider first the subgraph of H induced by F_H and S. By Observation 2 and Lemma 1 any vertex in F_H has degree at least 3 in this subgraph, and they form an independent set. Consider the inequality of Lemma 2. Any vertex in F_H contributes at least 2 units to the left-hand side while the right-hand side is $12|S| - 24$. This proves Claim (i) after dividing.

Now consider the full graph H. By Observation 2 and Lemma 1 any vertex in $F_H \cup T_H$ has degree at least 3 in H, and they form an independent set. Claim (ii) now follows from Lemma 2 as above. □

Corollary 1. *If the minimum degree is 3, then $|M| \geq \frac{7}{50}(n + 12)$.*

Proof. Adding Lemma 3(ii) six times to Lemma 3(i) gives

$$7|F_H| + 6|T_H| \leq 42|S| + 36|U| - 84 \leq 42|M_S| + 36|M_U| - 84.$$

Adding Claim 1 seven times and Claim 2 six times gives

$$7|F_C| + 7|F_H| + 6|T_B| + 6|T_H| \leq 42|M_S| + 36|M_U| + 7|M_C| + 6|M_B| - 84.$$

Since $|S| = |M_S| = |T_H| + |T_B|$, this simplifies to

$$7|F| = 7|F_H| + 7|F_C| \leq 36|M_S| + 36|M_U| + 7|M_C| + 6|M_B| - 84 \leq 36|M| - 84.$$

Therefore $2|M| = n - |F| \geq n + 12 - \frac{36}{7}|M|$ which gives the bound after rearranging. □

It is worth pointing out that this result (as well as Theorem 2 below) does not use 1-planarity of the graph except when using the bound in Lemma 2. Hence, similar bounds could be proved for any graph class where the size of independent sets can be upper-bounded relative to its minimum degree.

Doing the improvement from $\frac{7}{50}$ to $\frac{1}{7}$ will be done by improving Lemma 3(ii) slightly. We will show the following in Sect. 5:

Lemma 4. $|F_H| + |T_H| \leq 6|S| + 5|U| - 12.$

This then gives our main result:

Theorem 1. *Let G be a 1-planar graph with minimum degree 3, and let M be a matching in G that has no augmenting path of length 9 or less. Then $|M| \geq \frac{n+12}{7}$.*

Proof. Using $|S| = |M_S|$ and $|U| = |M_U|$ we have

$$|F_H| + |T_H| \leq 6|M_S| + 5|M_U| - 12 \qquad \text{from Lemma 4}$$
$$|F_C| \leq |M_C| \qquad \text{from Claim 1}$$
$$|T_B| \leq |M_B| \qquad \text{from Claim 2.}$$

Since $|T_H| + |T_B| = |M_S|$ this gives $|F| + |M_S| \leq |M_C| + |M_B| + 6|M_S| + 5|M_U| - 12$, therefore $|F| \leq 5|M| - 12$. This implies $2|M| = n - |F| \geq n - 5|M| + 12$ or $7|M| \geq n + 12$. □

Higher Minimum Degree. Since the bound for independent sets in 1-planar graphs gets smaller when the minimum degree is larger, we can prove better matching-bounds for higher minimum degree. The following is proved exactly like Lemma 3:

Lemma 5. *If the minimum degree is $\delta > 3$, then*

$$(i) \ |F_H| \leq \tfrac{4}{\delta-2}(|S| - 2) \quad and \quad (ii) \ |F_H| + |T_H| \leq \tfrac{4}{\delta-2}(|S| + |U| - 2).$$

Theorem 2. *Let G be a 1-planar graph with minimum degree δ. Let M be any matching in G without 9-augmenting path. Then*

- $|M| \geq \tfrac{3}{10}(n + 12)$ *for $\delta = 4$,*
- $|M| \geq \tfrac{1}{3}(n + 12)$ *for $\delta \geq 5$.*

Proof. Set $c = \tfrac{4}{\delta-2}$, so $|F_H| \leq c(|S| - 12)$ and $|F_H| + |T_H| \leq c(|S| + |U| - 12)$. Taking the former incquality once and adding the latter one c times gives

$$(c+1)|F_H|+c|T_H| \leq (c^2+c)|S|+c^2|U|-(c+1)12 = (c^2+c)|M_S|+c^2|M_U|-(c+1)12.$$

Adding Claim 1 $c + 1$ times and Claim 2 c times gives

$$(c+1)(|F_C|+|F_H|) + c(|T_B|+|T_H|)$$
$$\leq (c^2+c)|M_S| + c^2|M_U| + (c+1)|M_C| + c|M_B| - (c+1)12. \quad (1)$$

For $\delta = 4$ we have $c = 2$, and with $|T_B| + |T_H| = |M_S|$ Eq. 1 simplifies to

$$3|F| \leq 4|M_S| + 4|M_U| + 3|M_C| + 2|M_B| - 36 \leq 4|M| - 36.$$

Therefore $2|M| = n - |F| \geq n + 12 - \tfrac{4}{3}|M|$. For $\delta \geq 5$ we have $c^2 < c+1$ and so can only simplify Eq. 1 to $(c + 1)(|F_C| + |F_H|) \leq (c + 1)|M| - (c + 1)12$, hence $2|M| = n - |F| \geq n + 12 - |M|$. The bounds follow after rearranging. □

For $\delta = 4, 5$ these are close to the bounds of $\tfrac{1}{3}(n+4)$ (for $\delta = 4$) and $\tfrac{1}{5}(2n+3)$ (for $\delta = 5$) that we know to be the tight lower bounds on the maximum matching size [6]. Unfortunately we do not know how to improve Theorem 2 for $\delta > 3$; the techniques of Sect. 5 do not work for higher minimum degree.

Stopping Earlier? Currently we remove all augmenting paths up to length 9. Naturally one wonders whether one could stop earlier? We can show that it suffices to remove only 7-augmenting paths by inspecting the analysis. The details are not difficult but tedious and require even more notation; we omit them.

On the other hand, it is not enough to remove only 3-augmenting paths. Figure 4 shows a matching in a 1-planar graph that has no 3-augmenting paths, but only size $\tfrac{n+12}{8}$. We can show that this is tight.

Theorem 3. *Let G be a 1-planar graph with minimum degree 3 and let M be a matching without 3-augmenting paths. Then $|M| \geq \tfrac{n+12}{8}$.*

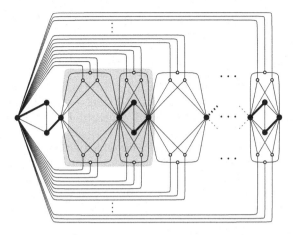

Fig. 4. A graph with a matching marked in thick edges of size $\frac{n+12}{8}$. No 3-augmenting path exists for the chosen matching, but there are 5-augmenting paths. The gray area marks an example of 16 vertices such that only 2 matching edges exist. Repeating this configuration gives the example for arbitrary n.

Proof. The proof is very similar to the one of Theorem 2 in [13] except that we use Lemma 2 rather than the edge-bound for planar bipartite graphs. We repeat it here for completeness, mimicking their notation. Let M_c be all those matching-edges (x, y) for which some free vertex $f \in F$ is adjacent to both x and y, and let F_c be all such free vertices. Vertex f is necessarily the only F-neighbour of x and y, else there would be a 3-augmenting path. Hence $|F_c| \leq |M_c|$.

Let M_o and F_o be the remaining matching-edges and free vertices. For each edge (x, y) in M_o, at most one of the ends can have F-neighbours, else (x, y) would be in M_c or there would be a 3-augmenting path. Let S be the ends of edges in M_o that have F-neighbours, and let G' be the auxiliary graph induced by F_o and S. Then $|F_o| \leq 6|S| - 12 \leq 6|M_o| - 12$ by Lemma 2.

Putting both together, $2|M| = n - |F| \geq n + 12 - |M_c| - 6|M_o| \geq n + 12 - 6|M|$ and the bound follows after rearranging.   □

5   Proof of Lemma 4

(Sketch; details are in the full paper [5].) Fix an arbitrary 1-planar drawing of H. We obtain a 1-planar drawing H^+ from H by inserting any potential kite-edge (t, x) with $t \in T_H$ and $x \in S \cup U$ that does not exist yet. If (t, x) exists, but has a crossing, then re-route it to become *uncrossed* (i.e., without crossing).

We split T_H-vertices and assign them as follows. If $t \in T_H$ has an uncrossed edge to a U-neighbour u, then assign t to u. Else, if t has three or more S-neighbours, then add t to a vertex set T_σ. Else assign t to an arbitrary U-neighbour u. In the first and third case we call (t, u) the *assignment-edge*. Let $U^{(d)}$ be the set of all those vertices $u \in U$ that have d incident assignment-edges.

Let $T^{(d)}$ be all those vertices in $T_H \setminus T_\sigma$ that have been assigned to a vertex in $U^{(d)}$. Since $|T^{(d)}| = d|U^{(d)}|$, we have:

Observation 3. $|T_0| = 0$ and $\sum_{d=1}^{5} |T^{(d)}| \leq 5 \sum_{d=0}^{5} |U^{(d)}|$.

Transform drawing H^+ as follows:

- Delete all vertices in $U^{(0)} \cup \cdots \cup U^{(5)}$ and $T^{(1)} \cup \cdots \cup T^{(5)}$ and all SU-edges.
- For any remaining $t \in T_H$, delete all edges to U-neighbours except the assignment-edge (if any).
- While there exists a vertex $t \in T_H \setminus T_\sigma$ for which either the assignment-edge (t, u) or the matching-edge (s, t) is uncrossed: Delete t and insert edge (s, u). Normally (s, u) is routed along the path s-t-u, which has at most one crossing. But if this leads to a crossing of (s, u) with an edge that ends at s or u, then instead draw (s, u) as a kite-edge of that crossing so that the drawing remains good.

For this proof sketch, let us assume that *all* vertices in $T_H \setminus T_\sigma$ get deleted. (This is not always the case, and those "remaining" vertices of T_H are a major difficulty to overcome; see [5].)

Assuming this to be the case, we have in the resulting drawing J the independent set $F_H \cup T_\sigma \cup \bigcup_{d \geq 6} U^{(d)}$ and the vertices of S_H. All vertices in $F_H \cup T_\sigma$ have degree at least 3. Vertex $u \in U^{(d)}$ (for $d \geq 6$) has degree at least d in J, because it was assigned to d T_H-vertices and therefore inherits edges to their d distinct matching-partners. Lemma 4 now holds by applying Lemma 2 to drawing J and combining it with Observation 3 as follows:

$$12|S_H| - 24 \geq 2|F_H| + 2|T_\sigma| + \sum_{d \geq 6}(3d-6)|U^{(d)}| \geq 2|F_H| + 2|T_\sigma| + \sum_{d \geq 6}(2d-10)|U^{(d)}|$$

$$\geq 2|F_H| + 2|T_\sigma| + 2\sum_{d \geq 6}|T^{(d)}| - 10\sum_{d \geq 6}|U^{(d)}| + 2\sum_{d \leq 5}|T^{(d)}| - 10\sum_{d \leq 5}|U^{(d)}|$$

$$\geq 2|F_H| + 2|T_H| - 10|U|$$

and hence $|F_H| + |T_H| \leq 6|S_H| + 5|U| - 12$.

6   Summary and Outlook

In this paper, we considered how to find a large matching in a 1-planar graph with minimum degree 3. We argued that any matching without augmenting paths of length up to 9 has size at least $\frac{n+12}{7}$, which is also the largest matching one can guarantee to exist in any 1-planar graph with minimum degree 3. Such a matching can easily be found in linear time, even if no 1-planar drawings is known, by stopping the matching algorithm by Micali and Vazirani after a constant number of rounds.

It remains open how to find large matchings in 1-planar graphs with minimum degree $\delta > 3$ that match the upper bounds. It would also be interesting to study other near-planar graph classes such as k-planar graphs (for $k > 1$); here we do not even know what tight matching-bounds exist and much less how to find matchings of that size in linear time.

References

1. Baker, B.S.: Approximation algorithms for NP-complete problems on planar graphs. J. ACM **41**(1), 153–180 (1994). https://doi.org/10.1145/174644.174650
2. Bast, H., Mehlhorn, K., Schäfer, G., Tamaki, H.: Matching algorithms are fast in sparse random graphs. Theor. Comput. Syst. **39**(1), 3–14 (2006). https://doi.org/10.1007/s00224-005-1254-y
3. Berge, C.: Graphs and Hypergraphs, 2nd edn. North-Holland (1976). Translated from Graphes et Hypergraphes, Dunod (1970)
4. Biedl, T., Bose, P., Demaine, E.D., Lubiw, A.: Efficient algorithms for Petersen's matching theorem. J. Algorithms **38**(1), 110–134 (2001). https://doi.org/10.1006/jagm.2000.1132
5. Biedl, T., Klute, F.: Finding large matchings in 1-planar graphs of minimum degree 3 (2020). CoRR 2002.11818. http://arxiv.org/abs/2002.11818
6. Biedl, T., Wittnebel, J.: Matchings in 1-planar graphs with large minimum degree (2019). CoRR 1911.04603. http://arxiv.org/abs/1911.04603
7. Booth, K.S., Lueker, G.S.: Testing for the consecutive ones property, interval graphs and graph planarity using PQ-tree algorithms. J. Comput. Syst. Sci. **13**, 335–379 (1976). https://doi.org/10.1016/S0022-0000(76)80045-1
8. Chiba, N., Nishizeki, T.: The Hamiltonian cycle problem is linear-time solvable for 4-connected planar graphs. J. Algorithms **10**(2), 187–211 (1989). https://doi.org/10.1016/0196-6774(89)90012-6
9. Cole, R., Ost, K., Schirra, S.: Edge-coloring bipartite multigraphs in $O(E \log D)$ time. Combinatorica **21**(1), 5–12 (2001). https://doi.org/10.1007/s004930170002
10. Demaine, E.D., Fomin, F.V., Hajiaghayi, M., Thilikos, D.M.: Subexponential parameterized algorithms on bounded-genus graphs and H-minor-free graphs. J. ACM **52**(6), 866–893 (2005). https://doi.org/10.1145/1101821.1101823
11. Diestel, R.: Graph Theory. Graduate Texts in Mathematics, vol. 173, 4th edn. Springer, Heidelberg (2012)
12. Edmonds, J.: Maximum matchings and a polyhedron with 0,1-vertices. J. Res. Natl. Bur. Stan. **69B**, 125–130 (1965)
13. Franke, R., Rutter, I., Wagner, D.: Computing large matchings in planar graphs with fixed minimum degree. Theor. Comput. Sci. **412**(32), 4092–4099 (2011). https://doi.org/10.1016/j.tcs.2010.06.012
14. Grigoriev, A., Bodlaender, H.L.: Algorithms for graphs embeddable with few crossings per edge. Algorithmica **49**(1), 1–11 (2007). https://doi.org/10.1007/s00453-007-0010-x
15. Hall, M., Jr.: An algorithm for distinct representatives. Am. Math. Mon. **63**, 716–717 (1956)
16. Hopcroft, J.E., Karp, R.M.: An $n^{5/2}$ algorithm for maximum matchings in bipartite graphs. SIAM J. Comput. **2**, 225–231 (1973). https://doi.org/10.1137/0202019
17. Hopcroft, J.E., Tarjan, R.E.: Efficient planarity testing. J. ACM **21**(4), 549–568 (1974). https://doi.org/10.1145/321850.321852
18. Kobourov, S.G., Liotta, G., Montecchiani, F.: An annotated bibliography on 1-planarity. Comput. Sci. Rev. **25**, 49–67 (2017). https://doi.org/10.1016/j.cosrev.2017.06.002
19. Lovász, L., Plummer, M.D.: Matching Theory. Annals of Discrete Mathematics, p. 29. North-Holland Publishing Co., Amsterdam (1986)

20. Micali, S., Vazirani, V.V.: An $O(\sqrt{|V|}|E|)$ algoithm for finding maximum matching in general graphs. In: Proceedings of the 21st Annual Symposium on Foundations of Computer Science (FOCS 1980), pp. 17–27. IEEE Computer Society (1980). https://doi.org/10.1109/SFCS.1980.12

21. Nishizeki, T., Baybars, I.: Lower bounds on the cardinality of the maximum matchings of planar graphs. Discrete Math. **28**(3), 255–267 (1979). https://doi.org/10.1016/0012-365X(79)90133-X

22. Petersen, J.: Die Theorie der regulären graphs (The theory of regular graphs). Acta Math. **15**, 193–220 (1891)

23. Schaefer, M.: Crossing Numbers of Graphs. CRC Press, Boca Raton (2017)

24. Tutte, W.T.: A theorem on planar graphs. Trans. Am. Math. Soc. **82**, 99–116 (1956). https://doi.org/10.2307/1992980

25. Vazirani, V.V.: A simplification of the MV matching algorithm and its proof. CoRR, abs/1210.4594 (2012). http://arxiv.org/abs/1210.4594

Strong Cliques in Diamond-Free Graphs

Nina Chiarelli[1,2], Berenice Martínez-Barona[3], Martin Milanič[1,2(✉)],
Jérôme Monnot[4], and Peter Muršič[1]

[1] FAMNIT, University of Primorska, Koper, Slovenia
[2] IAM, University of Primorska, Koper, Slovenia
{nina.chiarelli,martin.milanic,peter.mursic}@famnit.upr.si
[3] Departament d'Enginyeria Civil i Ambiental, Universitat Politècnica de Catalunya,
Barcelona, Spain
berenice.martinez@upc.edu
[4] LAMSADE, University Paris-Dauphine, Paris Cedex 16, France

Abstract. A strong clique in a graph is a clique intersecting all
inclusion-maximal stable sets. Strong cliques play an important role
in the study of perfect graphs. We study strong cliques in the class
of diamond-free graphs, from both structural and algorithmic points of
view. We show that the following five NP-hard or co-NP-hard problems
remain intractable when restricted to the class of diamond-free graphs:
Is a given clique strong? Does the graph have a strong clique? Is every
vertex contained in a strong clique? Given a partition of the vertex set
into cliques, is every clique in the partition strong? Can the vertex set
be partitioned into strong cliques?

On the positive side, we show that the following two problems whose
computational complexity is open in general can be solved in linear time
in the class of diamond-free graphs: Is every maximal clique strong? Is
every edge contained in a strong clique? These results are derived from
a characterization of diamond-free graphs in which every maximal clique
is strong, which also implies an improved Erdős Hajnal property for such
graphs.

Keywords: Maximal clique · Maximal stable set · Diamond-free
graph · Strong clique · Simplicial clique · CIS graph · NP-hard
problem · Linear-time algorithm · Erdős-Hajnal property

1 Introduction

Background and Motivation. Given a graph G, a *clique* in G is a set of pairwise
adjacent vertices, and a *stable set* (or *independent set*) is a set of pairwise non-
adjacent vertices. A clique (resp., stable set) is *maximal* if it is not contained in
any larger clique (resp., stable set). A clique is *strong* if intersects all maximal
stable sets and a strong stable set is defined analogously. The concepts of strong
cliques and strong stable sets in graphs play an important role in the study
of perfect graphs (see, e.g., [7]) and were studied in a number of papers (see,

© Springer Nature Switzerland AG 2020
I. Adler and H. Müller (Eds.): WG 2020, LNCS 12301, pp. 261–273, 2020.
https://doi.org/10.1007/978-3-030-60440-0_21

e.g., [1,2,4,5,9,10,18,19,21,23–25,28,30,31,37]). Several algorithmic problems related to strong cliques and stable sets in graphs are NP-hard or co-NP-hard, in particular:

- STRONG CLIQUE: Given a clique C in a graph G, is C strong?
- STRONG CLIQUE EXISTENCE: Given a graph G, does G have a strong clique?
- STRONG CLIQUE VERTEX COVER: Given a graph G, is every vertex contained in a strong clique?
- STRONG CLIQUE PARTITION: Given a graph G and a partition of its vertex set into cliques, is every clique in the partition strong?
- STRONG CLIQUE PARTITION EXISTENCE: Given a graph G, can its vertex set be partitioned into strong cliques?

The first problem in the above list is co-NP-complete (see [37]), the second one is NP-hard (see [23]), and the remaining three are co-NP-hard (see [25] for the third and the fourth problem[1] and [16,24,33] for the fifth one).

Another interesting property related to strong cliques is the one defining CIS graphs. A graph is said to be *CIS* if every maximal clique is strong, or equivalently, if every maximal stable set is strong, or equivalently, if every maximal clique intersects every maximal stable set. Although the name CIS (Cliques Intersect Stable sets) was suggested by Andrade et al. in a recent book chapter [5], this concept has been studied under different names since the 1990s [4,9,18,19,21,37,38] (see [37] for a historical overview). Several other graph classes studied in the literature can be defined in terms of properties involving strong cliques (see, e.g., [10,24,25,30,31]).

There are several intriguing open questions related to strong cliques and strong stable sets, for instance: (i) What is the complexity of determining whether every edge of a given graph is contained in a strong clique? (See, e.g., [1].) (ii) What is the complexity of recognizing CIS graphs? (See, e.g., [5].) (iii) Is there some $\varepsilon > 0$ such that every n-vertex CIS graph has either a clique or a stable set of size at least n^{ε}? (See [2].)

The main purpose of this paper is to study strong cliques in the class of diamond-free graphs. The *diamond* is the graph obtained by removing an edge from the complete graph on four vertices, and a graph G is said to be *diamond-free* if no induced subgraph of G is isomorphic to the diamond. Our motivation for focusing on the class of diamond-free graphs comes from several sources. First, no two maximal cliques in a diamond-free graph share an edge, which makes interesting the question to what extent this structural restriction is helpful for understanding strong cliques. Second, structural and algorithmic questions related to strong cliques in particular graph classes were extensively studied in the literature (see, e.g., [1,2,9,19,24,25,28]), so this work represents a natural continuation of this line of research. Finally, this work furthers the knowledge about diamond-free graphs. In 1984, Tucker proved the Strong Perfect Graph Conjecture, now Strong Perfect Graph Theorem, for diamond-free graphs [36],

---

[1] In [25], the authors state that the two problems are NP-hard, but their proof actually shows co-NP-hardness.

and more recently there has been an increased interest regarding the coloring problem and the chromatic number of diamond-free graphs and their subclasses (see, e.g., [11,17,27,29]). Diamond-free graphs also played an important role in a recent work of Chudnovsky et al. [15] who proved that there are exactly 24 4-critical P_6-free graphs.

Our Contributions. Our study of strong cliques in diamond-free graphs is done from several interrelated points of view. First, we give an efficiently testable characterization of diamond-free CIS graphs. A vertex v in a graph G is *simplicial* if its closed neighborhood is a clique in G. Any such clique will be referred to as a *simplicial clique*. We say that a graph G is *clique simplicial* if every maximal clique in G is simplicial. The characterization is as follows.

Theorem 1. *Let G be a connected diamond-free graph. Then G is CIS if and only if G is either clique simplicial, $G \cong K_{m,n}$ for some $m, n \geq 2$, or $G \cong L(K_{n,n})$ for some $n \geq 3$.*

Second, we derive several consequences of Theorem 1. A graph class $\mathcal{G}$ is said to satisfy the *Erdős-Hajnal property* if there exists some $\varepsilon > 0$ such that every graph $G \in \mathcal{G}$ has either a clique or a stable set of size at least $|V(G)|^\varepsilon$. The well-known Erdős-Hajnal Conjecture [20] asks whether for every graph F, the class of F-free graphs has the Erdős-Hajnal property. The conjecture is still open, but it has been confirmed for graphs F with at most 4 vertices (see, e.g., [14]). In the case when F is the diamond, a simple argument shows that the inequality holds with $\varepsilon = 1/3$ (see [22]), but it is not known whether this value is best possible. Theorem 1 implies the following improvement for the diamond-free CIS graphs.

Theorem 2. *Let G be a diamond-free CIS graph. Then $\alpha(G) \cdot \omega(G) \geq |V(G)|$. Consequently, G has either a clique or a stable set of size at least $|V(G)|^{1/2}$.*

Next, we develop a linear-time algorithm to test if every edge of a given diamond-free graph is in a simplicial clique. This leads to the following algorithmic consequence of Theorem 1.

Theorem 3. *There is a linear-time algorithm that determines whether a given diamond-free graph G is CIS.*

Theorem 3 implies a linear-time algorithm for testing if every edge of a given diamond-free graph is contained in a strong clique. Furthermore, as a consequence of Theorem 3 and other results in the literature, we report on the following partial progress on the open question about the complexity of recognizing CIS graphs.

Theorem 4. *For every graph F with at most 4 vertices, it can be determined in polynomial time whether a given F-free graph is CIS.*

Finally, we complement the above efficient characterizations with hardness results about several problems related to strong cliques when restricted to the class of diamond-free graphs. More specifically, using reductions from the 3-COLORABILITY problem in the class of triangle-free graphs we show the following.

Theorem 5. *When restricted to the class of diamond-free graphs, the* STRONG CLIQUE, STRONG CLIQUE EXISTENCE, STRONG CLIQUE VERTEX COVER, *and* STRONG CLIQUE PARTITION *problems are* co-NP-*complete, and the* STRONG CLIQUE PARTITION EXISTENCE *problem is* co-NP-*hard.*

Due to space restrictions, most proofs are omitted and will appear in the full version of this work.

2 Preliminaries

We consider only graphs that are finite and undirected. We refer to simple graphs as graphs and to graphs with multiple edges allowed as multigraphs. Let $G = (V, E)$ be a graph with vertex set $V(G) = V$ and edge set $E(G) = E$. For a subset of vertices $X \subseteq V(G)$, we will denote by $G[X]$ the subgraph of G induced by X, that is, the graph with vertex set X and edge set $\{\{u, v\} \mid \{u, v\} \in E(G); \ u, v \in X\}$. We denote the complete graph, the path, and the cycle graph of order n by K_n, P_n, and C_n, respectively. The graph K_3 will be also referred as a *triangle*. By $K_{m,n}$ we denote the complete bipartite graph with parts of size m and n. The fact that a graph G is isomorphic to a graph H will be denoted by $G \cong H$. We say that G is *H-free* if no induced subgraph of G is isomorphic to H.

The *neighborhood* of a vertex v in a graph G, denoted by $N_G(v)$ (or just $N(v)$ if the graph is clear from the context), is the set of vertices adjacent to v in G. The cardinality of $N_G(v)$ is the *degree* of v in G, denoted by $d_G(v)$ (or simply $d(v)$). The *closed neighborhood*, $N(v) \cup \{v\}$, is denoted by $N[v]$. Given a graph G and a set $X \subseteq V(G)$, we denote by $N_G(X)$ the set of vertices in $V(G) \setminus X$ having a neighbor in X. The *line graph* of a graph G, denoted with $L(G)$, is the graph with vertex set $E(G)$ and such that two vertices in $L(G)$ are adjacent if and only if their corresponding edges in G have a vertex in common. A *matching* in a graph G is a set of pairwise disjoint edges. A matching is *perfect* if every vertex of the graph is an endpoint of an edge in the matching. Given a graph G, we denote by $\alpha(G)$ the maximum size of a stable set in G and by $\omega(G)$ the maximum size of a clique in G.

We first recall a basic property of CIS graphs (see, e.g., [5]). An induced P_4, (a, b, c, d), in a graph G is said to be *settled* (in G) if G contains a vertex v adjacent to both b and c and non-adjacent to both a and d.

Proposition 6. *In every CIS graph each induced P_4 is settled.*

Given a clique C in G and a set $S \subseteq V(G) \setminus C$, we say that clique C is *dominated by* S if every vertex of C has a neighbor in S. Using this notion, simplicial cliques can be characterized as follows.

Lemma 7. *A clique C in a graph G is not simplicial if and only C is dominated by $V(G) \setminus C$.*

Proof. Note that C is simplicial if and only if there exists a vertex $v \in C$ such that $C = N[v]$. However, since C is a clique, we have $C \subseteq N[v]$ for all vertices $v \in C$. It follows that C is simplicial if and only if there exists a vertex $v \in C$ such that $N(v) \subseteq C$. Equivalently, C is not simplicial if and only every vertex in C has a neighbor outside C, that is, C is dominated by $V(G) \setminus C$. $\quad\square$

A similar characterization is known for strong cliques (see, e.g., the remarks following [24, Theorem 2.3]).

Lemma 8. *A clique C in a graph G is not strong if and only if it is dominated by a stable set $S \subseteq V(G) \setminus C$.*

3   A Characterization of Diamond-Free CIS Graphs

We first derive a property of diamond-free graphs in which every induced P_4 is settled.

Lemma 9. *Let G be a connected diamond-free graph in which every induced P_4 is settled. Then, either G is complete bipartite, or each edge $e \in E(G)$ that is a maximal clique in G is also a simplicial clique.*

Proof. Suppose that G is not complete bipartite. Suppose that there is an edge $uv \in E(G)$ such that $\{u, v\}$ is a non-simplicial maximal clique in G. Then, by Lemma 7, $N(v) \setminus \{u\} \neq \emptyset$ and $N(u) \setminus \{v\} \neq \emptyset$. If $x \in N(u) \setminus \{v\}$ and $y \in N(v) \setminus \{u\}$, then $xy \in E(G)$ otherwise the path (x, u, v, y) is a non-settled induced P_4 in G, which is a contradiction. Moreover, since G is diamond-free, for any two different vertices $x, x' \in N(u) \setminus \{v\}$ and two different vertices $y, y' \in N(v) \setminus \{u\}$ we have $xx', yy' \notin E(G)$. This, together with the fact that $N(u) \cap N(v) = \emptyset$ implies that $N(u)$ and $N(v)$ are stable sets. Then, neither u nor v are contained in a triangle. Hence, $G[N[u] \cup N[v]]$ is complete bipartite. Since G is connected and not complete bipartite, we can assume without loss of generality that there is a vertex $w \in V(G) \setminus (N[u] \cup N[v])$ such that $wx \in E(G)$ for some $x \in N(u) \setminus \{v\}$. Then, (v, u, x, w) is a non-settled induced P_4 in G, a contradiction. $\quad\square$

Next, we analyze the global properties of an induced subgraph H in a connected diamond-free CIS graph G such that $H \cong L(K_{n,n})$ for some $n \geq 3$.

Lemma 10. *Let G be a connected diamond-free CIS graph. Suppose that G contains an induced subgraph H such that $H \cong L(K_{n,n})$ for some $n \geq 3$. Then, either all the $2n$ copies of K_n in H are maximal cliques in G or none of them is a maximal clique in G. Furthermore, if all the $2n$ copies of K_n are maximal cliques in G, then $G = H$.*

We will also need the following technical lemma about diamond-free graphs.

Lemma 11. *Let G be a diamond-free graph. Let $C' = \{v_1, v_2, \ldots, v_\ell\}$ be a clique in G and let $C_1, \ldots, C_\ell$ be maximal cliques in G such that for every $i \in \{1, \ldots, \ell\}$ we have $C_i \cap C' = \{v_i\}$, $2 \leq \ell = |C'| \leq |C_i|$ and $C_i \cap C_j = \emptyset$ for any $j \neq i$. Then, there exists a stable set $S \subseteq \bigcup_{i=1}^{\ell} C_i$ in $G - C'$ of size $\ell - 1$ such that $|N_G(S) \cap C'| = |S|$. Furthermore, if $\max_{i=1}^{\ell} |C_i| > \ell$, then there exists a stable set $S \subseteq \bigcup_{i=1}^{\ell} C_i$ in $G - C'$ such that $|S| = \ell$ and S dominates C'.*

We can now characterize the connected diamond-free CIS graphs. The result shows that, apart from the clique simplicial diamond-free graphs, there are only two highly structured infinite families of connected diamond-free CIS graphs.

Theorem 1. *Let G be a connected diamond-free graph. Then G is CIS if and only if G is either clique simplicial, $G \cong K_{m,n}$ for some $m, n \geq 2$, or $G \cong L(K_{n,n})$ for some $n \geq 3$.*

Proof (sketch). It is not difficult to see that each of the three conditions is sufficient for G to be CIS. To show necessity, consider a connected diamond-free CIS graph G and suppose for a contradiction that it does not satisfy any of the three conditions. Let $C = \{v_1, \ldots, v_k\}$ be a smallest maximal clique in G that is not simplicial, and for each $i \in \{1, \ldots, k\}$, let $C_i \neq C$ be a maximal clique containing v_i. First observe that if $k = 2$, then by Proposition 6 and Lemma 9 we have that G is complete bipartite, a contradiction, thus $k \geq 3$. Next, we show using Lemmas 8 and 11 that all cliques C_i have size k. Finally we show that either Lemma 10 implies $G \cong L(K_{k,k})$ or by using Lemma 11 we can construct a stable set in $V(G) \backslash C$ dominating C, which leads to a contradiction by Lemma 8. □

4 Consequences

We now discuss several consequences of Theorem 1.

4.1 Large Cliques or Stable Sets in Diamond-Free CIS Graphs

Recall that every diamond-free graph G has either a clique or a stable set of size at least $|V(G)|^{1/3}$. For diamond-free CIS graphs, Theorem 1 leads to an improvement of the exponent to $1/2$.

Lemma 12. *Let G be a clique simplicial graph. Then $\alpha(G) \cdot \omega(G) \geq |V(G)|$.*

Proof. Let $C_1, \ldots, C_k$ be all the maximal cliques of G. Since G is clique simplicial, each C_i is a simplicial clique. Selecting one simplicial vertex from each C_i gives a stable set S of size k, and since every vertex is in some simplicial clique, we infer that $|V(G)| \leq \sum_{i=1}^{k} |C_i| \leq |S| \cdot \omega(G) \leq \alpha(G) \cdot \omega(G)$. □

Theorem 2. *Let G be a diamond-free CIS graph. Then $\alpha(G) \cdot \omega(G) \geq |V(G)|$. Consequently, G has either a clique or a stable set of size at least $|V(G)|^{1/2}$.*

Proof. First we show that it suffices to prove the statement for connected graphs. Let G be a disconnected diamond-free CIS graph, let C be a component of G and let $G' = G - V(C)$. Then C and G' are diamond-free CIS graphs, and by induction on the number of components we may assume that $\alpha(C) \cdot \omega(C) \geq |V(C)|$ and $\alpha(G') \cdot \omega(G') \geq |V(G')|$. Since $\alpha(G) = \alpha(C) + \alpha(G')$ and $\omega(G) = \max\{\omega(C), \omega(G')\}$, we obtain $|V(G)| = |V(C)| + |V(G')| \leq \alpha(C) \cdot \omega(G) + \alpha(G') \cdot \omega(G) = \alpha(G) \cdot \omega(G)$.

Now let G be a connected diamond-free CIS graph. By Theorem 1, G is either clique simplicial, $G \cong K_{m,n}$ for some $m, n \geq 2$, or $G \cong L(K_{n,n})$ for some $n \geq 3$. If G is clique simplicial, then Lemma 12 yields $|V(G)| \leq \alpha(G) \cdot \omega(G)$. If $G \cong K_{m,n}$ for some $m, n \geq 2$, then $|V(G)| = m + n \leq 2 \cdot \max\{m, n\} = \omega(G) \cdot \alpha(G)$. Finally, if $G \cong L(K_{n,n})$, then $\alpha(G)$ equals the maximum size of a matching in $K_{n,n}$, that is, $\alpha(G) = n$, and $\omega(G)$ equals the maximum degree of a vertex in $K_{n,n}$, that is, $\omega(G) = n$. Thus, in this case equality holds, $|V(G)| = n^2 = \alpha(G) \cdot \omega(G)$. □

4.2 Testing the CIS Property in the Class of Diamond-Free Graphs

Our next consequence is a linear-time algorithm for testing the CIS property in the class of diamond-free graphs. The bottleneck to achieve linearity is the recognition of the clique simplicial property. Instead of checking this property directly, we check whether the graph is *edge simplicial*, that is, every edge is contained in a simplicial clique. Clearly, every clique simplicial graph is edge simplicial. While the converse implication fails in general, the two properties are equivalent in the case of diamond-free graphs, where every edge is in a unique maximal clique.

Recognizing if a general graph $G = (V, E)$ is edge simplicial can be done in time $\mathcal{O}(|V| \cdot |E|)$ (see [13, 34]). The algorithm is based on the observation that within a simplicial clique, every vertex of minimum degree is a simplicial vertex (see [12]). We show that in the case of diamond-free graphs, the running time can be improved to $\mathcal{O}(|V| + |E|)$.

Given a graph G and a linear ordering $\sigma = (v_1, \ldots, v_n)$ of its vertices, the σ-*greedy stable set* is the stable set S in G computed by repeatedly adding to the initially empty set S the smallest σ-indexed vertex as long as the resulting set is still stable. A *degree-greedy stable set* is any σ-greedy stable set where $\sigma = (v_1, \ldots, v_n)$ satisfies $d(v_i) \leq d(v_j)$ for $i < j$. Note that a degree-greedy stable set does not need to coincide with a stable set computed by the greedy algorithm that iteratively selects a minimum degree vertex and deletes the vertex and all its neighbors: to compute a degree-greedy stable set, the vertex degrees are only considered in the original graph G, and not in the subgraphs obtained by deleting the already selected vertices and their neighbors.

Our first lemma analyzes the structure of a degree-greedy stable set in a graph in which the simplicial cliques cover all the vertices. In particular, it applies to edge simplicial graphs.

Lemma 13. *Let G be a graph in which every vertex belongs to a simplicial clique and let S be a degree-greedy stable set in G. Then, S consists of simplicial vertices only, one from each simplicial clique.*

Proof. Let $\sigma = (v_1, \ldots, v_n)$ be a linear ordering of $V(G)$ with $d(v_i) \leq d(v_j)$ if $i < j$ and such that S is the σ-greedy stable set. Let C be a simplicial clique in G and let v_i be the simplicial vertex in C with the smallest index. Then, all neighbors of v_i have indices larger than i and therefore v_i is selected to be in S. This shows that S contains a vertex from C. Clearly, S cannot contain two vertices from C since S is stable and C is a clique. Thus, S contains exactly one vertex from every simplicial clique in G. Finally, suppose that S contains some non-simplicial vertex v. By assumption on G, vertex v belongs to some simplicial clique C. Let w be the vertex from C contained in S. Then w is simplicial and thus $w \neq v$. This means that S contains two adjacent vertices v and w, which is in contradiction with the fact that S is a stable set. Thus, every vertex in S is simplicial.   □

Using Lemma 13 we now show the importance of degree-greedy stable sets for testing if a given diamond-free graph is edge simplicial. The characterization is based on the following auxiliary construction. Given a graph $G = (V, F)$ and a stable set S in G, we define the multigraph G_S where $V(G_S) = V \setminus S$ and the multiset of edges is given by

$$E(G_S) = \bigcup_{v \in S} \{xy \mid x \neq y \text{ and } x, y \in N_G(v)\}.$$

Note that since S is a stable set in G, every edge in $E(G_S)$ indeed has both endpoints in $V \setminus S = V(G_S)$.

Lemma 14. *Let S be a degree-greedy stable set in a diamond-free graph G. Then G is edge simplicial if and only if $G_S = G - S$.*

The condition given by Lemma 14 can be tested in linear time, see Algorithm 1. Together with Theorem 1, this leads to the following.

Theorem 3. *There is a linear-time algorithm that determines whether a given diamond-free graph G is CIS.*

A graph $G = (V, E)$ is *general partition* if there exists a set U and an assignment of vertices $x \in V$ to sets $U_x \subseteq U$ such that $xy \in E$ if and only if $U_x \cap U_y \neq \emptyset$ and for each maximal stable set S in G, the sets U_x, $x \in S$ form a partition of U. It is known that G is a general partition graph if and only if every edge of G is contained in a strong clique (see [30]). Clearly, every CIS graph is a general partition graph, and the two properties are equivalent in the class of diamond-free graphs. Therefore, Theorem 3 implies the following.

Corollary 15. *There is a linear-time algorithm that determines whether every edge of a given diamond-free graph G is in a strong clique.*

4.3 Testing the CIS Property in Classes of F-free graphs

No good characterization or recognition algorithm for CIS graphs is known. Recognizing CIS graphs is believed to be co-NP-complete [37], conjectured to

---

Algorithm 1. Diamond-Free Edge Simplicial Recognition

Input: A diamond-free graph G given by adjacency lists $(L_v : v \in V(G))$.
Output: "Yes" if G is edge simplicial, "No" otherwise.

1: compute a linear order $\sigma = (v_1, \ldots, v_n)$ of the vertices of G such that $d(v_1) \leq \ldots \leq d(v_n)$;
2: sort the adjacency list of each vertex in increasing order with respect to σ;
3: set $S = \emptyset$, all the vertices are unmarked;
4: **for** $i = 1, \ldots, n$ **do**
5:     **if** v_i is not marked **then**
6:         mark all vertices in $N(v_i)$, $S = S \cup \{v_i\}$;
7: compute the adjacency lists L_w^- of $G - S$ based on the order of $V \setminus S$ induced by σ;
8: **if** $\sum_{v \in S} \binom{d(v)+1}{2} > |E(G)|$ **then**
9:     **return** "No";
10: **for** $w \in V \setminus S$ **do**
11:     $L_w' = $ empty list;
12: **for** $v \in S$ **do**
13:     **for** $w \in N_G(v)$ **do**
14:         append each element of $N_G(v) \setminus \{w\}$ at the end of L_w';
15: **for** $w \in V \setminus S$ **do**
16:     **if** length$(L_w') \neq $ length(L_w^-) **then**
17:         **return** "No";
18: sort the adjacency lists L_w' based on the linear order of $V \setminus S$ induced by σ;
19: **for** $w \in V \setminus S$ **do**
20:     **if** $L_w' \neq L_w^-$ **then**
21:         **return** "No";
22: **return** "Yes";

---

be co-NP-complete [38], and conjectured to be polynomial [5]. Using Theorem 3 together with some known results from the literature (on strong cliques, resp. CIS graphs [2,5,24] along with [3,6,32,35]) implies that the CIS property can be recognized in polynomial time in any class of F-free graphs where F has at most 4 vertices.

Theorem 4. *For every graph F with at most 4 vertices, it can be determined in polynomial time whether a given F-free graph is CIS.*

5   Hardness Results

We consider five more decision problems related to strong cliques: STRONG CLIQUE, STRONG CLIQUE EXISTENCE, STRONG CLIQUE VERTEX COVER, STRONG CLIQUE PARTITION, and STRONG CLIQUE PARTITION EXISTENCE (see Sect. 1 for definitions). These problems were studied by Hujdurović et al. in [25], who determined the computational complexity of these problems in the classes of chordal graphs, weakly chordal graphs, line graphs and their complements, and graphs of maximum degree at most three.

In contrast with the problems of verifying whether every maximal clique is strong, or whether every edge is in a strong clique, we prove that all the above five problems are co-NP-hard in the class of diamond-free graphs. The hardness proofs are obtained using a reduction from the 3-COLORABILITY problem in the class of triangle-free graphs: Given a triangle-free graph G, can $V(G)$ be partitioned into three stable sets? As shown by Kamiński and Lozin in [26], this problem is NP-complete. Furthermore, it is clear that the problem remains NP-complete if we additionally assume that the input graph has at least five vertices and minimum degree at least three. Let $\mathcal{G}$ denote the class of all triangle-free graphs with at least five vertices and minimum degree at least 3.

Theorem 16. *The* 3-COLORABILITY *problem is* NP-*complete in the class* $\mathcal{G}$.

The reductions are based on the following construction. Given a graph $G \in \mathcal{G}$, we associate to it two diamond-free graphs G' and G'', defined as follows. The vertex set of G' is $V(G) \times \{0, 1, 2, 3\}$. For every $v \in V(G)$, the set of vertices of G' with value v in the first coordinate forms a clique (of size 4); we will refer to this clique as C_v. The set of vertices of G' with value 0 in the second coordinate forms a clique C (of size $|V(G)|$). For every $i \in \{1, 2, 3\}$ and every two distinct vertices $u, v \in V(G)$, vertices (u, i) and (v, i) are adjacent in G' if and only if u and v are adjacent in G. There are no other edges in G'. The graph G'' is obtained from the graph G' by adding, for each vertex $w \in V(G') \setminus C$, a new vertex w' adjacent only to w. See Fig. 1 for an example.

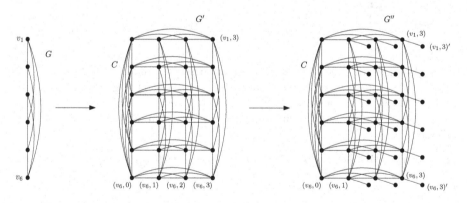

Fig. 1. Transforming G into G' and G''.

Proposition 17. *Let* $G \in \mathcal{G}$ *and let* G' *and* G'' *be the graphs constructed from* G *as described above. Then,* G' *and* G'' *are diamond-free and the following statements are equivalent.*

1. G *is not 3-colorable.*
2. C *is a strong clique in* G'.

3. G' has a strong clique.
4. C is a strong clique in G''.
5. Every vertex of G'' is contained in a strong clique.
6. Every clique from the following collection of cliques in G'' is strong:

$$\{C\} \cup \{\{w, w'\} \mid w \in V(G') \setminus C\}.$$

7. The vertex set of G'' can be partitioned into strong cliques.

Using Proposition 17 we derive the following hardness results.

Theorem 5. *When restricted to the class of diamond-free graphs, the* STRONG CLIQUE, STRONG CLIQUE EXISTENCE, STRONG CLIQUE VERTEX COVER, *and* STRONG CLIQUE PARTITION *problems are* co-NP-*complete, and the* STRONG CLIQUE PARTITION EXISTENCE *problem is* co-NP-*hard.*

6 Conclusion

We established the complexity of seven problems related to strong cliques in the class of diamond-free graphs. Five of these problems remain intractable and the remaining two become solvable in linear time. Our work refines the boundaries of known areas of tractability and intractability of algorithmic problems related to strong cliques in graphs. Besides the open problems of the complexity of testing whether every maximal clique is strong, or whether every edge is contained in a strong clique, many other interesting questions remain. For example, it is still open whether there exists a polynomial-time algorithm to recognize the class of strongly perfect graphs, introduced in 1984 by Berge and Duchet [8] and defined as graphs in which every induced subgraph has a strong stable set. To the best of our knowledge, the recognition complexity of strongly perfect graphs is also open when restricted to the class of diamond-free graphs.

Acknowledgements. The authors are grateful to Ademir Hujdurović for helpful discussions and the anonymous reviewers for constructive remarks. The second named author has been supported by the European Union's Horizon 2020 research and innovation programme under the Marie Sklodowska-Curie grant agreement No. 734922, by Mexico's CONACYT scholarship 254379/438356, by Erasmus+ for practices (SMT) Action KA103 - Project 2017/2019, by MICINN from the Spanish Government under project PGC2018-095471-B-I00, and by AGAUR from the Catalan Government under project 2017SGR1087. The first and third named authors are supported in part by the Slovenian Research Agency (I0-0035, research programs P1-0285 and P1-0404, and research projects J1-9110, N1-0102, N1-0160). Part of this work was done while the third named author was visiting LAMSADE, University Paris-Dauphine; their support and hospitality is gratefully acknowledged.

References

1. Alcón, L., Gutierrez, M., Kovács, I., Milanič, M., Rizzi, R.: Strong cliques and equistability of EPT graphs. Discret. Appl. Math. **203**, 13–25 (2016)
2. Alcón, L., Gutierrez, M., Milanič, M.: A characterization of claw-free CIS graphs and new results on the order of CIS graphs. Electron. Notes Theoret. Comput. Sci. **346**, 15–27 (2019)
3. Alekseev, V.E.: On the number of maximal independent sets in graphs from hereditary classes. In: Combinatorial-Algebraic Methods in Discrete Optimization, pp. 5–8. University of Nizhny Novgorod (1991). (in Russian)
4. Andrade, D.V., Boros, E., Gurvich, V.: Not complementary connected and not CIS d-graphs form weakly monotone families. Discret. Math. **310**(5), 1089–1096 (2010)
5. Andrade, D.V., Boros, E., Gurvich, V.: On graphs whose maximal cliques and stable sets intersect. In: Goldengorin, B. (ed.) Optimization Problems in Graph Theory. SOIA, vol. 139, pp. 3–63. Springer, Cham (2018). https://doi.org/10.1007/978-3-319-94830-0_2
6. Balas, E., Yu, C.S.: On graphs with polynomially solvable maximum-weight clique problem. Networks **19**(2), 247–253 (1989)
7. Berge, C., Chvátal, V. (eds.): Topics on Perfect Graphs. North-Holland Mathematics Studies, vol. 88. North-Holland Publishing Co., Amsterdam (1984)
8. Berge, C., Duchet, P.: Strongly perfect graphs. In: Topics on Perfect Graphs. North-Holland Mathematics Studies, vol. 88, pp. 57–61. North-Holland, Amsterdam (1984)
9. Boros, E., Gurvich, V., Milanič, M.: On CIS circulants. Discrete Math. **318**, 78–95 (2014)
10. Boros, E., Gurvich, V., Milanič, M.: On equistable, split, CIS, and related classes of graphs. Discret. Appl. Math. **216**(part 1), 47–66 (2017)
11. Brandstädt, A., Giakoumakis, V., Maffray, F.: Clique separator decomposition of hole-free and diamond-free graphs and algorithmic consequences. Discret. Appl. Math. **160**(4–5), 471–478 (2012)
12. Cheston, G.A., Hare, E.O., Hedetniemi, S.T., Laskar, R.C.: Simplicial graphs. In: 1988 19th Southeastern Conference on Combinatorics, Graph Theory, and Computing, Baton Rouge, LA, vol. 67, pp. 105–113 (1988)
13. Cheston, G.A., Jap, T.S.: A survey of the algorithmic properties of simplicial, upper bound and middle graphs. J. Graph Algorithms Appl. **10**(2), 159–190 (2006)
14. Chudnovsky, M.: The Erdös-Hajnal conjecture-a survey. J. Graph Theor. **75**(2), 178–190 (2014)
15. Chudnovsky, M., Goedgebeur, J., Schaudt, O., Zhong, M.: Obstructions for three-coloring graphs without induced paths on six vertices. J. Combin. Theory Ser. B **140**, 45–83 (2020)
16. Chvátal, V., Slater, P.J.: A note on well-covered graphs. In: Quo Vadis, Graph Theory? Annals of Discrete Mathematics, vol. 55, pp. 179–181. North-Holland, Amsterdam (1993)
17. Dabrowski, K.K., Dross, F., Paulusma, D.: Colouring diamond-free graphs. J. Comput. Syst. Sci. **89**, 410–431 (2017)
18. Deng, X., Li, G., Zang, W.: Proof of Chvátal's conjecture on maximal stable sets and maximal cliques in graphs. J. Combin. Theor. Ser. B **91**(2), 301–325 (2004)
19. Dobson, E., Hujdurović, A., Milanič, M., Verret, G.: Vertex-transitive CIS graphs. Eur. J. Combin. **44**(part A), 87–98 (2015)

20. Erdős, P., Hajnal, A.: Ramsey-type theorems. Discret. Appl. Math. **25**(1–2), 37–52 (1989). Combinatorics and complexity (Chicago, IL, 1987)
21. Gurvich, V.: On exact blockers and anti-blockers, Δ-conjecture, and related problems. Discret. Appl. Math. **159**(5), 311–321 (2011)
22. Gyárfás, A.: Reflections on a problem of Erdős and Hajnal. In: Graham, R.L., Nesetril, J. (eds.) The Mathematics of Paul Erdős, II. Algorithms and Combinatorics, vol. 14, pp. 93–98. Springer, New York (1997). https://doi.org/10.1007/978-1-4614-7254-4_11
23. Hoàng, C.T.: Efficient algorithms for minimum weighted colouring of some classes of perfect graphs. Discret. Appl. Math. **55**(2), 133–143 (1994)
24. Hujdurović, A., Milanič, M., Ries, B.: Graphs vertex-partitionable into strong cliques. Discret. Math. **341**(5), 1392–1405 (2018)
25. Hujdurović, A., Milanič, M., Ries, B.: Detecting strong cliques. Discret. Math. **342**(9), 2738–2750 (2019)
26. Kamiński, M., Lozin, V.: Coloring edges and vertices of graphs without short or long cycles. Contrib. Discret. Math. **2**(1), 61–66 (2007)
27. Karthick, T., Mishra, S.: On the chromatic number of $(P_6$, diamond)-free graphs. Graphs Combin. **34**(4), 677–692 (2018)
28. Kloks, T., Lee, C.-M., Liu, J., Müller, H.: On the recognition of general partition graphs. In: Bodlaender, H.L. (ed.) WG 2003. LNCS, vol. 2880, pp. 273–283. Springer, Heidelberg (2003). https://doi.org/10.1007/978-3-540-39890-5_24
29. Kloks, T., Müller, H., Vušković, K.: Even-hole-free graphs that do not contain diamonds: a structure theorem and its consequences. J. Combin. Theor. Ser. B **99**(5), 733–800 (2009)
30. McAvaney, K., Robertson, J., DeTemple, D.: A characterization and hereditary properties for partition graphs. Discret. Math. **113**(1–3), 131–142 (1993)
31. Milanič, M., Trotignon, N.: Equistarable graphs and counterexamples to three conjectures on equistable graphs. J. Graph Theor. **84**(4), 536–551 (2017)
32. Prisner, E.: Graphs with few cliques. In: Alavi, Y., Schwenk, A. (eds.) Graph Theory, Combinatorics, and Algorithms: Proceedings of 7th Quadrennial International Conference on the Theory and Applications of Graphs, Western Michigan University, pp. 945–956. Wiley, New York (1995)
33. Sankaranarayana, R.S., Stewart, L.K.: Complexity results for well-covered graphs. Networks **22**(3), 247–262 (1992)
34. Skowrońska, M., Sysło, M.M.: An algorithm to recognize a middle graph. Discret. Appl. Math. **7**(2), 201–208 (1984)
35. Tsukiyama, S., Ide, M., Ariyoshi, H., Shirakawa, I.: A new algorithm for generating all the maximal independent sets. SIAM J. Comput. **6**(3), 505–517 (1977)
36. Tucker, A.: Coloring perfect $(K_4 - e)$-free graphs. J. Combin. Theor. Ser. B **42**, 313–318 (1987)
37. Zang, W.: Generalizations of Grillet's theorem on maximal stable sets and maximal cliques in graphs. Discret. Math. **143**(1–3), 259–268 (1995)
38. Zverovich, I., Zverovich, I.: Bipartite bihypergraphs: a survey and new results. Discret. Math. **306**(8–9), 801–811 (2006)

Recognizing k-Clique Extendible Orderings

Mathew Francis[1($\boxtimes$)], Rian Neogi[2], and Venkatesh Raman[2]

[1] Indian Statistical Institute, Chennai Centre, Chennai, India
`mathew@isichennai.res.in`
[2] The Institute of Mathematical Sciences, HBNI, Chennai, India
`rianneogi@gmail.com`, `vraman@imsc.res.in`

Abstract. We consider the complexity of recognizing k-clique-extendible graphs (k-C-E graphs) introduced by Spinrad (Efficient Graph Representations, AMS 2003), which are generalizations of comparability graphs. A graph is k-clique-extendible if there is an ordering of the vertices such that whenever two k-sized overlapping cliques A and B have $k-1$ common vertices, and these common vertices appear between the two vertices $a, b \in (A \setminus B) \cup (B \setminus A)$ in the ordering, there is an edge between a and b, implying that $A \cup B$ is a $(k+1)$-sized clique. Such an ordering is said to be a k-C-E ordering. These graphs arise in applications related to modelling preference relations. Recently, it has been shown that a maximum sized clique in such a graph can be found in $n^{O(k)}$ time [Hamburger et al. 2017] when the ordering is given. When k is 2, such graphs are precisely the well-known class of comparability graphs and when k is 3 they are called triangle-extendible graphs. It has been shown that triangle-extendible graphs appear as induced subgraphs of visibility graphs of simple polygons, and the complexity of recognizing them has been mentioned as an open problem in the literature.

While comparability graphs (i.e. 2-C-E graphs) can be recognized in polynomial time, we show that recognizing k-C-E graphs is NP-hard for any fixed $k \geq 3$ and CO-NP-hard when k is part of the input. While our NP-hardness reduction for $k \geq 4$ is from the betweenness problem, for $k = 3$, our reduction is an intricate one from the 3-colouring problem. We also show that the problems of determining whether a given ordering of the vertices of a graph is a k-C-E ordering, and that of finding an ℓ-sized (or maximum sized) clique in a k-C-E graph, given a k-C-E ordering, are complete for the parameterized complexity classes CO-W[1] and W[1] respectively, when parameterized by k. However we show that the former is fixed-parameter tractable when parameterized by the treewidth of the graph.

1 Introduction and Motivation

An undirected graph is a comparability (or transitively orientable) graph if the edges can be oriented in a way that for any three vertices u, v, w whenever there is a (directed) edge from u to v and an edge from v to w, there is an edge from u

© Springer Nature Switzerland AG 2020
I. Adler and H. Müller (Eds.): WG 2020, LNCS 12301, pp. 274–285, 2020.
https://doi.org/10.1007/978-3-030-60440-0_22

to w. They are a well-studied class of graphs [3,6] and they can be recognized in polynomial time [5]. Spinrad [8] generalized this class of graphs and introduced the notion of k-clique-extendible orderings (abbr. k-C-E ordering) on the vertices of a graph defined as follows.

Definition 1 (k-C-E ordering, Spinrad [8]). *An ordering ϕ of the vertices of a graph $G = (V, E)$ is a k-clique-extendible ordering (or k-C-E ordering) of G if, whenever X and Y are two overlapping cliques of size k such that $|X \cap Y| = k - 1$, $X \setminus Y = \{a\}$, $Y \setminus X = \{b\}$, and all the vertices in $X \cap Y$ occur between a and b in ϕ, we have $(a, b) \in E(G)$ and hence $X \cup Y$ is a $(k + 1)$-clique.*

A graph G is said to be k-*clique-extendible* (k-C-E for short) if there exists a k-clique-extendible ordering ϕ of G. It can be observed that comparability graphs are exactly the 2-clique-extendible graphs. Spinrad [8] observed that 3-clique-extendible graphs, also called triangle-extendible graphs, arise in the visibility graphs of simple polygons and that a maximum clique can be found in polynomial time in such graphs if a 3-clique-extendible ordering is given. This result has been generalized to obtain an $n^{O(k)}$ algorithm for finding a maximum clique in k-C-E graphs (given with a k-C-E ordering) on n vertices [4]. The question of whether there is a polynomial time algorithm to recognise 3-C-E graphs has been mentioned as an open problem [8].

We believe that k-C-E graphs are natural generalizations of comparability graphs and our main contribution in this paper is a serious study of this class of graphs. Our results show that recognizing k-C-E graphs is NP-hard for any fixed $k \geq 3$ and also CO-NP-hard when k is part of the input. This solves the open problem regarding the complexity of recognizing 3-C-E graphs and we hope that our results will trigger further study of k-C-E graphs in general.

If an ordering of the vertices is given, then it is easy to get an $n^{O(k)}$ algorithm to determine whether it is a k-C-E ordering of the graph (see Sect. 4). We show that this problem is CO-NP-complete and also complete for the parameterized complexity class CO-W[1]. The reduction also implies that unless the Exponential Time Hypothesis fails, this problem does not have an $f(k)n^{o(k)}$ algorithm for any function f of k. However, we show that the problem is fixed-parameter tractable when parameterized by the treewidth of the graph, that is, there is an $f(tw)n^{O(1)}$ algorithm for the problem, where tw is the treewidth of the graph.

Organization of the Paper. In the next section, we give the necessary notation and definitions. In Sect. 3, we prove some results about k-C-E graphs which are used in our reductions in later sections. In Sect. 4, we show that the problem of checking whether a given ordering is a k-C-E ordering is CO-NP-complete and CO-W[1]-complete. In this section, we also show that the problem is fixed-parameter tractable when parameterized by the treewidth of the graph. In Sect. 5 we show that the $n^{O(k)}$ algorithm for finding maximum clique in a k-C-E graph [4] is likely optimal. Sect. 6 gives our main NP-hardness reductions for the problem of recognizing k-C-E graphs. We give two reductions, one for $k = 3$ and another for $k \geq 4$. We list some open problems in Sect. 7.

2   Preliminaries

All graphs considered in this paper are undirected and simple. Given a graph G, by $V(G)$ we denote the set of vertices in the graph and by $E(G)$ we denote the set of edges in the graph. Let G be a graph. For a subset of vertices $S \subseteq V(G)$, we define $G[S]$ as the induced subgraph of G having vertex set S.

Given a linear order ϕ of a set A, we write $a <_\phi b$ to mean that a and b are two elements of A such that a occurs before b in ϕ. Also, we write $\phi = (a_1, a_2, \ldots, a_n)$ to mean that $A = \{a_1, a_2, \ldots, a_n\}$ and $a_1 <_\phi a_2 <_\phi \cdots <_\phi a_n$. We say that a vertex b comes between vertices a and c in ϕ if $a <_\phi b <_\phi c$ or $c <_\phi b <_\phi a$. By ϕ^{-1} we denote the reverse of ϕ, that is, $a <_{\phi^{-1}} b$ if and only if $b <_\phi a$.

Given an ordering ϕ of a set V and a set $S \subseteq V$, we define $\phi|_S$ to be the ordering of the elements of S in the order in which they occur in ϕ. Further, we say that $a, b \in S$ are the endpoints of S if a is the first element of $\phi|_S$ and b is the last element of $\phi|_S$. Given two disjoint sets A and B, and orderings $\phi_1 = (a_1, a_2, \ldots, a_n)$ of the set A and $\phi_2 = (b_1, b_2, \ldots, b_m)$ of the set B, we define $\phi_1 + \phi_2 = (a_1, a_2, \ldots, a_n, b_1, b_2, \ldots, b_m)$ that is an ordering on the set $A \cup B$, that is, $+$ is the concatenation operator on orderings. We will abuse notation to allow sets to be used with the concatenation operator: if γ is an expression that is a concatenation of orderings and sets, we say that an ordering ϕ is of the form γ, if there exists an ordering for each set appearing in γ such that replacing each set with its corresponding ordering in γ yields the ordering ϕ.

A clique in a graph is a set of vertices that are pairwise adjacent in the graph. An independent set is a set of vertices that are pairwise non-adjacent. Given subsets $S, A, B \subseteq V(G)$, we say that S separates A and B if there is no path from A to B in $G[V(G) \setminus S]$. For a pair u, v of nonadjacent vertices of a graph, by identifying u with v, we mean adding the edges (u, w) for all $w \in N(v) \setminus N(u)$ and then deleting v.

We denote by K_n^- the graph obtained by removing an edge from the complete graph K_n on n vertices. Given an ordering ϕ of the vertices of a graph G, we say that an induced subgraph H of G is an *ordered* K_t^- in ϕ if $\phi|_{V(H)} = (h_1, h_2, \ldots, h_t)$ and $E(H) = \{(h_i, h_j) \mid 1 \leq i < j \leq t\} \setminus \{(h_1, h_t)\}$. It follows that an ordering of the vertices of a graph is a k-C-E ordering if and only if it contains no ordered K_{k+1}^-. We refer the reader to the full version of this paper [2] for a definition of fixed-parameter tractability and the notion of parameterized reductions, treewidth and the Exponential Time Hypothesis.

3   Basic Results

We start with the following observations which are used throughout the paper.

Observation 1. *An ordering $\phi = \{v_1, v_2, \ldots, v_n\}$ is a k-C-E ordering, if and only if its reverse ordering, $\phi^{-1} = \{v_n, v_{n-1}, \ldots, v_1\}$ is also a k-C-E ordering.*

Observation 2. *Given a graph G and an induced subgraph H of G, if an ordering ϕ is a k-C-E ordering of G, then $\phi|_{V(H)}$ is a k-C-E ordering of H. Thus every induced subgraph of a k-clique-extendible graph is also k-clique-extendible.*

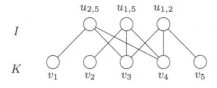

Fig. 1. Diagram depicting F_3. Edges in the clique are not shown, and only 3 of the $u_{i,j}$ vertices are shown to avoid visual clutter.

Observation 3. *If G is a k-colourable graph with colour classes $V_1, \ldots, V_k$, then any ordering ϕ of $V(G)$ of the form $V_1 + V_2 + \cdots + V_k$ is a k-C-E ordering of G. Thus, every k-colourable graph is k-clique-extendible.*

It is not difficult to see that any k-clique-extendible ordering of a graph is also a $(k+1)$-clique-extendible ordering of it. Thus, every k-clique-extendible graph is also a $(k+1)$-clique-extendible graph. Note that every graph on n vertices is trivially n-clique-extendible. So the notion of k-clique-extendibility gives rise to a hierarchy of graph classes starting with comparability graphs and ending with the entire set of graphs. This motivates the use of k as a graph parameter.

We prove a lemma that will help us construct a k-C-E ordering of a graph from k-C-E orderings of its subgraphs.

Lemma 1 ($\star$).[1] *For a graph G, let $V_1, V_2 \subseteq V(G)$ and let σ_1, σ_2 be k-C-E orderings of $G[V_1]$ and $G[V_2]$ respectively for any $k \geq 2$, such that (1.) $V_1 \cup V_2 = V(G)$, (2.) $V_1 \cap V_2$ separates V_1 and V_2, (3.) $\sigma_1|_{V_1 \cap V_2} = \sigma_2|_{V_1 \cap V_2}$ and (4.) if C is a $(k-1)$-clique in $V_1 \cap V_2$ and u, v are the endpoints of C in σ_1, then every vertex $a \in V_1 \setminus V_2$ that is adjacent to all of C satisfies $u <_{\sigma_1} a <_{\sigma_1} v$. Then G has a k-C-E ordering ϕ such that $\phi|_{V_1} = \sigma_1$ and $\phi|_{V_2} = \sigma_2$.*

Forbidden Subgraph. We construct a forbidden subgraph for the class of k clique extendible graphs which is used to build gadgets in our NP-hard reductions.

For a positive integer k, let $K = \{v_1, v_2, \ldots, v_{2k-1}\}$ be a $(2k-1)$ sized clique. For every pair of vertices v_i and v_j in K, add a vertex $u_{i,j}$ such that $u_{i,j}$ is adjacent to every vertex in K except v_i and v_j. Let $I = \{u_{i,j} \mid i, j \in [2k-1], i < j\}$ be the set of all such $u_{i,j}$ for every pair of vertices in K. Let F_k be the graph thus obtained having vertex set $K \cup I$. See Fig. 1 for an example that demonstrates the adjacencies between I and K when $k = 3$.

Lemma 2 ($\star$). *F_k is not k-clique-extendible.*

4   Verifying a k-C-E Ordering

In this section, we prove that even verifying whether an ordering is a k-clique-extendible ordering is hard (assuming k is considered as part of the input, rather than a constant).

---

[1] Refer to the full version of this paper [2] for the proofs of results marked with a $\star$.

VERIFY k-C-E ORDERING
Input: Graph G, integer k and an ordering ϕ of $V(G)$
Question: Is ϕ a k-C-E ordering of G?

VERIFY k-C-E ORDERING has a simple $n^{O(k)}$ algorithm as one can enumerate all $\binom{n}{k+1}$ subgraphs isomorphic to K^-_{k+1}, and check if any of them are ordered with respect to the ordering. We prove that the problem is CO-W[1]-complete and CO-NP-complete by a reduction from and to the CLIQUE problem, and that the problem also cannot have an $f(k)n^{o(k)}$ algorithm assuming ETH. The reduction maps the YES instances of VERIFY k-C-E ORDERING to the NO instances of CLIQUE and vice-versa. Hence showing that VERIFY k-C-E ORDERING is CO-NP-complete.

Theorem 1 ($\star$). VERIFY k-C-E ORDERING *is* CO-W[1]-*complete,* CO-NP-*complete and there is no* $f(k)n^{o(k)}$ *algorithm for it unless* ETH *fails.*

If all the k-cliques in a graph can be enumerated in time $O(f(k)poly(n))$ for some function f, then we can verify if an ordering is a k-C-E ordering in $O(f(k)poly(n))$ time by checking every pair of such cliques to see if they form an ordered K^-_{k+1}. We show that a similar situation happens if G has bounded treewidth and so the verification problem becomes easy.

Theorem 2 ($\star$). *Given an ordering of the vertices of a graph G on n vertices, we can verify whether it is a k-C-E ordering of G in time* $O(tw^{O(tw)}poly(n))$, *where* tw *is the treewidth of G.*

5   Hardness of Finding Clique

There exists an $n^{O(k)}$ algorithm for finding a maximum clique in a k-C-E graph [4] when a k-C-E ordering is given. In this section, we will prove that this is most likely optimal, that is, we prove that unless ETH fails, there is no $f(k)n^{o(k)}$ algorithm for finding a maximum clique in a k-C-E graph *even if the ordering is given.* We will reduce from the following problem.

MULTICOLOURED CLIQUE
Input: Graph G, a partition $V_1, \ldots, V_k$ of $V(G)$
Question: Does there exist a k-clique C in G such that $|C \cap V_i| = 1$ for each $i \in [k]$?

MULTICOLOURED CLIQUE is W[1]-hard and cannot be solved in time $f(k)n^{o(k)}$ unless ETH fails [1]. Given an instance $G, V_1, \ldots, V_k$ of MULTICOLOURED CLIQUE, we will first remove all edges that lie within each partition V_i. Hence the graph is now k-colourable with colour classes $V_1, \ldots, V_k$. Any k-colourable graph is also a k-C-E graph by Observation 3, and we can use an ordering ϕ of the form $V_1 + V_2 + \cdots + V_k$ to find the maximum clique size of G using an algorithm to find maximum clique in a k-C-E graph. If the clique size is equal to k, we output yes, otherwise output no. The following theorem follows.

Theorem 3. *Finding a maximum clique in a k-C-E graph, even if given a k-C-E ordering of the graph, is* NP-*hard,* W[1]-*hard and cannot be solved in time $f(k)n^{o(k)}$ unless* ETH *fails.*

6 Finding a k-C-E Ordering

In this section, we consider the following problem and prove the main result of the paper.

FIND k-C-E ORDERING
Input: Graph G, integer k
Question: Is G a k-C-E graph?

Note that this is possibly a harder problem than VERIFY k-C-E ORDERING, but still Theorem 1 doesn't immediately imply even CO-W[1]-hardness for this problem, as one may be able to determine whether G has a k-C-E ordering without even verifying an ordering. Our main result in this section is to show that FIND k-C-E ORDERING is NP-hard for each $k \geq 3$. First we will show that FIND k-C-E ORDERING is CO-W[1]-hard and CO-NP-hard. This result rules out algorithms running in time $f(k)n^{o(k)}$ assuming ETH (where as the NP-hardness rules out even $n^{f(k)}$ algorithms assuming P$\neq$NP).

Theorem 4 ($\star$). FIND k-C-E ORDERING *is* CO-W[1]-*hard and* CO-NP-*hard.*

6.1 NP-Hardness for $k \geq 4$

We now prove the NP-hardness of FIND k-C-E ORDERING by a reduction from BETWEENNESS defined below. The reduction strategy works for all $k \geq 4$ but not for $k = 3$ and so we give a different reduction for $k = 3$ in the next section.

BETWEENNESS
Input: Universe U of size n, and a set of triples $\mathcal{T} = \{t_1, \ldots, t_m\}$ where each $t_i = (a_i, b_i, c_i)$ is an ordered triple of elements in U
Question: Does there exist an ordering ϕ of U such that either $a_i <_\phi b_i <_\phi c_i$ or $c_i <_\phi b_i <_\phi a_i$ for each triple $(a_i, b_i, c_i) \in \mathcal{T}$?

BETWEENNESS is NP-hard [7]. To prove our reduction, we will require a gadget that takes as input a graph G and 3 vertices $x, y, z \in V(G)$ and converts them to a modified graph G' in such a way that either $x <_\phi y <_\phi z$ or $z <_\phi y <_\phi x$ for any k-C-E ordering ϕ of G'. Moreover, if ϕ is a k-C-E ordering of G such that $x <_\phi y <_\phi z$ or $z <_\phi y <_\phi x$ then ϕ is also a k-C-E of G'. Thus the gadget 'prunes' out the orderings of the graph where y does not lie between x and z in the ordering. The k-C-E orderings of G' are exactly the k-C-E orderings ϕ of G where either $x <_\phi y <_\phi z$ or $z <_\phi y <_\phi x$. Thus to construct the reduction, we will start with a graph where all $n!$ orderings are valid k-C-E orderings, and apply the gadget for each $(a_i, b_i, c_i) \in \mathcal{T}$. After applying the gadgets, we will

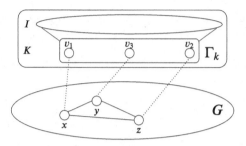

Fig. 2. The construction of the gadget. Dotted lines indicate vertices identified to each other.

have pruned out all the 'bad' orderings and we will remain with exactly the set of orderings in which b_i lies between a_i and c_i for each $i \in [m]$. To describe the construction of the gadget, first we need to define an auxiliary graph Γ_k.

Definition of the Auxiliary Graph. Recall the graph F_k, defined in Sect. 2 on the vertex set $K \cup I$ where $K = \{v_1, v_2, \ldots, v_{2k-1}\}$ induces a clique on $2k-1$ vertices, and every vertex in I is indexed by a pair of vertices of K to which the vertex is not adjacent. Pick arbitrary vertices v_1 and v_2 of K and let $u_{1,2}$ be the vertex of I that is adjacent to every vertex of K except v_1 and v_2. Define $\Gamma_k = F_k \setminus \{u_{1,2}\}$. Note that Γ_k has $O(k^2)$ many vertices.

Lemma 3 ($\star$). *In any k-C-E ordering ϕ of Γ_k, v_1 and v_2 are the endpoints of K. Furthermore, there exists a k-C-E ordering ϕ of Γ_k such that v_1 is the first element in ϕ and v_2 is the last.*

The Gadget. We will use Γ_k as a gadget to constrict the set of orderings a graph can have. Pick an arbitrary vertex $v_3 \in K$ such that $v_3 \neq v_1, v_2$. Given a graph G, applying the gadget on a triplet of vertices $x, y, z \in V(G)$ involves taking the disjoint union of G and Γ_k and identifying the vertices x with v_1, y with v_3 and z with v_2 (see Fig. 2). For technical reasons, we will only be applying the gadget on vertices x, y, z that induce a clique in G. Since Γ_k has $O(k^2)$ many vertices, the gadget will add $O(k^2)$ vertices to G, keeping it well within a polynomial factor. We use notation $G' = \mathcal{C}_k(G, x, y, z)$ to denote "G' is obtained by applying the gadget on G on vertices x, y, z". The valid k-C-E orderings of G' should exactly be the k-C-E orderings of G where y does not come between x and z. The following lemmas give us exactly that.

Lemma 4 ($\star$). *Let G be a graph and let $x, y, z \in V(G)$ be vertices of G. In any k-C-E ordering ϕ of $G' = \mathcal{C}_k(G, x, y, z)$, y comes between x and z.*

Lemma 5 ($\star$). *Let $k \geq 4$ and let G be a graph that has a k-C-E ordering ψ such that y comes between x and z for some three vertices $x, y, z \in V(G)$ that form a 3-clique in G, then $G' = \mathcal{C}_k(G, x, y, z)$ has a k-C-E ordering ϕ such that $\phi|_{V(G)} = \psi$.*

The Reduction. We are now ready to prove that the problem of checking whether a graph has a k-C-E ordering is NP-hard for each $k \geq 4$.

Theorem 5. FIND k-C-E ORDERING *is* NP-*hard for each* $k \geq 4$.

Proof. We will reduce from BETWEENNESS. Let $\mathcal{I} = (U, \mathcal{T})$ be the input BETWEENNESS instance. We want to construct a graph G' such that G' has a k-C-E ordering if and only if the BETWEENNESS instance is satisfiable. We will construct a graph with vertex set equal to the universe U and apply the gadget for every triple (a_i, b_i, c_i) in $\mathcal{T}$. We do this iteratively, that is, we first define G_0 to be the complete graph on vertex set U and then construct $G_i = \mathcal{C}_k(G_{i-1}, a_i, b_i, c_i)$ for each $i \in [m]$ (where m is the number of triples in $\mathcal{T}$). The final graph G' is equal to G_m. There are m many calls to the gadget and each gadget adds $O(k^2)$ vertices to G'. So the final size of G' is $O(n + mk^2)$ (where n is the size of U), which is polynomial in n and m.

Claim $(\star)$. G' has a k-C-E ordering if and only if $\mathcal{I}$ is a YES-instance.

The theorem follows from the above claim. $\square$

The above reduction shows that FIND k-C-E ORDERING is NP-hard. From Theorem 4, FIND k-C-E ORDERING is also CO-NP-hard. Thus it is unlikely that the problem is in NP or in CO-NP. Moreover, it is easy to verify that the problem lies in Σ_2, as one can simply guess the ordering ϕ and use a CO-NP machine (Theorem 1) to check whether ϕ is a k-C-E ordering. Thus it is an open question whether FIND k-C-E ORDERING is Σ_2-complete.

Remark (1). It is important to note that the problem is *not* CO-NP-hard when k is a fixed constant as opposed to it being given as a input. When k is fixed, the k-C-E ordering itself is an NP certificate for the problem, as given an ordering it is easy to check whether it is a k-C-E ordering for constant k. Thus, when k is constant, the problem is NP-complete. Indeed, the proof of CO-NP-hardness in Theorem 1 assumes that k is given as an input.

Remark (2). The reduction in Theorem 5 does not work for $k = 3$ due to technicalities that arise in order to satisfy the fourth condition of Lemma 1, due to which we require that $|V(G) \cap V(\Gamma_k)| \leq k - 1$ (see proof of Lemma 5). Since $|V(G) \cap V(\Gamma_k)| = |\{x, y, z\}| = 3$ in the gadgets we construct, this forces k to be at least 4. We give a separate proof for NP-hardness of $k = 3$ in the following section that uses some different ideas.

6.2   NP-Hardness for $k = 3$

In this section, we prove that the problem of finding a 3-C-E ordering is NP-hard. We will reduce from the 3-COLOURING problem. Given a graph G and an ordering ϕ of $V(G)$, we say that three edges $(u, v), (w, x), (y, z) \in E(G)$ form a *disjoint triple* in ϕ if $u <_\phi v \leq_\phi w <_\phi x \leq_\phi y <_\phi z$. Here $x \leq_\phi y$ means that either $x = y$ or $x <_\phi y$.

Observation 4 ($\star$). *Let G be a 3-colourable graph and let C_1, C_2, C_3 be a partition of $V(G)$ into three independent sets. Then any ordering ϕ of $V(G)$ of the form $C_1 + C_2 + C_3$ contains no disjoint triple.*

Observation 5 ($\star$). *Let G be any graph. If there is an ordering ϕ of $V(G)$ that contains no disjoint triple, then G is 3-colourable.*

It follows from Observations 4 and 5 that a graph G is 3-colourable if and only if there is an ordering of its vertex set containing no disjoint triple.

Another observation is that in any 3-C-E ordering ϕ of G, for any pair of non-adjacent vertices $u, v \in V(G)$, the vertices that are adjacent to both u and v and lie between u and v in ϕ must be an independent set in G. Indeed, if there is an edge (a, b) such that $u <_\phi a <_\phi b <_\phi v$ and a, b are adjacent to both u and v, then $G[\{u, a, b, v\}]$ is an ordered K_4^- in ϕ. This suggests a reduction from 3-COLOURING. The idea is that, associated to every edge $e = (u, v) \in E(G)$, we will add a vertex t_1^e, and a pair of adjacent vertices t_2^e and t_3^e. We will add edges so that the t_2 vertices and t_3 vertices together form a clique and the t_1 vertices form an independent set. We also add edges between all t_2, t_3 vertices and t_1 vertices. We will add a gadget to ensure that t_1^e, t_2^e, t_3^e all lie between u and v in any 3-C-E ordering of G'.

If G is not 3-colourable, then for any ordering ϕ of $V(G')$, there will be a disjoint triple in $\phi|_{V(G)}$. If the disjoint triple is formed by the edges $(u, v), (w, x), (y, z)$ of G, where $u <_\phi v \leq_\phi w <_\phi x \leq_\phi y <_\phi z$, then the vertices $t_1^{(u,v)}, t_2^{(w,x)}, t_3^{(w,x)}, t_1^{(y,z)}$ form an ordered K_4^- in ϕ, and hence there can be no 3-C-E ordering of G'. On the other hand, our construction makes sure that if G is a 3-colourable graph, then there exists a 3-C-E ordering for G'. We now describe the reduction in detail.

The Construction. Given a graph G, we construct a supergraph G' as explained below (also see Fig. 3). For subsets $A, B \subseteq V(G)$, by "join A and B", we mean that we add all possible edges between vertices in A and vertices in B. To construct the vertex set of G', we take the vertex set of G and add the following.

1. Add 4 sets of vertices $A = \{a, a_1, a_2, a_3\}$, $B = \{b, b_1, b_2, b_3\}$, $C = \{c, c_1, c_2, c_3\}$ and $D = \{d, d_1, d_2, d_3\}$
2. Add the sets of vertices $F = \{f_i^e \mid e \in E(G), i \in \{1, 2, \ldots, 6\}\}$ and $T = \{t_i^e \mid e \in E(G), i \in \{1, 2, 3\}\}$

To construct the edge set of G', we take the edge set of G and add the following.

1. Add edges to make A, B, C and D into cliques of size 4 each.
2. Add edges to make $t_i^e, f_{2i-1}^e, f_{2i}^e$ into a clique, for each edge $e \in E(G)$ and $i \in [3]$
3. Join $\{a_1, a_2, a_3\}$ and $\{b_1, b_2, b_3\}$
4. Join $\{b_1, b_2, b_3\}$ and $\{c_1, c_2, c_3\}$
5. Join $\{c_1, c_2, c_3\}$ and $\{d_1, d_2, d_3\}$

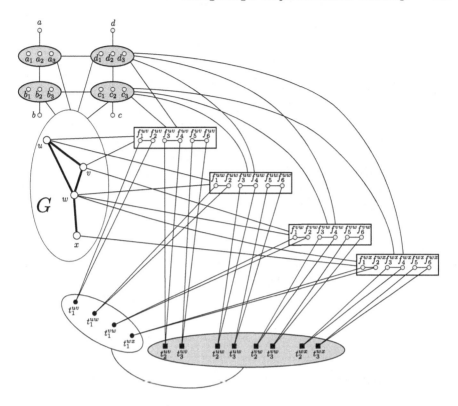

Fig. 3. The construction of G' from G. The vertices inside each shaded block form a clique. An edge between a vertex u and a block means that u is adjacent to every vertex in the block, and an edge between two blocks means that every vertex in one block is adjacent to every vertex in the other block. Note that an edge between vertices a and b is denoted as ab instead of (a, b) to reduce clutter.

6. Join $\{d_1, d_2, d_3\}$ and $\{a_1, a_2, a_3\}$
7. Join $\{a_1, a_2, a_3, b_1, b_2, b_3\}$ and $V(G)$
8. Join $\{c_1, c_2, c_3, d_1, d_2, d_3\}$ and $V(G) \cup F$
9. Add edges $(f_i^{(u,v)}, u)$ and $(f_i^{(u,v)}, v)$, for each $(u, v) \in E(G)$ and $i \in \{1, \ldots, 6\}$
10. Add edges to make $\bigcup_{e \in E(G)} \{t_2^e, t_3^e\}$ into a clique
11. Join $\bigcup_{e \in E(G)} t_1^e$ and $\bigcup_{e \in E(G)} \{t_2^e, t_3^e\}$

Lemma 6. *If G' has a 3-C-E ordering then G is 3-colourable.*

Proof. Suppose that ϕ is a 3-C-E ordering of G'. By Observation 5, we only need to show that there is no disjoint triple in the ordering $\phi|_{V(G)}$. We can assume without loss of generality that there exist distinct $i, j \in \{1, 2, 3\}$ such that in the ordering ϕ, we have $a_i <_\phi a_j <_\phi a$ (reversing the ordering ϕ if necessary; recall Observation 1). If there is a vertex $w \in V(G) \cup \{b_1, b_2, b_3, d_1, d_2, d_3\}$ such that $w <_\phi a_i$, then w, a_i, a_j, a form an ordered K_4^- in ϕ, which contradicts the fact that ϕ is a 3-C-E ordering. Therefore, we can assume without loss of generality that the vertex a_1 occurs before every vertex of $V(G) \cup \{b_1, b_2, b_3, d_1, d_2, d_3\}$ in the ordering ϕ. This also means that if there exist distinct $i, j \in \{1, 2, 3\}$ such that $b_i <_\phi b_j <_\phi b$, then a_1, b_i, b_j, b would form an ordered K_4^- in ϕ. Thus, we conclude that there exist distinct $i, j \in \{1, 2, 3\}$ such that $b <_\phi b_i <_\phi b_j$, and arguing as before, we assume without loss of generality that the vertex b_1 occurs after every vertex of $V(G) \cup \{a_1, a_2, a_3, c_1, c_2, c_3\}$ in the ordering ϕ. Now if there exist distinct $i, j \in \{1, 2, 3\}$ such that $c <_\phi c_i <_\phi c_j$, then c, c_i, c_j, b_1 form an ordered K_4^- in ϕ. Thus, there exist distinct $i, j \in \{1, 2, 3\}$ such that $c_i <_\phi c_j <_\phi c$, and reasoning as before, we can assume without loss of generality that c_1 occurs before every vertex in $V(G) \cup \{b_1, b_2, b_3, d_1, d_2, d_3\}$. Using similar arguments, we conclude that d_1 occurs after every vertex in $V(G) \cup \{a_1, a_2, a_3, c_1, c_2, c_3\}$ in ϕ.

Claim. For every edge $(u, v) \in E(G)$, all the vertices in $\{f_i^{(u,v)} : 1 \le i \le 6\}$ occur between u and v in ϕ.

Proof. Suppose that there exists $i \in \{1, 2, \ldots, 6\}$ such that $f_i^{(u,v)} <_\phi u <_\phi v$. Then the vertices $f_i^{(u,v)}, u, v, b_1$ form an ordered K_4^- in ϕ, which contradicts the fact that ϕ is a 3-C-E ordering. Similarly, if $u <_\phi v <_\phi f_i^{(u,v)}$ for some $i \in \{1, 2, \ldots, 6\}$, then $a_1, u, v, f_i^{(u,v)}$ form an ordered K_4^- in ϕ; again a contradiction. $\quad\square$

Claim. For every edge $e \in E(G)$ and $i \in \{1, 2, 3\}$, the vertex t_i^e occurs between f_{2i-1}^e and f_{2i}^e in ϕ.

Proof. Since in the ordering ϕ, c_1 occurs before every vertex in $V(G)$ and d_1 occurs after every vertex in $V(G)$, it follows from the above claim that c_1 occurs before every vertex in F and d_1 occurs after every vertex in F. Now suppose that for some $e \in E(G)$ and $i \in \{1, 2, 3\}$, we have $f_{2i-1}^e, f_{2i}^e <_\phi t_i^e$. Then the vertices $c_1, f_{2i-1}^e, f_{2i}^e, t_i^e$ form an ordered K_4^- in ϕ, which is a contradiction. Similarly, if $t_i^e <_\phi f_{2i-1}^e, f_{2i}^e$, then the vertices $t_i^e, f_{2i-1}^e, f_{2i}^e, d_1$ form an ordered K_4^- in ϕ, again a contradiction. $\quad\square$

From the above two claims, it follows that for any edge $(a, b) \in E(G)$ and $i \in \{1, 2, 3\}$, the vertex $t_i^{(a,b)}$ occurs between a and b in ϕ. Now suppose for the sake of contradiction that $(u, v), (w, x), (y, z) \in E(G)$ form a disjoint triple in $\phi|_{V(G)}$, where $u <_\phi v \le_\phi w <_\phi x \le_\phi y <_\phi z$. Then we have $u <_\phi t_1^{(u,v)} <_\phi v \le_\phi w <_\phi t_2^{(w,x)}, t_3^{(w,x)} <_\phi x \le_\phi y <_\phi t_1^{(y,z)} <_\phi z$. But then the vertices $t_1^{(u,v)}, t_2^{(w,x)}, t_3^{(w,x)}, t_1^{(y,z)}$ form an ordered K_4^- in ϕ, a contradiction. $\quad\square$

Lemma 7 ($\star$). *If G is 3-colourable then G' has a 3-C-E ordering.*

Lemma 6 and Lemma 7 prove the correctness of the reduction and thus we have the following theorem.

Theorem 6. FIND k-C-E ORDERING *is NP-hard for $k = 3$.*

7 Conclusion

We have shown that the problem of determining whether a given graph is a k-C-E is NP-hard for each $k \geq 3$ and CO-NP-hard for general k. Finding a maximum clique in a k-C-E graph on n vertices is known to have an $n^{O(k)}$ algorithm when a k-clique-extendible ordering is given, which we prove to be optimal. It is also an open problem mentioned before [8] whether we can find a maximum clique in a k-C-E graph in polynomial time for a fixed k, if given only the adjacency matrix of the graph. Finally, it would be interesting to know polynomial time solvable problems in k-C-E graphs, even for $k = 3$. As triangle free graphs and diamond-free graphs are 3-C-E graphs, we know that the independent set problem and the colouring problem are NP-hard in these classes of graphs.

It would also be interesting to study whether these graphs can be recognised approximately. There are two suitable notions for approximation. One is the following: An algorithm is said to be an α-factor approximation (for $\alpha \geq 1$) if, given a graph G and integer k, it either outputs a (αk)-C-E ordering or concludes that no k-C-E ordering exists for G. The second notion is the following: An algorithm is said to be a α-factor approximation (for $\alpha \leq 1$) if, given a graph G and integer k, outputs an ordering ϕ such that at most α fraction of the induced K_{k+1}^- in the graph are ordered in ϕ. Note that solving this problem for $\alpha = 0$ is equivalent to solving FIND k-C-E ORDERING.

For the second notion of approximation, there is an easy $(\frac{2}{k(k+1)})$-factor approximation. Simply output a random ordering of the vertices of G. The probability that any given induced K_{k+1}^- is ordered is $\frac{2}{k(k+1)}$. Thus by linearity of expectation, a $\frac{2}{k(k+1)}$ fraction of all the induced K_{k+1}^- in G will be ordered.

References

1. Cygan, M., et al.: Parameterized Algorithms. Springer, Cham (2015). https://doi.org/10.1007/978-3-319-21275-3
2. Francis, M., Neogi, R., Raman, V.: Recognizing k-clique extendible orderings. arXiv:2007.06060 (2020)
3. Golumbic, M.C.: Algorithmic Graph Theory and Perfect Graphs, vol. 57. Elsevier, Amsterdam (2004)
4. Hamburger, P., McConnell, R.M., Pór, A., Spinrad, J.P., Xu, Z.: Double threshold digraphs. In: 43rd International Symposium on Mathematical Foundations of Computer Science (MFCS 2018), vol. 117. Leibniz International Proceedings in Informatics (LIPIcs), pp. 69:1–69:12. Schloss Dagstuhl-Leibniz-Zentrum fuer Informatik (2018)
5. McConnell, R.M., Spinrad, J.P.: Linear-time transitive orientation. In: Proceedings of the Eighth Annual ACM-SIAM Symposium on Discrete Algorithms, pp. 19–25. Society for Industrial and Applied Mathematics (1997)
6. Möhring, R.H.: Algorithmic aspects of comparability graphs and interval graphs. In: Rival, I. (ed.) Graphs and Order. NATO ASI Series, pp. 41–101. Springer, Dordrecht (1985). https://doi.org/10.1007/978-94-009-5315-4_2
7. Opatrny, J.: Total ordering problem. SIAM J. Comput. 8(1), 111–114 (1979)
8. Spinrad, J.P.: Efficient Graph Representations. American Mathematical Society, Providence (2003)

Linear-Time Recognition
of Double-Threshold Graphs

Yusuke Kobayashi[1] , Yoshio Okamoto[2] , Yota Otachi[3]([⊠]) , and Yushi Uno[4]

[1] Kyoto University, Kyoto, Japan
yusuke@kurims.kyoto-u.ac.jp
[2] The University of Electro-Communications, Chofu, Tokyo, Japan
okamotoy@uec.ac.jp
[3] Nagoya University, Nagoya, Japan
otachi@i.nagoya-u.ac.jp
[4] Osaka Prefecture University, Osaka, Japan
uno@cs.osakafu-u.ac.jp

Abstract. A graph $G = (V, E)$ is a *double-threshold graph* if there exist a vertex-weight function $w \colon V \to \mathbb{R}$ and two real numbers $\mathtt{lb}, \mathtt{ub} \in \mathbb{R}$ such that $uv \in E$ if and only if $\mathtt{lb} \le \mathtt{w}(u) + \mathtt{w}(v) \le \mathtt{ub}$. In the literature, those graphs are studied as the pairwise compatibility graphs that have stars as their underlying trees. We give a new characterization of double-threshold graphs, which gives connections to bipartite permutation graphs. Using the new characterization, we present a linear-time algorithm for recognizing double-threshold graphs. Prior to our work, the fastest known algorithm by Xiao and Nagamochi [COCOON 2018] ran in $O(n^6)$ time, where n is the number of vertices.

Keywords: Double-threshold graph · Bipartite permutation graph · Recognition algorithm · Star pairwise compatibility graph

1 Introduction

A graph is a *threshold graph* if there exist a vertex-weight function and a real number called a weight lower bound such that two vertices are adjacent in the graph if and only if the associated vertex weight sum is at least the weight lower bound. Threshold graphs and their generalizations are well studied because of their beautiful structures and applications in many areas [5,12]. In particular, the edge-intersections of two threshold graphs, and their complements (i.e., the union of two threshold graphs) have attracted several researchers in the past, and recognition algorithms with running time $O(n^5)$ by Ma [11], $O(n^4)$ by Raschle

Partially supported by JSPS KAKENHI Grant Numbers JP16K16010, JP17K00017, JP18H04091, JP18H05291, JP18K11168, JP18K11169, and JST CREST Grant Number JPMJCR1402. The authors are grateful to Robert E. Jamison and Alan P. Sprague for sharing their manuscript [9]. The authors would also like to thank Martin Milanič, Gregory J. Puleo, and Vaidy Sivaraman for useful information.

© Springer Nature Switzerland AG 2020
I. Adler and H. Müller (Eds.): WG 2020, LNCS 12301, pp. 286–297, 2020.
https://doi.org/10.1007/978-3-030-60440-0_23

and Simon [15], and $O(n^3)$ by Sterbini and Raschle [17] have been developed, where n is the number of vertices.

In this paper, we study the class of double-threshold graphs, which is a proper generalization of threshold graphs and a proper specialization of the graphs that are edge-intersections of two threshold graphs. A graph is a *double-threshold graph* if there exist a vertex-weight function and two real numbers called lower and upper bounds of weight such that two vertices are adjacent if and only if their weight sum is at least the lower bound and at most the upper bound. To the best of our knowledge, this natural generalization of threshold graphs was not studied until quite recently. In 2018, Xiao and Nagamochi [18] studied this graph class under the different name of "star pairwise compatibility graphs" and presented an $O(n^6)$-time recognition algorithm.

Our main result is to give a new characterization of double-threshold graphs that gives a simple linear-time recognition algorithm. We first show that every double-threshold graph is a permutation graph (but not vice versa) and that a bipartite graph is a double-threshold graph if and only if it is a permutation graph. These facts imply that many NP-hard graph problems are polynomial-time solvable on (bipartite or non-bipartite) double-threshold graphs. We then show that a graph is a double-threshold graph if and only if an auxiliary graph constructed from the original graph is a bipartite permutation graph. This characterization gives a linear-time algorithm for recognizing double-threshold graphs.

Recently, we have realized that Jamison and Sprague [9] have independently showed that all double-threshold graphs are permutation graphs and that all bipartite permutation graphs are double-threshold graphs. Their proofs are based on a vertex-ordering characterization of permutation graphs and a BFS structure of bipartite permutation graphs, while ours are direct transformations between vertex weights and permutation diagrams. Note that in their paper Jamison and Sprague [9] used the term *bi-threshold graphs* instead of double-threshold graphs. However, the name of "bi-threshold graphs" is already used[1] by Hammer and Mahadev [7] for a different generalization of threshold graphs (see below). Thus, even though "bi-threshold" would sound better and probably more appropriate, we would like to keep our term "double-threshold" in this paper.

Other Generalizations of Threshold Graphs. There are many other generalizations of threshold graphs such as threshold signed graphs [3], threshold tolerance graphs [14], quasi-threshold graphs (also known as trivially perfect graphs) [19], weakly threshold graphs [1], paired threshold graphs [16], and mock threshold graphs [2]. We omit the definitions of these graph classes and only note that some small graphs show that these classes are incomparable to the class of double-threshold graphs (e.g., $3K_2$ and bull for threshold signed graphs, $2K_2$ and bull for threshold tolerance graphs, C_4 and $2K_3$ for quasi-threshold graphs, $2K_2$ and bull

---

[1] Strictly speaking, the names are not exactly the same. One is written with a hyphen, but the other is written without a hyphen.

for weakly threshold graphs, C_4 and bull for paired threshold graphs, $K_3 \cup C_4$ and bull for mock threshold graphs[2]). For the class of bithreshold graphs introduced by Hammer and Mahadev [7] (not the one introduced by Jamison and Sprague [9]), we can use $3K_2$ and bull to show that this class is incomparable to the class of double-threshold graphs.

Note that the concept of double-threshold digraphs introduced by Hamburger et al. [6] is concerned with directed acyclic graphs defined from a generalization of semiorders involving two thresholds and not related to threshold graphs or double-threshold graphs.

Pairwise Compatibility Graphs. Motivated by uniform sampling from phylogenetic trees in bioinformatics, Kearney, Munro, and Phillips [10] defined pairwise compatibility graphs. A graph $G = (V, E)$ is a *pairwise compatibility graph* if there exists a quadruple $(T, \mathtt{w}, \mathtt{lb}, \mathtt{ub})$, where T is a tree, $\mathtt{w} \colon E(T) \to \mathbb{R}$, and $\mathtt{lb}, \mathtt{ub} \in \mathbb{R}$, such that the set of leaves in T coincides with V and $uv \in E$ if and only if the (weighted) distance $d_T(u, v)$ between u and v in T satisfies $\mathtt{lb} \leq d_T(u, v) \leq \mathtt{ub}$.

Since its introduction, several authors have studied properties of pairwise compatibility graphs, but the existence of a polynomial-time recognition algorithm for that graph class has been open. The survey article by Calamoneri and Sinaimeri [4] proposed to look at the class of pairwise compatibility graphs defined on stars (i.e., star pairwise compatibility graphs), and asked for a characterization of star pairwise compatibility graphs. Recently, Xiao and Nagamochi [18] solved the open problem and gave an $O(n^6)$-time algorithm to recognize a star pairwise compatibility graph.

As we will see after the formal definition of double-threshold graphs, the star pairwise compatibility graphs are precisely the double-threshold graphs. Although the pairwise compatibility graphs are rather well studied, we study the double-threshold graphs in the context of threshold graphs and their generalizations. This is because, in our opinion, the double-threshold graphs and techniques used for them are more relevant in that context.

2   Preliminaries

All graphs in this paper are undirected, simple, and finite. A graph G is given by the pair of its vertex set V and its edge set E as $G = (V, E)$. The vertex set and the edge set of G are often denoted by $V(G)$ and $E(G)$, respectively. The *order* of a graph refers to the number of its vertices. For a vertex v in a graph $G = (V, E)$, its *neighborhood* is the set of vertices that are adjacent to v, and denoted by $N_G(v) = \{u \mid uv \in E\}$. When the graph G is clear from the context, we often omit the subscript. A linear ordering $\prec$ on a set S with

---

[2] K_n and C_n denote the complete graph and the cycle of n vertices, respectively. The disjoint union of two graphs G and H is denoted by $G \cup H$. For a graph G and an positive integer k, kG is the disjoint union of k copies of G. The graph *bull* is a five-vertex path with an additional edge connecting the 2nd and 4th vertices.

$|S| = n$ can be represented by a sequence $\langle s_1, s_2, \ldots, s_n \rangle$ of the elements in S, in which $s_i \prec s_j$ if and only if $i < j$. By abusing the notation, we sometimes write $\prec = \langle s_1, s_2, \ldots, s_n \rangle$.

2.1 Double-Threshold Graphs

A graph $G = (V, E)$ is a *threshold graph* if there exist a vertex-weight function $\mathtt{w} \colon V \to \mathbb{R}$ and a real number $\mathtt{lb} \in \mathbb{R}$ such that $uv \in E$ if and only if $\mathtt{lb} \leq \mathtt{w}(u) + \mathtt{w}(v)$. A graph $G = (V, E)$ is a *double-threshold graph* if there exist a vertex-weight function $\mathtt{w} \colon V \to \mathbb{R}$ and two real numbers $\mathtt{lb}, \mathtt{ub} \in \mathbb{R}$ with the following property: $uv \in E \iff \mathtt{lb} \leq \mathtt{w}(u) + \mathtt{w}(v) \leq \mathtt{ub}$. Then, we say that the double-threshold graph G is *defined* by $\mathtt{w}$, $\mathtt{lb}$ and $\mathtt{ub}$.

The definition of a double-threshold graph can be understood visually in the plane, by its so called *slab representation*. See Fig. 1 for an example. In the xy-plane, we consider the slab defined by $\{(x, y) \mid \mathtt{lb} \leq x + y \leq \mathtt{ub}\}$ that is illustrated in gray. Then, two vertices $u, v \in V$ are joined by an edge if and only if the point $(\mathtt{w}(u), \mathtt{w}(v))$ lies in the slab.

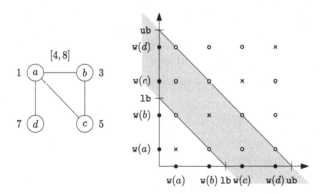

Fig. 1. (Left) A double-threshold graph. The weight of each vertex is given as $\mathtt{w}(a) = 1$, $\mathtt{w}(b) = 3$, $\mathtt{w}(c) = 5$, and $\mathtt{w}(d) = 7$; the lower bound is $\mathtt{lb} = 4$ and the upper bound is $\mathtt{ub} = 8$. (Right) The slab representation of the graph. A white dot represents the point $(\mathtt{w}(u), \mathtt{w}(v))$ for distinct vertices u, v, and a cross represents the point $(\mathtt{w}(v), \mathtt{w}(v))$ for a vertex v. Two distinct vertices u and v are joined by an edge if and only if the corresponding white dot lies in the gray slab.

Every threshold graph is a double-threshold graph as one can set a dummy upper bound $\mathtt{ub} > 2 \cdot \max\{\mathtt{w}(v) \mid v \in V\}$. We can easily see that the double-threshold graphs coincide with the star pairwise compatibility graphs ($\bigstar^3$).

The *threshold dimension* of a graph $G = (V, E)$ is the minimum integer k such that there are k threshold graphs $G_i = (V, E_i)$, $1 \leq i \leq k$, with $E = \bigcup_{1 \leq i \leq k} E_i$. A graph $G = (V, E)$ has *co-threshold dimension* k if its complement $\overline{G}$ has

---

[3] A star $\bigstar$ means that the proof is omitted. The ommited proofs can be found in the arXiv version https://arxiv.org/abs/1909.09371.

threshold dimension k. Since the class of threshold graphs is closed under taking complements [12], the co-threshold dimension of $G = (V, E)$ is the minimum integer k such that there are k threshold graphs $G_i = (V, E_i)$, $1 \leq i \leq k$, with $E = \bigcap_{1 \leq i \leq k} E_i$. Every double-threshold graph has co-threshold dimension at most 2 (★).

The next lemma allows us to use any values as lb and ub for defining a double-threshold graph. It also says that we do not have to consider degenerated cases, where some vertices have the same weight or some weight sum equals to the lower or upper bound.

Lemma 2.1 (★). *Let* $G = (V, E)$ *be a double-threshold graph defined by* $\mathtt{w}: V \to \mathbb{R}$ *and* $\mathtt{lb}, \mathtt{ub} \in \mathbb{R}$. *For every pair* $(\mathtt{lb}^*, \mathtt{ub}^*) \in \mathbb{R}^2$ *with* $\mathtt{lb}^* < \mathtt{ub}^*$, *there exists* $\mathtt{w}^*: V \to \mathbb{R}$ *defining* G *with* $\mathtt{lb}^*$ *and* $\mathtt{ub}^*$ *such that*

1. $\{v \mid \mathtt{lb}/2 \leq \mathtt{w}(v) \leq \mathtt{ub}/2\} = \{v \mid \mathtt{lb}^*/2 \leq \mathtt{w}^*(v) \leq \mathtt{ub}^*/2\}$,
2. $\mathtt{w}^*(u) + \mathtt{w}^*(v) \notin \{\mathtt{lb}^*, \mathtt{ub}^*\}$ *for all* $(u, v) \in V^2$, *and*
3. $\mathtt{w}^*(u) \neq \mathtt{w}^*(v)$ *if* $u \neq v$.

The following fact shown by Xiao and Nagamochi [18] allows us to consider bipartite components separately.

Lemma 2.2 ([18]). *A graph is a double-threshold graph if and only if it contains at most one non-bipartite component and all components are double-threshold graphs.*

2.2 Permutation Graphs

A graph $G = (V, E)$ is a *permutation graph* if there exist linear orderings $\prec_1$ and $\prec_2$ on V with the following property:

$$uv \in E \iff (u \prec_1 v \text{ and } v \prec_2 u) \text{ or } (u \prec_2 v \text{ and } v \prec_1 u).$$

We say that $\prec_1$ and $\prec_2$ *define* the permutation graph G. We call $\prec_1$ a *permutation ordering* of G if there exists a linear ordering $\prec_2$ satisfying the condition above. Since $\prec_1$ and $\prec_2$ play a symmetric role in the definition, $\prec_2$ is also a permutation ordering of G. Note that for a graph G and a permutation ordering $\prec_1$ of G, the other ordering $\prec_2$ that defines G together with $\prec_1$ is uniquely determined. Also note that if $\prec_1$ and $\prec_2$ define G, then $\prec_1^{\mathrm{R}}$ and $\prec_2^{\mathrm{R}}$ also define G, where $\prec^{\mathrm{R}}$ denotes the reversed ordering of $\prec$.

We often represent a permutation graph with a *permutation diagram*, which is drawn as follows. Imagine two horizontal parallel lines ℓ_1 and ℓ_2 on the plane. Then, we place the vertices in V on ℓ_1 from left to right according to the permutation ordering $\prec_1$ as distinct points, and similarly place the vertices in V on ℓ_2 from left to right according to $\prec_2$ as distinct points. The positions of $v \in V$ can be represented by x-coordinates on ℓ_1 and ℓ_2, which are denoted by $\mathtt{x}_1(v)$ and $\mathtt{x}_2(v)$, respectively. We connect the two points representing the same vertex with a line segment. The process results in a diagram (called a permutation diagram) with $|V|$ line segments. By definition, $uv \in E$ if and only if the line segments representing u and v cross in the permutation diagram, which is equivalent to the inequality $(\mathbf{x}_1(u) - \mathbf{x}_1(v))(\mathbf{x}_2(u) - \mathbf{x}_2(v)) < 0$.

Conversely, from a permutation diagram of G, we can extract linear orderings $\prec_1$ and $\prec_2$ as $\mathbf{x}_1(u) < \mathbf{x}_1(v) \iff u \prec_1 v$ and $\mathbf{x}_2(u) < \mathbf{x}_2(v) \iff u \prec_2 v$. When those conditions are satisfied, we say that the orderings of the x-coordinates on ℓ_1 and ℓ_2 are *consistent* with the linear orderings $\prec_1$ and $\prec_2$, respectively.

A graph is a *bipartite permutation graph* if it is a bipartite graph and a permutation graph. Although a permutation graph may have an exponential number of permutation orderings, it is essentially unique for a connected bipartite permutation graph in the sense of Lemma 2.3 below. For a graph $G = (V, E)$, linear orderings $\langle v_1, \ldots, v_n \rangle$ and $\langle v'_1, \ldots, v'_n \rangle$ on V are *neighborhood-equivalent* if $N(v_i) = N(v'_i)$ for all i.

Lemma 2.3 ([8]). *Let G be a connected bipartite permutation graph defined by $\prec_1$ and $\prec_2$. Then, every permutation ordering of G is neighborhood-equivalent to $\prec_1$, $\prec_2$, $\prec_1^R$, or $\prec_2^R$.*

The following lemma and corollary show that a bipartite permutation graph can be represented by a permutation diagram with a special property.

Lemma 2.4 (★). *Let $G = (X, Y; E)$ be a bipartite permutation graph. Then, G can be represented by a permutation diagram in which $\mathbf{x}_2(x) = \mathbf{x}_1(x) + 1$ for $x \in X$ and $\mathbf{x}_2(y) - \mathbf{x}_1(y) - 1$ for $y \in Y$.*

Corollary 2.5 (★). *Let $G = (X, Y; E)$ be a connected bipartite permutation graph defined by $\prec_1$ and $\prec_2$. If the first vertex in $\prec_1$ belongs to X, then G can be represented by a permutation diagram such that the orderings of the x-coordinates on ℓ_1 and ℓ_2 are consistent with $\prec_1$ and $\prec_2$, respectively, and that $\mathbf{x}_2(x) = \mathbf{x}_1(x) + 1$ for $x \in X$ and $\mathbf{x}_2(y) = \mathbf{x}_1(y) - 1$ for $y \in Y$.*

2.3 Double-Threshold Graphs as Permutation Graphs

We show that double-threshold graphs are strongly related to permutation graphs.

Lemma 2.6 (★). *Every double-threshold graph is a permutation graph.*

Lemma 2.7 (★). *Every bipartite permutation graph is a double-threshold graph.*

Corollary 2.8 (★). *The bipartite double-threshold graphs are exactly the bipartite permutation graphs.*

3 New Characterization

Let $G = (V, E)$ be a graph. From G and a vertex subset $M \subseteq V$, we construct an auxiliary bipartite graph $G'_M = (V', E')$ defined as $V' = \{v, \bar{v} \mid v \in V\}$ and $E' = \{u\bar{v} \mid uv \in E\} \cup \{v\bar{v} \mid v \in M\}$.

Lemma 3.1. *For a connected non-bipartite graph $G = (V, E)$ and a vertex subset $M \subseteq V$, G'_M is connected.*

Proof. For any $u, v \in V$, since G is connected and non-bipartite, G contains both an odd walk and an even walk from u to v. This shows that G'_M contains walks from u to v, from u to $\bar{v}$, from $\bar{u}$ to v, and from $\bar{u}$ to $\bar{v}$. Hence, G'_M is connected.   □

For the auxiliary graph $G'_M = (V', E')$ of $G = (V, E)$, a linear ordering on V' represented by $\langle w_1, w_2, \ldots, w_{2n} \rangle$ is *symmetric* if $w_i = v$ implies $w_{2n-i+1} = \bar{v}$ for any $v \in V$ and any $i \in \{1, 2, \ldots, 2n\}$.

The next is the key lemma for our characterization.

Lemma 3.2. *Let $G = (V, E)$ be a non-bipartite graph and $M \subseteq V$. The following are equivalent.*

1. *G is a double-threshold graph defined by $\mathtt{w}: V \to \mathbb{R}$ and $\mathtt{lb}, \mathtt{ub} \in \mathbb{R}$ such that $M = \{v \in V \mid \mathtt{lb}/2 \le \mathtt{w}(v) \le \mathtt{ub}/2\}$.*
2. *The auxiliary graph $G'_M = (V', E')$ can be represented by a permutation diagram in which both orderings $\prec_1$ and $\prec_2$ are symmetric.*

Proof $(1 \implies 2)$. An illustration is given in Fig. 2. Let G be a double-threshold graph defined by $\mathtt{w}: V \to \mathbb{R}$ and $\mathtt{lb}, \mathtt{ub} \in \mathbb{R}$ such that $M = \{v \in V \mid \mathtt{lb}/2 \le \mathtt{w}(v) \le \mathtt{ub}/2\}$. By Lemma 2.1, we can assume that $\mathtt{lb} = 0$ and $\mathtt{ub} = 2$, that $\mathtt{w}(u) + \mathtt{w}(v) \notin \{0, 2\}$ for every $(u, v) \in V^2$, and that $\mathtt{w}(u) \neq \mathtt{w}(v)$ if $u \neq v$. We construct a permutation diagram of G'_M as follows. Let ℓ_1 and ℓ_2 be two horizontal parallel lines. For each vertex $w \in V'$, we set the x-coordinates $\mathtt{x}_1(w)$ and $\mathtt{x}_2(w)$ on ℓ_1 and ℓ_2 as follows: for any $v \in V$,

$$\mathtt{x}_1(v) = \mathtt{w}(v) - 1, \qquad\qquad \mathtt{x}_1(\bar{v}) = 1 - \mathtt{w}(v),$$
$$\mathtt{x}_2(v) = \mathtt{w}(v), \qquad\qquad\quad \mathtt{x}_2(\bar{v}) = -\mathtt{w}(v).$$

Since $\mathtt{w}(u) + \mathtt{w}(v) \notin \{0, 2\}$ for every $(u, v) \in V^2$ and $\mathtt{w}(u) \neq \mathtt{w}(v)$ if $u \neq v$, the x-coordinates are distinct on ℓ_1 and on ℓ_2. By connecting $\mathtt{x}_1(w)$ and $\mathtt{x}_2(w)$ with a line segment for each $w \in V'$, we get a permutation diagram. The line segments corresponding to the vertices in V have negative slopes, and the ones corresponding to the vertices in $V' \setminus V$ have positive slopes. Thus, for any two vertices $u, v \in V$, the line segments corresponding to u and $\bar{v}$ cross if and only if both $\mathtt{x}_1(u) \le \mathtt{x}_1(\bar{v})$ and $\mathtt{x}_2(u) \ge \mathtt{x}_2(\bar{v})$ hold, which is equivalent to $0 \le \mathtt{w}(u) + \mathtt{w}(v) \le 2$, and thus to $u\bar{v} \in E'$. Similarly, the line segments corresponding to v and $\bar{v}$ cross if and only if $0 \le 2\mathtt{w}(v) \le 2$, i.e., $v \in M$. This shows that the obtained permutation diagram represents G'_M. Let $\prec_1$ be the ordering on V' defined by $\mathtt{x}_1$. Since $\mathtt{x}_1(v) = -\mathtt{x}_1(\bar{v})$ for each $v \in V$, if v is the ith vertex in $\prec_1$, then $\bar{v}$ is the ith vertex in $\prec_1^R$. This implies that $\bar{v}$ is the $(2n - i + 1)$st vertex in $\prec_1$, and thus $\prec_1$ is symmetric. In the same way, we can show that the ordering $\prec_2$ defined by $\mathtt{x}_2$ is symmetric.

$(2 \implies 1)$ Suppose we are given a permutation diagram of G'_M in which both $\prec_1$ and $\prec_2$ are symmetric. We may assume by symmetry that the first vertex

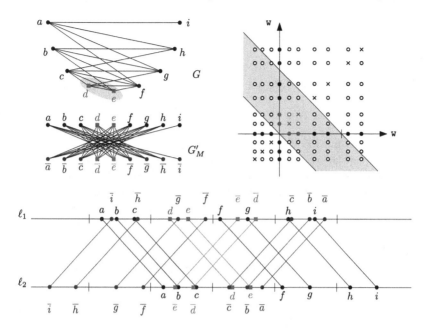

Fig. 2. An illustration of $(1 \implies 2)$ in Lemma 3.2. (Top left) A double-threshold graph G with $M = \{d, e\}$. The auxiliary bipartite graph G'_M is also depicted. (Top right) A slab representation of G. (Bottom) A permutation diagram of G'_M.

in $\prec_1$ belongs to V. Since G'_M is connected by Lemma 3.1, Corollary 2.5 shows that we can represent G'_M by a permutation diagram in which the x-coordinates x_1 and x_2 on ℓ_1 and ℓ_2 satisfy that

$$x_2(v) = x_1(v) + 1 \quad \text{and} \quad x_2(\bar{v}) = x_1(\bar{v}) - 1 \quad (v \in V) \tag{1}$$

and that the orderings of the x-coordinates on ℓ_1 and ℓ_2 are consistent with $\prec_1$ and $\prec_2$, respectively. Since $\prec_1$ is symmetric, if $u, v \in V$ are the ith and the jth vertices in $\prec_1$, then $\bar{u}, \bar{v}$ are the $(2n - i + 1)$st and the $(2n - j + 1)$st vertices in $\prec_1$. Since $i < 2n - j + 1$ is equivalent to $j < 2n - i + 1$, we have that $u \prec_1 \bar{v}$ if and only if $v \prec_1 \bar{u}$. As x_1 is consistent with $\prec_1$, it holds for $u, v \in V$ that $x_1(u) \leq x_1(\bar{v})$ if and only if $x_1(v) \leq x_1(\bar{u})$, and hence

$$x_1(u) \leq x_1(\bar{v}) \iff x_1(u) + x_1(v) \leq x_1(\bar{v}) + x_1(\bar{u}).$$

Similarly, we can show that for $u, v \in V$,

$$x_2(u) \geq x_2(\bar{v}) \iff x_2(u) + x_2(v) \geq x_2(\bar{v}) + x_2(\bar{u}).$$

Thus, for any two distinct vertices $u, v \in V$, it holds that

$$uv \in E \iff u\bar{v} \in E' \iff x_1(u) \leq x_1(\bar{v}) \text{ and } x_2(u) \geq x_2(\bar{v})$$
$$\iff x_1(u) + x_1(v) \leq x_1(\bar{v}) + x_1(\bar{u}) \text{ and } x_2(u) + x_2(v) \geq x_2(\bar{v}) + x_2(\bar{u}). \tag{2}$$

For each $v \in V$, define $\mathtt{w}(v) = (\mathtt{x}_2(v) - \mathtt{x}_2(\bar{v}))/2$. By (1), we can see that (2) is equivalent to $0 \leq \mathtt{w}(u) + \mathtt{w}(v) \leq 2$, which shows that $\mathtt{w}$, $\mathtt{lb} = 0$, and $\mathtt{ub} = 2$ define G. Furthermore, for any $v \in V$,

$$v \in M \iff v\bar{v} \in E' \iff \mathtt{x}_1(v) \leq \mathtt{x}_1(\bar{v}) \text{ and } \mathtt{x}_2(v) \geq \mathtt{x}_2(\bar{v}) \iff 0 \leq \mathtt{w}(v) \leq 1,$$

which shows that $M = \{v \in V \mid 0 \leq \mathtt{w}(v) \leq 1\}$.   □

To utilize Lemma 3.2, we need to find the set M of mid-weight vertices; that is, the vertices with weights in the range $[\mathtt{lb}/2, \mathtt{ub}/2]$. The first observation is that M has to be a clique as the weight sum of any two vertices in M is in the range $[\mathtt{lb}, \mathtt{ub}]$. In the following, we show that a special kind of maximum cliques can be chosen as M. To this end, we first prove that we only need to consider (inclusion-wise) maximal cliques.

Lemma 3.3 (★). *For a non-bipartite double-threshold graph $G = (V, E)$, there exist $\mathtt{w} \colon V \to \mathbb{R}$ and $\mathtt{lb}, \mathtt{ub} \subset \mathbb{R}$ defining G such that $\{v \in V \mid \mathtt{lb}/2 \leq \mathtt{w}(v) \leq \mathtt{ub}/2\}$ is a maximal clique of G.*

An *efficient maximum clique* K of a graph G is a maximum clique (i.e., a clique of the maximum size) that minimizes the degree sum $\sum_{v \in K} \deg_G(v)$. We show that every efficient maximum clique can be the set of mid-weight vertices.

Lemma 3.4. *Let G be a non-bipartite double-threshold graph. For every efficient maximum clique K of G, there exist $\mathtt{w} \colon V \to \mathbb{R}$ and $\mathtt{lb}, \mathtt{ub} \in \mathbb{R}$ defining G such that $K = \{v \in V \mid \mathtt{lb}/2 \leq \mathtt{w}(v) \leq \mathtt{ub}/2\}$.*

Proof. Let K be an efficient maximum clique of G. By Lemma 2.6, G is a permutation graph, and thus cannot contain an induced odd cycle of length 5 or more [5]. As G is non-bipartite, G contains K_3. This implies that $|K| \geq 3$.

By Lemma 3.3, there exist $\mathtt{w} \colon V \to \mathbb{R}$ and $\mathtt{lb}, \mathtt{ub} \subset \mathbb{R}$ defining G such that $M := \{v \in V \mid \mathtt{lb}/2 \leq \mathtt{w}(v) \leq \mathtt{ub}/2\}$ is a maximal clique of G. Assume that $\mathtt{w}$, $\mathtt{lb}$, and $\mathtt{ub}$ are chosen so that the size of the symmetric difference $|M \triangle K| = |M \setminus K| + |K \setminus M|$ is minimized. Assume that $K \neq M$ since otherwise we are done. This implies that $K \not\subseteq M$ and $K \not\supseteq M$ as both K and M are maximal cliques. Observe that $G - M$ is bipartite. This implies that $|K \setminus M| \in \{1, 2\}$ and that $K \cap M \neq \emptyset$ as $|K| \geq 3$. Since K is a maximum clique, $|M \setminus K| \leq |K \setminus M|$ holds.

Let $u \in K \setminus M$. By symmetry, we may assume that $\mathtt{w}(u) < \mathtt{lb}/2$. Note that no other vertex in K has weight less than $\mathtt{lb}/2$ as K is a clique. Let $v \in M$ be a nonneighbor of u that has the minimum weight among such vertices. Such a vertex exists since M is a maximal clique. Note that $v \in M \setminus K$.

We now observe that v has the minimum weight in M. If $w \in M$ is a nonneighbor of u, then $\mathtt{w}(v) \leq \mathtt{w}(w)$ follows from the definition of v. If $w \in M$ is a neighbor of u, then $\mathtt{w}(v) < \mathtt{w}(w)$ holds, since otherwise $\mathtt{w}(u) < \mathtt{lb}/2 \leq \mathtt{w}(w) \leq \mathtt{w}(v)$ and $uw, wv \in E$ imply that $uv \in E$ as $\mathtt{lb} \leq \mathtt{w}(u) + \mathtt{w}(w) \leq \mathtt{w}(u) + \mathtt{w}(v) < \mathtt{w}(w) + \mathtt{w}(v) \leq \mathtt{ub}$.

Claim 3.5 (★). $N(u) \cap \{x \mid \mathtt{w}(x) < \mathtt{lb}/2\} = N(v) \cap \{x \mid \mathtt{w}(x) < \mathtt{lb}/2\}$.

Claim 3.6 (★). $N(u) \cap M = N(v) \cap M$.

Claim 3.7 (★). $N(u) \cap \{x \mid \mathtt{w}(x) > \mathtt{ub}/2\} \supseteq N(v) \cap \{x \mid \mathtt{w}(x) > \mathtt{ub}/2\}$.

Claims 3.5, 3.6, and 3.7 imply that $N(v) \subseteq N(u)$. We now show that $N(v) = N(u)$ holds. Suppose to the contrary that $N(v) \subsetneq N(u)$. Let $K' = K \setminus \{u\} \cup \{v\}$. We first argue that K' is a (maximum) clique. If $K \setminus M = \{u\}$, then $K' = M$ is a clique. Assume that $K \setminus M = \{u, u'\}$ for some $u' \neq u$. Since $\mathtt{w}(u) < \mathtt{lb}/2$ and $u' \in K \setminus M$, we have $\mathtt{w}(u') > \mathtt{ub}/2$. Let $w \in K \cap M$. Then, $vw, wu' \in E$. Since $\mathtt{w}(v) \leq \mathtt{w}(w) \leq \mathtt{ub}/2 < \mathtt{w}(u')$, we have $vu' \in E$. Thus, K' is a clique. The assumption $N(v) \subsetneq N(u)$ implies that $\deg_G(v) < \deg_G(u)$, and thus, $\sum_{w \in K'} \deg_G(w) = (\sum_{w \in K} \deg_G(w)) - \deg_G(u) + \deg_G(v) < \sum_{w \in K} \deg_G(w)$. This contradicts that K is efficient. Therefore, we conclude that $N(v) = N(u)$.

Now, we define a weight function $\mathtt{w}' \colon V \to \mathbb{R}$ by setting $\mathtt{w}'(u) = \mathtt{w}(v)$, $\mathtt{w}'(v) = \mathtt{w}(u)$, and $\mathtt{w}'(x) = \mathtt{w}(x)$ for all $x \in V \setminus \{u, v\}$. Then, $\mathtt{w}'$, $\mathtt{lb}$, and $\mathtt{ub}$ define G and $M' := \{w \in V \mid \mathtt{lb}/2 \leq \mathtt{w}'(w) \leq \mathtt{ub}/2\} = M \cup \{u\} \setminus \{v\}$ as $N(u) = N(v)$. This contradicts the choice of $\mathtt{w}$ as $|M' \bigtriangleup K| < |M \bigtriangleup K|$. □

Next, we show that the symmetry required in Lemma 3.2 follows for free when M is a clique.

Lemma 3.8 (★). *Let $G = (V, E)$ be a connected non-bipartite graph and M be a clique of G. Then, G'_M is a permutation graph if and only if G'_M can be represented by a permutation diagram in which both orderings $\prec_1$ and $\prec_2$ are symmetric.*

By putting the above facts together, we obtain the following characterization of non-bipartite double-threshold graphs.

Theorem 3.9. *For a non-bipartite graph G, the following are equivalent.*

1. *G is a double-threshold graph.*
2. *For every efficient maximum clique M of G, G'_M is a permutation graph.*
3. *For some efficient maximum clique M of G, G'_M is a permutation graph.*

Proof. $2 \implies 3$ is trivial. To show $1 \implies 2$, assume G is a non-bipartite double-threshold graph. Let M be an efficient maximum clique of G. By Lemma 3.4, there exist $\mathtt{w} \colon V \to \mathbb{R}$ and $\mathtt{lb}, \mathtt{ub} \in \mathbb{R}$ defining G such that $M = \{v \in V \mid \mathtt{lb}/2 \leq \mathtt{w}(v) \leq \mathtt{ub}/2\}$. Now by Lemma 3.2, G'_M is a permutation graph. To show $3 \implies 1$, assume that for an efficient maximum clique M of a non-bipartite graph G, the graph G'_M is a permutation graph.

Let H be a non-bipartite component of G. Then, H contains an induced odd cycle of length $k \geq 3$. This means that, if H does not include M, then G'_M contains an induced cycle of length $2k \geq 6$. However, this is a contradiction as a permutation graph cannot contain an induced cycle of length at least 5. Thus, H includes M. Also, there is no other non-bipartite component in G as it does not intersect M. Since H includes M, H'_M is a component of G'_M. By Lemma 3.8,

H'_M can be represented by a permutation diagram in which both $\prec_1$ and $\prec_2$ are symmetric, and thus H is a double-threshold graph by Lemma 3.2.

Let B be a bipartite component of G (if one exists). Since B does not intersect M, G'_M contains two isomorphic copies of H as components. Since G'_M is a permutation graph, B is a permutation graph too. By Corollary 2.8, B is a double-threshold graph.

The discussion so far implies that all components of G are double-threshold graphs and exactly one of them is non-bipartite. By Lemma 2.2, G is a double-threshold graph.   □

4   Linear-Time Recognition Algorithm

Theorem 4.1. *There exists a linear-time algorithm that takes a graph as its input and outputs* YES *if and only if the graph is a double-threshold graph.*

Proof. The algorithm for Theorem 4.1 is given in Algorithm 1. By Corollary 2.8 and Theorem 3.9, the algorithm is correct since it returns YES if and only if G is a bipartite permutation graph, or G is a non-bipartite permutation graph and G'_M is a permutation graph, where M is an efficient maximum clique of G.

At Steps 1 and 4, deciding whether a graph is a permutation graph and, if so, computing a permutation ordering can be done in linear time [13]. At Step 2, bipartiteness can be checked in linear time by, e.g., depth-first search. Observe that $|V(G'_M)| = 2|V|$ and $|E(G'_M)| = 2|E| + |M|$. Thus, it suffices to show that an efficient maximum clique of a permutation graph can be computed in linear time at Step 3.

To find an efficient maximum clique of G, we set to each vertex $v \in V$ the weight $f(v) = n^2 - \deg_G(v)$, where $n = |V|$, and then find a maximum-weight clique M of G with respect to these weights. Using the permutation ordering of G computed before, we can find M in linear time [5, pp. 133–134]. We show that M is an efficient maximum clique of G. Let K be an efficient maximum clique of G. Since $\sum_{v \in K} f(v) \leq \sum_{v \in M} f(v)$, we have

$$|K| \cdot n^2 - \sum_{v \in K} \deg_G(v) \leq |M| \cdot n^2 - \sum_{v \in M} \deg_G(v). \tag{3}$$

Since $0 \leq \sum_{v \in S} \deg_G(v) < n^2$ for any $S \subseteq V$, it holds that $|K| \cdot n^2 - n^2 < |M| \cdot n^2$. Since $|M| \leq |K|$, this implies that $|K| = |M|$. It follows from (3) that $\sum_{v \in K} \deg_G(v) \geq \sum_{v \in M} \deg_G(v)$. Thus, M is an efficient maximum clique.   □

Algorithm 1. Decide if G is a double-threshold graph.

1: **if** G has a permutation ordering $\prec$ **then**
2:     **if** G is bipartite **then return** YES
3:     Find an efficient maximum clique M of G using $\prec$.
4:     **if** G'_M is a permutation graph **then return** YES
5: **return** NO

References

1. Barrus, M.D.: Weakly threshold graphs. Discret. Math. Theor. Comput. Sci. **20**(1), 22 p. (2018). https://doi.org/10.23638/DMTCS-20-1-15
2. Behr, R., Sivaraman, V., Zaslavsky, T.: Mock threshold graphs. Discret. Math. **341**(8), 2159–2178 (2018). https://doi.org/10.1016/j.disc.2018.04.023
3. Benzaken, C., Hammer, P.L., de Werra, D.: Threshold characterization of graphs with Dilworth number two. J. Graph Theor. **9**(2), 245–267 (1985). https://doi.org/10.1002/jgt.3190090207
4. Calamoneri, T., Sinaimeri, B.: Pairwise compatibility graphs: a survey. SIAM Rev. **58**(3), 445–460 (2016). https://doi.org/10.1137/140978053
5. Golumbic, M.C.: Algorithmic Graph Theory and Perfect Graphs, 2nd edn. North Holland, Amsterdam (2004)
6. Hamburger, P., McConnell, R.M., Pór, A., Spinrad, J.P., Xu, Z.: Double threshold digraphs. In: MFCS 2018. LIPIcs, vol. 117, pp. 69:1–69:12 (2018). https://doi.org/10.4230/LIPIcs.MFCS.2018.69
7. Hammer, P.L., Mahadev, N.V.R.: Bithreshold graphs. SIAM J. Algebraic Discret. Meth. **6**(3), 497–506 (1985). https://doi.org/10.1137/0606049
8. Heggernes, P., van 't Hof, P., Meister, D., Villanger, Y.: Induced subgraph isomorphism on proper interval and bipartite permutation graphs. Theor. Comput. Sci. **562**, 252–269 (2015). https://doi.org/10.1016/j.tcs.2014.10.002
9. Jamison, R.E., Sprague, A.P.: Bi-threshold permutation graphs (2019, in press)
10. Kearney, P., Munro, J.I., Phillips, D.: Efficient generation of uniform samples from phylogenetic trees. In: Benson, G., Page, R.D.M. (eds.) WABI 2003. LNCS, vol. 2812, pp. 177 189. Springer, Heidelberg (2003). https://doi.org/10.1007/978 3 540-39763-2_14
11. Ma, T.-H.: On the threshold dimension 2 graphs. Technical report, Institute of Information Science, Nankang, Taipei, Taiwan (1993)
12. Mahadev, N.V.R., Peled, U.N.: Threshold Graphs and Related Topics, 1st edn. North Holland, Amsterdam (1995)
13. McConnell, R.M., Spinrad, J.P.: Modular decomposition and transitive orientation. Discret. Math. **201**(1–3), 189–241 (1999). https://doi.org/10.1016/S0012-365X(98)00319-7
14. Monma, C.L., Reed, B.A., Trotter, W.T.: Threshold tolerance graphs. J. Graph Theor. **12**(3), 343–362 (1988). https://doi.org/10.1002/jgt.3190120307
15. Raschle, T., Simon, K.: Recognition of graphs with threshold dimension two. In: STOC 1995, pp. 650–661. ACM (1995). https://doi.org/10.1145/225058.225283
16. Ravanmehr, V., Puleo, G.J., Bolouki, S., Milenkovic, O.: Paired threshold graphs. Discret. Appl. Math. **250**, 291–308 (2018). https://doi.org/10.1016/j.dam.2018.05.008
17. Sterbini, A., Raschle, T.: An $O(n^3)$ time algorithm for recognizing threshold dimension 2 graphs. Inf. Process. Lett. **67**(5), 255–259 (1998). https://doi.org/10.1016/S0020-0190(98)00112-4
18. Xiao, M., Nagamochi, H.: Characterizing Star-PCGs. Algorithmica **82**, 1–25 (2020). https://doi.org/10.1007/s00453-020-00712-8
19. Yan, J., Chen, J., Chang, G.J.: Quasi-threshold graphs. Discret. Appl. Math. **69**(3), 247–255 (1996). https://doi.org/10.1016/0166-218X(96)00094-7

Characterization and Linear-Time Recognition of Paired Threshold Graphs

Yixin Cao[1]⊕, Guozhen Rong[2](✉), and Jianxin Wang[2]

[1] Department of Computing, Hong Kong Polytechnic University, Hong Kong, China
yixin.cao@polyu.edu.hk
[2] School of Computer Science and Engineering, Central South University,
Changsha, China
rongguozhen@csu.edu.cn

Abstract. In a paired threshold graph, each vertex has a weight, and two vertices are adjacent if and only if their weight sum is large enough and their weight difference is small enough. It generalizes threshold graphs and unit interval graphs, both very well studied. We present a vertex ordering characterization of this graph class, which enables us to prove that it is a subclass of interval graphs. Further study of clique paths of paired threshold graphs leads to a simple linear-time recognition algorithm for the graph class.

Keywords: (Paired) threshold graph · (Unit) interval graph · Interval model · Umbrella ordering · Interval ordering · Broom ordering · Clique path

1 Introduction

A graph is a *threshold graph* if one can assign positive weights to its vertices in a way that two vertices are adjacent if and only if the sum of their weights is not less than a certain threshold. Originally formulated from combinatorial optimization [1], threshold graphs found applications in many diversified areas. As one of the simplest nontrivial classes, the mathematical properties of threshold graphs have been thoroughly studied. They admit several nice characterizations, including inductive construction, degree sequences, forbidden induced subgraphs (Fig. 1), to name a few [12]. Relaxing these characterizations in one way or another, we end with several graph classes, e.g., cographs, split graphs, trivially perfect graphs, and double-threshold graphs [4,9,13,17]. Yet another closely related graph class are difference graphs, defined solely by weight differences [6].

Motivated by applications in social and economic interaction modeling, Ravanmehr et al. [15] introduced paired threshold graphs, another generalization

---

The full version of this paper, available at arXiv:1909.13029, contains all the proofs, some of which are omitted from this version due to the lack of space.

Y. Cao—Supported by RGC grants 15201317 and 15226116, and NSFC grant 61972330.
G. Rong and J. Wang—Supported by NSFC grants 61828205 and 61672536.

© Springer Nature Switzerland AG 2020
I. Adler and H. Müller (Eds.): WG 2020, LNCS 12301, pp. 298–309, 2020.
https://doi.org/10.1007/978-3-030-60440-0_24

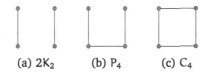

(a) $2K_2$ (b) P_4 (c) C_4

Fig. 1. Minimal forbidden induced subgraphs for threshold graphs.

of threshold graphs. A graph is a *paired threshold graph* if there exist a positive vertex weight assignment w and two positive thresholds, T_+ and T_-, such that two vertices are adjacent if and only if the sum of their weights are not less than T_+ and the difference of their weights are not greater than T_-.

An easy observation on a threshold graph is, vertices of small weights, less than half of the threshold to be specific, form an independent set, while the other vertices form a clique. (Hence, each threshold graph is a split graph.) Clearly, the first part remains true for paired threshold graphs, but not the second. Since the adjacency between a pair of high-weight vertices ($\geq \frac{T_+}{2}$) is only decided by their weight difference, they induce an indifference graph [16], which is more widely known as a unit interval graph. The crucial point is thus to understand the interaction between these two sets of vertices. For this purpose we may focus on paired threshold graphs that are neither threshold graphs nor unit interval graphs. Ravanmehr et al. [15] presented a distance decomposition for such a paired threshold graph G: If G is connected, then they are able to decompose $V(G)$ into a set X, which induces a threshold graph, and a sequence of cliques, where the vertices in a same clique have the same distance to X.

It is straightforward to show that paired threshold graphs are chordal: The vertex with the smallest weight is necessarily simplicial. Since interval graphs also contain all threshold graphs and all unit interval graphs, a natural question is on the relationship between interval graphs and paired threshold graphs.

Theorem 1. *All paired threshold graphs are interval graphs.*

Threshold graphs enjoy a very simple ordering characterization by the vertex degrees [1], while the ordering of the intervals gives a vertex ordering characterization for unit interval graphs, called an umbrella ordering [11]. On the other hand, interval graphs have a vertex ordering characterization with the so-called 3-vertex conditions [14]. We show that a paired threshold graph admits an interval ordering with the additional conditions: (1) it can be partitioned such that the first part induces an independent set and the second is an umbrella ordering; and (2) the neighborhood of every vertex from the first part is consecutive in this ordering. We call such a vertex ordering a broom ordering.

Theorem 2. *A graph is a paired threshold graph if and only if it admits a broom ordering.*

Unit interval graphs are interval graphs that can be represented using intervals of the same length. It is known that any threshold graph can be represented by intervals of at most two different lengths [10]. (But not all interval graphs with two-length representation are threshold graphs.) This is nevertheless not true for paired threshold graphs. For any $k > 0$, we are able to construct paired threshold graphs that cannot be represented by intervals of k different lengths. In other words, the class of paired threshold graphs is not a subclass of k-length interval graphs, defined by Klavík et al. [8]. Recall that unit interval graphs are also proper interval graphs, interval graphs that can be represented using intervals none of which properly contains the other. This has also been generalized by Klavík et al. [8], who defined the classes of k-nested interval graphs, for $k > 0$, to be the interval graphs that can be represented using intervals of which no k nested. Indeed, for any positive integer k, we can construct a paired threshold graph such that there must be k nested intervals in any interval model of this graph. On the other hand, it is easy to see that $K_{1,3} + K_{1,3}$, the disjoint union of two claws (the four-vertex tree with three leaves), is a 2-length interval graph but not a paired threshold graph. Therefore, the class of paired threshold graphs and the class of k-nested interval graphs are not comparable to each other.

Yet another graph class sandwiched between interval graphs and threshold graphs is the class of trivially perfect graphs [17]. It is not comparable to the class of paired threshold graphs or the class of k-nested interval graphs either. First, note that P_4 is a unit interval graph but not a trivially perfect graph. Second, the disjoin union of two claws is also a trivially perfect graph. Finally, for any positive integer k, we can recursively construct a trivially perfect graph that is not a k-nested interval graph as follows. Suppose that G is a trivially perfect graph but not a $(k - 1)$-nested interval graph. We take three disjoint copies of G, and add a universal vertex. See Fig. 2 for an overview of related graph classes.

Similar as threshold graphs and split graphs, the class of paired threshold graphs is not closed under taking disjoint union of subgraphs. If a paired threshold graph is not a unit interval graph, then in any assignment, there must be some vertex receiving weight $< T_+/2$ and some vertex receiving weight $\geq T_+/2$. (Note that an edgeless graph is trivially a unit interval graph.) From the definition it is easy to verify that at most one component can be a non–unit interval graphs, which is of course a connected paired threshold graph. This turns out to be also sufficient. (Since both the class of 2-length interval graphs and the class of trivially perfect graphs are closed under taking disjoint union of subgraphs, this also explains that they are not subclasses of paired threshold graphs. In particular, the claw is a trivially perfect graph and a 2-length interval graphs, but not a unit interval graph.)

For the recognition of paired threshold graphs, we may focus on connected non–unit interval graphs. For such a graph, we show that it is a paired threshold graph if and only if there is an induced subgraph with certain property. From this subgraph, we can produce two partitions of its vertex set, and it is a paired threshold graph if and only if one of them defines a broom ordering of this subgraph. Putting them together, we develop a linear-time algorithm for recognizing paired threshold graphs, improving the $O(|V(G)|^6)$-time algorithm of Ravanmehr et al. [15].

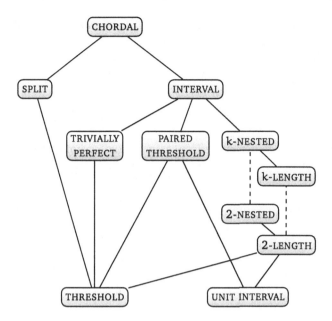

Fig. 2. A summary of related graph classes. Note that the three immediate subclasses of interval graphs are not comparable to each other.

Theorem 3. *Given a graph G, we can decide in $O(|V(G)| + |E(G)|)$ time whether G is a paired threshold graph.*

2   Preliminaries

All graphs discussed in this paper are undirected and simple. The vertex set and edge set of a graph G are denoted by, respectively, $V(G)$ and $E(G)$. For a subset $X \subseteq V(G)$, denote by $G[X]$ the subgraph of G induced by X, and by $G - X$ the subgraph $G[V(G) \setminus X]$; when X consists of a single vertex v, we use $G - v$ as a shorthand for $G - \{v\}$. The *neighborhood* $N(v)$ of a vertex $v \in V(G)$ comprises vertices adjacent to v, i.e., $N(v) = \{u \mid uv \in E(G)\}$, and its *closed neighborhood* is $N[v] = N(v) \cup \{v\}$. The *closed neighborhood* and the *neighborhood* of a set $X \subseteq V(G)$ of vertices are defined as $N[X] = \bigcup_{v \in X} N[v]$ and $N(X) = N[X] \setminus X$, respectively. We say that a vertex v is *simplicial* if $N[v]$ is a clique.

A graph G is a *threshold graph* if there exist a weight assignment $w : V(G) \to \mathbb{R}^+$ and a fixed threshold $T \in \mathbb{R}^+$ such that $uv \in E(G)$ if and only if $w(u) + w(v) \geq T$. Alternatively, a graph G is a threshold graph if and only if its vertices can be partitioned into a clique and an independent set I such that the neighborhoods of vertices in I form a total order under the containment relation. A graph G is a *paired threshold graph* if there exist a weight assignment $w : V(G) \to \mathbb{R}^+$ and two fixed thresholds $T_+, T_- \in \mathbb{R}^+$ such that $uv \in E(G)$ if and only if

$$w(u) + w(v) \geq T_+ \quad \text{and} \quad |w(u) - w(v)| \leq T_-.$$

Given an assignment w and thresholds T_+, T_- for a paired threshold graph, we may adjust all the vertex weights by the same value ϵ and T_+ by 2ϵ, while keeping T_- unchanged. It is easy to verify that it represents the same graph. Thus, we can always make the two thresholds equal.

Proposition 1. *A graph G is a paired threshold graph if and only if there exist a weight assignment $w : V(G) \to \mathbb{R}^+$ and a threshold $T_\pm \in \mathbb{R}^+$ such that $uv \in E(G)$ if and only if $w(u) + w(v) \geq T_\pm$ and $|w(u) - w(v)| \leq T_\pm$.*

In the rest of the paper we use the same value for both thresholds. We use $G(w, T_\pm)$ to denote a paired threshold graph with weight assignment w and threshold $T_\pm$.

In a threshold graph G with weight assignment w and threshold T, the weight $T/2$ defines a natural partition of the vertices: $\{v \mid w(v) < T/2\}$ forms an independent set, while $\{v \mid w(v) \geq T/2\}$ a clique. We can use $T_\pm/2$ to get a similar partition of the vertex set of a paired threshold graph $G(w, T_\pm)$. In particular, $\{v \mid w(v) < T_\pm/2\}$ remains an independent set. However, $\{v \mid w(v) \geq T_\pm/2\}$ is no longer a clique in general. Since the weight sum of any two such vertices in this set is at least $T_\pm$, it induces an indifference graph, or more widely known, a unit interval graph [16]. A graph is a *unit interval graph* if its vertices can be assigned to unit-length intervals on the real line such that two vertices are adjacent if and only if their corresponding intervals intersect.

Proposition 2. *In a paired threshold graph $G(w, T_\pm)$, the subgraph induced by $\{v \mid w(v) \geq T_\pm/2\}$ is a unit interval graph.*

As a matter of fact, all unit interval graphs are paired threshold graphs. This has been observed in [15], and we include a proof because the construction used in it will be exemplary in this paper. An interval model can be specified by the $2n$ endpoints for the n intervals: The interval for vertex v is denoted by $[\text{lp}(v), \text{rp}(v)]$, where $\text{lp}(v)$ and $\text{rp}(v)$ are the, respectively, left point and the right point of the interval.

Proposition 3. *A unit interval graph G is a paired threshold graph. Moreover, there is an assignment w such that $w(v) \geq T_\pm$ for all vertices $v \in V(G)$.*

The discussion above can be summarized as that a paired threshold graph can be partitioned into an independent set and a unit interval graph. Now that we have thus fully understood both parts, it is time to put their connection under scrutiny.

Proposition 4. *Let $G(w, T_\pm)$ be a paired threshold graph, and let $I = \{v \mid w(v) < T_\pm/2\}$. Then $N[I]$ induces a threshold graph.*

Although Proposition 4 is straightforward, we want to point out that the assignment w for vertices in $N[I]$ is not necessarily a threshold assignment for them with respect to threshold $T_\pm$; see Fig. 3 for an example. Moreover, Propositions 2 and 4 are not sufficient for a graph to be a paired threshold graph. Neither the net nor the tent is a paired threshold graph [15]. However, setting I to be a single simplicial vertex in the net nor the tent, we get a partition satisfying both propositions. There is a more technical condition on the connection in between.

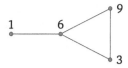

Fig. 3. A threshold graph. The weight assignment w shown in the graph and the threshold $T_\pm = 7$ witness that it is a paired threshold graph. However, this is not a threshold assignment because there is no edge between the two vertices with weights 1 and 9.

3 Characterization

An *ordering* σ of the vertex set of a graph G is a bijection from $V(G) \to \{1, \ldots, n\}$. We use $u <_\sigma v$ to denote that $\sigma(u) < \sigma(v)$. An ordering σ of the vertex set of a graph G is an *umbrella ordering* if for every triple of vertices x, y, z,

$$x <_\sigma y <_\sigma z \text{ and } xz \in E(G) \text{ imply } xy, yz \in E(G).$$

Looges and Olariu [11] showed that a graph is a unit interval graph if and only if it admits an umbrella ordering. Indeed, given a unit interval model of a unit interval graph G, ordering the vertices by their left endpoints produces an umbrella ordering of G. Likewise interval graphs can be characterized by the so called *interval orderings* σ [14]: for every triple of vertices x, y, z,

$$x <_\sigma y <_\sigma z \text{ and } xz \in E(G) \text{ imply } xy \in E(G).$$

We now formally define the broom ordering.

Definition 1. *An ordering σ of the vertex set of a graph G is a* broom ordering *if its reversal is an interval ordering of G and there exists p with $0 \le p \le n$ such that*

(i) for each of the first p vertices of σ, its neighborhood is after p and appears consecutively in σ; and
(ii) the sub-ordering of the last $n - p$ vertices is an umbrella ordering.

Two remarks on this definition are in order. For each of the last $n-p$ vertices, its closed neighborhood appears consecutively in σ. A graph having a broom ordering has also an interval ordering, hence an interval graph. Therefore, the following lemma implies Theorem 1.

Lemma 1. *Let $G(T_\pm, w)$ be a paired threshold graph. The ordering of $V(G)$ decided by w, with ties broken arbitrarily, is a broom ordering.*

Although the class of paired threshold graphs is not closed under taking disjoint union of subgraphs, the following proposition focuses us upxson connected graphs.

Lemma 2. *A graph G is a paired threshold graph if and only if one component is a connected paired threshold graph and all the others are unit interval graphs.*

Another way to characterize paired threshold graphs is through partition, with the focus on the connection between two parts.

Definition 2. *A partition $I \uplus U$ of a graph G is a* paired threshold partition *of G if*

(i) *I is an independent set, $G[U]$ is a unit interval graph, and $N[I]$ induces a threshold graph;*

(ii) *$G[U]$ has an umbrella ordering σ in which $N(v)$ appears consecutively for each $v \in I$; and*

(iii) *$N(I) \subseteq N[u]$, where u is the first vertex of σ.*

The following theorem implies Theorem 2.

Theorem 4. *The following are equivalent on a graph G.*

(1) G is a paired threshold graph.

(2) G admits a broom ordering.

(3) G has a paired threshold partition.

The paired threshold partition $I \uplus U$ of a paired threshold graph is not necessarily unique: For example, if the first vertex in U is nonadjacent to any vertex in I, then we may move it from U to I to make a new partition. We say that a paired threshold partition $I \uplus U$ is a *canonical partition* if I is maximal in all paired threshold partitions,—i.e., $(I \cup \{v\}) \uplus (U \setminus \{v\})$ is not a valid paired threshold partition for any $v \in U$. The following characterizes canonical partitions.

Proposition 5. *Let $I \uplus U$ be a canonical partition of a paired threshold graph G. If a vertex $v \in U \setminus N(I)$ is simplicial in G, then*

(1) $N(I) \nsubseteq N(v)$; and

(2) the subgraph induced by $N[I \cup \{v\}]$ is not a threshold graph.

If we drop the unit-length from the definition of unit interval graphs, then we end with interval graphs,—i.e., it allows intervals of arbitrary lengths. A graph G is an interval graph if and only if its maximal cliques can be arranged in a sequence such that for every vertex $v \in V(G)$, the set of maximal cliques containing v occur consecutively in the sequence [5]. This sequence is called a *clique path* of G. It is known that a connected unit interval graph has a unique clique path, up to full reversal [2,3].

By Theorem 1, we know that all paired threshold graphs are interval graphs. This leads us to consider paired threshold graphs on the perspective of clique path. A paired threshold graph admits a clique path with some important properties to be used in our recognition algorithm.

Theorem 5. *Let G be a connected paired threshold graph, and let $I \uplus U$ be a canonical partition of G.*

(1) *There exists a clique path $C_1, \ldots, C_\ell$ of G such that for $1 \le i \le |I|$, the only vertex in $I \cap C_i$ is a simplicial vertex of G, and $C_i \setminus I \subseteq C_{|I|}$;*

(2) *In any clique path of G, the maximal cliques disjoint from I appear consecutively, and they appear in the same order, up to full reversal;*

(3) *If G remains connected after all universal vertices removed, then for any clique path of G, vertices in I appear in the first $|I|$ or the last $|I|$ cliques.*

A set of k intervals $[\texttt{lp}(v_1), \texttt{rp}(v_1)], \ldots, [\texttt{lp}(v_k), \texttt{rp}(v_k)]$ is *nested* if

$$[\texttt{lp}(v_1), \texttt{rp}(v_1)] \subset \cdots \subset [\texttt{lp}(v_k), \texttt{rp}(v_k)].$$

Lemma 3. *For any positive integer k, there exists a paired threshold graph G such that there are k nested intervals in any interval model of G.*

4 Recognition

It is well known that interval graphs, unit interval graphs, and threshold graphs can be recognized in linear time [1,2,7]. For (unit) interval graphs, the recognition algorithms return an interval model, from which we can retrieve a clique path or an umbrella ordering in the same time. We say that two vertices u, v are *true twins* if $N[u] = N[v]$. A set of vertices is a *true-twin class* if it is a maximal set of vertices that are pairwise true twins. A graph has a unique partition into true-twin classes. The following proposition is from Deng et al. [3], and it is the core idea of Corneil [2].

Proposition 6. *Let G be a unit interval graph and let σ be an umbrella ordering of G.*

(1) *For each set of true-twin class T of G, vertices in T appear consecutively in σ, and this subsequence can be replaced by an arbitrary ordering of T.*

(2) *If G does not contain true twins, then it has a unique umbrella ordering, up to full reversal.*

We may assume that the input graph G is a connected interval graph. If it is not an interval graph, then we may return "no" by Theorem 1. If it is not connected, then we may remove all the components that are unit interval graphs, and return "no" if more than one component is left by Lemma 2.

The way we handle a connected interval graph G is to use a clique path $\mathcal{P}$ of G. We try to find a canonical partition from $\mathcal{P}$. According to Theorem 5(2), if G is a paired threshold graph, then we can find the partition $I \uplus U$ with I from the two ends of the clique path. However, a problem is that we do not know how many simplicial vertices we can take from each end. This can be simplified by Theorem 5(3): If G remains connected after all universal vertices removed, then it suffices to search only one end of $\mathcal{P}$. We hence proceed dependent upon whether G contains a universal vertex.

Proposition 7 ([1]). *A nonempty threshold graph contains either an isolated vertex or a universal vertex.*

In other words, a threshold graph can always be made empty by exhaustively removing isolated vertices and universal vertices. It is easy to see that paired threshold graphs are closed under adding isolated vertices, but not necessarily universal vertices. The following result is an extension of Proposition 7 to paired threshold graphs. Recall that minimal forbidden induced subgraphs of threshold graphs are $2K_2$, P_4, and C_4 (Fig. 1).

Lemma 4. *Let G be a connected graph, and let G' be obtained from G by exhaustively removing universal vertices and isolated vertices. If $G' \neq G$, then G is a paired threshold graph if and only if one of the following conditions holds:*

(1) G' is an empty graph;
(2) G' has two components, one being a complete graph, and the other a threshold graph;
(3) G' is connected, and it remains a paired threshold graph after adding a universal vertex.

We then remove exhaustively universal vertices and isolated vertices from G and study the resulting graph G'. By Lemma 4, we return "yes" if G' is empty, and return "no" if G' contains more than two components. If G' has precisely two components, then we return whether they satisfy Lemma 4(2), i.e., one component being a complete graph and the other a threshold graph. In the rest G' is connected; By Lemma 4(3), if $G' \neq G$, then we add a universal vertex to G'. Note that it is an induced subgraph of G. We build a clique path $K_1, \ldots, K_p$ for this subgraph. Let I_L be greedily obtained as follows. From $i = 1, \ldots, p$, if there is no simplicial vertex in K_i, then return I_L; otherwise, we pick a simplicial vertex of K_i and add it to I_L, as long as $N[I_L]$ still induces a threshold graph. Let I_R constructed in a similar way, but from K_p to K_1. By Proposition 5 and Theorem 5(3), G is a paired threshold graph if and only if one of I_L and I_R defines a paired threshold partition of this subgraph. It remains to verify whether one of the partitions, (i.e., $I_L \uplus (V(G') \setminus I_L)$ or $I_R \uplus (V(G') \setminus I_R)$,) is a paired threshold partition. As usual, n and m denote the number of vertices and the number of edges, respectively. Given a partition $I \uplus U$ of $V(G)$, we can check whether it satisfies Definition 2(1) in $O(m + n)$ time: to recognize an independent set is easy, and there exist $O(m + n)$ time algorithms for recognizing unit interval graphs and threshold graphs [3,12]. The remaining is to check whether it satisfies Definition 2(2–3) as well.

Lemma 5. *Let G be a connected interval graph. Given a partition $I \uplus U$ of $V(G)$ that satisfies Definition 2(1), we can check in $O(m + n)$ time whether it satisfies Definition 2(2–3) as well.*

Proof. We may number vertices in I in a way that $I = \{u_1, \ldots, u_{|I|}\}$ and $N(u_1) \subseteq \ldots \subseteq N(u_{|I|})$; this is possible because I is an independent set and $N[I]$ induces a threshold graph. Note that $d(u_1) \leq \ldots \leq d(u_{|I|})$. We call the procedure given in Fig. 4.

In the first two steps, it starts with finding an umbrella ordering σ of $G - I$, and then lists the true-twin classes of $G - I$ in their order of occurrences in σ.

INPUT: a connected interval graph G, and
    a partition I ⊎ U satisfying Definition 2(1).
OUTPUT: whether the partition satisfies Definition 2(2–3) or not.

1.   compute an umbrella ordering σ of G − I;
2.   let $\mathcal{T} = T_1, \ldots T_t$ be the true-twin classes of G − I in the order of σ;
3.   If $N(I) \not\subseteq N[T_1]$ and $N(I) \not\subseteq N[T_t]$ then return "no";
4.   If $N(I) \not\subseteq N[T_1]$ then reverse the sequence $\mathcal{T}$;
5.   for i ← 1, ..., |I| do
5.1.   find the first set T_ℓ in $\mathcal{T}$ intersecting $N(u_i)$;
5.2.   find the last set T_r in $\mathcal{T}$ intersecting $N(u_i)$;
5.3.   if $T \not\subseteq N(u_i)$ for any set T between T_ℓ and T_r in $\mathcal{T}$ then return "no";
5.4.   replace T_ℓ by $T_\ell \setminus N(u_i)$ and $T_\ell \cap N(u_i)$ in order;
5.5.   replace T_r by $T_r \cap N(u_i)$ and $T_r \setminus N(u_i)$ in order;
6.   return "yes."

Fig. 4. Verifying whether a partition satisfies the conditions in Definition 2.

By Proposition 6, the first vertex of any umbrella ordering of $G - I$ has to be from T_1 or T_t. If $N(I) \not\subseteq N[T_1]$ and $N(I) \not\subseteq N[T_t]$, then Definition 2(3) cannot be satisfied by any umbrella ordering of $G - I$. This justifies step 3. Note that it is possible $N(I) \subseteq N[T_t]$ and $N(I) \subseteq N[T_1]$, and in this case $N(I)$ are universal vertices of $G - I$. This is the trivial case. Otherwise we make sure that $N(I) \subseteq N(v)$ for some vertex in the first set of $\mathcal{T}$ in step 4.

It now enters step 5. The sets T_ℓ and T_r exist because G is connected and I is an independent set. The focus is on step 5.3. Suppose that $T \not\subseteq N(u_i)$. Note that T was not split from the same twin class as T_ℓ or T_r: Otherwise, $T \subseteq N(u_j)$ for some $j < i$, but then $T \subseteq N(u_i)$ as $N(u_j) \subseteq N(u_i)$. By Proposition 6, vertices in T have to be between vertices of T_ℓ and T_r in any umbrella ordering of $G - I$. Therefore, Definition 2(2) cannot be satisfied by any umbrella ordering of $G - I$. This justifies Step 5.3. We prove the correctness of step 6 by arguing that if the procedure passes step 5, then it satisfies Definition 2(2–3). We use the umbrella ordering of $G - I$ from $\mathcal{T}$ by replacing each set by an arbitrary ordering. Condition (2) is satisfied for vertex u_i after the ith iteration, and the sets containing $N(u_i)$ would never be touched after that. On the other hand, step 5 never switches the order of two sets, and hence $N(I) \subseteq N[v]$ for any vertex in the first set, which is a subset of T_1. Hence, condition (3) is satisfied as well.

It remains to show that the algorithm can be implemented in $O(n + m)$ time. It is straightforward for steps 1–4, and hence we focus on step 5. For each set in $\mathcal{T}$, we maintain a doubly linked list; further, we connect these lists into another doubly linked list. This allows us, among others, to split in time proportional to the number of elements to be split from a set. We also maintain an array of size n, of which the ith element points to the position of the ith vertex in the doubly

1. **for each** component C of G **do**
       if C is an unit interval graph **then** remove C from G;
2. **if** G is empty **then return** "yes";
3. **if** G has more than one component **then return** "no";
4. **while** G has a universal or isolated vertex v **then** $G \leftarrow G - v$;
5. **if** G is empty **then return** "yes";
6. **if** G is not connected **then**
       if there are more than two components **then return** "no";
       if one component is a clique, and the other is a threshold graph
       then return "yes";
       else return "no";
7. **if** a vertex is deleted in step 4, **then** add a universal vertex to G;
8. build a clique path $K_1, \ldots, K_p$ of G;
9. $I_L \leftarrow \emptyset$; $I_R \leftarrow \emptyset$;
10. **for** $i \leftarrow 1, \ldots, p$ **do**
        if there does not exists a simplicial vertex of G in K_i **then goto** 11;
        $x \leftarrow I_L \cup$ a simplicial vertex of G in K_i;
        if $N[I_L \cup \{x\}]$ does not induce a threshold graph **then goto** 11;
        $I_L \leftarrow I_L \cup \{x\}$;
11. **if** I_L defines a partition satisfying Definition 2 **then return** "yes";
12. **for** $i \leftarrow p, \ldots, 1$ **do**
        if there does not exists a simplicial vertex of G in K_i **then goto** 13;
        $x \leftarrow I_L \cup$ a simplicial vertex of G in K_i;
        if $N[I_R \cup \{x\}]$ does not induce a threshold graph **then goto** 13;
        $I_R \leftarrow I_R \cup \{x\}$;
13. **if** I_R defines a partition satisfying Definition 2 **then return** "yes";
14. **return** "no."

Fig. 5. The algorithm for recognizing paired threshold graphs.

linked lists. With these data structures it is straightforward to implement step 5 in $O(n + m)$ time. In the first iteration, we go through all the n vertices to find T_ℓ and T_r; after that we scan from the T_ℓ and T_r of the previous iteration to the left and to the right respectively. Hence, in the ith iteration with $1 < i \leq |I|$, we scan only $O(d(u_i))$ vertices. This completes the proof. □

We summarize our algorithm in Fig. 5.

Proof (Proof of Theorem 3). We use the algorithm described in Fig. 5. We first prove its correctness. Steps 1–3 follow from Proposition 3 and Lemma 2. Steps 4–7 follow from Lemma 4. Steps 8–10 follow from Proposition 5 and Theorem 5. We leave the time analysis to the full version of the paper. □

References

1. Chvátal, V., Hammer, P.L.: Aggregation of inequalities in integer programming. In: Hammer, P.L., Johnson, E.L., Korte, B.H., Nemhauser, G.L. (eds.) Annals of Discrete Mathematics. Studies in Integer Programming, vol. 1, pp. 145–162. Elsevier (1977). https://doi.org/10.1016/S0167-5060(08)70731-3
2. Corneil, D.G.: A simple 3-sweep LBFS algorithm for the recognition of unit interval graphs. Discret. Appl. Math. **138**(3), 371–379 (2004). https://doi.org/10.1016/j.dam.2003.07.001
3. Deng, X., Hell, P., Huang, J.: Linear-time representation algorithms for proper circular-Arc graphs and proper interval graphs. SIAM J. Comput. **25**(2), 390–403 (1996). https://doi.org/10.1137/S0097539792269095
4. Foldes, S., Hammer, P.L.: Split graphs. In: Proceedings of the 8th Southeastern Conference on Combinatorics, Graph Theory and Computing, Congressus Numerantium, vol. XIX, pp. 311–315 (1977)
5. Gilmore, P.C., Hoffman, A.J.: A characterization of comparability graphs and of interval graphs. Can. J. Math. **16**, 539–548 (1964). https://doi.org/10.4153/CJM-1964-055-5
6. Hammer, P.L., Peled, U.N., Sun, X.: Difference graphs. Discret. Appl. Math. **28**(1), 35–44 (1990). https://doi.org/10.1016/0166-218X(90)90092-Q
7. Hsu, W.L., Ma, T.H.: Fast and simple algorithms for recognizing chordal comparability graphs and interval graphs. SIAM J. Comput. **28**(3), 1004–1020 (1999). https://doi.org/10.1137/S0097539792224814
8. Klavík, P., Otachi, Y., Sejnoha, J.: On the classes of interval graphs of limited nesting and count of lengths. In: Hong, S. (ed.) ISAAC 2016. LIPIcs, vol. 64, pp. 45:1–45:13. Schloss Dagstuhl - Leibniz-Zentrum fuer Informatik (2016). https://doi.org/10.4230/LIPIcs.ISAAC.2016.45
9. Kobayashi, Y., Okamoto, Y., Otachi, Y., Uno, Y.: Linear-time recognition of double-threshold graphs (2020, this volume)
10. Leibowitz, R.: Interval Counts and Threshold Numbers of Graphs. Ph.D. thesis, Rutgers University, New Brunswick, New Jersey (1978)
11. Looges, P.J., Olariu, S.: Optimal greedy algorithms for indifference graphs. Comput. Math. Appl. **25**(7), 15–25 (1993). https://doi.org/10.1016/0898-1221(93)90308-I
12. Mahadev, N.V., Peled, U.N.: Threshold graphs and related topics. In: Annals of Discrete Mathematics, vol. 56. North-Holland Publishing Co., Amsterdam, The Netherlands (1995). https://doi.org/10.1016/S0167-5060(06)80001-4
13. Nikolopoulos, S.D.: Recognizing cographs and threshold graphs through a classification of their edges. Inf. Proc. Lett. **74**(3–4), 129–139 (2000). https://doi.org/10.1016/S0020-0190(00)00041-7
14. Ramalingam, G., Rangan, C.P.: A unified approach to domination problems on interval graphs. Inf. Proc. Lett. **27**(5), 271–274 (1988). https://doi.org/10.1016/0020-0190(88)90091-9
15. Ravanmehr, V., Puleo, G.J., Bolouki, S., Milenkovic, O.: Paired threshold graphs. Discret. Appl. Math. **250**, 291–308 (2018). https://doi.org/10.1016/j.dam.2018.05.008
16. Roberts, F.S.: Indifference graphs. In: Harary, F. (ed.) Proof Techniques in Graph Theory (Proceedings of the 2nd Ann Arbor Graph Theory Conference 1968), pp. 139–146. Academic Press, New York (1969)
17. Wolk, E.S.: The comparability graph of a tree. P. Amer. Math. Soc. **13**, 789–795 (1962). https://doi.org/10.1090/S0002-9939-1962-0172273-0

Drawing Graphs as Spanners

Oswin Aichholzer[1], Manuel Borrazzo[2], Prosenjit Bose[3], Jean Cardinal[4],
Fabrizio Frati[2(✉)], Pat Morin[3], and Birgit Vogtenhuber[1]

[1] Graz University of Technology, Graz, Austria
{oaich,bvogt}@ist.tugraz.at
[2] Roma Tre University, Rome, Italy
{borrazzo,frati}@dia.uniroma3.it
[3] Carleton University, Ottawa, Canada
{jit,morin}@scs.carleton.ca
[4] Université Libre de Bruxelles (ULB), Brussels, Belgium
jcardin@ulb.ac.be

Abstract. We study the problem of embedding graphs in the plane as good geometric spanners. That is, for a graph G, the goal is to construct a straight-line drawing Γ of G in the plane such that, for any two vertices u and v of G, the ratio between the minimum length of any path from u to v and the Euclidean distance between u and v is small. The maximum such ratio, over all pairs of vertices of G, is the *spanning ratio* of Γ.

First, we show that deciding whether a graph admits a straight-line drawing with spanning ratio 1, a proper straight-line drawing with spanning ratio 1, and a planar straight-line drawing with spanning ratio 1 are NP-complete, $\exists\mathbb{R}$-complete, and linear-time solvable problems, respectively. Second, we prove that, for every $\epsilon > 0$, every (planar) graph admits a proper (resp. planar) straight-line drawing with spanning ratio smaller than $1 + \epsilon$. Third, we note that our drawings with spanning ratio smaller than $1 + \epsilon$ have large edge-length ratio, that is, the ratio between the lengths of the longest and of the shortest edge is exponential. We show that this is sometimes unavoidable. More generally, we identify having bounded toughness as the criterion that distinguishes graphs that admit straight-line drawings with constant spanning ratio and polynomial edge-length ratio from graphs that do not.

1 Introduction

Let P be a set of points in the plane and let $\mathcal{G}$ be a geometric graph whose vertex set is P. We say that $\mathcal{G}$ is a *t-spanner* if, for every pair of points p and q in P, there exists a path from p to q in $\mathcal{G}$ whose total edge length is at most t times the Euclidean distance $\|pq\|$ between p and q. The *spanning ratio* of $\mathcal{G}$ is the smallest real number t such that $\mathcal{G}$ is a t-spanner. The problem of constructing, for a given set P of points in the plane, a sparse (and possibly planar) geometric

Partially supported by the MSCA-RISE project "CONNECT", N° 734922, by the NSERC of Canada, and by the MIUR-PRIN project "AHeAD", N° 20174LF3T8.

I. Adler and H. Müller (Eds.): WG 2020, LNCS 12301, pp. 310–324, 2020.
https://doi.org/10.1007/978-3-030-60440-0_25

graph whose vertex set is P and whose spanning ratio is small has received considerable attention; see, e.g., [12–15, 18, 21, 23, 39, 50, 51].

In this paper we look at the construction of geometric graphs with small spanning ratio from a different perspective. Namely, the problem we consider is whether it is possible to embed a given abstract graph in the plane as a geometric graph with small spanning ratio. That is, for a given graph, we want to construct a straight-line drawing with small spanning ratio, where the spanning ratio of a straight-line drawing is the maximum ratio, over all pairs of vertices u and v, between the total edge length of a shortest path from u to v and $\|uv\|$.

Graph embeddings in which every pair of vertices is connected by a path satisfying certain geometric properties have been the subject of intensive research. As a notorious example, a *greedy* drawing of a graph [6, 8, 19, 24, 31, 36, 40, 42, 43, 48] is such that, for every pair of vertices u and v, there is a path from u to v that monotonically decreases the distance to v at every vertex. Further examples are *self-approaching* and *increasing-chord* drawings [4, 20, 41], *angle-monotone* drawings [10, 20, 37], *monotone* drawings [5, 7, 29, 30, 32, 35] and *strongly-monotone* drawings [5, 25, 35]. While greedy, monotone, and strongly-monotone drawings might have unbounded spanning ratio, self-approaching, increasing-chord, and angle-monotone drawings are known to have spanning ratio at most 5.34 [33], at most 2.1 [44], and at most 1.42 [10], respectively. However, not all graphs, and not even all trees [36, 40], admit such drawings.

Our results are the following.

First, we look at straight-line drawings with spanning ratio equal to 1, which is clearly the smallest attainable value by any graph. We prove that deciding whether a graph admits a straight-line drawing, a proper straight-line drawing (in which no vertex-vertex or vertex-edge overlaps are allowed), and a planar straight-line drawing with spanning ratio 1 are NP-complete, $\exists\mathbb{R}$-complete, and linear-time solvable problems, respectively.

Second, we show that allowing each shortest path to have a total edge length slightly larger than the Euclidean distance between its end-vertices makes it possible to draw all graphs. Namely, for every $\epsilon > 0$, every graph has a proper straight-line drawing with spanning ratio smaller than $1 + \epsilon$ and every planar graph has a planar straight-line drawing with spanning ratio smaller than $1 + \epsilon$.

Third, we address the issue that our drawings with spanning ratio smaller than $1 + \epsilon$ have poor resolution. That is, the *edge-length ratio* of these drawings, i.e., the ratio between the lengths of the longest and of the shortest edge, might be super-polynomial in the number of vertices of the graph. We show that this is sometimes unavoidable, as stars have exponential edge-length ratio in any straight-line drawing with constant spanning ratio. More in general, we present graph families for which any straight-line drawing with constant spanning ratio has edge-length ratio which is exponential in the inverse of the toughness. On the other hand, we prove that graph families with constant toughness admit proper straight-line drawings with polynomial edge-length ratio and constant spanning ratio. Finally, we prove that bounded-degree trees admit planar straight-line drawings with polynomial edge-length ratio and constant spanning ratio.

Full versions of sketched or omitted proofs can be found in [3].

2   Preliminaries

For a graph G and a set S of vertices of G, we denote by $G - S$ the graph obtained from G by removing the vertices in S and their incident edges. The subgraph of G *induced by* S is the graph whose vertex set is S and whose edge set consists of every edge of G that has both its end-vertices in S. The *toughness* of a graph G is the largest real number $t > 0$ such that, for any set S such that $G - S$ consists of $k \geq 2$ connected components, we have $|S| \geq t \cdot k$.

A *drawing* of a graph maps each vertex to a distinct point in the plane and each edge to a Jordan arc between its end-vertices. A *straight-line* drawing maps each edge to a straight-line segment. Let Γ be a straight-line drawing of a graph G. The *length of a path* in Γ is the sum of the lengths of its edges. We denote by $\|uv\|_\Gamma$ (by $\pi_\Gamma(u, v)$) the Euclidean distance (resp. the length of a shortest path) between two vertices u and v in Γ; we sometimes drop the subscript Γ when the drawing we refer to is clear from the context. The *spanning ratio* of Γ is the real value $\max\limits_{u,v} \frac{\pi_\Gamma(u,v)}{\|uv\|_\Gamma}$, where the maximum is over all pairs of vertices u and v of G.

A drawing is *planar* if no two edges intersect, except at common end-vertices. A planar drawing partitions the plane into connected regions, called *faces*; the bounded faces are *internal*, while the unbounded face is the *outer face*. A graph is *planar* if it admits a planar drawing. A planar graph is *maximal* if adding any edge to it violates its planarity. In any planar drawing of a maximal planar graph every face is delimited by a 3-cycle. The *bounding box* $\mathcal{B}(\Gamma)$ of a drawing Γ is the smallest axis-parallel rectangle containing Γ in the closure of its interior. The *width* and *height* of Γ are the width and height of $\mathcal{B}(\Gamma)$.

3   Drawings with Spanning Ratio 1

In this section we study drawings with spanning ratio equal to 1.

Theorem 1. *Recognizing whether a graph admits a straight-line drawing with spanning ratio equal to 1 is an NP-complete problem.*

Proof Sketch: The core of the proof consists of showing that a graph has a straight-line drawing with spanning ratio 1 if and only if it contains a Hamiltonian path (then the theorem follows from the NP-completeness of the problem of deciding whether a graph contains a Hamiltonian path [27,28]). In particular, let Γ be a straight-line drawing with spanning ratio 1 of a graph G and assume w.l.o.g. that no two vertices have the same x-coordinate in Γ. Let $v_1, v_2, \ldots, v_n$ be the vertices of G, ordered by increasing x-coordinates. Then, for $i = 1, 2, \ldots, n - 1$, we have that G contains the edge $v_i v_{i+1}$, as any other path between v_i and v_{i+1} would be longer than $\|v_i v_{i+1}\|$. Hence, G contains the Hamiltonian path $(v_1, v_2, \ldots, v_n)$.   □

The *existential theory of the reals* problem asks whether real values exist for n variables such that a quantifier-free formula, consisting of polynomial equalities and inequalities on such variables, is satisfied. The class of problems that are complete for the existential theory of the reals is denoted by $\exists\mathbb{R}$ [45]. It is known that NP $\subseteq \exists\mathbb{R} \subseteq$ PSPACE [16], however it is not known whether $\exists\mathbb{R} \subseteq$ NP. Many geometric problems are $\exists\mathbb{R}$-complete, see, e.g., [1,38].

Theorem 2. *Recognizing whether a graph admits a proper straight-line drawing with spanning ratio equal to 1 is an $\exists\mathbb{R}$-complete problem.*

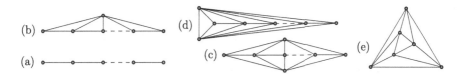

Fig. 1. The five graph classes defined in [22].

Proof Sketch: Let Γ be a proper straight-line drawing with spanning ratio 1 of a graph G. Let S be the set of points at which the vertices of G are drawn. It is easy to prove that the *point visibility graph* G_S of S is isomorphic to G, where the point visibility graph G_P of a point set $P \subset \mathbb{R}^2$ has a vertex for each point $p \in P$ and has an edge between two vertices if and only if the straight-line segment between the corresponding points does not contain any point of P in its interior. The theorem follows from the fact that recognizing point visibility graphs is a problem that is $\exists\mathbb{R}$-complete [17].   □

Theorem 3. *Recognizing whether a graph admits a planar straight-line drawing with spanning ratio equal to 1 is a linear-time solvable problem.*

Proof: Dujmović et al. [22] characterized the graphs that admit a planar straight-line drawing with a straight-line segment between every two vertices as the graphs in the five graph classes in Fig. 1. Since a straight-line drawing has spanning ratio 1 if and only if every two vertices are connected by a straight-line segment, the theorem follows from the fact that recognizing whether a graph belongs to such five graph classes can be easily done in linear time.   □

4   Drawings with Spanning Ratio $1 + \epsilon$

In this section we study straight-line drawings with spanning ratio arbitrarily close to 1. Most of the section is devoted to a proof of the following result.

Theorem 4. *For every $\epsilon > 0$, every connected planar graph admits a planar straight-line drawing with spanning ratio smaller than $1 + \epsilon$.*

Let G be an n-vertex maximal planar graph with $n \geq 3$, let $\mathcal{G}$ be a planar drawing of G, and let (u, v, z) be the cycle delimiting the outer face of G in $\mathcal{G}$. A *canonical ordering* [9,26,34] for G is a total ordering $[v_1, \ldots, v_n]$ of its vertex set such that the following hold for $k = 3, \ldots, n$: (i) $v_1 = u$, $v_2 = v$, and $v_n = z$;

(ii) the subgraph G_k of G induced by $v_1, \ldots, v_k$ is 2-connected and the cycle $\mathcal{C}_k$ delimiting its outer face in $\mathcal{G}$ consists of the edge $v_1 v_2$ and of a path $\mathcal{P}_k$ between v_1 and v_2; and (iii) v_k is incident to the outer face of G_k in $\mathcal{G}$. The following lemma generalizes the concept of canonical orderings to non-maximal connected planar graphs.

Lemma 1. *Let H be an n-vertex connected planar graph. There exist an n-vertex maximal planar graph G and a canonical ordering $[v_1, \ldots, v_n]$ for G such that, for each $k \in \{1, \ldots, n\}$, the subgraph H_k of H induced by $\{v_1, \ldots, v_k\}$ is connected.*

Fig. 2. Construction for the case in which $a(v) = b(v)$.

Proof Sketch: For each $k = 2, \ldots, n$, we let G_k be the subgraph of G induced by $v_1, \ldots, v_k$ and L_k be the graph composed of G_k and of the vertices and edges of H that are not in G_k. Further, we define $v_1, \ldots, v_k$ and G_k so that H_k is connected, G_k is 2-connected, and L_k admits a planar drawing $\mathcal{L}_k$ such that:

1. the outer face of the planar drawing $\mathcal{G}_k$ of G_k in $\mathcal{L}_k$ is delimited by a cycle $\mathcal{C}_k$ composed of the edge $v_1 v_2$ and of a path $\mathcal{P}_k$ between v_1 and v_2;
2. v_k is incident to the outer face of $\mathcal{G}_k$;
3. every internal face of $\mathcal{G}_k$ is delimited by a 3-cycle; and
4. the vertices and edges of H that are not in G_k lie in the outer face of $\mathcal{G}_k$.

If $k = 2$, then construct any planar drawing $\mathcal{L}_2$ of H and define v_1 and v_2 as the end-vertices of any edge $v_1 v_2$ incident to the outer face of $\mathcal{L}_2$. Properties 1–4 are then trivially satisfied (in this case the path $\mathcal{P}_2$ is the single edge $v_1 v_2$).

If $2 < k < n$, assume that $v_1, \ldots, v_{k-1}$ and G_{k-1} have been defined so that H_{k-1} is connected, G_{k-1} is 2-connected, and L_{k-1} admits a planar drawing $\mathcal{L}_{k-1}$ satisfying Properties 1–4. Let $\mathcal{P}_{k-1} = (u = w_1, w_2, \ldots, w_x = v)$, where $x \geq 2$.

Consider any vertex v in $L_{k-1} \setminus G_{k-1}$. By Properties 1 and 4 of $\mathcal{L}_{k-1}$, all the neighbors of v in G_{k-1} lie in $\mathcal{P}_{k-1}$. We say that v is a *candidate* (to be designated as v_k) *vertex* if, for some $1 \leq i \leq x$, there exists an edge $w_i v$ such that $w_i v$ immediately follows the edge $w_i w_{i-1}$ in clockwise order around w_i or immediately follows the edge $w_i w_{i+1}$ in counter-clockwise order around w_i.

For each candidate vertex v, let $w_{a(v)}$ and $w_{b(v)}$ be the neighbors of v in $\mathcal{P}_{k-1}$ such that $a(v)$ is minimum and $b(v)$ is maximum (possibly $a(v) = b(v)$). If $a(v) < b(v)$, let the *reference cycle $\mathcal{C}(v)$ of v* be composed of the edges $w_{a(v)} v$ and $w_{b(v)} v$ and of the subpath of $\mathcal{P}_{k-1}$ between $w_{a(v)}$ and $w_{b(v)}$. Define the *depth* of v as 0 if $a(v) = b(v)$ or as the number of candidate vertices that lie inside $\mathcal{C}(v)$ in $\mathcal{L}_{k-1}$ otherwise. We select as $v_k := v$ any candidate vertex v with depth 0.

If $a(v) = b(v)$, as in Fig. 2, assume that $w_{a(v)}v$ immediately follows the edge $w_{a(v)}w_{a(v)+1}$ in counter-clockwise order around $w_{a(v)}$; the other case is symmetric. Define G_k as G_{k-1} plus the vertex v and the edges $w_{a(v)}v$ and $w_{a(v)+1}v$. Further, construct $\mathcal{L}_k$ by drawing the edge $w_{a(v)+1}v$ so that the cycle $(w_{a(v)}, w_{a(v)+1}, v)$ does not contain any vertex or edge in its interior.

If $a(v) < b(v)$, as in Fig. 3, redraw each biconnected component of L_{k-1} whose vertices different from v lie inside $\mathcal{C}(v)$ planarly so that it now lies outside $\mathcal{C}(v)$; after this modification, no vertex of L_{k-1} lies inside $\mathcal{C}(v)$. Then define G_k as G_{k-1} plus the vertex v and the edges $w_{a(v)}v, w_{a(v)+1}v, \ldots, w_{b(v)}v$. Further, construct $\mathcal{L}_k$ by drawing the edges among $w_{a(v)}v, w_{a(v)+1}v, \ldots, w_{b(v)}v$ not in H so that they all lie inside $\mathcal{C}(v)$ and so that the edges $w_{a(v)}v, w_{a(v)+1}v, \ldots, w_{b(v)}v$ appear consecutively and in this counter-clockwise order around v.

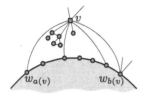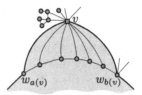

Fig. 3. Construction for the case in which $a(v) < b(v)$.

In both cases, it is easy to see that H_k is connected, G_k is 2-connected, and $\mathcal{L}_k$ satisfies Properties 1–4. See [3] for details.

If $k = n$, the construction is similar to the one described for the case $2 < k < n$, however further edges are added to G_n so to ensure that the outer face of $\mathcal{G}_n$ is delimited by the 3-cycle (v_1, v_2, v_n). Setting $G := G_n$ concludes the proof. $\square$

Lemma 2. *For every $k = 3, \ldots, n$ and for every $\epsilon > 0$, there exists a planar straight-line drawing Γ_k of G_k such that: (1) the outer face of Γ_k is delimited by the cycle $\mathcal{C}_k$; further, the path $\mathcal{P}_k$ is x-monotone and lies above the edge uv, except at u and v; and (2) the restriction Ξ_k of Γ_k to the vertices and edges of H_k is a drawing with spanning ratio smaller than $1 + \epsilon$.*

Proof Sketch: The proof is by induction on k. The base case $k = 3$ is trivial.

Assume that, for some $k = 4, \ldots, n$, a planar
straight-line drawing Γ_{k-1} of G_{k-1} has been constructed satisfying Properties 1 and 2; see Fig. 4. Let
δ be the diameter of a disk D enclosing Γ_{k-1}. We
construct Γ_k from Γ_{k-1} as follows. Let $\mathcal{P}_{k-1} = (u =
w_1, w_2, \ldots, w_x = v)$. As proved in [26], the neighbors
of v_k in G_{k-1} are the vertices in a sub-path $(w_p, \ldots, w_q)$
of $\mathcal{P}_{k-1}$, where $p < q$. By Property 1 of Γ_{k-1}, we
have $x(w_p) < x(w_q)$. We then place v_k at any point
in the plane satisfying the following conditions: (i)
$x(w_p) < x(v_k) < x(w_q)$; (ii) for every $i = p, \ldots, q - 1$,
the y-coordinate of v_k is larger than those of the intersection points between the line through $w_i w_{i+1}$ and the
vertical lines through w_p and w_q; and (iii) the distance
between v_k and the point of D closest to v_k is a real
value $d > \frac{k\delta}{\epsilon}$.

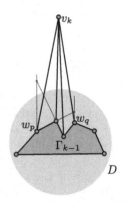

Fig. 4. Construction of Γ_k from Γ_{k-1}.

Since $\mathcal{P}_k$ is obtained from $\mathcal{P}_{k-1}$ by substituting the path $(w_p, w_{p+1}, \ldots, w_q)$
with the path (w_p, v_k, w_q), Condition (i) and the x-monotonicity of $\mathcal{P}_{k-1}$ imply
that $\mathcal{P}_k$ is x-monotone. Condition (ii), the x-monotonicity of $\mathcal{P}_{k-1}$, and the
planarity of Γ_{k-1} imply that Γ_k is planar. We now prove that the spanning ratio
of Ξ_k is smaller than $1 + \epsilon$. Consider any two vertices v_i and v_j. If $i < k$ and
$j < k$, then $\frac{\pi_{\Xi_k}(v_i, v_j)}{\|v_i v_j\|_{\Xi_k}} \leq \frac{\pi_{\Xi_{k-1}}(v_i, v_j)}{\|v_i v_j\|_{\Xi_{k-1}}} < 1 + \epsilon$. If $i = k$, then $\|v_k v_j\|_{\Xi_k} \geq d$,
by Condition (iii). Consider the path $P(v_k, v_j)$ composed of any edge $v_k v_\ell$ in
H_k incident to v_k (which exists since H_k is connected) and of any path in H_{k-1}
between v_ℓ and v_j (which exists since H_{k-1} is connected). The length of $P(v_k, v_j)$
is at most $d + \delta$ (by Condition (iii) and by the triangular inequality, this is an
upper bound on $\|v_k v_\ell\|_{\Xi_k}$) plus $(k - 2) \cdot \delta$ (this is an upper bound on the length
of any path in H_{k-1}). Hence, $\frac{\pi_{\Xi_k}(v_k, v_j)}{\|v_k v_j\|_{\Xi_k}} < \frac{d + k\delta}{d} < 1 + \epsilon$. This completes the
induction and the proof of the lemma.   □

Lemmata 1 and 2 imply Theorem 4. Namely, for a connected planar graph
H, by Lemma 1 we can construct a maximal planar graph G that, by Lemma 2
(with $k = n$) and for every $\epsilon > 0$, admits a planar straight-line drawing whose
restriction to H is a drawing with spanning ratio smaller than $1 + \epsilon$.

The following can be obtained by means of techniques similar to (and simpler
than) the ones in the proof of Theorem 4; the proof can be found in [3].

Theorem 5. *For every $\epsilon > 0$, every connected graph admits a proper straight-
line drawing with spanning ratio smaller than $1 + \epsilon$.*

5   Drawings with Small Spanning and Edge-Length Ratios

In this section we study straight-line drawings with small spanning ratio and
edge-length ratio. Our main result is the following.

Theorem 6. *For every $\epsilon > 0$ and every $\tau > 0$, every n-vertex graph with toughness τ admits a proper straight-line drawing whose spanning ratio is at most $1 + \epsilon$ and whose edge-length ratio is in $\mathcal{O}\left(n^{\frac{\log_2(2 + \lceil 2/\epsilon \rceil)}{\log_2(2 + \lceil 1/\tau \rceil) - \log_2(1 + \lceil 1/\tau \rceil)}} \cdot 1/\epsilon\right)$. Further, for every $0 < \tau < 1$, there is a graph G with toughness τ whose every straight-line drawing with spanning ratio at most s has edge-length ratio in $2^{\Omega(1/(\tau \cdot s^2))}$.*

In order to prove Theorem 6, we study straight-line drawings of bounded-degree trees. This is because there is a strong connection between the toughness of a graph and the existence of a spanning tree with bounded degree. Indeed, if a graph G has toughness τ, then it has a spanning tree with maximum degree $\lceil 1/\tau \rceil + 2$ [49]. Further, a tree has toughness equal to the inverse of its maximum degree. We start by proving the following lower bound.

Theorem 7. *For any $s \geq 1$, any straight-line drawing with spanning ratio at most s of a tree with a vertex of degree d has edge-length ratio in $2^{\Omega(d/s^2)}$.*

Proof: For any $s \geq 1$, let Γ be any straight-line drawing of T with spanning ratio at most s; refer to Fig. 5(a). Let u_T be a vertex of degree d. Assume w.l.o.g. up to a scaling (which does not alter the edge-length ratio and the spanning ratio of Γ) that the length of the shortest edge incident to u_T in Γ is 1. For any integer $i \geq 0$, let $\mathcal{C}_i$ be the circle centered at u_T whose radius is $r_i = 2^i$. Further, for any integer $i > 0$, let $\mathcal{A}_i$ be the closed annulus delimited by $\mathcal{C}_{i-1}$ and $\mathcal{C}_i$. By assumption, no neighbor of u_T lies inside the open disk delimited by $\mathcal{C}_0$. We claim that, for any integer $i > 0$ and for some constant c, there are at most $c \cdot s^2$ neighbors of u_T inside $\mathcal{A}_i$. This implies that at most $k \cdot c \cdot s^2$ neighbors of u_T lie inside the closed disk delimited by $\mathcal{C}_k$. Hence, if $d > k \cdot c \cdot s^2$, e.g., if $k = \lfloor \frac{d-1}{c \cdot s^2} \rfloor$, then there is a neighbor v_T of u_T outside $\mathcal{C}_k$. Then $\|u_T v_T\| > 2^k \in 2^{\Omega(d/s^2)}$. Hence, the theorem follows from the claim.

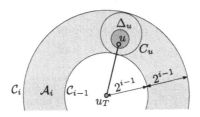

Fig. 5. Illustration for the proof of Theorem 7.

It remains to prove the claim. For each neighbor u of u_T inside $\mathcal{A}_i$, let Δ_u be a closed disk such that: (i) u lies inside Δ_u; (ii) Δ_u lies inside $\mathcal{A}_i$; and (iii) the diameter of Δ_u is $\delta_i = 2^{i-2}/s$. The existence of Δ_u can be proved as follows. Consider the circle $\mathcal{C}_u$ whose antipodal points are the intersection points of $\mathcal{C}_{i-1}$ and $\mathcal{C}_i$ with the ray from u_T through u. Note that $\mathcal{C}_u$ lies inside $\mathcal{A}_i$ and has diameter $2^{i-1} > \delta_i = 2^{i-2}/s$. Then Δ_u is any disk with diameter δ_i that contains u and that lies inside the closed disk delimited by C_u.

Suppose, for a contradiction, that there exist two neighbors u and v of u_T inside $\mathcal{A}_i$ such that the disks Δ_u and Δ_v intersect. Then $\pi_\Gamma(u,v) \geq 2^i$, since both the edges uu_T and vu_T are longer than $r_{i-1} = 2^{i-1}$. By the triangular inequality, $\|uv\|_\Gamma \leq 2 \cdot \delta_i = 2^{i-1}/s$. Hence $\frac{\pi_\Gamma(u,v)}{\|uv\|_\Gamma} \geq 2s$, while the spanning ratio of Γ is at most s. This contradiction proves that, for any two neighbors u and v of u_T inside $\mathcal{A}_i$, the disks Δ_u and Δ_v do not intersect. The area of $\mathcal{A}_i$ is $\pi \cdot (r_i^2 - r_{i-1}^2) = \pi \cdot (2^{2i} - 2^{2i-2}) = 3\pi \cdot (2^{2i-2})$. Since each disk Δ_u lying inside $\mathcal{A}_i$ has area $\pi \cdot (2^{2i-6}/s^2)$ and does not intersect any different disk Δ_v, it follows that $\mathcal{A}_i$ contains at most $\frac{3\pi \cdot (2^{2i-2}) \cdot s^2}{\pi \cdot 2^{2i-6}} = 48 \cdot s^2$ distinct disks Δ_u and hence at most $48 \cdot s^2$ neighbors of u_T. This proves the claim and hence the theorem. $\square$

Corollary 1. *Let S be an n-vertex star. For any $s \geq 1$, any straight-line drawing of S with spanning ratio at most s has edge-length ratio in $2^{\Omega(n/s^2)}$.*

The lower bound of Theorem 6 follows from Theorem 7 and from the fact that a tree with maximum degree d has toughness $1/d$. On the other hand, the upper bound of Theorem 6 is obtained by means of the following.

Theorem 8. *For every $\epsilon > 0$, every n-vertex tree T with maximum degree d admits a proper straight-line drawing such that no three vertices are collinear, the spanning ratio is at most $1+\epsilon$, the distance between any two vertices is at least 1, and the width, the height, and the edge-length ratio are in $\mathcal{O}\left(n^{\frac{\log_2(2+\lceil 2/\epsilon \rceil)}{\log_2(d/(d-1))}} \cdot 1/\epsilon\right)$.*

Theorem 8 proves the upper bound in Theorem 6 and hence concludes its proof. Namely, let G be an n-vertex graph with toughness τ and let $\epsilon > 0$; then G has a spanning tree T with maximum degree $d = \lceil 1/\tau \rceil + 2$ [49]. Apply Theorem 8 to construct a straight-line drawing Γ_T of T. Construct a straight-line drawing Γ_G of G from Γ_T by drawing the edges of G not in T as straight-line segments. Then Γ_G is proper, as no three vertices are collinear in Γ_T. Further, the spanning ratio of Γ_G is at most the one of Γ_T, hence it is at most $1 + \epsilon$. Finally, the edge-length ratio of Γ_G is in $\mathcal{O}\left(n^{\frac{\log_2(2+\lceil 2/\epsilon \rceil)}{\log_2(d/(d-1))}} \cdot 1/\epsilon\right)$, given that the distance between any two vertices in Γ_T (and hence in Γ_G) is at least 1 and given that the width and height of Γ_T (and hence of Γ_G) are in $\mathcal{O}\left(n^{\frac{\log_2(2+\lceil 2/\epsilon \rceil)}{\log_2(d/(d-1))}} \cdot 1/\epsilon\right)$.

We defer the proof of Theorem 8 to the full version of the paper [3] and present a proof of Theorem 9, in which it is shown that trees with bounded maximum degree even admit *planar* straight-line drawings with constant spanning ratio and polynomial edge-length ratio. The cost of planarity is found in the dependence on the maximum degree, which is worse than in Theorem 8.

Theorem 9. *For every $\epsilon > 0$, every n-vertex tree T with maximum degree d admits a planar straight-line drawing whose spanning ratio is at most $1 + \epsilon$ and whose edge-length ratio is in $\mathcal{O}\left((2n)^{2+(d-2) \cdot \log_2(1+\lceil \frac{2}{\epsilon} \rceil)} \cdot \log_2 n\right)$.*

Proof Sketch: Let $\gamma = \lceil \frac{2}{\epsilon} \rceil$. If $d \leq 2$, then T is a path and the desired drawing is trivially constructed. We can hence assume that $d \geq 3$. Root T at any leaf r; this ensures that every vertex of T has at most $d - 1$ children. In order to avoid some technicalities in the upcoming algorithm, we also assume that every non-leaf vertex of T has at least two children. This is obtained by inserting a new child for each vertex of T with just one child; note that the *size* of the tree, i.e., its number of vertices, is less than doubled by this modification. We again call T the tree after this modification and by n its size.

Our construction is a "well-spaced" version of an algorithm by Shiloach [47]. We construct a planar straight-line drawing Γ of T in which (i) r is at the top-left corner of $\mathcal{B}(\Gamma)$, and (ii) for every vertex u of T, the path from u to r in T is (non-strictly) xy-monotone.

If $n = 1$, then Γ is obtained by placing r at any point in the plane. If $n > 1$, then let $r_1, \ldots, r_k$ be the children of r, where $k \leq d - 1$, let $T_1, \ldots, T_k$ be the subtrees of T rooted at $r_1, \ldots, r_k$, whose sizes are $n_1, \ldots, n_k$, respectively. Assume, w.l.o.g. up to a relabeling, that $n_1 \leq \cdots \leq n_k$; hence, $n_i \leq n/2$ for $i = 1, 2, \ldots, k - 1$. Refer to Figure 6. Place r at any point in the plane. Inductively construct planar straight-line drawings $\Gamma_1, \ldots, \Gamma_k$ of $T_1, \ldots, T_k$, respectively. Position Γ_1 so that r_1 is on the same vertical line as r, one unit below it; let d_1 be the width of Γ_1. Then, for $i = 2, \ldots, k$, position Γ_i so that r_i is one unit below r and $\gamma \cdot (d_{i-1} + \log_2 n)$ units to the right of the right side of $\mathcal{B}(\Gamma_{i-1})$; denote by d_i the width of the bounding box of the drawings $\Gamma_1, \ldots, \Gamma_i$. Finally, move Γ_k one unit above, so that r_k is on the same horizontal line as r.

We now analyze the properties of Γ. By construction Γ is a straight-line drawing. The planarity of Γ is easily proved by exploiting the fact that r_i is at the top-left corner of $\mathcal{B}(\Gamma_i)$ and that $r_1, r_2, \ldots, r_{k-1}$ all lie one unit below r.

Height. Let $h(n)$ be the maximum height of a drawing of an n-vertex tree constructed by the algorithm. The same analysis as in [47] shows that $h(n) \leq \log_2 n$, given that $h(1) = 0$ and $h(n) < \max\{h(\frac{n}{2}) + 1, h(n-1)\}$ for $n \geq 2$.

Spanning Ratio. We prove that, for any two vertices u and v that do not belong to the same subtree T_i, it holds true that $\frac{\pi_\Gamma(u,v)}{\|uv\|_\Gamma} \leq \frac{\gamma+2}{\gamma}$. This suffices to prove that the spanning ratio of Γ is at most $\frac{\gamma+2}{\gamma}$. Suppose w.l.o.g. that u belongs to a subtree T_i and v belongs to a subtree T_j, with $i < j$.

First, we have $\|uv\| \geq x_v + \gamma \cdot (d_{j-1} + \log_2 n)$, where x_v denotes the distance between v and the left side of $\mathcal{B}(\Gamma_j)$, while the second term is the distance between the left side of $\mathcal{B}(\Gamma_j)$ and the right side of $\mathcal{B}(\Gamma_{j-1})$.

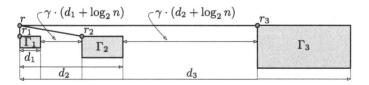

Fig. 6. Inductive construction of Γ. In this example $k = 3$.

Clearly, we have $\pi_\Gamma(u,v) = \pi_\Gamma(u,r) + \pi_\Gamma(r,v)$. The path between u and r (between v and r) is xy-monotone, hence $\pi_\Gamma(u,r)$ (resp. $\pi_\Gamma(v,r)$) is upper bounded by the horizontal distance plus the vertical distance between u and r (resp. between v and r). The vertical distance between u and r (between v and r) is at most $\log_2(n)$, since the height of Γ is at most $\log_2(n)$. The horizontal distance between u and r is at most $d_i \leq d_{j-1}$, while the one between v and r is $x_v + \gamma \cdot (d_{j-1} + \log_2 n) + d_{j-1}$. Hence, $\pi_\Gamma(u,v) \leq (d_{j-1} + \log_2 n) + (x_v + \gamma \cdot (d_{j-1} + \log_2 n) + d_{j-1} + \log_2 n) = x_v + (\gamma + 2) \cdot (d_{j-1} + \log_2 n)$. Thus:

$$\frac{\pi_\Gamma(u,v)}{\|uv\|_\Gamma} \leq \frac{(\gamma+2) \cdot (\frac{x_v}{\gamma} + d_{j-1} + \log_2 n)}{\gamma \cdot (\frac{x_v}{\gamma} + d_{j-1} + \log_2 n)} \leq \frac{\gamma+2}{\gamma} \leq 1 + \epsilon.$$

Width. Let $w_1, \ldots, w_k$ be the widths of $\Gamma_1, \ldots, \Gamma_k$. By construction, $d_1 = w_1$ and, for each $j = 2, \ldots, k$, we have $d_j = d_{j-1} + \gamma \cdot (d_{j-1} + \log_2 n) + w_j = (\gamma+1) \cdot d_{j-1} + \gamma \cdot \log_2 n + w_j$. Hence, by induction on j, we have $d_j = (\gamma+1)^{j-1} \cdot w_1 + (\gamma+1)^{j-2} \cdot w_2 + \ldots + (\gamma+1) \cdot w_{j-1} + w_j + ((\gamma+1)^{j-1} - 1) \cdot \log_2 n$. In particular, the width of Γ is equal to d_k and hence to:

$$\sum_{i=1}^{k} ((\gamma+1)^{k-i} \cdot w_i) + ((\gamma+1)^{k-1} - 1) \cdot \log_2 n. \tag{1}$$

Let $w(n)$ be the maximum width of a drawing of an n-vertex tree constructed by the algorithm. By construction $w(1) = 0$. For $n \geq 2$, by Equality 1, we get:

$$w(n) \leq (\gamma+1)^{d-2} \cdot \sum_{i=1}^{k-1} w(n_1) + w(n_k) + (\gamma+1)^{d-2} \cdot \log_2 n. \tag{2}$$

Recall that $n_1, \ldots, n_{k-1} \leq n/2$. On the other hand, n_k might be larger than $n/2$; if that is so, Inequality 2 is used to replace the term $w(n_k)$ into Inequality 2 itself. The repetition of this substitution eventually results in the following (see [3] for details):

$$w(n) \leq (\gamma+1)^{d-2} \cdot \sum_{i,j} w(n_{i,j}) + (\gamma+1)^{d-2} \cdot (n-1) \cdot \log_2 n, \tag{3}$$

where the terms $n_{i,j}$ denote the sizes of distinct subtrees of T (hence $\sum n_{i,j} \leq n - 1$), each of which has at most $n/2$ nodes (hence $n_{i,j} \leq n/2$).

We prove, by induction on n, that $w(n) \leq f(n) := ((\gamma+1)^{d-2})^{\log_2 n} \cdot n^2 \cdot \log_2 n$. This is trivial when $n = 1$, given that $w(1) = 0$. Assume now that $n > 1$. By Inequality 3 and by induction, we get $w(n) \leq (\gamma+1)^{d-2} \cdot \sum_{i,j} \left(((\gamma+1)^{d-2})^{\log_2 n_{i,j}} \cdot n_{i,j}^2 \cdot \log_2 n_{i,j} \right) + (\gamma+1)^{d-2} \cdot (n-1) \cdot \log_2 n$. Since $n_{i,j} \leq n/2 < n$, we get $w(n) \leq (\gamma+1)^{d-2} \cdot ((\gamma+1)^{d-2})^{\log_2(n/2)} \cdot \sum_{i,j} n_{i,j}^2 \cdot \log_2 n + (\gamma+1)^{d-2} \cdot (n-1) \cdot \log_2 n = ((\gamma+1)^{d-2})^{\log_2 n} \cdot \sum_{i,j} n_{i,j}^2 \cdot \log_2 n + (\gamma+1)^{d-2} \cdot (n-1) \cdot \log_2 n.$

Since $\sum n_{i,j} \le n-1$, we have $\sum\limits_{i,j} n_{i,j}^2 \le (n-1)^2$. Thus, $w(n) \le \left((\gamma+1)^{d-2}\right)^{\log_2 n}$. $((n-1)^2 + (n-1)) \cdot \log_2 n \le \left((\gamma+1)^{d-2}\right)^{\log_2 n} \cdot n^2 \cdot \log_2 n$. This completes the induction and the analysis of the width of Γ.

Edge-Length Ratio. By construction, the length of each edge connecting r to a child is larger than or equal to 1, hence the same is true for every edge of T. Thus, the edge-length ratio of Γ is upper bounded by the maximum length of an edge of T. In turn, this is at most the sum of the height plus the width of Γ, which is in $\mathcal{O}\left(\left((\gamma+1)^{d-2}\right)^{\log_2 n} \cdot n^2 \cdot \log_2 n\right)$, as proved above. The factor $\left((\gamma+1)^{d-2}\right)^{\log_2 n}$ can be rewritten as $n^{(d-2)\cdot\log_2(\gamma+1)}$. The bound claimed in the statement is then obtained by substituting $\gamma = \lceil\frac{2}{\epsilon}\rceil$ and by observing that the value of n used in the calculations is at most twice the size of the initial tree. $\square$

6   Open Problems

Our research raises a number of open problems which might be worth studying.

First, it would be interesting to tighten the bounds in Theorem 6 relating the toughness to the edge-length ratio of a drawing with constant spanning ratio.

Second, there is still much to be understood about the edge-length ratio of planar straight-line drawings with constant spanning ratio. Theorem 9 shows that planar straight-line drawings with constant spanning ratio and polynomial edge-length ratio exist for bounded-degree trees. We also observe that every n-vertex 2-connected outerplanar graph G admits a planar straight-line drawing with spanning ratio at most $\sqrt{2}$ and edge-length ratio in $\mathcal{O}(n^{1.5})$; this can be achieved by placing the vertices of G, in the order given by the Hamiltonian cycle of G, at the vertices of a lattice xy-monotone polygonal curve; see, e.g., [2]. Further, Schnyder drawings are known to be 2-spanners [13]. Hence, n-vertex 3-connected planar graphs admit planar straight-line drawings with spanning ratio at most 2 and edge-length ratio in $\mathcal{O}(n)$ [11,46]; do they admit planar straight-line drawings with spanning ratio smaller than 2 (and possibly arbitrarily close to 1) and polynomial edge-length ratio? Can Theorem 6 be extended to prove that a planar straight-line drawing with constant spanning ratio and polynomial edge-length ratio exists for planar graphs with bounded toughness?

References

1. Abrahamsen, M., Adamaszek, A., Miltzow, T.: The art gallery problem is $\exists\mathbb{R}$-complete. In: Diakonikolas, I., Kempe, D., Henzinger, M. (eds.) 50th Annual Symposium on Theory of Computing (STOC 2018), pp. 65–73. ACM (2018)
2. Acketa, D.M., Zunic, J.D.: On the maximal number of edges of convex digital polygons included into an $m \times m$-grid. J. Comb. Theory Ser. A **69**(2), 358–368 (1995)
3. Aichholzer, O., et al.: Drawing graphs as spanners. CoRR abs/2002.05580 (2020). https://arxiv.org/abs/2002.05580

4. Alamdari, S., Chan, T.M., Grant, E., Lubiw, A., Pathak, V.: Self-approaching Graphs. In: Didimo, W., Patrignani, M. (eds.) GD 2012. LNCS, vol. 7704, pp. 260–271. Springer, Heidelberg (2013). https://doi.org/10.1007/978-3-642-36763-2_23

5. Angelini, P., Colasante, E., Di Battista, G., Frati, F., Patrignani, M.: Monotone drawings of graphs. J. Graph Algorithms Appl. **16**(1), 5–35 (2012)

6. Angelini, P., Di Battista, G., Frati, F.: Succinct greedy drawings do not always exist. Networks **59**(3), 267–274 (2012)

7. Angelini, P., et al.: Monotone drawings of graphs with fixed embedding. Algorithmica **71**(2), 233–257 (2015)

8. Angelini, P., Frati, F., Grilli, L.: An algorithm to construct greedy drawings of triangulations. J. Graph Algorithms Appl. **14**(1), 19–51 (2010)

9. Badent, M., Brandes, U., Cornelsen, S.: More canonical ordering. J. Graph Algorithms Appl. **15**(1), 97–126 (2011)

10. Bonichon, N., Bose, P., Carmi, P., Kostitsyna, I., Lubiw, A., Verdonschot, S.: Gabriel triangulations and angle-monotone graphs: local routing and recognition. In: Hu, Y., Nöllenburg, M. (eds.) GD 2016. LNCS, vol. 9801, pp. 519–531. Springer, Cham (2016). https://doi.org/10.1007/978-3-319-50106-2_40

11. Bonichon, N., Felsner, S., Mosbah, M.: Convex drawings of 3-connected plane graphs. Algorithmica **47**(4), 399–420 (2007)

12. Bose, P., Devroye, L., Löffler, M., Snoeyink, J., Verma, V.: Almost all Delaunay triangulations have stretch factor greater than $\pi/2$. Comput. Geom. Theory Appl. **44**(2), 121–127 (2011)

13. Bose, P., Fagerberg, R., van Renssen, A., Verdonschot, S.: Competitive routing in the half-θ_6-graph. In: Rabani, Y. (ed.) 23rd Annual ACM-SIAM Symposium on Discrete Algorithms (SODA 2012), pp. 1319–1328 (2012)

14. Bose, P., Fagerberg, R., van Renssen, A., Verdonschot, S.: On plane constrained bounded-degree spanners. Algorithmica **81**(4), 1392–1415 (2019)

15. Bose, P., Smid, M.H.M.: On plane geometric spanners: a survey and open problems. Comput. Geom. Theory Appl. **46**(7), 818–830 (2013)

16. Canny, J.F.: Some algebraic and geometric computations in PSPACE. In: Simon, J. (ed.) 20th Annual ACM Symposium on Theory of Computing (STOC 1988), pp. 460–467. ACM (1988)

17. Cardinal, J., Hoffmann, U.: Recognition and complexity of point visibility graphs. Discrete Comput. Geom. **57**(1), 164–178 (2017)

18. Chew, P.: There are planar graphs almost as good as the complete graph. J. Comput. Syst. Sci. **39**(2), 205–219 (1989)

19. Da Lozzo, G., D'Angelo, A., Frati, F.: On planar greedy drawings of 3-connected planar graphs. In: Aronov, B., Katz, M.J. (eds.) 33rd International Symposium on Computational Geometry (SoCG 2017). LIPIcs, vol. 77, pp. 33:1–33:16. Schloss Dagstuhl - Leibniz-Zentrum fuer Informatik (2017)

20. Dehkordi, H.R., Frati, F., Gudmundsson, J.: Increasing-chord graphs on point sets. J. Graph Algorithms Appl. **19**(2), 761–778 (2015)

21. Dobkin, D.P., Friedman, S.J., Supowit, K.J.: Delaunay graphs are almost as good as complete graphs. Discrete Comput. Geom. **5**, 399–407 (1990)

22. Dujmović, V., Eppstein, D., Suderman, M., Wood, D.R.: Drawings of planar graphs with few slopes and segments. Comput. Geom. Theory Appl. **38**(3), 194–212 (2007)

23. Dumitrescu, A., Ghosh, A.: Lower bounds on the dilation of plane spanners. Int. J. Comput. Geom. Appl. **26**(2), 89–110 (2016)

24. Eppstein, D., Goodrich, M.T.: Succinct greedy geometric routing using hyperbolic geometry. IEEE Trans. Comput. **60**(11), 1571–1580 (2011)

25. Felsner, S., Igamberdiev, A., Kindermann, P., Klemz, B., Mchedlidze, T., Scheucher, M.: Strongly monotone drawings of planar graphs. In: 32nd International Symposium on Computational Geometry (SoCG 2016). LIPIcs, vol. 51, pp. 37:1–37:15. Schloss Dagstuhl - Leibniz-Zentrum fuer Informatik (2016)
26. de Fraysseix, H., Pach, J., Pollack, R.: How to draw a planar graph on a grid. Combinatorica 10(1), 41–51 (1990)
27. Garey, M.R., Johnson, D.S.: Computers and Intractability: A Guide to the Theory of NP-Completeness. W. H. Freeman, New York (1979)
28. Garey, M.R., Johnson, D.S., Tarjan, R.E.: The planar Hamiltonian circuit problem is NP-complete. SIAM J. Comput. 5(4), 704–714 (1976)
29. He, D., He, X.: Optimal monotone drawings of trees. SIAM J. Discrete Math. 31(3), 1867–1877 (2017)
30. He, X., He, D.: Monotone drawings of 3-connected plane graphs. In: Bansal, N., Finocchi, I. (eds.) ESA 2015. LNCS, vol. 9294, pp. 729–741. Springer, Heidelberg (2015). https://doi.org/10.1007/978-3-662-48350-3_61
31. He, X., Zhang, H.: On succinct greedy drawings of plane triangulations and 3-connected plane graphs. Algorithmica 68(2), 531–544 (2014)
32. Hossain, M.I., Rahman, M.S.: Good spanning trees in graph drawing. Theoret. Comput. Sci. 607, 149–165 (2015)
33. Icking, C., Klein, R., Langetepe, E.: Self-approaching curves. Math. Proc. Cambridge Philos. Soc. 125(3), 441–453 (1999)
34. Kant, G.: Drawing planar graphs using the canonical ordering. Algorithmica 16(1), 4–32 (1996)
35. Kindermann, P., Schulz, A., Spoerhase, J., Wolff, A.: On monotone drawings of trees. In: Duncan, C., Symvonis, A. (eds.) GD 2014. LNCS, vol. 8871, pp. 488–500. Springer, Heidelberg (2014). https://doi.org/10.1007/978-3-662-45803-7_41
36. Leighton, T., Moitra, A.: Some results on greedy embeddings in metric spaces. Discrete Comput. Geom. 44(3), 686–705 (2010)
37. Lubiw, A., Mondal, D.: Construction and local routing for angle-monotone graphs. J. Graph Algorithms Appl. 23(2), 345–369 (2019)
38. Mnev, N.E.: The universality theorems on the classification problem of configuration varieties and convex polytopes varieties. In: Viro, O.Y., Vershik, A.M. (eds.) Topology and Geometry — Rohlin Seminar. LNM, vol. 1346, pp. 527–543. Springer, Heidelberg (1988). https://doi.org/10.1007/BFb0082792
39. Mulzer, W.: Minimum dilation triangulations for the regular n-gon. Master's thesis, Freie Universität Berlin (2004)
40. Nöllenburg, M., Prutkin, R.: Euclidean greedy drawings of trees. Discrete Comput. Geom. 58(3), 543–579 (2017)
41. Nöllenburg, M., Prutkin, R., Rutter, I.: On self-approaching and increasing-chord drawings of 3-connected planar graphs. J. Comput. Geom. 7(1), 47–69 (2016)
42. Papadimitriou, C.H., Ratajczak, D.: On a conjecture related to geometric routing. Theoret. Comput. Sci. 344(1), 3–14 (2005)
43. Rao, A., Papadimitriou, C.H., Shenker, S., Stoica, I.: Geographic routing without location information. In: Johnson, D.B., Joseph, A.D., Vaidya, N.H. (eds.) 9th Annual International Conference on Mobile Computing and Networking (MOBICOM 2003), pp. 96–108. ACM (2003)
44. Rote, G.: Curves with increasing chords. Math. Proc. Cambridge Philos. Soc. 115(1), 1–12 (1994)
45. Schaefer, M.: Complexity of some geometric and topological problems. In: Eppstein, D., Gansner, E.R. (eds.) GD 2009. LNCS, vol. 5849, pp. 334–344. Springer, Heidelberg (2010). https://doi.org/10.1007/978-3-642-11805-0_32

46. Schnyder, W.: Embedding planar graphs on the grid. In: Johnson, D.S. (ed.) 1st Annual ACM-SIAM Symposium on Discrete Algorithms (SODA 1990), pp. 138–148 (1990)
47. Shiloach, Y.: Linear and planar arrangements of graphs. Ph.D. thesis, Weizmann Institute of Science (1976)
48. Wang, J.J., He, X.: Succinct strictly convex greedy drawing of 3-connected plane graphs. Theoret. Comput. Sci. **532**, 80–90 (2014)
49. Win, S.: On a connection between the existence of k-trees and the toughness of a graph. Graphs Comb. **5**(1), 201–205 (1989)
50. Xia, G.: The stretch factor of the Delaunay triangulation is less than 1.998. Comput. Geom. Theory Appl. **42**(4), 1620–1659 (2013)
51. Xia, G., Zhang, L.: Toward the tight bound of the stretch factor of Delaunay triangulations. In: 23rd Annual Canadian Conference on Computational Geometry (CCCG 2011) (2011)

Inserting One Edge into a Simple Drawing Is Hard

Alan Arroyo[1]📷, Fabian Klute[2]📷, Irene Parada[3]📷, Raimund Seidel[4]📷, Birgit Vogtenhuber[5(✉)]📷, and Tilo Wiedera[6]📷

[1] IST Austria, Klosterneuburg, Austria
alanmarcelo.arroyoguevara@ist.ac.at
[2] Utrecht University, Utrecht, The Netherlands
f.m.klute@uu.nl
[3] TU Eindhoven, Eindhoven, The Netherlands
i.m.de.parada.munoz@tue.nl
[4] Universität des Saarlandes, Saarbrücken, Germany
rseidel@cs.uni-saarland.de
[5] Graz University of Technology, Graz, Austria
bvogt@ist.tugraz.at
[6] Osnabrück University, Osnabrück, Germany
tilo.wiedera@uos.de

Abstract. A *simple drawing* $D(G)$ of a graph G is one where each pair of edges share at most one point: either a common endpoint or a proper crossing. An edge e in the complement of G can be *inserted* into $D(G)$ if there exists a simple drawing of $G + e$ extending $D(G)$. As a result of Levi's Enlargement Lemma, if a drawing is rectilinear (pseudolinear), that is, the edges can be extended into an arrangement of lines (pseudolines), then any edge in the complement of G can be inserted. In contrast, we show that it is NP-complete to decide whether one edge can be inserted into a simple drawing. This remains true even if we assume that the drawing is pseudocircular, that is, the edges can be extended to an arrangement of pseudocircles. On the positive side, we show that, given an arrangement of pseudocircles $\mathcal{A}$ and a pseudosegment σ, it can be decided in polynomial time whether there exists a pseudocircle Φ_σ extending σ for which $\mathcal{A} \cup \{\Phi_\sigma\}$ is again an arrangement of pseudocircles.

A. Arroyo—Funded by the Marie Skłodowska-Curie grant agreement No 754411.

F. Klute—This work was conducted while Fabian Klute was a member of the "Algorithms and Complexity Group" at TU Wien.

I. Parada—Partially supported by the Austrian Science Fund (FWF): W1230 and within the collaborative DACH project *Arrangements and Drawings* as FWF project I 3340-N35.

B. Vogtenhuber—Partially supported by Austrian Science Fund (FWF) within the collaborative DACH project *Arrangements and Drawings* as FWF project I 3340-N35.

T. Wiedera—Supported by the German Research Foundation (DFG) grant CH 897/2-2.

© Springer Nature Switzerland AG 2020
I. Adler and H. Müller (Eds.): WG 2020, LNCS 12301, pp. 325–338, 2020.
https://doi.org/10.1007/978-3-030-60440-0_26

1 Introduction

A *simple drawing* of a graph G (also known as *good drawing* or as *simple topological graph* in the literature) is a drawing $D(G)$ of G in the plane such that every pair of edges shares at most one point that is either a proper crossing or a common endpoint. In particular, no tangencies between edges are allowed, edges must not contain any vertices in their relative interior, and no three edges intersect in the same point. Simple drawings have received a great deal of attention in various areas of graph drawing, for example in connection with two long-standing open problems: the crossing number of the complete graph [30] and Conway's thrackle conjecture [26].

In this work, we study the problem of inserting an edge into a simple drawing of a graph. Given a simple drawing $D(G)$ of a graph $G = (V, E)$ and an edge e of the complement $\overline{G}$ of G we say that e can be *inserted* into $D(G)$ if there exists a simple drawing of $G' = (V, E \cup \{e\})$ that contains $D(G)$ as a subdrawing.

A *pseudoline arrangement* is an arrangement of simple biinfinite arcs, called *pseudolines*, such that every pair of pseudolines intersects in a single point that is a proper crossing. Similarly, an *arrangement of pseudocircles* is an arrangement of simple closed curves, called *pseudocircles*, such that every pair of pseudocircles intersects in either zero or two points, where in the latter case, both intersection points are proper crossings. A simple drawing $D(G)$ is called *pseudolinear* if the drawing of every edge can be extended to a pseudoline such that the extended drawing forms a pseudoline arrangement. Likewise, $D(G)$ is called *pseudocircular* if the drawing of every edge can be extended to a pseudocircle such that the extended drawing forms an arrangement of pseudocircles.

Pseudoline arrangements were introduced by Levi [24] in 1926 and have since been extensively studied; see for example [13]. One of the most fundamental results on pseudoline arrangements, nowadays well known as Levi's Enlargement Lemma, stems from Levi's original paper[1]. It states that, for any given pseudoline arrangement $\mathcal{L}$ and any two points p and q not on the same pseudoline of $\mathcal{L}$, it is always possible to insert a pseudoline through p and q into $\mathcal{L}$ such that the resulting arrangement is again a valid pseudoline arrangement.

From Levi's Enlargement Lemma, it immediately follows that given any pseudolinear drawing $D(G)$ and any set E^* of edges from $\overline{G}$, it is always possible to insert all edges from E^* into $D(G)$ such that the resulting drawing is again pseudolinear. To the contrary, as shown by Kynčl [23], this is in general not the case for simple drawings, not even if G is a matching plus two isolated vertices which are the endpoints of the edge to be inserted [22]. The latter implies that an analogous statement to Levi's Enlargement Lemma is not true for arrangements of pseudosegments (simple arcs that pairwise intersect at most once). Moreover, Arroyo, Derka, and Parada [2] recently showed that given a simple drawing $D(G)$ and a set E^* of edges from $\overline{G}$, it is NP-complete to decide whether E^* can be inserted into $D(G)$ (such that the resulting drawing is again simple). However,

---

[1] Also known as Levi's Extension Lemma. Several different proofs of Levi's Enlargement Lemma have been published since then [3,14,31–33].

the cardinality of E^* required for their hardness proof is linear in the size of the constructed graph. The main open problem posed in [2] is the complexity of deciding whether one single given edge e of $\overline{G}$ can be inserted into $D(G)$.

In this work, we show that this decision problem is NP-complete, even if G is a matching plus two isolated vertices which are the endpoints of e. This implies that, given an arrangement $\mathcal{S}$ of pseudosegments and two points p and q not on the same pseudosegment, it is NP-complete to decide whether it is possible to insert a pseudosegment from p to q into $\mathcal{S}$ such that the resulting arrangement is again a valid arrangement of pseudosegments (Sect. 2). On the positive side, we observe that the decision problem is fixed-parameter tractable (FPT) in the number of crossings of the drawing (Sect. 4).

Snoeyink and Hershberger [32] showed the following analogon to Levi's Enlargement Lemma for arrangements of pseudocircles: For any arrangement $\mathcal{A}$ of pseudocircles and any three points p, q, and r, not all of them on one pseudocircle of $\mathcal{A}$, there exists a pseudocircle Φ through p, q, and r such that $\mathcal{A} \cup \{\Phi\}$ is again an arrangement of pseudocircles. Refining our hardness proof, we show that the edge-insertion decision problem remains NP-complete when $D(G)$ is a pseudocircular drawing, regardless of whether the resulting drawing is required to be again pseudocircular or allowed to be any simple drawing. This holds even if we are in addition given an arrangement of pseudocircles extending $D(G)$. On the positive side, we show that, given an arrangement $\mathcal{A}$ of pseudocircles and a pseudosegment σ, it can be decided in polynomial time whether there exists an extension Φ_σ of σ to a simple closed curve such that $\mathcal{A} \cup \{\Phi_\sigma\}$ is again an arrangement of pseudocircles (Sect. 3).

One of the implications of the results presented in this paper concerns so-called saturated drawings [22]. A simple drawing $D(G)$ of a graph G is called *saturated* if no edge e from $\overline{G}$ can be inserted into $D(G)$. It is known that there are saturated simple drawings with a linear number of edges [16]. A natural question is to determine the complexity of deciding whether a simple drawing is saturated. Our hardness result implies that the straight-forward idea of testing whether $D(G)$ is saturated by checking for every edge in $\overline{G}$ whether it can be inserted into $D(G)$ is not feasible unless P = NP.

The problem of inserting an edge (or multiple edges or a star) into a planar graph has been extensively studied in the contexts of determining the crossing number of the resulting graph [6,29] and of finding a drawing of the resulting graph in which the original planar graph is drawn crossing-free and the drawing of the resulting graph has as few crossings as possible [10,11,15,28]. In relation to our work, a main difference is that we consider inserting edges into some given non-plane drawing of a graph. Furthermore, the question considered in this paper is strongly related to work on extending partial representations of graphs. Here, we are usually given a representation of a part of the graph G and are asked to extend it into a full representation of G such that the partial representation is a sub-representation of the full one. Recent years have seen a plethora of results in this topic [1,4,5,7–9,12,17–21,25,27].

Proofs of statements marked with ⋆ are deferred to the full version of this work.

2   Inserting One Edge into a Simple Drawing Is Hard

Theorem 1. *Given a simple drawing $D(G)$ of a graph $G = (V, E)$ and an edge uv of $\overline{G}$, it is NP-complete to decide whether uv can be inserted into $D(G)$, even if $V \setminus \{u, v\}$ induces a matching in G and u and v are isolated vertices.*

It is straightforward to verify that the problem is in NP (see Arroyo et al. [2] for a combinatorial description of our problem using the dual of the planarization of the drawing). We show NP-hardness via a reduction from 3SAT. Let $\phi(x_1, \ldots x_n)$ be a 3SAT-formula with *variables* $x_1, \ldots, x_n$ and set of *clauses* $\mathcal{C} = \{C_1, \ldots, C_m\}$. An occurrence of a variable x_i in a clause $C_j \in \mathcal{C}$ is called a *literal*. For convenience, we assume that in $\phi(x_1, \ldots, x_n)$, each clause has three (not necessarily different) literals. In a preprocessing step, we eliminate clauses with only positive or only negative literals via the transformation from Lemma 1.

Lemma 1 ($\star$). *The following transformation of a clause with only positive or only negative literals, respectively, preserves the satisfiability of the clause (y is a new variable and* false *is the constant value false):*

$$x_i \vee x_j \vee x_k \Rightarrow \begin{cases} x_k \vee y \vee \texttt{false} & \textit{(i)} \\ x_i \vee x_j \vee \neg y & \textit{(ii)} \end{cases} \qquad \neg x_i \vee \neg x_j \vee \neg x_k \Rightarrow \begin{cases} \neg x_i \vee \neg x_j \vee y & \textit{(iii)} \\ \neg x_k \vee \neg y \vee \texttt{false} & \textit{(iv)} \end{cases}$$

After the preprocessing, we have a *transformed* 3SAT-formula where each clause is of one of the following four types: Type (i) two positive literals and one constant false; Type (ii) one negative and two positive literals; Type (iii) one positive and two negative literals, and finally, Type (iv) two negative literals and one constant false.

Given a transformed 3SAT-formula $\phi = \phi(x_1, \ldots, x_n)$ with set of clauses $\mathcal{C} = \{C_1, \ldots, C_m\}$, satisfiability of ϕ will correspond to being able to insert a given edge uv into a simple drawing D of a matching constructed from the formula ϕ. The main idea of the reduction is that the variable and clause gadgets in D act as "barriers" inside a simple closed region R of D, in which we need to insert a simple arc γ from one side to the other to connect u and v. Crossing a barrier in some way imposes constraints on how or whether we can cross other barriers afterwards.

To simplify the description, we first focus our attention to the inside of the simple closed region R. We assume that γ cannot cross the boundary of R. In the following we use two lines, named λ and μ, to bound the regions in which a variable and clause gadget will be placed. Particularly, these lines will be identified with opposite segments on R's boundary.

Variable Gadget. A variable gadget W is bounded from above by a horizontal line λ and from below by a horizontal line μ. Additionally, it contains a vertical segment κ between λ and μ, a set P of pairwise non-crossing arcs (parts of later-defined edges), each with one endpoint on κ and the other endpoint on μ, and a set N of pairwise non-crossing arcs, each with one endpoint on κ and the other endpoint on λ. On κ, all the endpoints of arcs in P lie above all the endpoints of arcs in N, implying that every arc in P crosses every arc in N. Finally, we choose two points u and v such that u is to the left of all arcs in W and v is to the right of them; see Fig. 1 for an illustration. The arcs in P and N correspond to positive and negative appearances of the variable, respectively.

Lemma 2 ($\star$). *Let W be a variable gadget. Any arc between the horizontal lines λ and μ that connects u and v crosses either all arcs in P or all arcs in N.*

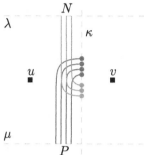

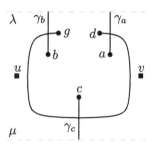

Fig. 1. Variable gadget. The orange arcs belong to N, the green ones to P. (Color figure online)

Fig. 2. Clause gadget.

Clause Gadget. Similar to a variable gadget, a clause gadget K is bounded from above and below by two horizontal lines λ and μ, respectively. Additionally, it contains three horizontal arcs (parts of later-defined edges) γ_a, γ_b, and γ_c, where the former two have one endpoint on λ and the latter has one endpoint on μ. On λ, the endpoint of γ_a lies to the right of the one of γ_b. The other endpoints of γ_a, γ_b, and γ_c are called a, b, and c, respectively. None of these three arcs cross. Moreover, K contains two points d and g and an edge dg that crosses γ_a, γ_c, and γ_b in that order when traversed from d to g. Notice that we do not require any specific rotation of the crossings of dg with γ_a and γ_b (where the rotation is the clockwise order of the endpoints of the crossing arcs). However, to simplify the description, we assume that the rotations of the crossings are as in Fig. 2. The rotation of the crossing of dg with γ_c is forced by the order of the crossings along dg. Finally, we again choose two points u and v such that u is to the left of all arcs in K and v is to the right of them; see Fig. 2 for an illustration.

Lemma 3 ($\star$). *Let K be a clause gadget. Any arc uv between the horizontal lines λ and μ that connects u and v crosses either dg twice or at least one of the arcs γ_a, γ_b, and γ_c.*

The Reduction. Let $\phi(x_1, \ldots, x_n)$ be a transformed 3SAT-formula with clause set $C = \{C_1, \ldots, C_m\}$ (each clause being of one of the four types identified above). To build our reduction we need one more gadget. First, we introduce the following simple drawing introduced by Kynčl et al. [22, Figure 11] and depicted in Fig. 3. Here, we denote this drawing by ⓞ. Following the notation by Kynčl et al., we denote its six arcs by a_1, a_2, a_3, b_1, b_2,

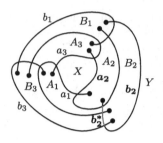

Fig. 3. The simple drawing ⓞ presented in [22].

and b_3; and its eight cells by X, A_1, A_2, A_3, B_1, B_2, B_3, and Y; see Fig. 3 for an illustration. The core property $\mathcal{P}$ of ⓞ is that it is not possible to insert an edge between a point in cell X and another point in cell Y such that the result is a simple drawing [22, Lemma 15].

For our reduction, we first choose two arbitrary points u and v in the cells X and B_2 and insert them as vertices into ⓞ. Let ⓞ$'$ be the obtained drawing. Further, let b_2^* be the part of the arc b_2 between the crossing point of b_2 and a_2 and the crossing point of b_2 and b_3, see again Fig. 3.

Lemma 4 (⋆). *The edge uv cannot be inserted into* ⓞ$'$ *without crossing b_2^*.*

The final piece we need for our reduction is a set F of $m^I + m^{IV} + 4$ arcs that we insert into ⓞ$'$, where m^I is the number of clauses of Type (i) and m^{IV} the number of clauses of Type (iv). For an arc $f \in F$ we will place one of its endpoints on a horizontal line κ_F inside A_2 and the other one inside B_2. The only crossings of f with ⓞ$'$ are with the arcs a_2, a_1, b_3, and b_2, in that order, when traversing f from its endpoint on κ_F to its endpoint in B_2. Furthermore, when f is traversed in that direction, it crosses from A_2 to A_1, from A_1 to B_3, from B_3 to Y, and from Y to B_2.

Consider the $m^I + m^{IV} + 4$ endpoints on κ_F sorted from left to right. We denote by f_j the arc in F incident with the j-th such endpoint. When traversing b_2 from its endpoint in A_2 to its endpoint in B_1, the crossings of arcs in F with b_2 appear in the same order as their endpoints on κ_F. More precisely, the crossings of b_2, when b_2 is traversed in that direction, are with a_2, a_1, b_3, f_1, f_2, $\ldots$, $f_{|F|}$, and b_1, in that order.

The arcs f_{m^I+1}, f_{m^I+2}, f_{m^I+3}, and f_{m^I+4} will behave differently than the other arcs in F. In the following, we denote these four arcs by r_2, r_1, ℓ_1, and ℓ_2, respectively. There are only two crossings between arcs in F, namely, between r_1 and r_2, and between ℓ_1 and ℓ_2, and both these crossings are inside B_2. These four crossing arcs divide B_2 into three regions. Let R denote the region with b_2^* on its boundary; let R_r denote the (other) region incident with the crossing between r_1 and r_2; and let R_ℓ denote the (other) region incident with the crossing between ℓ_1 and ℓ_2. Arcs r_1, r_2, ℓ_1, and ℓ_2 must be drawn such that the vertex v lies in R; see the red arcs in Fig. 4 for an illustration. The precise endpoints of the edges in $F \setminus \{r_1, r_2, \ell_1, \ell_2\}$ will be fixed when we insert the clause gadgets.

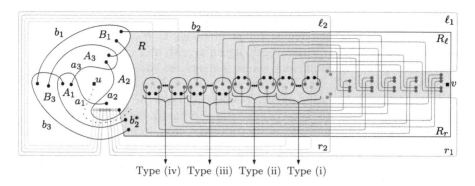

Type (iv) Type (iii) Type (ii) Type (i)

Fig. 4. Illustration of the reduction. (Color figure online)

Lemma 5 (⋆)**.** *The edge uv cannot be inserted into $\text{\textcircled{\scriptsize ◎}}'$ without crossing every arc in F in the closure of A_1 or of B_3.*

It remains to insert inside R the clause and variable gadgets and precisely define the endpoints of arcs in $F \setminus \{\ell_1, \ell_2, r_1, r_2\}$. For simplicity, we first insert the variable gadgets and then the clause gadgets. The idea is that each clause and variable gadget is inserted in R separating b_2^* from v. This is done by identifying the endpoints that were lying on λ or μ with points on ℓ_1, ℓ_2, r_1, r_2, or b_2. As a result, Lemmas 2 and 3 can be applied to the arc that we insert connecting u and v in the final drawing, since it has to cross b_2^* by Lemma 4.

We now insert the variable gadgets into R. Let $W^{(i)}$ be the variable gadget corresponding to variable x_i. For a gadget $W^{(i)}$, the arcs in N are drawn such that the endpoints on λ lie on the part of ℓ_1 that bounds R. The arcs in P are drawn similarly, but with the endpoints on μ lying on the part of r_1 that bounds R. Moreover, we identify vertex v in the gadget with vertex v in $\text{\textcircled{\scriptsize ◎}}'$. Gadgets corresponding to different variables are inserted without crossing each other. We now specify how they are inserted relative to each other. As we traverse ℓ_1 from its endpoint on κ_F to its endpoint in R, we encounter the endpoints of arcs in $W^{(i)}$ before the endpoints of arcs in $W^{(i+1)}$. Analogously, as we traverse r_1 from its endpoint on κ_F to its endpoint in R, we encounter the endpoints of arcs in $W^{(i)}$ before the endpoints of arcs in $W^{(i+1)}$. See Fig. 4 for an illustration.

In a similar way we insert the clause gadgets. Let $K^{(j)}$ be the clause gadget corresponding to clause C_j. If C_j is of Type (i), $K^{(j)}$ is inserted such that the endpoints on λ lie on the part of ℓ_2 that bounds R. If C_j is the j'-th clause of Type (i), we identify c with the endpoint of the arc $f_{j'}$. Similarly, if C_j is of Type (iv), $K^{(j)}$ is inserted such that the endpoints on λ lie on the part of r_2 that bounds R. If C_j is the j'-th clause of Type (iv), we identify c with the endpoint of the arc $f_{m'+4+j'}$. If C_j is of Type (ii), $K^{(j)}$ is inserted such that the endpoints on λ lie on the part of ℓ_2 that bounds R and the endpoint on μ lies on the part of r_2 that bounds R. Similarly, if C_j is of Type (iii), $K^{(j)}$ is inserted such that the endpoint on μ lies on the part of ℓ_2 that bounds R and the endpoints on λ lie on the part of r_2 that bounds R. The crossings in R of arcs from different clause gadgets are of arcs with an endpoint in r_2 with arcs in $\{f_j : 1 \leq j \leq m^I\}$.

We now specify how different clause gadgets are inserted relative to each other. As we traverse ℓ_2 from its endpoint on κ_F to its endpoint in R, we first encounter the endpoints of arcs corresponding to Type (iii) clauses, followed by the ones corresponding to Type (ii) clauses, and finally the ones corresponding to Type (i) clauses. Analogously, as we traverse r_2 from its endpoint on κ_F to its endpoint in R, we first encounter the endpoints of arcs corresponding to Type (iv) clauses, followed by the ones corresponding to Type (iii) clauses, and finally the ones corresponding to Type (ii) clauses. Moreover, as we traverse ℓ_2 and r_2 in the specified directions, the endpoints of arcs corresponding to the j'-th clause of a certain type are encountered before the endpoints of arcs corresponding to the $(j' - 1)$-st clause of this type. An illustration can be found in Fig. 4.

Finally, we connect arcs from variable and clause gadgets inside the regions R_ℓ and R_r. This is done such that if a literal in a clause is x_k then the corresponding arc in the clause gadget, that has an endpoint on ℓ_2, is connected with an arc in N of the gadget $W^{(k)}$, that has an endpoint on ℓ_1. Thus, these connections can lie in R_ℓ. Analogously, if a literal in a clause is $\neg x_k$ then the corresponding arc in the clause gadget, that has an endpoint on r_2, is connected with an arc in P of the gadget $W^{(k)}$, that has an endpoint on r_1. Thus, these connections can lie in R_r. Since, without loss of generality, we can assume that R_ℓ and R_r are convex regions and the endpoints we want to connect are pairwise distinct points on the boundaries of those regions, the connections can be drawn as straight-line segments. (For clarity, in Fig. 4, these connections have one bend per arc.) Therefore, there is at most one crossing between each pair of connecting arcs.

Each connecting arc is concatenated with the arcs in a variable and in a clause gadget that it joins. These concatenated arcs are edges in our drawing that have one endpoint in a variable gadget and the other one in a clause gadget. By construction, each such edge corresponds to a literal in the formula ϕ and each pair of them crosses at most once. Similarly, the arcs in $F \setminus \{\ell_1, \ell_2, r_1, r_2\}$ have one endpoint in a clause gadget and also define edges in our final drawing that we denote by the same names as the corresponding arcs.

We now have all the pieces that constitute our final drawing. It consists of (i) the simple drawing $\circledcirc'$; (ii) the edges $f_i \in F$ drawn as the described arcs (with their endpoints as vertices); (iii) the edges corresponding to literals (with their endpoints as vertices); and (iv) the edges dg in each clause gadget (with d and g as vertices). Observe that the constructed drawing is a simple drawing, as it is the drawing of a matching (plus the vertices u and v) and, by construction, any two edges cross at most once.

It remains to show that the presented construction is a valid reduction.

Lemma 6 ($\star$). *The above construction is a poly-time reduction from* 3SAT *to the problem of deciding whether an edge can be inserted into a simple drawing.*

Remarks and Extensions. As our reduction from 3SAT constructs a simple drawing $D(G)$ of a matching, the general problem is NP-hard even if G is as sparse as possible. We remark that if we do not require G to be a matching, our variable gadget can be simplified by identifying all the vertices on κ and removing the crossings between edges in N and P. Moreover, from the constructed drawing $D(G)$, one can produce an equivalent instance that is connected: This is done by inserting an apex vertex into an arbitrary cell of the drawing, and then subdividing its incident edges so that the resulting drawing D^* is simple. If uv can be inserted into $D(G)$ then it can be inserted also into D^*. Finally, it is possible to show that the simple drawings produced by our reduction are pseudocircular implying the following result.

Corollary 1 ($\star$). *Given a pseudocircular drawing $D(G)$ of a graph $G = (V, E)$ and an edge uv of $\overline{G}$, it is* NP*-complete to decide whether uv can be inserted into $D(G)$, even if an arrangement of pseudocircles extending the drawing of the edges in $D(G)$ is provided.*

3   Extending an Arrangement of Pseudocircles Is Easy

In the previous section we proved that deciding whether an edge can be inserted into a pseudocircular drawing such that the result is a simple (or a pseudocircular) drawing is hard. In this section we focus on extending arrangements instead of drawings of graphs. Snoeyink and Hershberger [32] showed that given an arrangement $\mathcal{A}$ of pseudocircles and three points, not all three on the same pseudocircle, one can find a pseudocircle Φ through the three points such that $\mathcal{A} \cup \{\Phi\}$ is again an arrangement of pseudocircles. Now, given any arrangement $\mathcal{A}$ and a pseudosegment σ intersecting each pseudocircle in $\mathcal{A}$ at most twice, it is not always possible to extend σ to a pseudocircle $\Phi_\sigma \supset \sigma$ such that $\mathcal{A} \cup \{\Phi_\sigma\}$ is again an arrangement of pseudocircles. Two examples are shown in Figs. 5 and 6. In either, any pseudocircle Φ_σ extending σ crosses one red or blue pseudocircle at least four times. However, we show in the following that the extension decision question can be answered in polynomial time:

Theorem 2. *Given an arrangement $\mathcal{A}$ of n pseudocircles and a pseudosegment σ intersecting each pseudocircle in $\mathcal{A}$ at most twice, it can be decided in time polynomial in n whether there exists an extension of σ to a pseudocircle Φ_σ such that $\mathcal{A} \cup \{\Phi_\sigma\}$ is an arrangement of pseudocircles.*

Proof. Throughout this proof we write $\overline{R} := \mathbb{R}^2 \setminus R$ for the *complement* of a set $R \subseteq \mathbb{R}^2$. An arrangement (of pseudocircles) partitions the plane into *vertices* (0-dimensional cells), *edges* (1-dimensional cells), and *faces* (2-dimensional cells). Since tangencies are not allowed, all vertices are proper crossings. Two arrangements are *combinatorially equivalent* (or, *isomorphic*) if the corresponding cell complexes are isomorphic, that is, if there is an incidence- and dimension-preserving bijection between their cells. By possibly transforming $\mathcal{A}$ into an isomorphic arrangement while preserving the incidences of σ, we can assume without loss of generality that σ is a horizontal segment. Let u and v be the left and right endpoints of σ, respectively. Further, we can assume that u is incident with the unbounded cell and that the intersection points of σ with the pseudocircles in $\mathcal{A}$ are all proper crossings. Our algorithm aims to compute a pseudocircle $\Phi_\sigma = \sigma \cup \sigma'$ such that $\mathcal{A} \cup \{\Phi_\sigma\}$ is an arrangement of pseudocircles, or determine that no such σ' exists. We call σ' an *extension* of σ.

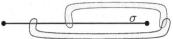

Fig. 5. Obstruction where all pseudo-circles intersect σ twice.

Fig. 6. Obstruction where one pseudo-circle intersects σ only once.

We partition the set of pseudocircles of $\mathcal{A}$ into three sets $\mathcal{C}_0$, $\mathcal{C}_1$, and $\mathcal{C}_2$, where for each $i \in \{0,1,2\}$, $\mathcal{C}_i$ is the set of pseudocircles in $\mathcal{A}$ crossing σ exactly i times. Note that u lies outside all pseudocircles $\phi \in \mathcal{A}$ while v lies outside of all $\phi \in \mathcal{C}_0 \cup \mathcal{C}_2$ and inside all $\phi \in \mathcal{C}_1$, that is, each $\phi \in \mathcal{C}_1$ separates u and v. Further, an extension σ' must not cross any $\phi \in \mathcal{C}_2$, it needs to cross every $\phi \in \mathcal{C}_1$ exactly once, and it can cross each $\phi \in \mathcal{C}_0$ either twice or not at all.

The idea is to construct a finite sequence $R_0 \subset R_1 \subset \ldots$ of closed subsets of $\mathbb{R}^2$, each consisting of cells of $\mathcal{A} \cup \sigma$ that cannot be reached by σ'. Each set R_i will be a simply connected closed region of $\mathbb{R}^2$ with both u and v on its boundary. Further, for each R_i and each $\phi \in \mathcal{C}_0$, we will maintain the invariant that $\text{int}(\phi) \cap \overline{R_i}$ is either a connected region or empty, where $\text{int}(\phi)$ denotes the interior of the bounded area enclosed by ϕ. (Note that $\text{int}(\phi) \cap \overline{R_i}$ is connected if and only if $R_i \setminus \text{int}(\phi)$ is connected.) The construction will either end by determining that σ cannot be extended, or with a set R_m such that routing σ' closely along the boundary of R_m gives a valid extension of σ.

Let R_0' be the union of σ and all the closed disks bounded by the pseudo-circles in $\mathcal{C}_2$ and consider the faces induced by R_0'. Since u is incident with the unbounded cell of R_0', and since σ' must not intersect the interior of R_0', σ' cannot reach any bounded face of R_0'. Let R_0 be the closure of the union of these bounded faces and σ. We may assume that $v \in \partial R_0$, as otherwise no extension σ' exists and we are done.

To see that the invariant holds for R_0, assume that there exists a pseudocircle $\phi \in \mathcal{C}_0$ such that $R_0 \setminus \text{int}(\phi)$ is not connected. As ϕ does not intersect σ, there exists a component D of $R_0 \setminus \text{int}(\phi)$ that is disjoint from σ. Further, as $\text{int}(\phi)$ is simply connected, $D \cap \partial R_0 \neq \emptyset$. Moreover, any point x on $\partial D \cap \partial R_0$ lies on some circle $\phi_x \in \mathcal{C}_2$. On the other hand, any path from a point of σ to x must enter and leave $\text{int}(\phi)$ and hence intersect ϕ at least twice. As ϕ_x intersects σ twice and lies in R_0, we get that ϕ_x intersects ϕ in at least four points, a contradiction.

For the iterative step, consider the arrangement $\mathcal{A}_i^\phi$ formed by ∂R_i and a pseudocircle $\phi \in \mathcal{C}_0 \cup \mathcal{C}_1$, and the cells of it that lie in $\overline{R_i}$. If $\phi \in \mathcal{C}_1$ and an extension σ' exists, then the only two such cells that can be intersected by σ' are the ones incident to u and v, respectively. Similarly, if $\phi \in \mathcal{C}_0$, then σ' can only intersect the cell(s) incident to u and v, plus the (by the invariant) unique cell $\text{int}(\phi) \cap \overline{R_i}$. In both cases, all other cells of this arrangement should be added to the forbidden area. We denote all cells $\mathcal{A}_i^\phi \cap \overline{R_i}$ that can possibly be intersected by σ' as *reachable* (by σ') and all other cells as *unreachable* (by σ').

Assume that there exists some pseudocircle $\phi \in \mathcal{C}_0 \cup \mathcal{C}_1$ such that the arrangement $\mathcal{A}_i^\phi$ of ϕ and ∂R_i contains unreachable cells. Then we obtain R'_{i+1} by adding all those cells to R_i. If v lies in a bounded face of $\overline{R'_{i+1}}$, then no extension σ' exists and we are done. Otherwise, $R_{i+1} = R'_{i+1}$ is a simply connected region that has both u and v on its boundary. It remains to show that the invariant is still maintained for R_{i+1}.

Lemma 7 ($\star$). *If R_i fulfills the invariant and u and v both lie in the unbounded region of R'_{i+1} then R_{i+1} also fulfills the invariant.*

Now assume that both u and v lie on the boundary of all sets R_i constructed in this way. Then the iterative process stops with a set R_m where for each $\phi \in \mathcal{C}_0 \cup \mathcal{C}_1$, all cells in the arrangement $\mathcal{A}_m^\phi$ of ϕ and ∂R_m that are contained in $\overline{R_m}$ are reachable by σ'. Note that $m = O(n^4)$ as $\mathcal{A}$ has $\Theta(n^4)$ cells, as in every iteration i, at least one cell of $\mathcal{A}$ has been added to R_i, and as each cell of $\mathcal{A}$ is added at most once. Consider a path P from u to v in $\overline{R_m}$ that is routed closely along the boundary ∂R_m (note that there are two different such paths). Then for any $\phi \in \mathcal{C}_1$, P intersects exactly two cells of $\mathcal{A}_m^\phi$, namely, the ones incident to u and v, respectively. Hence P crosses ϕ exactly once. Similarly, for any $\phi \in \mathcal{C}_0$, the path P intersects at most three cells of $\mathcal{A}_m^\phi$, namely, the one(s) incident to u and v plus possibly the cell $\text{int}(\phi) \cap \overline{R_m}$, which is one cell by the invariant. Hence P crosses ϕ at most twice. Thus $\sigma' = P$ is a valid extension for σ, which completes the correctness proof.

Note that computing R_0 and σ' (in case that the algorithm didn't terminate with a negative answer before) can be done in poly-time. Also, for each R_i and each $\phi \in \mathcal{C}_0 \cup \mathcal{C}_1$, the set of unreachable cells of $\mathcal{A}_i^\phi$ can be determined in poly-time. As we have $O(n^4)$ iteration steps, we can hence compute R_m from R_0 (or determine that σ is not extendible) in poly-time, which concludes the proof.

As an immediate consequence of Theorem 2 we have the following result:

Corollary 2. *Given an arrangement $\mathcal{A}$ of n pseudocircles and a pseudosegment σ, it can be decided in polynomial time whether σ can be extended to a pseudocircle $\Phi_\sigma \supset \sigma$ such that $\mathcal{A} \cup \{\Phi_\sigma\}$ is an arrangement of pseudocircles.*

4   FPT-Algorithm for Bounded Number of Crossings

In this section we show that for drawings with a bounded number of crossings it can be decided in FPT-time whether an edge can be inserted. Given a simple drawing $D(G)$ with k crossings, one can construct a *kernel* of size $O(k)$ by exhaustively removing isolated vertices and uncrossed edges from $D(G)$. For a simple drawing $D(G)$ of a graph $G = (V, E)$ and $e \in E$, let $D(G - e)$ be the subdrawing of $D(G)$ without the drawing of e. Similarly, for an isolated vertex $u \in V$ let $D(G - u)$ be the subdrawing of $D(G)$ without the drawing of u.

Observation 1. *Given a simple drawing $D(G)$ of a graph $G = (V, E)$ and an isolated vertex $w \in V$, an edge uv of $\overline{G}$ can be inserted into $D(G)$ if and only if uv can be inserted into $D(G - w)$.*

Lemma 8. ($\star$). *Given a simple drawing $D(G)$ of a graph $G = (V, E)$ and an edge $e \in E$ that is uncrossed in $D(G)$, an edge uv of $\overline{G}$ can be inserted into $D(G)$ if and only if uv can be inserted into $D(G - e)$.*

Theorem 3. ($\star$). *Given a simple drawing $D(G)$ of a graph $G = (V, E)$ and an edge uv of $\overline{G}$, there is an FPT-algorithm in the number k of crossings in $D(G)$ for deciding whether uv can be inserted into $D(G)$.*

Acknowledgements. This work was started during the 6th Austrian-Japanese-Mexican-Spanish Workshop on Discrete Geometry in Juni 2019 in Austria. We thank all the participants for the good atmosphere as well as discussions on the topic. Also, we thank Jan Kynčl for sending us remarks on the first arXiv version of this work and an anonymous referee for further helpful comments.

References

1. Angelini, P., et al.: Testing planarity of partially embedded graphs. ACM Trans. Algorithms **11**(4), 32:1–32:42 (2015). https://doi.org/10.1145/2629341
2. Arroyo, A., Derka, M., Parada, I.: Extending simple drawings. In: Archambault, D., Tóth, C.D. (eds.) GD 2019. LNCS, vol. 11904, pp. 230–243. Springer, Cham (2019). https://doi.org/10.1007/978-3-030-35802-0_18
3. Arroyo, A., McQuillan, D., Richter, R.B., Salazar, G.: Levi's lemma, pseudolinear drawings of K_n, and empty triangles. J. Graph Theory **87**(4), 443–459 (2018). https://doi.org/10.1002/jgt.22167
4. Bagheri, A., Razzazi, M.: Planar straight-line point-set embedding of trees with partial embeddings. Inf. Process. Lett. **110**(12–13), 521–523 (2010). https://doi.org/10.1016/j.ipl.2010.04.019
5. Brückner, G., Rutter, I.: Partial and constrained level planarity. In: Proceedings of the 28th Annual ACM-SIAM Symposium on Discrete Algorithms (SODA 2017), pp. 2000–2011. SIAM (2017). https://doi.org/10.1137/1.9781611974782.130
6. Cabello, S., Mohar, B.: Adding one edge to planar graphs makes crossing number and 1-planarity hard. SIAM J. Comput. **42**(5), 1803–1829 (2013). https://doi.org/10.1137/120872310
7. Chaplick, S., Dorbec, P., Kratochvíl, J., Montassier, M., Stacho, J.: Contact representations of planar graphs: extending a partial representation is hard. In: Kratsch, D., Todinca, I. (eds.) WG 2014. LNCS, vol. 8747, pp. 139–151. Springer, Cham (2014). https://doi.org/10.1007/978-3-319-12340-0_12
8. Chaplick, S., Fulek, R., Klavík, P.: Extending partial representations of circle graphs. J. Graph Theory **91**(4), 365–394 (2019). https://doi.org/10.1002/jgt.22436

9. Chaplick, S., Guśpiel, G., Gutowski, G., Krawczyk, T., Liotta, G.: The partial visibility representation extension problem. Algorithmica **80**(8), 2286–2323 (2017). https://doi.org/10.1007/s00453-017-0322-4

10. Chimani, M., Gutwenger, C., Mutzel, P., Wolf, C.: Inserting a vertex into a planar graph. In: Proceedings of the 20th ACM-SIAM Symposium on Discrete Algorithms (SODA 2009), pp. 375–383. SIAM (2009). https://doi.org/10.1137/1.9781611973068.42

11. Chimani, M., Hliněný, P.: Inserting multiple edges into a planar graph. In: Proceedings of the 32nd International Symposium on Computational Geometry (SoCG 2016), vol. 51, pp. 30:1–30:15. Schloss Dagstuhl - Leibniz-Zentrum für Informatik (2016). https://doi.org/10.4230/LIPIcs.SoCG.2016.30

12. Da Lozzo, G., Di Battista, G., Frati, F.: Extending upward planar graph drawings. In: Friggstad, Z., Sack, J.-R., Salavatipour, M.R. (eds.) WADS 2019. LNCS, vol. 11646, pp. 339–352. Springer, Cham (2019). https://doi.org/10.1007/978-3-030-24766-9_25

13. Felsner, S., Goodman, J.E.: Pseudoline arrangements. In: Tóth, C.D., O'Rourke, J., Goodman, J.E. (eds.) Handbook of Discrete and Computational Geometry, 3rd edn., pp. 125–157. CRC Press, Boca Raton. (2017)

14. Grünbaum, B.: Arrangements and spreads. AMS (1972)

15. Gutwenger, C., Mutzel, P., Weiskircher, R.: Inserting an edge into a planar graph. Algorithmica **41**(4), 289–308 (2005). https://doi.org/10.1007/s00453-004-1128-8

16. Hajnal, P., Igamberdiev, A., Rote, G., Schulz, A.: Saturated simple and 2-simple topological graphs with few edges. J. Graph Algorithms Appl. **22**(1), 117–138 (2018). https://doi.org/10.7155/jgaa.00460

17. Jelínek, V., Kratochvíl, J., Rutter, I.: A Kuratowski-type theorem for planarity of partially embedded graphs. Comput. Geom. Theory Appl. **46**(4), 466–492 (2013). https://doi.org/10.1016/j.comgeo.2012.07.005. Special Issue on the 27th Annual Symposium on Computational Geometry (SoCG'11)

18. Klavík, P., Kratochvíl, J., Krawczyk, T., Walczak, B.: Extending partial representations of function graphs and permutation graphs. In: Epstein, L., Ferragina, P. (eds.) ESA 2012. LNCS, vol. 7501, pp. 671–682. Springer, Heidelberg (2012). https://doi.org/10.1007/978-3-642-33090-2_58

19. Klavík, P., et al.: Extending partial representations of proper and unit interval graphs. Algorithmica **77**(4), 1071–1104 (2016). https://doi.org/10.1007/s00453-016-0133-z

20. Klavík, P., Kratochvíl, J., Otachi, Y., Saitoh, T.: Extending partial representations of subclasses of chordal graphs. Theor. Comput. Sci. **576**, 85–101 (2015). https://doi.org/10.1016/j.tcs.2015.02.007

21. Klavík, P., Kratochvíl, J., Otachi, Y., Saitoh, T., Vyskočil, T.: Extending partial representations of interval graphs. Algorithmica **78**(3), 945–967 (2016). https://doi.org/10.1007/s00453-016-0186-z

22. Kynčl, J., Pach, J., Radoičić, R., Tóth, G.: Saturated simple and k-simple topological graphs. Comput. Geom. Theory Appl. **48**(4), 295–310 (2015). https://doi.org/10.1016/j.comgeo.2014.10.008

23. Kynčl, J.: Improved enumeration of simple topological graphs. Discrete Comput. Geom. **50**(3), 727–770 (2013). https://doi.org/10.1007/s00454-013-9535-8

24. Levi, F.: Die Teilung der projektiven Ebene durch Gerade oder Pseudogerade. Berichte über die Verhandlungen der Sächsischen Akademie der Wissenschaften zu Leipzig, Mathematisch-Physische Klasse **78**, 256–267 (1926). (in German)

25. Mchedlidze, T., Nöllenburg, M., Rutter, I.: Extending convex partial drawings of graphs. Algorithmica **76**(1), 47–67 (2015). https://doi.org/10.1007/s00453-015-0018-6

26. Pach, J., Brass, P., Moser, W.O.J.: Research Problems in Discrete Geometry. Springer, New York (2005). https://doi.org/10.1007/0-387-29929-7

27. Patrignani, M.: On extending a partial straight-line drawing. Int. J. Found. Comput. Sci. **17**(5), 1061–1070 (2006). https://doi.org/10.1142/S0129054106004261

28. Radermacher, M., Rutter, I.: Inserting an edge into a geometric embedding. In: Biedl, T., Kerren, A. (eds.) GD 2018. LNCS, vol. 11282, pp. 402–415. Springer, Cham (2018). https://doi.org/10.1007/978-3-030-04414-5_29

29. Riskin, A.: The crossing number of a cubic plane polyhedral map plus an edge. Studia Scientiarum Mathematicarum Hungarica **31**(4), 405–414 (1996)

30. Schaefer, M.: Crossing Numbers of Graphs. CRC Press, Boca Raton (2018)

31. Schaefer, M.: A proof of Levi's extension lemma. ArXiv e-Prints (2019)

32. Snoeyink, J., Hershberger, J.: Sweeping arrangements of curves. In: Discrete and Computational Geometry: Papers from the DIMACS Special Year, vol. 6, pp. 309–350. DIMACS/AMS (1991). https://doi.org/10.1090/dimacs/006/21

33. Sturmfels, B., Ziegler, G.M.: Extension spaces of oriented matroids. Discrete Comput. Geom. **10**(1), 23–45 (1993). https://doi.org/10.1007/BF02573961

Bitonic st-Orderings for Upward Planar Graphs: The Variable Embedding Setting

Patrizio Angelini[1] (ID), Michael A. Bekos[2(✉)] (ID), Henry Förster[2] (ID),
and Martin Gronemann[3] (ID)

[1] Department of Mathematics and Applied Sciences, John Cabot University,
Rome, Italy
pangelini@johncabot.edu

[2] Department of Computer Science, University of Tübingen, Tübingen, Germany
{bekos,foersth}@informatik.uni-tuebingen.de

[3] Theoretical Computer Science, Osnabrück University, Osnabrück, Germany
martin.gronemann@uni-osnabrueck.de

Abstract. Bitonic st-orderings for st-planar graphs were recently introduced as a method to cope with several graph drawing problems. Notably, they have been used to obtain the best-known upper bound on the number of bends for upward planar polyline drawings with at most one bend per edge. For an st-planar graph that does not admit a bitonic st-ordering, one may split certain edges such that for the resulting graph such an ordering exists. Since each split is interpreted as a bend, one is usually interested in splitting as few edges as possible. While this optimization problem admits a linear-time algorithm in the fixed embedding setting, it remains open in the variable embedding setting. We close this gap in the literature by providing a linear-time algorithm that optimizes over all embeddings of the input st-planar graph.

The best-known lower bound on the number of required splits of an st-planar graph with n vertices is $n-3$. However, it is possible to compute a bitonic st-ordering without any split for the st-planar graph obtained by reversing the orientation of all edges. In terms of upward planar polyline drawings, the former translates into $n-3$ bends, while the latter into no bends. We show that this idea cannot always be exploited by describing an st-planar graph that needs at least $n-5$ splits in both orientations.

Keywords: Upward planar graphs · Bitonic st-orderings · Planar polyline drawings · Bend minimization

1 Introduction

Incremental drawing algorithms have a long history in the field of Graph Drawing. The central result of de Fraysseix, Pach and Pollack [8], who showed that every planar graph admits a planar straight-line drawing within quadratic area, marks the beginning of this line of research. They introduced the concept of

© Springer Nature Switzerland AG 2020
I. Adler and H. Müller (Eds.): WG 2020, LNCS 12301, pp. 339–351, 2020.
https://doi.org/10.1007/978-3-030-60440-0_27

canonical ordering, an ordering of the vertices that is used to drive their incremental drawing algorithm. In each step, one vertex at a time is placed, while it is ensured that certain invariants are satisfied. Another important result with respect to canonical orderings is by Kant [13]. While the original ordering is only defined for maximal planar graphs, he generalizes this concept to triconnected planar graphs. However, Kant's ordering is no longer a vertex ordering, instead it is an ordered partition of vertices. Later on, Harel and Sardas [12] show how one may further extend canonical orderings to the biconnected case.

Another type of vertex ordering that has its origins not in Graph Drawing, but finds its applications there [3,16], is the so-called *st-ordering* [7]. However, *st*-orderings are not restricted to planar graphs, hence, the ordering is not related directly to the embedding of the underlying planar graph. This relation between a planar embedding and the ordering itself is established by the *bitonic st-orderings*, which have been used to solve various graph drawing problems, e.g., T-contact representations [9], L-drawings [4], finding universal slope sets [1]. Besides being a proper *st* ordering, the definition takes the embedding into account and ensures that the vertex ordering has similar properties to a canonical ordering. Initially introduced for undirected graphs in [9], where it is shown that for every biconnected planar graph a bitonic *st*-ordering can be found in linear time, the concept has been extended to directed graphs [10].

The idea that led initially to the extension to directed graphs, namely the *st*-planar graphs, is rather simple. By slightly modifying the original algorithm of de Fraysseix, Pach and Pollack, one may use a bitonic *st*-ordering to obtain a planar straight-line drawing. Combined with the observation that a vertex is always drawn above its predecessors in the ordering, the resulting drawing is upward planar straight-line. However, not every *st*-planar graph admits such a bitonic *st*-ordering, but a full characterization is given in [10] that is based on the existence of so-called *forbidden configurations*. These configurations, however, can be eliminated by splitting certain edges in the graph, such that for the resulting graph one can then obtain the desired ordering. This technique is used to prove that every upward planar graph with n vertices, admits an upward planar polyline drawing with at most one bend per edge within quadratic area. Moreover, the number of bends is at most $n-3$, which is the best-known bound so far. Hereby, each bend corresponds to a dummy vertex that has been introduced by splitting an edge. Note that in [10] an example is given that requires exactly $n-3$ splits, which shows that this bound is tight.

In practice, one is interested in splitting as few edges as possible. In [10], a simple linear-time algorithm is described that finds the optimal set of edges to split. This algorithm assumes the embedding of the underlying *st*-planar graph to be part of the input, hence, it is only optimal in the fixed embedding scenario. Changing the embedding, however, may have a big impact on the required number of splits. Chaplick et al. [4] take a first step towards the variable embedding scenario by describing an SPQR-tree based recognition algorithm. Namely, if an *st*-planar graph admits a bitonic *st*-ordering in any of its embeddings, then their algorithm computes such an embedding and a corresponding bitonic *st*-ordering.

If no such ordering exists, one has to fall back to the fixed embedding splitting algorithm. In this work, we close this gap by describing a linear-time algorithm to compute an optimal set of edges to split over all possible embeddings.

Theorem 1. *Let $G = (V, E)$ be an st-planar graph with n vertices. It is possible to compute in $O(n)$ time a set of edges $E' \subseteq E$ of minimum cardinality such that the graph G' obtained from G by splitting each edge in E' once is bitonic.*

Having settled the problem of finding the smallest set of edges to split, we turn our attention to another idea which might enable us to improve the result for upward planar polyline drawings: Instead of only considering the original input graph, we can reverse all edges of this graph, obtain an upward planar drawing for this reversed graph, and then mirror this drawing vertically to obtain an upward planar drawing for the original graph. This idea stems from the observation that the example given in [10], which requires $n-3$ splits, does not require any split at all when all edges have been reversed. Hence, we may just choose the orientation with the minimum number of splits. Recently, Rettner [15] investigated this approach in his Bachelor's thesis, where he constructed an instance that requires at least $3/4n-3$ splits in each of the two orientations. The question that arises is whether one of the two orientations always requires significantly less than $n-3$ splits. We answer this question negatively with the following theorem.

Theorem 2. *For every integer $k \geq 1$, there is an st-planar graph $G_k = (V_k, E_k)$ with $n = 2k + 3$ vertices such that for every set $E' \subset E_k$ with less than $n-5$ edges neither the graph G'_k obtained from G_k by splitting each edge in E' once nor the graph $\tilde{G}'_k$ obtained by reversing the direction of all edges in G'_k is bitonic.*

We give preliminaries in Sect. 2, then we devote Sects. 3 and 4 to proving Theorems 1 and 2, respectively. We conclude with open problems in Sect. 5.

2   Preliminaries

Graph Drawings and Upward Planarity. A *drawing* Γ of a graph G maps the vertices of G to distinct points in the plane and the edges of G to simple Jordan arcs between their endpoints. Drawing Γ is *planar* if no two edges share an interior point. Planar drawings partition the plane into regions, called *faces*, whose boundaries consist of edges. The unbounded face is the *outer face*. An *embedding* is a class of drawings defining the sets of faces with the same boundaries.

A drawing Γ of a directed acyclic graph G is *upward planar* if for every edge (u, v), vertex u lies below v in Γ and (u, v) is drawn as a y-monotone curve in Γ; accordingly graph G is *upward planar* if it admits an upward planar drawing. A vertex v of a directed graph is a *source* (*sink*, resp.), if it only has outgoing (incoming, resp.) edges. A directed graph G is *st-planar* if it has a unique source s and a unique sink t such that there is an upward planar drawing of G, where s and t are incident to the outer face of it. In our definition, we assume that the edge (s, t) exists and is incident to the outer face. An embedding of an st-planar graph induces a left-to-right ordering of the incoming and outgoing edges of each vertex. We call the left-to-right ordered sequence of the neighbors of a vertex v connected with outgoing edges of v the *successor list* of v. Note that the faces of an st-planar graph have a unique source and a unique sink [17] connected by two paths. If one of these paths is a single edge, we call it *transitive*.

st-Orderings. An *st-ordering* of an st-planar graph is a linear ordering of its vertices with a prescribed vertex s being the first and a prescribed vertex t being the last vertex such that for every directed edge (u, v), it holds that u precedes v [14]. Given an st-ordering of an embedded st-planar graph, the successor list of a vertex u is *monotonically increasing* (*decreasing*, resp.) if the outgoing neighbors of u appear in this successor list in the same (opposite, resp.) order as they appear in the st-ordering. Further, the successor list of u is *bitonic* if there exists an outgoing neighbor h of u, called *apex* of u, such that the successor list of u is monotonically increasing from the beginning up to h and monotonically decreasing from h up to the end. Note that a monotonically increasing (decreasing, resp.) successor list is bitonic with the rightmost (leftmost, resp.) outgoing neighbor being its apex. We call an embedding $\mathcal{E}$ of an st-planar graph G *monotonic* (*bitonic*, resp.) if there exists an st-ordering of G such that the successor lists of all vertices defined by $\mathcal{E}$ are monotonically increasing/decreasing (bitonic, resp.); we call the corresponding st-ordering *monotonic* (*bitonic*, resp.). Further, we say that an st-planar graph G is *monotonic* (*bitonic*, resp.) if G admits a *monotonic* (*bitonic*, resp.) embedding.

Forbidden Configurations for Bitonic st-Orderings. Consider an embedding and an st-ordering of an st-planar graph G. Let u be a vertex and h the outgoing neighbor of u with largest rank in the st-ordering. Note that h is the only possible apex for the successor list of u. Then, the successor list of u is not bitonic if and only if there exist two vertices v, w such that v precedes w in the st-ordering and v appears between w and h in the successor list. We call this configuration a *conflict*. It has been shown [10] that for a given embedding of an st-planar graph there exists an st-ordering without conflicts if and only if the embedding does not contain any *forbidden configuration*, where a forbidden configuration is formed by two faces $f_1 = \langle u, v_{i+1}, \ldots, v_i \rangle$ and $f_2 = \langle u, v_j, \ldots, v_{j+1} \rangle$ such that the successor list of u contains $v_i, v_{i+1}, v_j, v_{j+1}$ in this order and $(v_{i+1}, \ldots, v_i)$ and $(v_j, \ldots, v_{j+1})$ are directed paths in G; see Fig. 1. In order to obtain bitonic embeddings even in the presence of forbidden configurations, Gronemann [10] proposed to *split* at least one of the transitive edges (u, v_i) and (u, v_{j+1}). More specifically, if we split edge (u, v_i), we obtain two new edges (u, v_i') and (v_i', v_i) with dummy vertex v_i'. Note that v_i' then replaces v_i in the successor list of u in the obtained graph. Since there exists no directed path from v_{i+1} to v_i', the forbidden configuration has been resolved.

Connectivity and SPQR-Trees. A graph is *connected* if for any pair of vertices there is a path connecting them. A graph is *k-connected* if the removal of any set of $k - 1$ vertices leaves it connected. A 2- or 3-connected graph is also

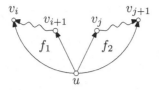

Fig. 1. Forbidden configuration that prevents a bitonic successor list for u.

referred to as *biconnected* or *triconnected*, respectively. Note that a triconnected planar graph has a unique embedding up to the choice of the outer face. Also note that st-planar graphs are always biconnected.

The *SPQR-tree* $\mathcal{T}$ of an st-planar graph G is a labeled tree representing the decomposition of G into its triconnected components [5,6]. Every triconnected component of G is associated with a node μ in $\mathcal{T}$. The two vertices separating the component associated with μ from the rest of the graph are called the *poles* s_μ and t_μ of μ. The *skeleton* of μ, denoted by $skel(\mu)$, is an st-planar graph where $s = s_\mu$ and $t = t_\mu$ whose edges are called *virtual edges*. In particular, there exists a virtual edge for every child ν of μ in $\mathcal{T}$ plus a *parent virtual edge* (s_μ, t_μ) that has a counterpart in the skeleton of its parent. A node $\mu \in \mathcal{T}$ can be of one of four different types: (i) *S-node*, if $skel(\mu)$ is composed of the parent virtual edge and a directed path of length at least 2 from s_μ to t_μ; (ii) *P-node*, if $skel(\mu)$ is a bundle of at least three parallel edges from s_μ to t_μ; (iii) *Q-node*, if $skel(\mu)$ consists of two parallel edges, one being the parent virtual edge and the other one being the corresponding edge in G; (iv) *R-node*, if $skel(\mu)$ is a simple triconnected st-planar graph with $s = s_\mu$ and $t = t_\mu$.

The set of leaves of $\mathcal{T}$ coincides with the set of Q-nodes, except for the Q-node ρ corresponding to edge (s,t), which is selected as the root of $\mathcal{T}$. Also, neither two S-nodes, nor two P-nodes are adjacent in $\mathcal{T}$. The subtree $\mathcal{T}_\mu$ of $\mathcal{T}$ rooted at μ induces a subgraph $pert(\mu)$ of G, called *pertinent*, which is described by $\mathcal{T}_\mu$ in the decomposition. In particular, $pert(\mu)$ is obtained from $skel(\mu)$ by recursively identifying each virtual edge with the corresponding parent virtual edge in the corresponding child node. We assume that the parent virtual edge of μ is not part of $pert(\mu)$. All embeddings of $pert(\mu)$ can be described by a permutation of the parallel virtual edges in each P-node in $\mathcal{T}_\mu$ and a flip of the skeleton of each R-node in $\mathcal{T}_\mu$. SPQR-tree $\mathcal{T}$ is unique, and can be computed in linear time [11].

3   Number of Splits in the Variable Embedding Setting

In this section, we present an algorithm to compute a set of edges E' as in Theorem 1 when graph G' is required to be bitonic. The algorithm for the monotonic case is analogous (and simpler), and will be discussed at the end of the section. Our goal is to construct an embedding $\mathcal{E}$ of G such that the embedding $\mathcal{E}'$ of G' obtained from $\mathcal{E}$ by splitting the edges in E' admits a bitonic ordering π'.

To compute $\mathcal{E}$, we adopt an SPQR-tree approach similar to the one by Chaplick et al. [4] to test whether an st-planar graph is bitonic. In contrast, however, we do not augment the graph explicitly. Instead, we specify the embedding $\mathcal{E}$ and a labeling of the edges describing whether they will eventually be split.

Let $\mathcal{T}$ be the SPQR-tree of G, rooted at the edge (s, t). We associate each node μ of $\mathcal{T}$, with poles s_μ and t_μ, with two costs $c_b(\mu)$ and $c_m(\mu)$, and with two embeddings $\mathcal{E}_b(\mu)$ and $\mathcal{E}_m(\mu)$ of $pert(\mu)$, whose edges are labeled as split or non-split, such that the following invariants hold:

I.1 $c_m(\mu)$ is the minimum number of splits to make $pert(\mu)$ bitonic, with the additional requirement that the successor list of s_μ is monotonically decreasing, and $\mathcal{E}_m(\mu)$ is an embedding of $pert(\mu)$ achieving this cost;

I.2 $c_b(\mu)$ is the minimum number of splits to make $pert(\mu)$ bitonic with no additional requirement, and $\mathcal{E}_b(\mu)$ is an embedding of $pert(\mu)$ achieving this cost;

I.3 a. an edge e in $\mathcal{E}_m(\mu)$ is labeled as split if and only if e contributes to $c_m(\mu)$,
   b. an edge e in $\mathcal{E}_b(\mu)$ is labeled as split if and only if e contributes to $c_b(\mu)$;

I.4 if the edge (s_μ, t_μ) exists in $pert(\mu)$, then t_μ is the apex of s_μ in $\mathcal{E}_m(\mu)$ and the edge (s_μ, t_μ) is labeled as non-split.

Observe that, by definition, it holds that $c_b(\mu) \leq c_m(\mu)$.

We perform a bottom-up traversal of $\mathcal{T}$ and compute for each node μ the costs $c_b(\mu)$ and $c_m(\mu)$, the two embeddings $\mathcal{E}_b(\mu)$ and $\mathcal{E}_m(\mu)$ of $pert(\mu)$, and the labeling of their edges, so that I.2–I.4 hold, assuming that these invariants hold for all the children of μ. We distinguish cases based on the type of node μ.

Node μ is a Q-node that is a Leaf of $\mathcal{T}$. We set both $\mathcal{E}_b(\mu)$ and $\mathcal{E}_m(\mu)$ to the unique embedding of $pert(\mu)$, which consists only of edge (s_μ, t_μ). Since this embedding is monotonic, we set both costs $c_m(\mu)$ and $c_b(\mu)$ to 0, and we label (s_μ, t_μ) as non-split in both embeddings. Hence, I.2–I.4 are satisfied.

Node μ is a P-node: Let $\nu_1, \ldots, \nu_k$ denote the children of μ. W.l.o.g., assume that if μ has a Q-node child then this child is ν_1. We construct both $\mathcal{E}_m(\mu)$ and $\mathcal{E}_b(\mu)$ by ordering the children of μ in clockwise order around s_μ from ν_1 to ν_k. Then, we choose embeddings and flips for the pertinent graphs of the children of μ in order to obtain $\mathcal{E}_m(\mu)$ and $\mathcal{E}_b(\mu)$, as follows.

In order to construct $\mathcal{E}_m(\mu)$, we choose the monotonic embedding $\mathcal{E}_m(\nu_i)$ for each child ν_i and perform no flip. We set the monotonic cost $c_m(\mu)$ for μ to $\sum_{i=1}^k c_m(\nu_i)$, satisfying I.1. The labeling of the edges in $\mathcal{E}_m(\mu)$ is inherited from the corresponding ones of $\mathcal{E}_m(\nu_1), \ldots, \mathcal{E}_m(\nu_k)$, which ensures I.3a and I.4.

To specify $\mathcal{E}_b(\mu)$, we select one of the children of μ to contain the apex of s_μ, so that the resulting bitonic cost $c_b(\mu)$ for μ is minimized. For this, we select the child ν_h, with $1 \leq h \leq k$, such that the difference $c_m(\nu_h) - c_b(\nu_h)$ is maximum. If this difference is 0 for all children of μ, we set ν_h to be ν_1. Then, we select the bitonic embedding $\mathcal{E}_b(\nu_h)$ for ν_h and the monotonic embedding $\mathcal{E}_m(\nu_i)$ for each child $\nu_i \neq \nu_h$. Finally, we flip the embeddings $\mathcal{E}_m(\nu_1), \ldots, \mathcal{E}_m(\nu_{h-1})$ of the pertinent graphs of $\nu_1, \ldots, \nu_{h-1}$. Note that the flip of these embeddings results in a monotonically increasing successor list at s_μ for each of them, and hence

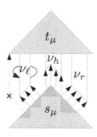

Fig. 2. Bitonic embedding $\mathcal{E}_b(\mu)$ for a P-node μ. One child ν_h uses its bitonic embedding $\mathcal{E}_b(\nu_h)$. Embedding $\mathcal{E}_m(\nu_\ell)$ of child ν_ℓ appearing left of ν_h is flipped.

guarantees that $\mathcal{E}_b(\mu)$ is bitonic, except when ν_1 is a Q-node and $\nu_h \neq \nu_1$; see Fig. 2. To guarantee that $\mathcal{E}_b(\mu)$ is bitonic also in the latter case, edge (s_μ, t_μ) must be split; note that, by I.4, edge $(s_\mu, t_\mu) = (s_{\nu_1}, t_{\nu_1})$ is labeled `non-split` in the embedding $\mathcal{E}_m(\nu_1)$ of ν_1. So, to guarantee I.2, we set $c_b(\mu) = c_m(\mu) - c_m(\nu_h) + c_b(\nu_h)$ in the former case and $c_b(\mu) = c_m(\mu) - c_m(\nu_h) + c_b(\nu_h) + 1$ otherwise.

To guarantee I.3b, we inherit the labeling of the edges in $\mathcal{E}_b(\mu)$ from the embeddings $\mathcal{E}_m(\nu_1), \ldots, \mathcal{E}_m(\nu_{h-1}), \mathcal{E}_b(\nu_h), \mathcal{E}_m(\nu_{h+1}), \ldots, \mathcal{E}_m(\nu_k)$. Further, in the special case in which ν_1 is a Q-node and $\nu_h \neq \nu_1$, we label edge (s_μ, t_μ) as `split`.

Node μ is an S-node: Let $\nu_1, \ldots, \nu_k$ denote the children of μ, where $s_\mu = s_{\nu_1}$ and $t_\mu = t_{\nu_k}$. To compute $\mathcal{E}_m(\mu)$ and $\mathcal{E}_b(\mu)$, we use $\mathcal{E}_m(\nu_1)$ and $\mathcal{E}_b(\nu_1)$ for child ν_1, respectively, and the bitonic embeddings $\mathcal{E}_b(\nu_2), \ldots, \mathcal{E}_b(\nu_k)$ for children $\nu_2, \ldots, \nu_k$ in both cases, without performing any flip. To guarantee I.1 and I.2, we set $c_m(\mu)$ and $c_b(\mu)$ to $c_m(\nu_1) + \sum_{i=2}^{k} c_b(\nu_i)$ and $\sum_{i=1}^{k} c_b(\nu_i)$, respectively. To guarantee I.3a and I.3b, the labeling of the edges in $\mathcal{E}_b(\mu)$ and $\mathcal{E}_m(\mu)$ is inherited from the corresponding ones in the chosen embeddings of the children. Finally, I.4 is satisfied since edge (s_μ, t_μ) does not exist in $pert(\mu)$.

Node μ is an R-node: Since $skel(\mu)$ is triconnected, it has a unique embedding; we will construct both $\mathcal{E}_m(\mu)$ and $\mathcal{E}_b(\mu)$ based on such an embedding and select for each child ν of μ a suitable embedding of $pert(\nu)$ and a flip. Since each edge is outgoing for only one of its end-vertices, we consider every vertex in $skel(\mu)$ independently, together with its outgoing virtual edges, similar to [10].

Let u be a vertex of $skel(\mu)$, and let $(u, v_1), \ldots, (u, v_k)$ be the outgoing virtual edges of u, as they appear consecutively clockwise around u, and let $\nu_1, \ldots, \nu_k$ be the corresponding children of μ. If $u \neq s_\mu$, we can construct a bitonic successor list for u in both $\mathcal{E}_m(\mu)$ and $\mathcal{E}_b(\mu)$. Otherwise, we may need to perform different choices when constructing $\mathcal{E}_m(\mu)$ and $\mathcal{E}_b(\mu)$, to guarantee I.1 and I.2, respectively.

Suppose first that $u \neq s_\mu$. Similar to the P-node case, we determine a child ν_h of μ, with $1 \leq h \leq k$, to contain the apex of u that minimizes the number of splits of the outgoing edges of u to make the successor list of u bitonic; we denote this number by $c_b(u)$. To determine ν_h, we consider each child ν_j, for $j = 1, \ldots, k$, to be candidate for ν_h, and compute the required number of splits for this choice, denoted by $c_b(u, j)$. We then obtain $c_b(u) = \min\{c_b(u, j) \mid j = 1, \ldots, k\}$.

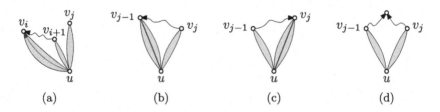

Fig. 3. (a) An unavoidable conflict if v_j is selected as apex of u. (b)–(d) Different cases, that arise, when computing $c_b(u, j)$.

In contrast to the P-node case, we cannot conclude that $c_b(u, j) = c_b(v_j) + \sum_{i \neq j} c_m(v_i)$, since the choice of v_j and the structure of $skel(\mu)$ may result in new conflicts. Namely, consider a child v_i of μ with $i < j$ and assume that the edge (u, v_i) exists in $pert(\mu)$ and that there is a directed path in $skel(\mu)$ from v_{i+1} to v_i; see e.g. Fig. 3a. This implies that there is a directed path from a successor w of u in $pert(v_{i+1})$ to v_i. Since $i < j$, vertex w should however not be a predecessor of v_i and hence edge (u, v_i) must be split. Note that, by I.4 edge (u, v_i) is labeled as non-split in $\mathcal{E}_m(v_i)$. Analogously, if $i > j$, a conflict may arise when there exists a directed path in $skel(\mu)$ from v_{i-1} to v_i. Denote by $c_s(u, j)$ the total number of these additional splits when v_j contains the apex of u. Thus,

$$c_b(u, j) = \sum_{i=1}^{j-1} c_m(v_i) + c_b(v_j) + \sum_{i=j+1}^{k} c_m(v_i) + c_s(u, j).$$

The computation of $c_b(u, j)$ for all $j = 1, \ldots, k$ can be done in quadratic time with respect to the number of outgoing virtual edges of u. Next, we make use of ideas of the fixed-embedding algorithm [10] to achieve linear time. Namely, we first compute $c_b(u, 1)$. Then, for each $j = 2, \ldots, k$, we can compute $c_b(u, j)$ from $c_b(u, j - 1)$. For $c_b(u, j)$, we assume the apex of u to be contained in child v_j, while assuming that we already computed $c_b(u, j - 1)$. In this transition, $pert(v_{j-1})$ changes its embedding from $\mathcal{E}_b(v_{j-1})$ to $\mathcal{E}_m(v_{j-1})$, while $pert(v_j)$ changes its embedding from $\mathcal{E}_m(v_j)$ to $\mathcal{E}_b(v_j)$; the pertinent graphs of the remaining children maintain their monotonic embedding. We take this change into account by considering the corresponding difference $\delta_j(u) = c_b(v_j) - c_m(v_j) + c_m(v_{j-1}) - c_b(v_{j-1})$. In addition, we must also take into account the difference between $c_s(u, j - 1)$ and $c_s(u, j)$, whose computation can be done again by only considering the children v_{j-1} and v_j. More precisely, if (u, v_{j-1}) is an edge in $pert(\mu)$, then it did not need to be split when the apex of u was in v_{j-1}, but it has to be split when moving the apex to v_j, if there is a directed path from v_j to v_{j-1}; see Fig. 3b. On the other hand, if (u, v_j) is an edge of $pert(\mu)$ and it had to be split when the apex of u was in v_{j-1}, i.e. there is a directed path from v_{j-1} to v_j, then edge (u, v_j) does not need to be split any longer when the apex is in v_j; see Fig. 3c. Note that, if there is no directed path between v_{j-1} and v_j, then neither of the two cases occurs, and edges (u, v_{j-1})

and (u, v_j) (if they exist) do not need to be split; see Fig. 3d. In either case, we conclude that the difference between $c_s(u, j-1)$ and $c_s(u, j)$ is at most 1. Depending on which of the three cases arises, we compute $c_b(u, j)$ as follows:

$$c_b(u, j) = c_b(u, j-1) + \delta_j(u) + \begin{cases} 1 & \text{if } \exists \text{ directed path from } v_j \text{ to } v_{j-1}, \text{ and} \\ & (u, v_{j-1}) \text{ is an edge in } pert(\mu) \\ -1 & \text{if } \exists \text{ directed path from } v_{j-1} \text{ to } v_j, \text{ and} \\ & (u, v_j) \text{ is an edge in } pert(\mu) \\ 0 & \text{otherwise} \end{cases}$$

Once $c_b(u, j)$ has been computed for all $j = 1, \ldots, k$, we choose ν_h, with $1 \leq h \leq k$, so that $c_b(u, h)$ is minimum among all $c_b(u, j)$ and define $c_b(u) = c_b(u, h)$. Hence, in order to construct $\mathcal{E}_m(\mu)$ and $\mathcal{E}_b(\mu)$, we select the bitonic embedding $\mathcal{E}_b(\nu_h)$ for $pert(\nu_h)$ and the monotonic embedding $\mathcal{E}_m(\nu_i)$ for the pertinent graph $pert(\nu_i)$ for $i \in \{1, \ldots, h-1, h+1, \ldots, k\}$. We further flip the embeddings $\mathcal{E}_m(\nu_1), \ldots, \mathcal{E}_m(\nu_{h-1})$, as in the P-node case. We inherit the labeling of the edges from the embeddings $\mathcal{E}_m(\nu_1), \ldots, \mathcal{E}_m(\nu_{h-1}), \mathcal{E}_b(\nu_h), \mathcal{E}_m(\nu_{h+1}), \ldots, \mathcal{E}_m(\nu_k)$, except for the edges that contribute to $c_s(u, h)$, which we label as split. We repeat the above operations for every vertex u of $skel(\mu)$ with $u \neq s_\mu$.

Consider now the case $u = s_\mu$. We distinguish two cases, based on which embedding of $pert(\mu)$ we are going to compute. Namely, for $\mathcal{E}_b(\mu)$ we perform the same operations as for any other vertex of $skel(\mu)$, since in this embedding we can have a bitonic successor list for s_μ. This guarantees I.2 and I.3b. In order to also guarantee I.1 and I.3b, we have to slightly adjust our approach. In particular, we have to obtain a monotonic successor list for s_μ in $\mathcal{E}_m(\mu)$. To achieve this, we first choose the monotonic embeddings for $pert(\nu_1), \ldots, pert(\nu_k)$. Then, we have to choose whether ν_1 or ν_k contains the apex of s_μ. In order to perform this choice, we have to consider the conflicts that are created due to the presence of directed paths in $skel(\mu)$, as in the bitonic case. Thus, we compute $c_s(s_\mu, 1)$ and $c_s(s_\mu, k)$ and choose the minimum of the two. We choose the corresponding child to contain the apex of s_μ and label edges as split such that all conflicts are resolved and inherit the labeling of the remaining edges from embeddings $\mathcal{E}_m(\nu_1), \ldots, \mathcal{E}_m(\nu_k)$. Note that if ν_k contains the apex of s_μ, we also have to flip all the embeddings $\mathcal{E}_m(\nu_1), \ldots, \mathcal{E}_m(\nu_k)$ and the resulting embedding of the entire R-node μ so to obtain a monotonically decreasing successor list for s_μ. Thus,

$$c_m(\mu) = \sum_{\substack{u \in V_\mu \\ u \neq s_\mu}} c_b(u) + \underbrace{\sum_{i=1}^k c_m(\nu_i) + \min\{c_s(s_\mu, 1), c_s(s_\mu, k)\}}_{u = s_\mu} \text{ and } c_b(\mu) = \sum_{u \in V_\mu} c_b(u)$$

where V_μ denotes the vertex set of $skel(\mu)$. Note that I.4 is trivially satisfied since edge (s_μ, t_μ) does not exist in $pert(\mu)$.

Node μ is a Q-node that is the Root of $\mathcal{T}$: Note that this case arises at the end of the traversal of $\mathcal{T}$. Since we seek to compute a bitonic embedding for G, we only have to compute $\mathcal{E}_b(\mu)$ and satisfy I.2 and I.3b (i.e., I.1, I.3a and I.4 can be safely neglected). Consider the unique child ν of μ. Assume first that $c_b(\nu) < c_m(\nu)$. We claim that the apex of $s_\mu = s_\nu$ in $\mathcal{E}_b(\nu)$ is not incident to any face that contains both s_μ and t_μ. Indeed, if this is not the case either s_ν has already a monotonic successor list in $\mathcal{E}_b(\nu)$ or ν is a P-node and it is possible to obtain a monotonic successor list of s_ν by only reordering and flipping the embeddings of the pertinent graphs of its children. In both cases, however, $c_b(\nu) = c_m(\nu)$ holds, which is a contradiction. By our claim, it follows that any possible embedding obtained from $\mathcal{E}_b(\nu)$ by adding the edge (s_μ, t_μ) would violate the bitonicity of the successor list of s_μ. Thus, we have to label (s_μ, t_μ) as split, and inherit the labeling of the remaining edges from $\mathcal{E}_b(\nu)$. Consider now the case $c_b(\nu) = c_m(\nu)$. Here, we use the monotonic embedding $\mathcal{E}_m(\nu)$ and label the edges according to the labeling of $\mathcal{E}_m(\nu)$ while labeling (s_μ, t_μ) as non-split. In both cases, edge (s_μ, t_μ) is embedded on the outer face of the embedding of $pert(\nu)$ such that t_μ is the leftmost successor of s_μ, which guarantees I.2. I.3b is satisfied by the way we treat edge (s_μ, t_μ) and by inheriting the labeling of the remaining edges of $pert(\mu)$ from the chosen embedding of $pert(\nu)$.

Proof (of Theorem 1). The correctness of our algorithm follows from the fact that at the end of the traversal of $\mathcal{T}$, I.2 is satisfied by the bitonic embedding $\mathcal{E}_b(\rho)$ of the root ρ of $\mathcal{T}$. Further, since the labeling of the edges of $\mathcal{E}_b(\rho)$ satisfies I.3b, we can set E' to be the set of edges that are labeled as split in $\mathcal{E}_b(\rho)$, which guarantees that E' is of minimum cardinality and that the graph G' obtained from G by splitting once each edge in E' is bitonic. Note that we can obtain an actual bitonic st-numbering for G' using the fixed-embedding algorithm [10] on the embedding of G' obtained from $\mathcal{E}_b(\rho)$ by splitting each edge of E'.

As for the time complexity, the construction of the SPQR-tree can be done in $O(n)$ time [11]. Then, at each step of the algorithm, we consider a node μ of $\mathcal{T}$ and we perform a set of operations in time linear to the size of $skel(\mu)$. This is clear for the Q-, S-, and P-node cases. In the R-node case, this follows from our analysis and the fact that the fixed-embedding algorithm [10] is linear in the size of the input embedding. Since the sum of the sizes of the skeletons over all the nodes of $\mathcal{T}$ is $O(n)$ [2], the time complexity of the algorithm follows.   □

Remark 1. Our algorithm can be adjusted so that the resulting graph G' is monotonic. To achieve that, in the S- and R-node cases, we apply to all vertices of $skel(\mu)$ the same procedure as we applied to s_μ when computing $\mathcal{E}_m(\mu)$. In this way, we guarantee that the successor lists of all vertices are in fact monotonic.

Remark 2. Every series-parallel graph, oriented consistently with the series-parallel structure, is monotonic. This is because, in the absence of R-nodes, there is no need to split when computing a monotonic embedding.

4   Lower Bound on the Number of Splits

In this section we prove Theorem 2. Graph $G_k = (V_k, E_k)$ is as follows. For $k = 1$, we set $V_1 = \{s_0, s_1, t_0, t_1, t_2\}$, and $E_1 = \{(s_0, t_0), (s_0, t_1), (s_0, t_2), (s_1, t_0),$

$(s_1, t_1), (s_1, t_2), (s_1, s_0), (t_0, t_1), (t_1, t_2)\}$; see Fig. 4a. Since G_1 is triconnected, it admits a unique embedding up to the choice of the outer face, which allows to embed (s_1, t_2) as leftmost or rightmost edge of s_1. Since G_1 contains a Hamiltonian path $(s_1, s_0, t_0, t_1, t_2)$, it has a unique st-ordering with $s = s_1$ and $t = t_2$.

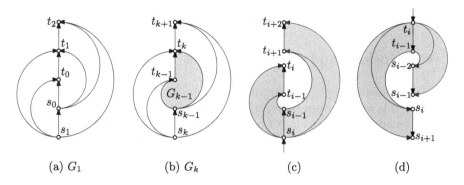

(a) G_1 (b) G_k (c)   (d)

Fig. 4. (a)–(b) Construction of graph G_k. (c)–(d) Illustration of forbidden configurations and edges that must be split in both orientations. (Color figure online)

For $k > 1$, graph G_k is constructed based on the triconnected Hamiltonian planar st-graph G_{k-1} with $s - s_{k-1}$ and $t - t_k$ as follows. Its vertex set is $V_k = V_{k-1} \cup \{s_k, t_{k+1}\}$, while its edge set is $E_k = E_{k-1} \cup \{(s_k, t_{k-1}), (s_k, t_k), (s_k, t_{k+1}),$ $(s_{k-1}, t_{k+1}), (s_k, s_{k-1}), (t_k, t_{k+1})\}$; see Fig. 4b. Since graph G_k is a triconnected Hamiltonian planar st-graph with $s = s_k$ and $t = t_{k+1}$, it has a unique st-ordering and a unique embedding up to the choice of the outer face, which simply allows us to embed edge (s_k, t_{k+1}) as the leftmost or rightmost edge of vertex s_k. As a side note, we also mention that G_k has pathwidth 3.

We describe forbidden configurations that inevitably appear in G_k, and then estimate the number of edge splits that are required to eliminate them. In particular, for vertex s_i with $1 \leq i \leq k-1$, each of the two faces $\langle s_i, s_{i-1}, t_{i-1}\rangle$ and $\langle s_i, t_{i-1}, t_i\rangle$ form a forbidden configuration with each of the faces $\langle s_i, s_{i-1}, t_{i+1}\rangle$ and $\langle s_i, t_{i+1}, t_{i+2}\rangle$; see the blue and red colored faces in Fig. 4c, respectively. In order to eliminate these forbidden configurations, at least one of the two pairs of edges $(s_i, t_{i-1}), (s_i, t_i)$ and $(s_i, t_{i+1}), (s_i, t_{i+2})$ must be split; see the blue and red edges in Fig. 4c, respectively. Forbidden configurations involving s_k are avoidable by embedding (s_k, t_{k+1}) as the leftmost edge of s_k.

Similar arguments apply when considering $\tilde{G}_k$. In particular, for vertex t_i with $2 \leq i \leq k-1$, each of the two faces $\langle t_i, t_{i-1}, s_i\rangle$ and $\langle t_i, s_i, s_{i+1}\rangle$ form a forbidden configuration with each of the two faces $\langle t_i, t_{i-1}, s_{i-2}\rangle$ and $\langle t_i, s_{i-2}, s_{i-1}\rangle$; see the blue and red colored faces in Fig. 4d, respectively. Hence, at least one of the two pairs of edges $(t_i, s_i), (t_i, s_{i+1})$ and $(t_i, s_{i-2}), (t_i, s_{i-1})$ must be split; see the blue and red edges in Fig. 4d, respectively. Note that forbidden configurations that involve vertex t_{k+1} can be avoided by embedding edge (t_{k+1}, s_k) as the leftmost edge of t_{k+1}. Moreover, for vertex t_1 (t_k, resp.), face $\langle t_1, t_0, s_0\rangle$

$(\langle t_k, t_{k-1}, s_k \rangle$, resp.) forms forbidden configurations with both faces $\langle t_1, t_0, s_1 \rangle$ and $\langle t_1, s_1, s_2 \rangle$ ($(\langle t_k, t_{k-1}, s_{k-1} \rangle$ and $\langle t_k, s_{k-2}, s_{k-1} \rangle$, resp.). Hence, for each of t_1 and t_k at least one more incident edge must be split.

We conclude that for both G_k and $\tilde{G}_k$, a set of edges E' of cardinality at least $2(k-1) = n-5$ has to be split to eliminate all the forbidden configurations discussed above. Note that in $\tilde{G}_1$ there is already one unavoidable forbidden configuration (at vertex t_1), while this is not the case for G_1.

5   Conclusions and Open Problems

We conclude with some open problems raised by our work. (i) In view of Remark 2, it is worth investigating other meaningful subclasses of upward planar graphs that admit improved upper bounds on the required number of splits; note that our lower bound example already imposes strong restrictions (e.g., Hamiltonicity, low pathwidth).(ii) We know that an upper bound on the number of splits provides an upper bound on the number of bends needed to compute a 1-bend upward planar drawing in quadratic area. Understanding the implication of a lower bound on the number of splits on the corresponding number of bends is an interesting research direction. (iii) An experimental evaluation of our algorithm would allow to estimate the required number of splits in practice.

Acknowledgments. The authors would like to thank Michael Kaufmann and Antonios Symvonis for fruitful discussions.

References

1. Angelini, P., Bekos, M.A., Liotta, G., Montecchiani, F.: Universal slope sets for 1-bend planar drawings. Algorithmica **81**(6), 2527–2556 (2019). https://doi.org/10.1007/s00453-018-00542-9
2. Bertolazzi, P., Di Battista, G., Didimo, W.: Computing orthogonal drawings with the minimum number of bends. IEEE Trans. Comput. **49**(8), 826–840 (2000). https://doi.org/10.1109/12.868028
3. Biedl, T.C., Kant, G.: A better heuristic for orthogonal graph drawings. Comput. Geom. **9**(3), 159–180 (1998). https://doi.org/10.1016/S0925-7721(97)00026-6
4. Chaplick, S., et al.: Planar L-drawings of directed graphs. In: Frati, F., Ma, K.-L. (eds.) GD 2017. LNCS, vol. 10692, pp. 465–478. Springer, Cham (2018). https://doi.org/10.1007/978-3-319-73915-1_36
5. Di Battista, G., Tamassia, R.: On-line maintenance of triconnected components with SPQR-trees. Algorithmica **15**(4), 302–318 (1996). https://doi.org/10.1007/BF01961541
6. Di Battista, G., Tamassia, R.: On-line planarity testing. SIAM J. Comput. **25**(5), 956–997 (1996). https://doi.org/10.1137/S0097539794280736
7. Even, S., Tarjan, R.E.: Computing an st-numbering. Theor. Comput. Sci. **2**(3), 339–344 (1976). https://doi.org/10.1016/0304-3975(76)90086-4
8. de Fraysseix, H., Pach, J., Pollack, R.: How to draw a planar graph on a grid. Combinatorica **10**(1), 41–51 (1990). https://doi.org/10.1007/BF02122694

9. Gronemann, M.: Bitonic *st*-orderings of biconnected planar graphs. In: Duncan, C., Symvonis, A. (eds.) GD 2014. LNCS, vol. 8871, pp. 162–173. Springer, Heidelberg (2014). https://doi.org/10.1007/978-3-662-45803-7_14

10. Gronemann, M.: Bitonic *st*-orderings for upward planar graphs. In: Hu, Y., Nöllenburg, M. (eds.) GD 2016. LNCS, vol. 9801, pp. 222–235. Springer, Cham (2016). https://doi.org/10.1007/978-3-319-50106-2_18

11. Gutwenger, C., Mutzel, P.: A linear time implementation of SPQR-trees. In: Marks, J. (ed.) GD 2000. LNCS, vol. 1984, pp. 77–90. Springer, Heidelberg (2001). https://doi.org/10.1007/3-540-44541-2_8

12. Harel, D., Sardas, M.: An algorithm for straight-line drawing of planar graphs. Algorithmica **20**(2), 119–135 (1998). https://doi.org/10.1007/PL00009189

13. Kant, G.: Drawing planar graphs using the canonical ordering. Algorithmica **16**(1), 4–32 (1996). https://doi.org/10.1007/BF02086606

14. Otten, R.H.J.M., van Wijk, J.G.: Graph representations in interactive layout design. In: Proceedings of the IEEE International Symposium on Circuits and Systems, pp. 914–918 (1978)

15. Rettner, C.: Flipped bitonic st-orderings of upward plane graphs. Bachelor's thesis, University of Würzburg (2019). https://www1.pub.informatik.uni-wuerzburg.de/pub/theses/2019-rettner-bachelor.pdf

16. Rosenstiehl, P., Tarjan, R.E.: Rectilinear planar layouts and bipolar orientations of planar graphs. Discrete Comput. Geom. **1**(4), 343–353 (1986). https://doi.org/10.1007/BF02187706

17. Tamassia, R., Tollis, I.G.: A unified approach to visibility representations of planar graphs. Discrete Comput. Geom. **1**(4), 321–341 (1986). https://doi.org/10.1007/BF02187705

2.5-Connectivity: Unique Components, Critical Graphs, and Applications

Irene Heinrich[1]($\boxtimes$) (iD), Till Heller[2,3] (iD), Eva Schmidt[3] (iD), and Manuel Streicher[3] (iD)

[1] Algorithms and Complexity Group, TU Kaiserslautern, 67663 Kaiserslautern, Germany
irene.heinrich@cs.uni-kl.de

[2] Fraunhofer-Institut für Techno- und Wirtschaftsmathematik, Fraunhofer-Platz 1, 67663 Kaiserslautern, Germany
till.heller@itwm.fraunhofer.de

[3] Optimization Research Group, TU Kaiserslautern, 67663 Kaiserslautern, Germany
{eva.schmidt,streicher}@mathematik.uni-kl.de

Abstract. If a biconnected graph stays connected after the removal of an arbitrary vertex and an arbitrary edge, then it is called 2.5-connected. We prove that every biconnected graph has a canonical decomposition into 2.5-connected components. These components are arranged in a tree-structure. We also discuss the connection between 2.5-connected components and triconnected components and use this to present a linear time algorithm which computes the 2.5-connected components of a graph. We show that every critical 2.5-connected graph other than K_4 can be obtained from critical 2.5-connected graphs of smaller order using simple graph operations. Furthermore, we demonstrate applications of 2.5-connected components in the context of cycle decompositions and cycle packings.

Keywords: Mixed connectivity · Triconnected components · Critical graphs

1 Introduction

Over the years, connectivity has become an indispensable notion of graph theory. A tremendous amount of proofs start with a reduction which says "The main result holds for all graphs if it holds for all *sufficiently connected* graphs". Here, *sufficiently connected* stands for some measure of connectedness as, for example, biconnected, 4-edge-connected, vertex-edge-connected, or just connected. Usually, first a reduction from the desired statement for general graphs to sufficiently connected graphs is proven. Then the subsequent section starts with a sentence of the following manner: "From now on, all considered graphs are

This research was partially funded by the European Research Council (ERC) under the European Union's Horizon 2020 research and innovation programme (EngageS: grant agreement No. 820148), and the Federal Ministry of Education and Research.

I. Adler and H. Müller (Eds.): WG 2020, LNCS 12301, pp. 352–363, 2020.
https://doi.org/10.1007/978-3-030-60440-0_28

sufficiently connected." For example, it is shown in [14] that the Tutte polynomial is multiplicative over the biconnected components of a considered graph. Another reduction to components of higher connectivity is that finding a planar embedding can be reduced to embedding the triconnected components of the graph, cf. [12].

By far the largest part of the existing literature treats either *k-connectivity* (where, loosely speaking, graphs which stay connected even if $k - 1$ vertices are removed are considered) or *k-edge connectivity* (where graphs which stay connected even if $k - 1$ edges are removed are considered). We speak of *mixed connectivity*, when graphs are regarded which stay connected after k vertices and l edges are removed. This measure of connectivity has only rarely been studied. We refer the reader to [2] for a brief survey of mixed connectivity. In [10] it is shown that the behaviour of cycle decompositions is preserved under splits at vertex-edge separators (that is, a vertex and an edge whose removal disconnects the graph).

Our Contribution. We introduce a canonical decomposition of a graph into its 2.5-connected components, where a graph is 2.5-connected if it is biconnected and the removal of a vertex and an edge does not disconnect the graph. We prove the following decomposition theorem.

Theorem 1 (Decomposition into 2.5-connected components). *Let G be a biconnected graph. The 2.5-connected components of G are unique and can be computed in linear time.*

Furthermore, we demonstrate that the behaviour of critical 2.5-connected graphs is preserved in their triconnected components. We obtain a result similar to Tutte's decomposition theorem for 3-connected 3-regular graphs: all critical 2.5-connected graphs other than K_4 can be obtained from critical 2.5-connected graphs of smaller order by simple graph operations.

Finally, we show that the minimum (maximum) cardinality of a cycle decomposition of an Eulerian graph can be obtained from the minimum (maximum) cardinalities of the cycle decompositions of its 2.5-connected component. This gives new insights into a long standing conjecture of Hajós.

Most of the proofs have been omitted due to space restrictions. We refer to [9] for the full version of this paper.

Techniques. We demonstrate that 2.5-connected components can be defined in the same manner as triconnected components. The novel underlying idea of the present article is a red-green-colouring of the virtual edges of the triconnected components (a virtual edge of a component is not part of the original graph but stores the information where the components need to be glued together in order to obtain the host graph). The colouring is assigned to the virtual edges during the process of carrying out splits that give the triconnected components. It preserves the information whether a virtual edge could arise in a sequence of 2.5-splits (those corresponding to a vertex-edge separator). If so, the edge is coloured green, otherwise red. We prove that this colouring can be assigned to the virtual edges of the triconnected components (without knowledge of the splits that led there) in linear time. It can be exploited to obtain 2.5-connected components: glue the red edges. We show that the uniqueness of the red-green-colouring implies the uniqueness of the 2.5-connected components.

Further Related Work. We refer to [15] as a standard book on graph connectivity. The same topic is considered from an algorithmic point of view in [13]. A short overview on mixed connectivity with strong emphasis on partly raising Menger's theorem to mixed separators can be found in Chapter 1.4 of [2]. Grohe [7] introduces a new decomposition of a graph into quasi-4-connected components and discusses the relation of the quasi-4-connected components to triconnected components.

The importance of triconnected components for planarity testing was already observed in [12]. Hopcroft and Tarjan [11] proved that these components are tree-structured and exploited this algorithmically. On this basis, Battista and Tamassia [1] developed the notion of SPQR-trees. Gutwenger and Mutzel [8] used this result and the results of [11] for a linear-time algorithm that computes the triconnected components of a given graph and their tree-structure (SPQR-tree).

Outline. Preliminary results and definitions are introduced in the next section. In particular, Hopcroft and Tarjan's notions of triconnected components and virtual edges (cf. [11]) are explained. In Sect. 3 we adapt the definition of triconnected components in order to give a natural definition of 2.5-connected components. We prove that these are unique and show how they can be obtained from the triconnected components. We exploit this knowledge in Sect. 4 in order to give a linear time algorithm which computes the 2.5-connected components of a given graph. We characterize the critical 2.5-connected graphs in Sect. 5. Finally, some applications of 2.5-connected graphs are discussed in Sect. 6.

2   Preliminaries

If not stated otherwise, we use standard graph theoretic notation as can be found in [3]. Graphs are finite and may contain multiple edges but no loops. A graph of order 2 and size $k \geq 2$ is a *multiedge* (or k-edge). In this article a graph G is equipped with an injective labelling $\ell_G := E_V \to \mathbb{N}$ where E_V is a (possibly empty) subset of $E(G)$. We call $E_V(G) := E_V$ the *virtual edges* of G. If G is described without a labelling, then we implicitly assume $E_V(G) = \emptyset$.

Most of the notation and all of the results in this paragraph are borrowed from [11]. A connected graph is *biconnected* if for each triple of distinct vertices $(u, v, w) \in V(G)^3$ there exists a u-v-path P in G with $w \notin V(P)$.[1]

---

[1] This differs from the definition of 2-connected graphs as can be found in [3]. Connected graphs of order 2 are biconnected but not 2-connected.

Let u and v be two vertices of a biconnected graph G. We divide $E(G)$ into equivalence classes $E_1, E_2, \ldots, E_k$ such that two edges lie in the same class if and only if they are edges of a (possibly closed) subpath of G which neither contains u nor v internally. The classes E_i are the *separation classes* of G with respect to $\{u, v\}$. The set $\{u, v\}$ is a *separation pair* if there exists a set $I \subsetneq \{1, \ldots, k\}$ such that $E' := \bigcup_{i \in I} E_i$ satisfies $\min\{|E'|, |E(G) \setminus E'|\} \geq 2$. In this case, let $G_1 := G[E'] + e_1$ and $G_2 := G[E(G) \setminus E'] + e_2$, where both, e_1 and e_2, are new edges with endvertices u and v. Fix some $x \in \mathbb{N} \setminus l_G(E_V(G))$. For $i \in \{1, 2\}$ let $\ell_{G_i} \colon (E_V(G) \cap E(G_i)) \cup \{e_i\} \to \mathbb{N}$ be the labelling with $\ell_{G_i}(e) = \ell_G(e)$ for $e \in E_V(G) \cap E(G_i)$ and $\ell_{G_i}(e_i) = x$. Replacing G by G_1 and G_2 is a *split*. The virtual edges e_1 and e_2 *correspond* to each other. Vice versa, if G_1 and G_2 can be obtained by a split from G, then G is the *merge graph* of G_1 and G_2. Replacing G_1 and G_2 by G is a *merge*. A biconnected graph without a separation pair is *triconnected*.

Suppose a multigraph G is split, the split graphs are split, and so on, until no more splits are possible. (Each graph remaining is triconnected). The graphs constructed this way are called *split components of G*.

We say that two graphs H and H' are *equivalent*, if H' can be obtained from H by renaming and relabelling the virtual edges in $E_V(H)$. Two sets of graphs $\{G_1, \ldots, G_k\}$ and $\{G'_1, \ldots, G'_k\}$ are *equivalent* if the elements can be ordered in such a way that G_i is equivalent to G'_i for all $i \in \{1, \ldots, k\}$ and the correspondence of the virtual edges is preserved by the according renaming and relabelling maps. Two sets of split components of the same graph are not equivalent in general. Consider for example a cycle of length 4. The two possible separation pairs yield different partitions of the edge set of the cycle.

Split components of G are of one of the following types:

$$\text{triangles,} \quad \text{3-edges,} \quad \text{and other triconnected graphs.}$$

Denote the latter set by $\mathcal{T}$. Merge the triangles of the split components as much as possible to obtain a set of cycles $\mathcal{C}$. Further, merge the 3-edges as much as possible to obtain a set of multiedges $\mathcal{M}$. The set $\mathcal{C} \cup \mathcal{M} \cup \mathcal{T}$ is the set of *triconnected components* of G. Indeed, it is accurate to speak of *the* triconnected components as the following statement of Hopcroft and Tarjan [11] shows:

Theorem 2 (Uniqueness of triconnected components [11]). *If $\mathcal{I}$ and $\mathcal{I}'$ are two sets of triconnected components of the same biconnected graph, then $\mathcal{I}$ and $\mathcal{I}'$ are equivalent.*

The following statement is crucial for the proof of Theorem 2, cf. [11]. We will discuss in the next section how a variation of Lemma 1 serves us in proving the uniqueness of 2.5-connected components.

Lemma 1 ([11]). *Let $\mathcal{I}$ be a set of graphs obtained from a biconnected graph G by a sequence of splits and merges.*

(a) The graph $S(\mathcal{I})$ with

$$V(S(\mathcal{I})) = \mathcal{I}, \quad E(S(\mathcal{I})) = \{st \colon s \text{ and } t \text{ contain corresponding virtual edges}\}$$

is a tree.

(b) The set $\mathcal{I}$ can be produced by a sequence of splits.

3   2.5-Connectivity

In the following, we transfer the above notation of [11] to *mixed connectivity*, where separators may contain both, vertices and edges. Given a biconnected graph which is not a triangle, a tuple $(c, uv) \in V(G) \times E(G)$ is a *vertex-edge-separator* if $G - uv - c$ is disconnected.

Lemma 2 ([10]).   *Let G be a biconnected graph. If (c, uv) is a vertex-edge-separator of G, then $G - uv - c$ has exactly two components, each containing a vertex of $\{u, v\}$. Let $a \in \{u, v\}$. Denote the component containing a by C_a and set $G_a := G[V(C_a) \cup \{c\}]$. Then $G_a + ca$ is biconnected.*

With the same notation as in Lemma 2, it holds $\max_{a \in \{u,v\}}\{|E(G_a)|\} \geq 2$. Let $a \in \{u, v\}$. If $|E(G_a)| \geq 2$, then $\{c, a\}$ is a separation pair of G. Let b denote the vertex in $\{u, v\} \setminus \{a\}$. Now $G_a + ac$, $G_b + ba + ac$ are split graphs of $\{c, a\}$ with virtual edges ac. We say that $\{a, c\}$ *supports* the vertex-edge-separator (c, uv) or that $\{a, c\}$ is *supporting*. Replacing G by the two graphs $G_a + ac$ and $G_b + ba + ac$ is called *2.5-split* of G at (c, uv) *with support* $\{a, c\}$. The graphs $G_a + ac$ and $G_b + ba + ac$ are *the 2.5-split graphs* of G at (c, uv) with support $\{a, c\}$. A *non-supporting* split is a split which is not of this form for any vertex-edge separator. Observe that a vertex-edge-separator has at least one and at most two separation pairs in its support.

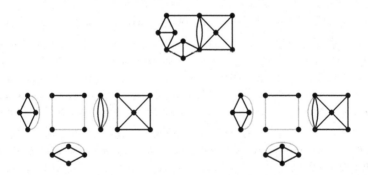

Fig. 1. A graph with its triconnected (left) and 2.5-connected (right) components. Non-virtual edges are black. Virtual edges are green if they can be obtained by a sequence of 2.5-splits and red otherwise. (Color figure online)

If G is biconnected and no tuple $(v, e) \in V(G) \times E(G)$ is a vertex-edge-separator, then G is *2.5-connected*. In analogy to the notion of *triconnected components* of Hopcroft and Tarjan [11], we define 2.5-connected components of G, see also Fig. 1.

Suppose a 2.5-split is carried out on G, the 2.5-split graphs are split by 2.5-splits, and so on, until no more 2.5-splits are possible. (Each graph remaining is 2.5-connected). The graphs constructed this way are called *2.5-split components of G*.

Observe that 2.5-split components of a biconnected graph are not unique (a cycle with more than 3-edges serves again as an example).

The 2.5-split components of a given graph are of the following types:

triangles,   multiedges of size at least 3,   and other 2.5-connected graphs.

Let G be a graph. Consider a decomposition of G into 2.5-split components, where $\mathcal{T}$ denotes the subset of triangles, $\mathcal{M}$ the set of multiedges and $\mathcal{H}$ denotes the set of other 2.5-connected graphs in the decomposition. Now merge the triangles in $\mathcal{T}$ as much as possible and leave the multiedges and 2.5-connected graphs unchanged. Replace $\mathcal{T}$ in the split components by the set $\mathcal{C}$ of cycles obtained this way. The components $\mathcal{C} \cup \mathcal{M} \cup \mathcal{H}$ obtained this way are the *2.5-connected components* of G.

Lemma 3. *Let $\mathcal{I}$ be a set of graphs obtained from a biconnected graph G by a sequence of 2.5-splits and merges.*

(a) The graph $S(\mathcal{I})$ is a tree.[2]
(b) The set $\mathcal{I}$ can be produced by a sequence of splits.
(c) If $\mathcal{I}$ is a set of 2.5-connected components of G, then $\mathcal{I}$ can be produced by a sequence of 2.5-splits.

Lemma 4 (cf. [10]). *Let G_1 and G_2 be split graphs of a biconnected graph G with respect to some separation pair. If (c, e) is a vertex-edge-separator of G_1 and $e \in E(G)$, then (c, e) is a vertex-edge-separator of G.*

Lemma 5. *Let $\mathcal{I}$ be a set of 2.5-connected components of a biconnected graph G. The triconnected components of G can be obtained from $\mathcal{I}$ by a sequence of splits.*

Let H be a graph and $e \in E(H)$. Recall that the *ear* of e in H is the maximal (possibly closed) path in H that contains e such that all its internal vertices are of degree two in H. A subgraph P of H is called an *ear* if it is an ear of some edge of H. If both endvertices of e are of degree at least 3 in H, then the ear of e in H is *trivial*, that is, the ear is the length-1 path containing e. Otherwise it is called *non-trivial*.

Lemma 6. *Let $s = s_1 \ldots s_k$ be a sequence of splits of a biconnected graph G such that the resulting graphs are the triconnected components of G.*

(a) None of the separation pairs that correspond to the splits in s contains a vertex which is of degree 2 when the split is carried out.

---

[2] Recall the definition of $S(\mathcal{I})$ from Lemma 1.

(b) Let $i \in \{1, \ldots, k\}$. Consider the graphs $G_1, \ldots, G_{i+1}$ obtained from carrying out $s_1, \ldots, s_i$ on G. Let $e, e' \in \bigcup_{j=1}^{i+1} E(G_j)$ be corresponding virtual edges. If e_1 lies on a non-trivial ear, then e_2 lies on a trivial ear.

(c) Let H_1 and H_2 be triconnected components of G containing corresponding virtual edges $e_1 \in E(H_1)$ and $e_2 \in E(H_2)$. If the ear of e_1 in H_1 is non-trivial, then H_1 is a cycle and the ear of e_2 in H_2 is trivial.

Theorem 3 (Unique colouring of virtual edges). Let $s = s_1 s_2 \ldots s_k$ be a sequence of splits that is carried out on a graph G such that the obtained graphs are the triconnected components of G. We define a 2-colouring of the virtual edges of the triconnected components starting from the uncoloured graph G. For $i \in \{1, \ldots, k\}$:

- If s_i is a non-supporting split, then the respective virtual edges are coloured red.
- If s_i is a 2.5-split for some vertex-edge-separator (c, e) where e is a green virtual edge or a non-virtual edge, then let $e^\star$ and $e^{\star\star}$ be the virtual edges arising from s_i. Colour all virtual edges with labels that appear in the ear of $e^\star$ and $e^{\star\star}$ green.
- If s_i is a 2.5-split only for vertex-edge-separators (c, e) with e red, then the virtual edges corresponding to s_i are coloured red.

The colouring of the virtual edges of the triconnected components obtained this way is independent of the choice of s.

See Fig. 1 for an example of the above colouring.

Proof. We prove the following more general statement:

Claim 1: Let $i \in \{0, 1, \ldots, l\}$ and let $G_1, G_2, \ldots, G_{i+1}$ be the graphs that are obtained from carrying out the splits $s_1, s_2, \ldots, s_i$. It holds for each virtual edge $e^\star \in \bigcup_{j=1}^{i+1} E(G_j)$ that $e^\star$ is coloured green if and only if $e^\star$ or its corresponding edge is contained in an ear with at least one non-virtual edge. Otherwise it is red.

Internal vertices of ears are of degree 2. Thus by Lemma 6, internal vertices are never contained in a separation pair that corresponds to one of the splits $s_1, \ldots, s_k$, that is,

$$\text{ears are never split by the sequence } s_1 s_2 \ldots s_k. \tag{1}$$

We prove Claim 1 by induction on i. If $i = 0$, then no split is carried out and, hence, there are no virtual edges to consider and Claim 1 satisfied.

Now let $i \geq 1$. By induction, Claim 1 holds for the graphs $G'_1, G'_2, \ldots, G'_i$ obtained from carrying out $s_1, \ldots, s_{i-1}$. Without loss of generality, s_i splits G'_i into G_i and G_{i+1}. Let $e_i^\star \in E(G_i)$ and $e_{i+1}^\star \in E(G_{i+1})$ be the new virtual edges.

First assume that s_i is a non-supporting split. The edges $e_i^\star$ and $e_{i+1}^\star$ are red and have trivial ears since s_i is non-supporting. Thus, $e_i^\star$ and $e_{i+1}^\star$ satisfy Claim 1. Other virtual edges and their ears remain unchanged by s_i.

Now assume that s_i supports a vertex-edge-separator (c, e) of G'_i. We may assume that $e \in E(G_i)$. In particular,

$$e \text{ and } e_i^\star \text{ lie on the same ear } P \text{ of } G_i. \tag{2}$$

By Lemma 6(b) and (2) it holds that

$$\text{all virtual edges that correspond to an edge of } P \text{ lie on a trivial ear.} \tag{3}$$

If an ear in G'_i is lengthened by s_i, then the ear is a subpath of P according to (2) and Lemma 6(b). In particular, it suffices to prove Claim 1 for the virtual edges of P.

If e is non-virtual, then all virtual edges of P and their corresponding edges are coloured green and Claim 1 is satisfied. If e is green, then by induction e or its corresponding edge lie in a non-trivial ear of G'_i which contains a non-virtual edge. This ear is a subpath of P by Lemma 6(b) and, hence, Claim 1 is satisfied. If e is red, then the ear P' of e in G'_i solely consists of red edges by induction. If s_i supports a vertex-edge-separator with a green or non-virtual edge, then one of the above cases applies. Otherwise, P is the union of the trivial ears $G'_i[e_i^\star]$, P', and possibly one additional ear that contains a red edge of a vertex-edge-separator supported by s_i. All of the ears consist solely of virtual red edges. This settles the claim.

Corollary 1. *Let G be a biconnected graph and let e and e' be corresponding virtual edges of the triconnected components of G. Apply the edge-colouring of Theorem 3. The following statements are equivalent:*

- *e is red.*
- *e' is red.*
- *The ears of e and e' in the triconnected components are both trivial, or, one of the two ears is a cycle solely consisting of virtual edges and the other ear is trivial.*

In Chapter 4 we will exploit Corollary 1 to develop a linear time algorithm that computes the 2.5-connected components of a given graph.

Theorem 4 (Uniqueness of 2.5-connected components) *If $\mathcal{I}$ and $\mathcal{I}'$ are two sets of 2.5-connected components of the same biconnected graph, then $\mathcal{I}$ and $\mathcal{I}'$ are equivalent. With respect to the colouring described in Theorem 3, the 2.5-connected components of G are obtained from the unique triconnected components by merging all red edges of the triconnected components of G.*

Corollary 2. *Let G be a graph. If $\mathcal{I}$ denotes the 2.5-connected components of G and $\mathcal{I}'$ denotes the triconnected components of G, then $S(\mathcal{I})$ is a minor of $S(\mathcal{I}')$.[3]*

Corollary 3. *A biconnected graph is 2.5-connected if and only if no cycle of its triconnected components contains a non-virtual edge.*

---

[3] Recall the definition of $S(\mathcal{I})$ from Lemma 1.

4   A Linear Time Algorithm for 2.5-Connected Components

Based on the work of Hopcroft and Tarjan [11] Gutwenger and Mutzel [8] showed that the triconnected components of a given graph can be computed in linear time. In this section, we provide a linear-time algorithm which computes the 2.5-connected components of a graph given its triconnected components. It follows that the 2.5-connected components of a given graph can be computed in linear time. The main idea is again, to exploit the red-green colouring of the virtual edges in order to obtain the 2.5-connected components from the triconnected components.

Theorem 5. *The 2.5-connected components of a biconnected graph can be computed in linear time.*

5   Critical 2.5-Connected Graphs

In this chapter, we provide novel decomposition techniques for critical 2.5-connected graphs. In analogy to Tutte's well-known decomposition theorem (Theorem 9) we show that critical 2.5-connected graphs which are not isomorphic to the K_4 can be reduced to critical 2.5-connected graphs of smaller order using simple graph operations.

Let G be a biconnected graph. A vertex-2-edge-separator of G is a triple $(c, e_1, e_2) \in V(G) \times E(G)^2$ such that $G - e_1 - e_2 - c$ is disconnected. A graph G is *critical 2.5-connected* if G is 2.5-connected and for every edge $e \in E(G)$ it holds that $G - e$ is not 2.5-connected, that is, e is contained in a vertex-2-edge-separator of G. If $u \in V(G)$ is a degree-3 vertex with incident edges e_0, e_1, e_2, then the vertex-2-edge-separator (c, e_1, e_2) is *degenerate*, where c denotes the vertex that is joined to u by e_0. A critical 2.5-connected graph is *degenerate* if every vertex-2-edge-separator is degenerate. Consider prisms of order at least 8 or complete bipartite graphs isomorphic to $K_{3,n}$ with $n \geq 3$ as examples for infinite families of degenerate graphs.

Theorem 6. *A 2.5-connected graph G with triconnected components $\mathcal{I}$ is critical if and only if the following conditions are satisfied:*

(a) every k-edge $M \in \mathcal{I}$ containing a non-virtual edge is a 3-edge that contains exactly one virtual edge and the unique neighbour of M in $\mathcal{S}(\mathcal{I})$ is a cycle,
(b) every other component $H \in \mathcal{I}$ satisfies that each non-virtual edge of H lies on a vertex-2-edge-separator of H with both edges non-virtual.

Theorem 7. *Let G be a 3-connected graph that contains a non-degenerate vertex-2-edge-separator (c, e_1, e_2).*

(a) If c is incident to an edge $e_0 \in E(G)$ such that $G - e_0 - e_1 - e_2$ is disconnected with components C_1 and C_2, then let G_1 (G_2) be the graph constructed by adding a new vertex x_1 (x_2) and the edges $u_i x_1$ $(v_i x_2)$ for $i \in \{0, 1, 2\}$, where u_i (v_i) denotes the endvertex of e_i in C_1 (C_2).

(b) *Otherwise, there are exactly two components C_1 and C_2 of $G - e_1 - e_2 - c$. Let G_1 (G_2) be the graph constructed from $G[V(C_1) \cup \{c\}]$ ($G[V(C_2) \cup \{c\}]$) by adding a new vertex x_1 (x_2) and the edges u_1x_1, u_2x_1, and cx_1 (v_1x_2, v_2x_2, and cx_2), where u_i (v_i) is the endvertex of e_i in C_1 (C_2).*

If G is critical 2.5-connected, then G_1 and G_2 are critical 2.5-connected 3-connected graphs of smaller order than G.

The only 3-connected critical graphs which cannot be decomposed into critical graphs of smaller order using operations above are degenerate graphs.

Theorem 8. *Let G be a degenerate 3-connected graph which is not 3-regular and let $u \in V(G)$ with $\deg_G(u) = 3$. Denote the neighbours of u by v_1, v_2, and v_3.*

(a) If $\deg_G(v_i) \geq 4$ for $i \in \{1, 2, 3\}$, then set $G' := G - u$.
(b) If $\deg_G(v_1) = 3$ and $\deg_G(v_3) \geq 4$, then set $G' := G - u + v_1v_2$.

The graph G' is critical 2.5-connected with $|V(G')| < |V(G)|$.

We have shown in this chapter that critical 2.5-connected graphs can be reduced using simple operations until the obtained graphs are 3-regular and 3-connected. Then we may apply the following theorem of Tutte.

Theorem 9 ([16], **cf.** [15])**.** *Each simple 3-connected 3-regular graph other than a complete graph on four vertices can be obtained from a 3-connected 3-regular graph H by subdividing two distinct edges of H and connecting the subdivision vertices with a new edge. Conversely, each graph obtainable in this way is 3-connected.*

The graph H in Theorem 9 is 3-regular and 3-connected and, hence, H is critical 2.5-connected. We close this chapter with an example. Consider Figure 2.

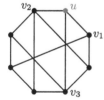 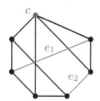 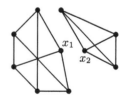

Fig. 2. Reduction of a degenerate graph.

The left graph is degenerate and not 3-regular. We apply Theorem 8(b) to obtain the graph in the middle. This graph is critical 2.5-connected and contains a non-degenerate separator (c, e_1, e_2). We obtain the isomorphic copies of the $K_{3,3}$ and the K_4 on the right by carrying out the construction of Theorem 7 (b). Observe that both of the graphs on the right are 3-regular and 3-connected. We may now apply Theorem 9 to reduce the bipartite graph further while the critical 2.5-connectivity is preserved.

6   Application to Extremal Cycle Decomposition

In this section, we prove that the problem of finding an extremal cycle decomposition of an Eulerian graph can be reduced to finding an extremal cycle decomposition for its 2.5-connected components. Furthermore, we show how Hajós' conjecture can be reduced to considering 2.5-connected components. A *decomposition* of a graph G is a set of subgraphs $\mathcal{C}$ of G such that each edge of G is contained in exactly one of the subgraphs. We say that G can be *decomposed* into the elements of $\mathcal{C}$. If all of the subgraphs in $\mathcal{C}$ are cycles, then $\mathcal{C}$ is a *cycle decomposition*. For an Eulerian graph G we set

$$c(G) := \min\{k \colon G \text{ can be decomposed into } k \text{ cycles}\} \text{ and}$$
$$\nu(G) := \max\{k \colon G \text{ can be decomposed into } k \text{ cycles}\}.$$

A cycle decomposition of G with $c(G)$ ($\nu(G)$) cycles is *minimal* (*maximal*). Let G_1 and G_2 be obtained from carrying out a 2.5-split on G. It is proven in [10] that $c(G) = c(G_1) + c(G_2) - 1$ and $\nu(G) = \nu(G_1) + \nu(G_2) - 1$. The theorem below follows.

Theorem 10. *Let G be a biconnected Eulerian graph and $G_1, G_2, \ldots, G_k$ its 2.5-connected components.*

(a) $c(G) = \sum_{i=1}^{k} c(G_i) - k + 1$,
(b) $\nu(G) = \sum_{i=1}^{k} \nu(G_i) - k + 1$.

Hajós' conjecture asserts that an Eulerian graph can be decomposed into at most $1/2(|V(G)| + m(G) - 1)$ cycles, where $m(G)$ denotes the minimal number of edges that need to be removed from G in order to obtain a simple graph.[4] The only progress made towards a verification of Hajós' conjecture concerns graphs that contain vertices of degree at most 4 (cf. [4]), very sparse graphs (cf. [5]) and, very dense graphs (cf. [6]).

Theorem 11. *Let G be a biconnected graph. If all 2.5-connected components of G satisfy Hajós' conjecture, then G satisfies Hajós' conjecture.*

In particular, the conjecture of Hajós' is satisfied if and only if all 2.5-connected graphs satisfy Hajós' conjecture.

7   Conclusion

We provide a canonical decomposition of a biconnected graph into its unique 2.5-connected components. Furthermore, we show how these components can be constructed from the triconnected components of the graph. This overall gives a linear-time algorithm for the 2.5-connected components. We show that all critical 2.5-connected except complete graphs on four vertices can be reduced to smaller critical 2.5-connected graphs. Finally, we prove that it suffices to verify Hajós' conjecture for all 2.5-connected graphs in order to verify the conjecture for all graphs.

---

[4] Originally, Hajós conjectured that at most $1/2|V(G)|$ cycles are needed. This equivalent reformulation is due to Fan and Xu, cf. [4].

References

1. di Battista, G., Tamassia, R.: On-line planarity testing. SIAM J. Comput. **25**(5), 956–997 (1996). https://doi.org/10.1137/s0097539794280736
2. Beineke, L.W., Wilson, R.J., Oellermann, O.R.: Topics in Structural Graph Theory. Cambridge Univ. Press, Cambridge (2012)
3. Diestel, R.: Graph Theory. Graduate Texts in Mathematics. Springer, Heidelberg (2000)
4. Fan, G., Xu, B.: Hajós' conjecture and projective graphs. Discrete Math. **252**(1), 91–101 (2002). https://doi.org/10.1016/s0012-365x(01)00290-4
5. Fuchs, E., Gellert, L., Heinrich, I.: Cycle decompositions of pathwidth-6 graphs. J. Graph Theory (2019). https://doi.org/10.1002/jgt.22516
6. Girão, A., Granet, B., Kühn, D., Osthus, D.: Path and cycle decompositions of dense graphs. arXiv preprint arXiv: 1911.05501 (2019)
7. Grohe, M.: Quasi-4-connected components. In: 43rd International Colloquium on Automata, Languages, and Programming (ICALP 2016) (2016). https://doi.org/10.4230/LIPIcs.ICALP.2016.8
8. Gutwenger, C., Mutzel, P.: A linear time implementation of SPQR-trees. In: Marks, J. (ed.) GD 2000. LNCS, vol. 1984, pp. 77–90. Springer, Heidelberg (2001). https://doi.org/10.1007/3-540-44541-2_8
9. Heinrich, I., Heller, T., Schmidt, E., Streicher, M.: 2.5-connectivity: unique components, critical graphs, and applications. arXiv preprint arXiv: 2003.01498 (2020)
10. Heinrich, I., Streicher, M.: Cycle decompositions and constructive characterizations. Electron. J. Graph Theory Appl. **7**(2), 411–428 (2019). https://doi.org/10.5614/ejgta.2019.7.2.15
11. Hopcroft, J.E., Tarjan, R.E.: Dividing a graph into triconnected components. SIAM J. Comput. **2**, 135–158 (1973). https://doi.org/10.1137/0202012
12. Mac Lane, S.: A structural characterization of planar combinatorial graphs. Duke Math. J. **3**(3), 460–472 (1937). https://doi.org/10.1215/S0012-7094-37-00336-3
13. Nagamochi, H., Ibaraki, T.: Algorithmic Aspects of Graph Theroy. Cambridge University Press, Cambridge (2008)
14. Tutte, W.T.: A contribution to the theory of chromatic polynomials. Can. J. Math. **6**, 80–81 (1954). https://doi.org/10.4153/cjm-1954-010-9
15. Tutte, W.T.: Connectivity in Graphs. University of Toronto Press (1966). https://doi.org/10.3138/9781487584863
16. Wormald, N.C.: Classifying K-connected cubic graphs. In: Horadam, A.F., Wallis, W.D. (eds.) Combinatorial Mathematics VI. LNM, vol. 748, pp. 199–206. Springer, Heidelberg (1979). https://doi.org/10.1007/BFb0102696

Stable Structure on Safe Set Problems in Vertex-Weighted Graphs II –Recognition and Complexity–

Shinya Fujita[1] , Boram Park[2] , and Tadashi Sakuma[3]([⊠])

[1] School of Data Science, Yokohama City University, Yokohama 236-0027, Japan
fujita@yokohama-cu.ac.jp
[2] Department of Mathematics, Ajou University, Suwon 16499, Republic of Korea
borampark@ajou.ac.kr
[3] Faculty of Science, Yamagata University, Yamagata 990-8560, Japan
sakuma@sci.kj.yamagata-u.ac.jp

Abstract. Let G be a graph, and let w be a non-negative real-valued weight function on $V(G)$. For every subset X of $V(G)$, let $w(X) = \sum_{v \in X} w(v)$. A non-empty subset $S \subset V(G)$ is a *weighted safe set* of (G, w) if for every component C of the subgraph induced by S and every component D of $G - S$, we have $w(C) \geq w(D)$ whenever there is an edge between C and D. If the subgraph of G induced by a weighted safe set S is connected, then the set S is called a *connected weighted safe set* of (G, w). The *weighted safe number* $\mathrm{s}(G, w)$ and *connected weighted safe number* $\mathrm{cs}(G, w)$ of (G, w) are the minimum weights $w(S)$ among all weighted safe sets and all connected weighted safe sets of (G, w), respectively. It is easy to see that for every pair (G, w), $\mathrm{s}(G, w) \leq \mathrm{cs}(G, w)$ by their definitions. In [*Journal of Combinatorial Optimization*, **37**:685–701, 2019], the authors asked which pair (G, w) satisfies the equality $\mathrm{s}(G, w) = \mathrm{cs}(G, w)$ and it was shown that every weighted cycle satisfies the equality. In the companion paper [*European Journal of Combinatorics*, in press] of this paper, we give a complete list of connected bipartite graphs G such that $\mathrm{s}(G, w) = \mathrm{cs}(G, w)$ for every weight function w on $V(G)$. In this paper, as is announced in the companion paper, we show that, for any graph G in this list and for any weight function w on $V(G)$, there exists an FPTAS for calculating a minimum connected safe set of (G, w). In order to prove this result, we also prove that for any tree T and for any weight function w' on $V(T)$, there exists an FPTAS for calculating a minimum connected safe set of (T, w'). This gives a complete answer to a question posed by Bapat et al. [*Networks*, **71**:82–92, 2018] and disproves a conjecture by Ehard and Rautenbach [*Discrete Applied Mathematics*, **281**:216–223, 2020]. We also show that determining whether a graph is in the above list or not can be done in linear time.

Supported by grant fundings from JSPS KAKENHI (No. 19K03603), National Research Foundation of Korea (NRF-2018R1C1B6003577), and JSPS KAKENHI (No. 26400185, No. 16K05260 and No. 18K03388).

© Springer Nature Switzerland AG 2020
I. Adler and H. Müller (Eds.): WG 2020, LNCS 12301, pp. 364–375, 2020.
https://doi.org/10.1007/978-3-030-60440-0_29

Keywords: Safe set · Connected safe set · Vertex-weight function · Safe-finite · Network majority · Network vulnerability · Subgraph component polynomial

1   Introduction

We use [7] for terminology and notation not defined here. Only finite, simple (undirected) graphs are considered. For a graph G, the subgraph of G induced by a subset $S \subseteq V(G)$ is denoted by $G[S]$. We often abuse/identify terminology and notation for subsets of the vertex set and subgraphs induced by them. In particular, a component is sometimes treated as a subset of the vertex set. For a subset X of $V(G)$, we denote $G[V(G) \setminus S]$ by $G - S$. For a graph G, when A and B are disjoint subsets of $V(G)$, the set of edges joining some vertex of A and some vertex of B is denoted by $E_G(A, B)$. If $E_G(A, B) \neq \emptyset$, then A and B are said to be *adjacent*. A (vertex) weight function w on $V(G)$ means a mapping associating each vertex in $V(G)$ with a non-negative real number. We call (G, w) a weighted graph. For every subset X of $V(G)$, let $w(X) = \sum_{v \in V(G)} w(v)$, and note that we also allow to use the notation $w(G[X])$ for $w(X)$.

Let G be a connected graph. A non-empty subset $S \subseteq V(G)$ is a *safe set* if, for every component C of $G[S]$ and every component D of $G - S$, we have $|C| \geq |D|$ whenever $E_G(C, D) \neq \emptyset$. If $G[S]$ is connected, then S is called a *connected safe set*. In [4], those notions are extended on (vertex) weighted graphs. Let w be a weight function on $V(G)$. A non-empty subset $S \subset V(G)$ is a *weighted safe set* of (G, w) if, for every component C of $G[S]$ and every component D of $G - S$, we have $w(C) \geq w(D)$ whenever $E_G(C, D) \neq \emptyset$. The *weighted safe number* of (G, w) is the minimum weight $w(S)$ among all weighted safe sets of (G, w), that is, $\mathrm{s}(G, w) := \min\{w(S) \mid S \text{ is a weighted safe set of } (G, w)\}$. If S is a weighted safe set of (G, w) and $w(S) = \mathrm{s}(G, w)$, then S is called a *minimum weighted safe set*. Similar to connected safe sets, if S is a weighted safe set of (G, w) and $G[S]$ is connected, then S is called a *connected weighted safe set* of (G, w). The *connected weighted safe number* of (G, w) is defined by $\mathrm{cs}(G, w) := \min\{w(S) \mid S \text{ is a connected weighted safe set of } (G, w)\}$, and a *minimum connected weighted safe set* is a connected weighted safe set S of (G, w) such that $w(S) = (G, w)$. It is easy to see that for every pair (G, w), $\mathrm{s}(G, w) \leq \mathrm{cs}(G, w)$ by their definitions.

Throughout this paper, we often drop 'weighted' to call a weighted safe set or a connected weighted safe set when it is clear from the context.

Fujita, MacGillivray and Sakuma [11] introduced the notion of a safe set with motivation to use the concept for facility location problems of safe evacuation plans. Kang, Kim and Park [13] explored the safe number of the Cartesian product of two complete graphs. This notion was extended to weighted graphs (G, w) due to Bapat et al. [4]. For a real application, we can regard (G, w) as a kind of network. In such a network, it is important to gain control of a "majority" so that we can control the network consensus. As pointed out by Bapat et al. [4], a significant feature of the safe set is that the minimum size of this parameter can be used as both of majority and (which coincides with) vulnerability measures of the network. The authors in [4] noticed that the minimum weight of a subnetwork which attains the above majority role for a given network (G, w) coincides with the safe number of (G, w).

Since the concept of a safe set can be thought as a suitable measure of network vulnerability, and hence it has some deep relation to other relevant graph invariants. One of the intensively studied measures of vulnerability of a graph network (G, w) may be the *graph integrity*, which is defined as $I(G, w) := \min_{S \subseteq V(G)} \{w(S) + \max\{w(H) : H \text{ is a component of } G - S\}\}$ (e.g., see [2, 3, 6, 15]). For every graph network (G, w), it is not difficult to see that the inequality $I(G, w) \leq 2s(G, w)$ holds. Furthermore, if a set $S(\subseteq V(G))$ attains the number $I(G, w)$ and the induced subgraph $G[S]$ is connected, then we also have the inequality $cs(G, w) \leq I(G, w) \leq 2cs(G, w)$. For the case of unweighted graphs, Fujita and Furuya [9] gave a tight lower bound on $I(G, w)$ in terms of the (connected) safe number of a graph.

We also remark that a common property in terms of the weighted safe number sometimes yields a characterization of graphs. Indeed, Fujita, Jensen, Park and Sakuma [10] showed that a graph G is a cycle or a complete graph if and only if $s(G, w) \geq w(G)/2$ for any weight function w on $V(G)$.

Motivated by those applications, weighted safe set problems in graphs attract much attention, especially in the algorithmic aspect. Fujita, MacGillivray and Sakuma [11] showed that computing the connected safe number in the case (G, w) with a constant weight function w is $\mathcal{NP}$-hard in general. On the other hand, when G is a tree and w is a constant weight function, they constructed a greedy algorithm for computing the connected safe number of G in linear time. Águeda et al. [1] constructed an efficient algorithm for computing the safe number of an unweighted graph with bounded treewidth. Somewhat surprisingly, Bapat et al. [4] showed that computing the connected weighted safe number in a tree is $\mathcal{NP}$-hard even if the underlying tree is restricted to be a star. They also constructed an efficient algorithm computing the safe number for a weighted path. Furthermore, the authors in [10] constructed a linear time algorithm computing the safe number for a weighted cycle. Bapat et al. [4] also gave a polynomial time 2-approximation algorithm for finding a minimum weighted connected safe set of a weighted tree, and asked whether there exist more accurate approximation algorithms or not. Ehard and Rautenbach [8] answered this question and gave a polynomial-time approximation scheme (PTAS) for finding a minimum weighted connected safe set of a weighted tree. Unfortunately, their algorithm is not a fully polynomial-time approximation scheme (FPTAS), and they conjectured that there exists no such algorithm. Contrary to their expectation, in this paper, we give an FPTAS for computing a minimum connected safe set of a weighted tree. Hence this result will be a complete answer to the above question of Bapat et al. [4]. We also note that, the parameterized complexity of some safe set enumeration problems was investigated by Belmonte et al. [5].

In contrast with the above algorithmic approach, in the companion paper[12] of this paper, we focus on a theoretical aspect on weighted safe set problems.

Since the safe number of a graph is no more and sometimes less than the connected safe number of it in general, it would be an important and natural question to ask which class of graphs satisfies the property that these two values are the same. On the other hand, even if we consider a safe set S of a graph G whose vertex weight function is uniform, after we contract each component of $G[S]$ and $G[V - S]$, the resultant weighted bipartite graph possibly has a non-constant weight function. Furthermore, some applications may require the robustness of the equality $s(G, w) = cs(G, w)$ against variations of weight functions w on $V(G)$. Taking account of these, one of the most essential problem is to characterize the class of unweighted bipartite graphs G such that, for every vertex weight function w, the weighted safe number $s(G, w)$ is equal to the weighted connected safe number $cs(G, w)$. In the paper [12] we resolve the above problem (Theorem 1), which settles the principal sub-case of an open problem in the paper [10].

In this paper, as is announced in the companion paper [12], we give an FPTAS to find a minimum (connected) safe set for any weighted graph (G, w) of such a bipartite graph G and its non-negative vertex-weight function w (Theorem 4). Note that a star is one of such bipartite graphs (see Theorem 1), and recall that it is $\mathcal{NP}$-hard to find a minimum connected safe set of a given weighted star. In order to prove the above FPTAS result, first we prove that, for any tree T and for any weight function w on $V(T)$, there exists an FPTAS for calculating a minimum connected safe set of (T, w) (Theorem 3). It might be useful if we can easily determine whether a given bipartite graph G satisfies the equation $s(G, w) = cs(G, w)$ for every vertex weight function w or not. Here we also provide an algorithm recognizing such graphs in linear time (Theorem 2).

2   Our Results

The following open problem was proposed by [10].

Problem 1 ([10]). Determine the family $\mathcal{G}^{cs}$ of all graphs G such that $s(G, w) = cs(G, w)$ for every weight function w on $V(G)$.

In the companion paper [12], we completely characterize all chordal graphs and all bipartite graphs in $\mathcal{G}^{cs}$. The following, one of the main results of that paper [12], gives the complete list of the connected bipartite graphs in $\mathcal{G}^{cs}$. A *double star* is a tree with diameter at most three. A *dominating clique* is a domination set which is a clique, that is, it induces a complete graph and every vertex v not in the clique has a neighbor in this clique.

Definition 1. *Let m, n, p, q be non-negative integers. Let $D(m, n; p, q)$ (resp. $D^*(m, n; p, q)$) be a connected bipartite graph with bipartition $(X_1 \cup X_2 \cup P, Y_1 \cup Y_2 \cup Q)$, where the unions are disjoint, satisfying (1) $\sim$ (4):*

(1) $|X_1| = m$, $|Y_1| = m + 1$, $|X_2| = n + 1$, $|Y_2| = n$, $|P| = p$, and $|Q| = q$;
(2) Both $G[X_1 \cup Y_1]$ and $G[X_2 \cup Y_2]$ are complete bipartite graphs;

(3) The vertices in P are pendant vertices which are adjacent to a vertex $y \in Y_1$ and the vertices in Q are pendant vertices which are adjacent to a vertex $x \in X_2$;

(4) $E_G(X_1, Y_2) = \emptyset$ and $G[X_2 \cup Y_1]$ is a complete bipartite graph (resp. a double star with a dominating edge xy).

Note that each of $D(m,n;p,q)$ and $D^*(m,n;p,q)$ has a dominating edge xy ($x \in X_2$ and $y \in Y_1$), where a dominating edge is a dominating clique of size two. See Fig. 1 for examples.

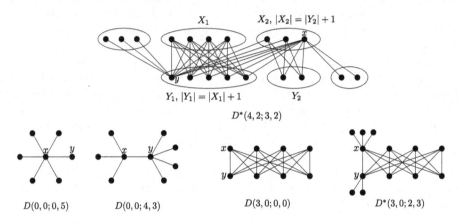

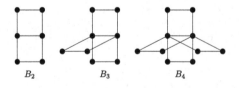

Fig. 1. Graphs $D(m,n;p,q)$ or $D^*(m,n;p,q)$

The *m-book graph*, denoted by B_m, is the Cartesian product of a star $K_{1,m}$ and a path P_2. See Fig. 2.

The following theorem gives a full list of graphs in $\mathcal{G}^{\mathrm{cs}}$ for the bipartite case.

Fig. 2. Book graphs

Theorem 1 ([12]). *A connected bipartite graph G belongs to $\mathcal{G}^{\mathrm{cs}}$ if and only if G is one of the following:*

(I) an even cycle C_{2n} with $n \geq 2$;

(II) a double star;

(III) a book graph B_n with $n \geq 1$;

(IV) a graph obtained from $K_{3,3}$ by deleting an edge;
(V) $D(m, n; p, q)$ or $D^*(m, n; p, q)$, with $m \geq 2$, $n \neq 1$ and $p, q \geq 0$.

Note that, from the above theorem, we see that if a bipartite graph G belongs to $\mathcal{G}^{cs}$, then G is an even cycle or G has a dominating edge. By using this fact, we can recognize the graphs in Theorem 1 in linear time.

Theorem 2. *There exists a linear time algorithm to decide whether a given graph is in the list of Theorem 1 or not.*

Now, let us move onto our main results. First we will consider the following problem:

Problem 2 (Finding a rooted subtree of a given weight).

Instance: An n-vertex tree T with its vertex-weight function $w : V(T) \rightarrow \mathbb{Z}_{\geq 0}$, a vertex (root) r of the tree T, and a non-negative integral value W.
  Task: Find a sub-tree T' of T containing the vertex r such that the weight $w(T')$ is closest to the value W subject to the condition that $w(T') \geq W$.

It is a folklore that the above problem has a simple DP based pseudo-polynomial algorithm whose running time is in $O(C^2 n)$, where the parameter C denotes the total weight $w(T)$. Quite recently, Kumabe, Maehara and Sin'ya [14] gave a more efficient pseudo-polynomial algorithm whose running time is in $O(Cn)$. From their result, we can easily obtain the following FPTAS for Problem 2:

Lemma 1. *For Problem 2, a sub-tree T' of T containing the vertex r such that $w(T') \leq (1 + \epsilon) \cdot \mathrm{OPT}(T, w)$ holds can be detected in $O(\frac{n^2}{\epsilon})$-time.*

By using Lemma 1, we also obtain an FPTAS for the following problem:

Problem 3 (Minimum Connected Safe Set Problem on Weighted Trees).

Instance: An n-vertex tree T with its vertex-weight function $w : V(T) \rightarrow \mathbb{Z}_{\geq 0}$.
  Task: Find a minimum connected safe set of (T, w).

Theorem 3. *For Problem 3, a connected safe set S of a weighted tree (T, w) such that $w(S) \leq (1 + \epsilon) \cdot \mathrm{cs}(T, w)$ holds can be detected in $O(\frac{n^4}{\epsilon})$-time.*

Proof. We find a connected safe set S such that $w(S) \leq (1 + \epsilon) \cdot \mathrm{cs}(T, w)$ holds using the following steps:
(Step 1). Guess a vertex r of T which is in the connected safe set S of (T, w). Note that the number of such candidates is at most n.
(Step 2). Guess an edge e_M of T such that the component $X(e_M, r)$ of $T - e_M$ not containing the vertex r is one of the heaviest component of $T - S$. Note that the number of such candidates is $n - 1$ for each fixed vertex r.
(Step 3). Starting from the vertex r, check all the edges e of T in order of a breadth-first search and, if the weight of the component $X(e, r)$ of $G - e$ not

containing r is strictly more than $w(X(e_M, r))$ then paint the edge e red. For each fixed vertex r, by using the same method in the algorithm CONNECTED SAFE SET of the paper [11], we can calculate and list the weights $w(X(e, r))$ for all edges e of T in $O(n)$-time.

(Step 4). Contract the subtree $R(e_M, r)$ which consists of all the red edges to a single vertex $v(R(e_M, r))$ and set its weight to be $w(R(e_M, r))$. Remove $X(e_M, r)$ from the resulting tree and denote it by $H(e_M, r)$.

(Step 5). By using Lemma 1, calculate a subtree $Y(e_M, r)$ of $H(e_M, r)$ containing the vertex $v(R(e_M, r))$ such that the weight $w(Y(e_M, r))$ is closest to the value $w(X(e_M, r))$ subject to the condition that $w(Y(e_M, r)) \geq w(X(e_M, r))$.

(Step 6). Return a tree with minimum weight in the set of candidate trees $\{Y(e_M, r) : (e_M, r) \in E(T) \times V(T)\}$.

It is clear that the most time-consuming part is Step 5, and its running time is in $O(\frac{n^2}{\epsilon})$ for each fixed graph $H(e_M, r)$. Since the number of the graphs $H(e_M, r)$ is $n(n-1)$, the total running time of this algorithm is in $O(\frac{n^4}{\epsilon})$.   □

Corollary 1. *If the input tree T of Problem 3 is restricted to a double-star, then a connected safe set S such that $w(S) \leq (1 + \epsilon) \cdot cs(T, w)$ holds can be detected in $O(\frac{n^3}{\epsilon})$-time.*

Now we are ready to prove the existence of an FPTAS for the following problem:

Problem 4 (Minimum Safe Set Problem on Bipartite Graphs in $\mathcal{G}^{cs}$)

Instance: An n-vertex bipartite graph G in the list of Theorem 1 and its vertex-weight function $w : V(G) \to \mathbb{Z}_{\geq 0}$.
Task: Find a minimum safe set of (G, w).

For a connected graph G and $S \subset V(G)$, let us denote by $\beta(G, S)$ the graph whose vertices are the components of $G[S]$ and of $G - S$, and two vertices A and B are adjacent in $\beta(G, S)$ if and only if $E_G(A, B) \neq \emptyset$. Note that $\beta(G, S)$ is always a bipartite graph.

Theorem 4. *For Problem 4, a connected safe set S of a weighted graph (G, w) such that $w(S) \leq (1 + \epsilon) \cdot cs(G, w)$ holds can be detected in $O(\frac{n^7}{\epsilon^2})$-time.*

Proof. If the input graph G is either in the cases (I) or (IV) of Theorem 1, then the statement of Theorem 4 is trivially true. And also if, the graph G is in the case (II) of Theorem 1, then our statement is true because of Corollary 1.

Now suppose that G is a book graph (i.e., the case (III)) and let xy be its unique dominating edge. If a minimum safe set $S_{\min}$ of (G, w) contains x while $S_{\min}$ does not contain y, then $G - S_{\min}$ is also connected. Hence, in order to detect $S_{\min}$, we should find a star subgraph T of $G - y$ containing x such that it has the closest to the value $w(G)/2$ subject to the condition that $w(T) \geq w(G)/2$. Hence we can apply Lemma 1 to calculate the candidate of S corresponding to this subcase. On the contrary, if a minimum connected safe set $S_{\min}$ of (G, w) contains both x and y, then let us contract the edge xy to a single vertex $v(xy)$ and let us denote the resultant graph by H. Set the weight of the new vertex $v(xy)$ of H to be $w(x) + w(y)$. Then w can be extended to a weight function on $V(H)$, and $\beta(G, S_{\min})$ can be thought as $\beta(H, S'_{\min})$ for some minimum connected safe set $S'_{\min}$ of (H, w). Now let us replace each triangle of H, say $v(xy) - u_i - v_i - v(xy)$, such that $w(u_i) \leq w(v_i)$ holds, with the arm $v(xy) - p_i - q_i - u_i$ where $w(p_i) := w(u_i)$ and $w(q_i) := w(v_i) - w(u_i)$. Then we obtain the resultant substitutional weighted tree (T, w) for (H, w). That is, under the assumption that any minimum safe set contains x (i.e., except for a very special case such that $w(q_i) \geq w(V(G) - u_i)$ holds for some i), there exists a natural one to one correspondence between the minimum connected safe sets of (T, w) and the minimum connected safe sets of (G, w). Hence we can apply Theorem 3 to calculate the candidate of S corresponding to this subcase. In any case, by comparing these two candidates corresponding the two subcases, we can detect a desired safe set S in $O(\frac{n^4}{\epsilon})$-time.

From now on, suppose that our input n-vertex graph G is in the case (V) of Theorem 1. Let us use the notations $X_1, X_2, Y_1, Y_2, P, Q, x, y$ as in Definition 1. Especially, let xy be the dominating edge of G. Let S denote a desired connected safe set of (G, w).

We divide the rest of the proof into several cases and subcases. For each subcase, we provide an FPTAS for calculating a candidate of our desired safe set S in such subcase. By using these FPTAS's, we develop a list of potential candidates and return the lightest set among all the candidates.

First we show the following lemma.

Lemma 2. *Let H be an n-vertex graph satisfying the following conditions:*

1. *H has a vertex u in the desired safe set S.*
2. *U denotes the set of pendant vertices adjacent to u.*
3. *H has two distinct vertices v_1, v_2 in $G \setminus S$.*
4. *$N[v_1] \cap N[v_2] = \{u\}$ and $N[v_1] \cup N[v_2] = N[u] \setminus U$.*

Then, for any non-negative integral-valued function w on $V(H)$, a connected safe set S of a weighted tree (H, w) such that $w(S) \leq (1 + \epsilon) \cdot cs(H, w)$ can be detected in $O(\frac{n^3}{\epsilon^2})$-time.

Corollary 2. *Let H be an n-vertex graph satisfying the following conditions:*

1. *H has a vertex u in the desired safe set S.*
2. *U denotes the set of pendant vertices adjacent to u.*
3. *H has a vertex v in $G \setminus S$.*
4. *$N[v] = N[u] \setminus U$.*

Then, for any non-negative integral-valued function w on $V(H)$, a connected safe set S of a weighted tree (H, w) such that $w(S) \leq (1 + \epsilon) \cdot cs(H, w)$ can be detected in $O(\frac{n^3}{\epsilon})$-time. If the set U is empty, then the above time-complexity is compressed to $O(\frac{n^2}{\epsilon})$-time.

In the following, we transform our graph (G, w) repeatedly, and we sometimes abuse our notation (G, w) to denote the transformed graph. In each step of the transformation, we contract some vertex set $A \subseteq V(G)$ into a single vertex $v(A)$. The weight of the new vertex $v(A)$ in the contracted graph is always defined to be $w(A)$, unless otherwise specified.

Let us start with the case of $D(a, 0, p, q)(= D^*(a, 0; p, q))$:

Lemma 3. *Let $X, Y, P, Q, \{x\}, \{y\}$ be disjoint sets. and let $n := |X| + |Y| + |P| + |Q| + 2$. Let G be a bipartite graph defined as follows: $V(G) := X \cup Y \cup P \cup Q \cup \{x\} \cup \{y\}$, $E(G) := \{\{u, v\} | u \in X \cup \{x\}, v \in Y \cup \{y\}\} \cup \{\{p, y\} | p \in P\} \cup \{\{x, q\} | q \in Q\}$. Then, for any weight function w on $V(G)$, a connected safe set S of the weighted graph (G, w) such that $w(S) \leq (1 + \epsilon) \cdot cs(G, w)$ holds can be detected in $O(\frac{n^5}{\epsilon})$-time.*

Proof of Lemma 3. First, suppose that there exist 4 distinct vertices x_1, x_2, y_1 and y_2 such that both $x_1 y_1$ and $x_2 y_2$ are edges of G and that $\{x_1, y_1\} \subseteq S$, $\{x_2, y_2\} \subseteq G - S$, $x \in \{x_1, x_2\} \subseteq X \cup \{x\}$ and $y \in \{y_1, y_2\} \subseteq Y \cup \{y\}$ hold. In this case, let us contract the edges $x_1 y_1$ and $x_2 y_2$ into single vertices $v(x_1 y_1)$ and $v(y_2 x_2)$ and denote the resulting weighted graph by (G, w) again.

If $x = x_1$ and $y = y_1$ then the above G satisfies the condition of Corollary 2 so that $H := G$, $u := v(x_1 y_1)$, $v := v(x_2 y_2)$, and $U := P \cup Q$. Hence we obtain the desired safe set S in $O(\frac{n^3}{\epsilon})$-time.

If $x = x_2$ and $y = y_2$ then we can assume $P \cup Q \subset G - S$, and hence we can contract the set $P \cup Q \cup \{v(x_2 y_2)\}$ to a single vertex $v(PQx_2 y_2)$. Let us denote the resulting weighted graph by (G, w) again. Then G satisfies the condition of Corollary 2 so that $H := G$, $u := v(x_1 y_1)$, $v := v(PQx_2 y_2)$, and $U := \emptyset$. Hence we obtain the desired safe set S in $O(\frac{n^2}{\epsilon})$-time.

If $x = x_1$ and $y = y_2$ then we can assume $P \subset G - S$, and hence we can contract the set $P \cup \{v(x_2 y_2)\}$ to a single vertex $v(Px_2 y_2)$. Let us denote the resulting weighted graph by (G, w) again. Then G satisfies the condition of Corollary 2 so that $H := G$, $u := v(x_1 y_1)$, $v := v(Px_2 y_2)$, and $U := Q$. Hence we obtain the desired safe set S in $O(\frac{n^3}{\epsilon})$-time. By symmetry, the same is true for the case that $x = x_2$ and $y = y_1$ hold.

The number of candidates of $\{x_1, x_2, y_1, y_2\} \setminus \{x, y\}$ is in $O(n^2)$ and hence the time-complexity so far is $O(\frac{n^5}{\epsilon})$.

In the other cases, at least one of the four conditions $X \cup \{x\} \subseteq S$, $X \cup \{x\} \subseteq G - S$, $Y \subseteq S$ and $Y \subseteq G - S$ holds.

Suppose that $X \cup \{x\} \subseteq S$ holds. Since we can assume that $G[S]$ is a connected, there exists at least one vertex $v_1 \in Y \cap S$. Guess the vertex v. Then let us contract the set $X \cup \{x\} \cup \{v_1\}$ to a single vertex $v(v_1 x X)$. Let us denote the resulting weighted graph by (G, w) again. Then G is a double star, and we can apply Corollary 1 here. Since the number of candidates of v_1 is at most n, we obtain the desired safe set S in $O(\frac{n^4}{\epsilon})$-time. By symmetry, the same is true for the case of $Y \subseteq S$.

Next suppose that $X \cup \{x\} \subseteq G - S$ holds. Note that every connected component of $G[Y \cup P \cup Q]$ is a singleton except for $G[P \cup \{y\}]$. Since we can assume that $G[S]$ is connected, we have $|Y \cap S| \leq 1$ and hence there exists a vertex $v_2 \in Y \cap (G - S)$. Then let us contract the set $X \cup \{x\} \cup \{v_2\}$ to a single vertex $v(v_2 x X)$. Let us denote the resulting weighted graph by (G, w) again. Then G is a double star again, and we can also apply Corollary 1 here. Since the number of candidates of v_2 is at most n, we obtain the desired safe set S in $O(\frac{n^4}{\epsilon})$-time. By symmetry, the same is true for the case that $Y_1 \subseteq G - S$ holds.

Thus we have that the total time-complexity is in $O(\frac{n^5}{\epsilon})$.   □

From now on, let us assume that our input n-vertex graph G is $D(a, b; p, q)$ or $D^*(a, b; p, q)$ with no collateral conditions of the parameters a, b, p, q. Let us use the notations $X_1, X_2, Y_1, Y_2, P, Q, x, y$ as in Definition 1.

(Case 1): $\{x, y\} \subseteq G - S$.

(Subcase 1–1): G is $D^*(a, b; p, q)$.

In this case, the set $\{x, y\}$ separates $X_1 \cup (Y_1 \setminus \{y\})$ from $(X_2 \setminus \{x\}) \cup Y_2$, and hence at least one of the sets $\{x\} \cup X_1 \cup Y_1$ and $X_2 \cup Y_2 \cup \{y\}$ is in the same component of $G - S$.

If $\{x\} \cup X_1 \cup Y_1$ is in the same component of $G - S$, then let us contract the set Y_1 to a single vertex $v(Y_1)$. Let us denote the resulting weighted graph by (G, w) again. Since this graph G satisfies the conditions in Lemma 3, we obtain the desired safe set S in $O(\frac{n^5}{\epsilon})$-time. By symmetry, the same is true for the case that $X_2 \cup Y_2 \cup \{y\}$ is in the same component of $G - S$.

(Subcase 1–2): G is $D(a, b; p, q)$.

First suppose that $Y_1 \subseteq G - S$ holds. Since Y_1 separates X_1 from $X_2 \cup Y_2$, at least one of the two sets $\{x\} \cup X_1 \cup Y_1$ and $X_2 \cup Y_1 \cup Y_2$ is in the same component of $G - S$. If $\{x\} \cup X_1 \cup Y_1$ is in the same component of $G - S$, then let us contract the set Y_1 to a single vertex $v(Y_1)$. Let us denote the resulting weighted graph by (G, w) again. Since the graph G satisfies the conditions in Lemma 3, we obtain the desired safe set S in $O(\frac{n^5}{\epsilon})$-time. By symmetry, the same is true for the case that $X_2 \cup Y_1 \cup Y_2$ is in the same component of $G - S$.

Also note that, if we assume $X_2 \subseteq G - S$ instead of $Y_1 \subseteq G - S$, again, by symmetry, we obtain the desired safe set S in $O(\frac{n^5}{\epsilon})$-time.

In the other cases, there exist a vertex $y_1 \in Y_1$ and a vertex $x_2 \in X_2$ such that $\{x_2, y_1\} \subseteq S$ holds. And the set $P \cup Q \cup \{x, y\}$ is in a component of $G - S$. Then let us contract the sets $\{x_2, y_1\}$ and $P \cup Q \cup \{x, y\}$ to single vertices $v(x_2 y_1)$ and $v(x y P Q)$. Let us denote the resulting weighted graph by (G, w) again. This G satisfies the condition of Corollary 2 so that $H := G$, $u := v(x_2 y_1)$, $v := v(x y P Q)$, and $U := \emptyset$. Since the number of candidates of the pair (x_2, y_1) is in $O(n^2)$, we obtain the desired safe set S in $O(\frac{n^4}{\epsilon})$-time.

(Case 2): $\{x, y\} \subseteq S$.

If at least one of the four set X_1, X_2, Y_1, Y_2 is a subset of S, then, as seen many times before, our input graph is reduced to the graphs satisfying the conditions in Lemma 3. Hence we obtain the desired safe set S in $O(\frac{n^5}{\epsilon})$-time.

In the other cases, there exist a vertex $x_1 \in X_1$ and a vertex $x_2 \in X_2 \setminus \{x\}$ and a vertex $y_1 \in Y_1 \setminus \{y\}$ and a vertex $y_2 \in Y_2$ such that $\{x_1, x_2, y_1, y_2\} \subseteq G - S$ holds. Note that $x_1 y_1$ and $x_2 y_2$ are edges of G. And the set $P \cup Q \cup \{x, y\}$ is in a component of $G - S$. Then let us contract the sets $\{x_1, y_1\}$, $\{x_2, y_2\}$ and $P \cup Q \cup \{x, y\}$ to single vertices $v(x_1 y_1)$, $v(x_2 y_2)$ and $v(xyPQ)$, respectively. Then let us denote the resulting weighted graph by (G, w) again.

If $v(x_1 y_1)$ is not adjacent to $v(x_2 y_2)$, then G satisfies the conditions of H in Lemma 2. Hence we obtain the desired safe set S in $O(\frac{n^3}{\epsilon^2})$-time.

If $v(x_1 y_1)$ is adjacent to $v(x_2 y_2)$, then let us contract the set $\{x_1, x_2, y_1, y_2\}$ to a single vertex $v(x_1 x_2 y_1 y_2)$. Let us denote the resulting weighted graph by (G, w) again. then G satisfies the conditions of the second graph in Corollary 2. Hence we obtain the desired safe set S in $O(\frac{n^2}{\epsilon})$-time. Since the number of candidates of the pair of edges $(x_1 y_1, x_2 y_2)$ is in $O(n^4)$, the total time-complexity is in $O(\frac{n^7}{\epsilon^2})$-time, so far.

(Case 3): $|\{x, y\} \cap S| = 1$.

By symmetry, we can assume without loss of generality that $x \in S$ and $y \in G - S$ hold.

In this case, the set $\{y\} \cup Q$ is in a component of $G - S$. Thus let us contract the set $\{y\} \cup Q$ to the single vertex y and set $w(y) := w(y) + w(Q)$. Let us denote the resulting weighted graph by (G, w) again.

If $Y_1 \subseteq G - S$ then, since Y_1 separate X_1 from x, $X_1 \cup Y_1$ is in a component of $G - S$. In this case, we can contract the set $X_1 \cup Y_1$ to the single vertex y and set $w(y) := w(X_1) + w(Y_1)$. Let us denote the resulting weighted graph by (G, w) again. This graph G satisfies the conditions of the first graph in Corollary 2. Hence we obtain the desired safe set S in $O(\frac{n^3}{\epsilon})$-time.

If $X_2 \subset S$, then, since X_2 is an independent set, there exists a vertex $y_1 \in (Y_1 \cup Y_2) \cap S$ such that $X_2 \subseteq N(y_1)$ holds. Then we can contract X_2 to the vertex x and set $w(x) := w(X_2)$. Let us denote the resulting weighted graph by (G, w) again. This graph G satisfies the conditions of H in Lemma 3. Since the number of candidates of y_1 is at most n, we obtain the desired safe set S in $O(\frac{n^6}{\epsilon})$-time. In the same way, for the case of $X_1 \subset S$, we can reach the same conclusion.

In the other cases, there exists a vertex $x_2 \in X_2 \setminus S$. Hence if $Y_2 \subset G - S$ then the set $\{x_2\} \cup Y_2$ is in a component of $G - S$. Then we can contract the set Y_2 to the vertex y and set $w(y) := w(y) + w(Y_2)$. Let us denote the resulting weighted graph by (G, w) again. This graph G satisfies the conditions of H in Lemma 3. Since the number of candidates of x_2 is at most n, we obtain the desired safe set S in $O(\frac{n^6}{\epsilon})$-time. In the same way, for the case of $X_1 \subset S$, we can reach the same conclusion.

The only remaining case is that there exist $x_1 \in X_1 \setminus S$ and $x_2 \in X_2 \setminus S$ and $y_1 \in Y_2 \cap S$ and $y_2 \in Y_2 \cap S$. In this case, we can contract the sets $\{x, y_1, y_2\}$ and $\{x_1, x_2, y\}$ to the vertices x and y respectively and set $w(x) := w(x) + w(y_1) + w(y_2)$ and $w(y) := w(y) + w(x_1) + w(x_2)$. Let us denote the resulting weighted graph by (G, w) again. Then this graph G satisfies the conditions of the first graph in Corollary 2. Since the number of candidates of the four-tuple (x_1, x_2, y_1, y_2) is in $O(n^4)$, we obtain the desired safe set S in $O(\frac{n^7}{\epsilon})$-time.   □

References

1. Águeda, R., et al.: Safe sets in graphs: graph classes and structural parameters. J. Comb. Optim. **36**(4), 1221–1242 (2017). https://doi.org/10.1007/s10878-017-0205-2
2. Atici, M., Ernst, C.: On the range of possible integrities of graphs $g(n, k)$. Graphs Comb. **27**, 475–485 (2011). https://doi.org/10.1007/s00373-010-0990-1
3. Bagga, K.S., Beineke, L.W., Goddard, W.D., Lipman, M.J., Pippert, R.E.: A survey of integrity. Discrete Appl. Math. **37**(38), 13–28 (1992)
4. Bapat, R.B., et al.: Weighted safe set problem on trees. Networks **71**, 81–92 (2018)
5. Belmonte, R., Hanaka, T., Katsikarelis, I., Lampis, M., Ono, H., Otachi, Y.: Parameterized complexity of safe set. In: Heggernes, P. (ed.) CIAC 2019. LNCS, vol. 11485, pp. 38–49. Springer, Cham (2019). https://doi.org/10.1007/978-3-030-17402-6_4
6. Benko, D., Ernst, C., Lanphier, D.: Asymptotic bounds on the integrity of graphs and separator theorems for graphs. SIAM J. Discrete Math. **23**, 265–277 (2009)
7. Chartrand, G., Lesniak, L., Zhang, P.: Graphs and Digraphs, 5th edn. Chapman and Hall, London (2011)
8. Ehard, S., Rautenbach, D.: Approximating connected safe sets in weighted trees. Discrete Appl. Math. **281**, 216–223 (2020)
9. Fujita, S., Furuya, M.: Safe number and integrity of graphs. Discrete Appl. Math. **247**, 398–406 (2018)
10. Fujita, S., Jensen, T., Park, B., Sakuma, T.: On weighted safe set problem on paths and cycles. J. Comb. Optim. **37**, 685–701 (2019). https://doi.org/10.1007/s10878-018-0316-4
11. Fujita, S., MacGillivray, G., Sakuma, T.: Safe set problem on graphs. Discrete Appl. Math. **215**, 106–111 (2016)
12. Fujita, S., Park, B., Sakuma, T.: Stable structure on safe set problems in vertex-weighted graphs. Eur. J. Comb. 103211 (2020). A preprint version. arXiv:1909.02718
13. Kang, B., Kim, S.-R., Park, B.: On the safe sets of Cartesian product of two complete graphs. Ars Comb. **141**, 243–257 (2018)
14. Kumabe, S., Maehara, T., Sin'ya, R.: Linear pseudo-polynomial factor algorithm for automaton constrained tree knapsack problem. In: Das, G.K., Mandal, P.S., Mukhopadhyaya, K., Nakano, S. (eds.) WALCOM 2019. LNCS, vol. 11355, pp. 248–260. Springer, Cham (2019). https://doi.org/10.1007/978-3-030-10564-8_20
15. Vince, A.: The integrity of a cubic graph. Discrete Appl. Math. **140**, 223–239 (2004)

The Linear Arboricity Conjecture for 3-Degenerate Graphs

Manu Basavaraju[1], Arijit Bishnu[2], Mathew Francis[3(✉)],
and Drimit Pattanayak[2]

[1] National Institute of Technology Karnataka, Surathkal, Mangalore, India
manub@nitk.ac.in
[2] Indian Statistical Institute, Kolkata, India
arijit@isical.ac.in, drimitpattanayak@gmail.com
[3] Indian Statistical Institute, Chennai Centre, Chennai, India
mathew@isichennai.res.in

Abstract. A k-*linear coloring* of a graph G is an edge coloring of G with k colors so that each color class forms a *linear forest*—a forest whose each connected component is a path. The *linear arboricity* $\chi'_l(G)$ of G is the minimum integer k such that there exists a k-linear coloring of G. Akiyama, Exoo and Harary conjectured in 1980 that for every graph G, $\chi'_l(G) \leq \left\lceil \frac{\Delta(G)+1}{2} \right\rceil$ where $\Delta(G)$ is the maximum degree of G. We prove the conjecture for 3-degenerate graphs. This establishes the conjecture for graphs of treewidth at most 3 and provides an alternative proof for the conjecture for triangle-free planar graphs. Our proof also yields an $O(n)$-time algorithm that partitions the edge set of any 3-degenerate graph G on n vertices into at most $\left\lceil \frac{\Delta(G)+1}{2} \right\rceil$ linear forests. Since $\chi'_l(G) \geq \left\lceil \frac{\Delta(G)}{2} \right\rceil$ for any graph G, the partition produced by the algorithm differs in size from the optimum by at most an additive factor of 1.

1 Introduction

All graphs considered in this paper are finite, simple and undirected. For a graph G, we let $V(G)$ and $E(G)$ denote its vertex set and edge set, respectively. The *neighborhood* of a vertex u in G, denoted by $N_G(u)$, is the set $\{v \colon uv \in E(G)\}$. We abbreviate it to just $N(u)$ when the graph G is clear from the context. Given a graph G, the *degree* of a vertex $u \in V(G)$ is $|N(u)|$ and is denoted by $d_G(u)$. The *maximum degree* of a graph G, denoted by $\Delta(G)$, is defined to be $\max\{d_G(u) \colon u \in V(G)\}$. When the graph G under consideration is clear, we sometimes abbreviate $\Delta(G)$ to just Δ. For any terms not defined here, please refer [8].

Given a graph G on n vertices and an integer t, an ordering $v_1, v_2, \ldots, v_n$ of the vertices of G such that for each $1 \leq i \leq n$, $|\{v_j \colon v_j \in N(v_i) \text{ and } j > i\}| \leq t$ is called a *t-degeneracy ordering* of G. A graph G is said to be *t-degenerate* if

© Springer Nature Switzerland AG 2020
I. Adler and H. Müller (Eds.): WG 2020, LNCS 12301, pp. 376–387, 2020.
https://doi.org/10.1007/978-3-030-60440-0_30

it has a t-degeneracy ordering. Equivalently, a graph G is t-degenerate if every subgraph of G has minimum degree at most t. The class of 3-degenerate graphs is well studied in the literature and contains many well known graph classes like triangle-free planar graphs and graphs of treewidth at most 3 (also called partial 3-trees, this class contains outerplanar graphs and series-parallel graphs). Note that even though a 3-degenerate graph can contain only as many edges as a planar graph on n vertices—at most $3n - 6$ edges—they contain a large number of non-planar graphs as well.

An *edge coloring* of a graph G using the colors $\{1, 2, \ldots, k\}$ is a mapping $c : E(G) \rightarrow \{1, 2, \ldots, k\}$. Given an edge coloring using colors $\{1, 2, \ldots, k\}$, the *color class* i, for some $i \in \{1, 2, \ldots, k\}$, is the set of edges $c^{-1}(i) = \{e \in E(G) \colon c(e) = i\}$.

Graphs without cycles are known as forests. The *arboricity* of a graph is the minimum integer k such that its edge set can be partitioned into k forests. A *linear forest* is a forest whose each connected component is a path. The *linear arboricity* of a graph is the minimum number of linear forests into which its edge set can be partitioned.

A k-*linear coloring* of a graph G is an edge coloring of G such that each color class is a linear forest. Or in other words, it is an edge coloring in which every vertex has at most two edges of the same color incident with it and there is no cycle in the graph whose edges all receive the same color. The linear arboricity of a graph G is clearly the smallest integer k such that it has a k-linear coloring and is denoted by $\chi'_l(G)$. The parameter $\chi'_l(G)$ was introduced by Harary [14]. The linear arboricity conjecture, first stated by Akiyama, Exoo and Harary [1], is as follows.

Conjecture 1 (Linear Arboricity Conjecture). For every graph G,

$$\chi'_l(G) \leq \left\lceil \frac{\Delta(G) + 1}{2} \right\rceil.$$

1.1 Brief History

Note that for any graph G, $\chi'_l(G) \geq \left\lceil \frac{\Delta(G)}{2} \right\rceil$, since in any linear coloring of G, there can be at most 2 edges of the same color incident with any vertex. In fact, as noted by Harary [14], if G is a $\Delta(G)$-regular graph, then $\chi'_l(G) \geq \left\lceil \frac{\Delta(G)+1}{2} \right\rceil$. The linear arboricity conjecture suggests that this lower bound for regular graphs is tight. The conjecture has been proven for all graphs G such that $\Delta(G) \in \{3, 4, 5, 6, 8, 10\}$ [1, 2, 9, 12] and was shown to be true for planar graphs by Wu and Wu [18, 19]. Cygan et al. [7] proved that the linear arboricity of planar graphs which have $\Delta \geq 10$ is $\left\lceil \frac{\Delta}{2} \right\rceil$. Works of Alon [3], Alon and Spencer [4], and Ferber et al. [10] show that the conjecture holds asymptotically—in particular, for any $\epsilon > 0$ there exists a Δ_0 such that $\chi'_l(G) \leq (\frac{1}{2} + \epsilon)\Delta(G)$ whenever $\Delta(G) \geq \Delta_0$. From Vizing's Theorem [17], which says that any graph can be properly edge colored with $\Delta + 1$ colors, we get that $\chi'_l(G) \leq \Delta(G) + 1$ for any graph G. The best known general bound for linear arboricity is $\lceil \frac{3\Delta}{5} \rceil$ when Δ is even and $\lceil \frac{3\Delta+2}{5} \rceil$ for Δ odd, obtained by Guldan [12,13].

Although arboricity can be computed in polynomial time [11], computing linear arboricity is NP-hard [16]. As $\chi'_l(G) \geq \lceil \frac{\Delta}{2} \rceil$ for any graph G, a 2-factor approximation algorithm for computing linear arboricity can be obtained using Vizing's Theorem. Cygan et al. [7] showed an $O(n \log n)$ algorithm that produces a linear coloring of every planar graph on n vertices with the optimum number of colors when $\Delta(G) \geq 9$. Linear arboricity has applications in File Retrieval Systems [15].

1.2   Our Result

It is known that the conjecture is true for 2-degenerate graphs from the fact that the acyclic chromatic index of 2-degenerate graphs is at most $\Delta + 1$ [6]. (The acyclic chromatic index $\chi'_a(G)$ of a graph G is the minimum number of colors required to *properly* color the edges of G—i.e. no two incident edges get the same color—such that the union of any two color classes is a forest. Since the union of any two color classes in such a coloring will always be a linear forest, we get that $\chi'_l(G) \leq \left\lceil \frac{\chi'_a(G)}{2} \right\rceil$).

We prove the following theorem which shows that the linear arboricity conjecture is true for 3-degenerate graphs.

Theorem 1. *Let G be a 3-degenerate graph having $\Delta(G) \leq 2k - 1$, where k is a positive integer. Then $\chi'_l(G) \leq k$.*

Our proof also serves as an alternative (we believe, simpler) proof for the the result of Akiyama, Exoo and Harary [1] that every cubic graph has a 2-linear coloring. Graphs having treewidth at most 3, also called partial 3-trees, are 3-degenerate graphs, and hence our result establishes the linear arboricity conjecture for this class of graphs. Our result provides an alternative proof for the conjecture for some other classes of 3-degenerate graphs like triangle-free planar graphs and Halin graphs.

We convert the proof to a linear time algorithm that computes a $\left\lceil \frac{\Delta(G)+1}{2} \right\rceil$-linear coloring for any input 3-degenerate graph G. For triangle-free planar graphs or partial 2-trees, our algorithm has better asymptotic runtime complexity than the algorithm for planar graphs given in [7], with the caveat that our algorithm may produce a linear coloring using one more color than the optimum number of required colors.

2   Notation and Preliminaries

Given a graph G and a set $S \subseteq V(G)$, we denote by $G - S$ the graph obtained by removing the vertices in S from G, i.e. $V(G - S) = V(G) \setminus S$ and $E(G - S) = E(G) \setminus \{uv \colon u \in S\}$. When $S \subseteq E(G)$, we abuse notation to let $G - S$ denote the graph obtained by removing the edges in S from G; i.e. $V(G - S) = V(G)$

and $E(G - S) = E(G) \setminus S$. In both cases, if $S = \{s\}$, we sometimes denote $G - S$ by just $G - s$.

Let G be a t-degenerate graph. A *pivot* in G is a vertex that has at most t neighbors of degree more than t. A *pivot edge* in G is an edge between a pivot and a vertex with degree at most t.

Observation 1. *Every t-degenerate graph G has at least one pivot edge.*

Proof. If $\Delta(G) \leq t$, then every vertex of G is a pivot, and every edge of G is a pivot edge. If $\Delta(G) > t$, then the graph $G' = G - \{u: d_G(u) \leq t\}$ contains at least one vertex. Since G' is also t-degenerate (as every subgraph of a t-degenerate graph is also t-degenerate), there exists a vertex $v \in V(G')$ such that $d_{G'}(v) \leq t$. It can be seen that $\{u \in N_G(v): d_G(u) > t\} = N_{G'}(v)$. As $|N_{G'}(v)| = d_{G'}(v) \leq t$, we have that v is a pivot in G. Also, as $d_G(v) > t$, there exists $u \in N_G(v)$ having $d_G(u) \leq t$. Then uv is a pivot edge in G.   □

Alternatively, given a t-degeneracy ordering of a graph G, consider the first vertex v such that it has a neighbor u before it in the ordering. It is not difficult to see that uv is a pivot edge of G.

The following observation about linear forests is easy to see.

Observation 2. *Let H be a linear forest and let P_1 and P_2 be two paths in H having end vertices u_1, v_1 and u_2, v_2 respectively such that $u_1 \neq v_1$, $u_2 \neq v_2$, $\{u_1, v_1\} \neq \{u_2, v_2\}$ and $V(P_1) \cap V(P_2) \neq \emptyset$. Then at least one of u_1, v_1, u_2, v_2 has degree 2 in H.*

Identification of Vertices: Given a graph G and vertices $u, v \in V(G)$ such that $uv \notin E(G)$ and $N(u) \cap N(v) = \emptyset$, we let $G/(u,v)$ denote the graph obtained by "identifying" the vertex v with u. That is, $V(G/(u,v)) = V(G) \setminus \{v\}$ and $E(G/(u,v)) = E(G - v) \cup \{ux: x \in N(v)\}$. Note that given a k-linear coloring of $G/(u,v)$, the vertex u can be "split back" into the vertices u and v so as to obtain the graph G together with a k-linear coloring of it. The following observation states this fact.

Observation 3. *Let G be a graph and $u, v \in V(G)$ such that $uv \notin E(G)$ and $N(u) \cap N(v) = \emptyset$. If c is a k-linear coloring of $G/(u,v)$, then*

$$c'(e) = \begin{cases} c(e) & \text{if e is not incident with v} \\ c(ux) & \text{if $e = vx$} \end{cases}$$

is a k-linear coloring of G.

Definition 1. *Let c be a k-linear coloring of a graph G. For a vertex $x \in V(G)$, we define Colors(x) to be the set of colors in $\{1, 2, \ldots, k\}$ that appear on the edges incident with x. Further, we define Missing(x) to be the colors in $\{1, 2, \ldots, k\}$ that do not appear on any edge incident with x, Twice(x) to be the set of colors that appear on two edges incident with x, and Once(x) to be the set of colors that appear on exactly one edge incident with x.*

Note that for any vertex $x \in V(G)$, $|\text{Missing}(x)| + |\text{Once}(x)| + |\text{Twice}(x)| = k$ and also that the degree of x in G is $|\text{Once}(x)| + 2|\text{Twice}(x)|$.

3   Proof of Theorem 1

We prove the theorem by induction on $|E(G)|$. We assume that all 3-degenerate graphs having maximum degree at most $2k - 1$ and less than $|E(G)|$ edges can be linearly colored using k colors. We shall show that G can be linearly colored using k colors.

Let uv be a pivot edge of G such that $d_G(u) \leq 3$ and v is a pivot.

Lemma 1. *If $d_G(v) < 2k - 1$, then G has a k-linear coloring.*

Proof. Let $H = G - uv$. By the induction hypothesis, there is a k-linear coloring c of H.

Claim. If either

1. $|\mathrm{Once}(v)| \geq 3$, or
2. $|\mathrm{Missing}(v)| \geq 1$ and $|\mathrm{Missing}(v) \cup \mathrm{Once}(v)| \geq 2$,

then G has a k-linear coloring.

First, suppose that $|\mathrm{Once}(v)| \geq 3$. Since $d_H(u) \leq 2$, we have $|\mathrm{Colors}(u)| \leq 2$, and therefore there exists a color $i \in \mathrm{Once}(v) \setminus \mathrm{Colors}(u)$. Then, by coloring uv with i, we can get a k-linear coloring of G, and we are done. Next, suppose that $|\mathrm{Missing}(v)| \geq 1$ and $|\mathrm{Missing}(v) \cup \mathrm{Once}(v)| \geq 2$. Then there exist $i \in \mathrm{Missing}(v)$ and $j \in (\mathrm{Missing}(v) \cup \mathrm{Once}(v)) \setminus \{i\}$. If $j \notin \mathrm{Colors}(u)$, then we can color uv with j to obtain a k-linear coloring of G. Otherwise, since $|\mathrm{Colors}(u)| \leq 2$, we have $i \notin \mathrm{Twice}(u)$, implying that we can color uv with i to obtain a k-linear coloring of G. This proves the claim.

Observe that $d_H(v) = 2|\mathrm{Twice}(v)| + |\mathrm{Once}(v)|$. From the above claim, if $|\mathrm{Missing}(v)| + |\mathrm{Once}(v)| \geq 3$, then we are done. Therefore, we shall assume that $|\mathrm{Missing}(v)| + |\mathrm{Once}(v)| \leq 2$. As $|\mathrm{Twice}(v)| + |\mathrm{Once}(v)| + |\mathrm{Missing}(v)| = k$, this means that $|\mathrm{Twice}(v)| \geq k - 2$. Since $d(v) \leq 2k - 2$, we have $d_H(v) \leq 2k - 3$, implying that $|\mathrm{Twice}(v)| \leq k - 2$. Thus, we have $|\mathrm{Twice}(v)| = k - 2$. Since $2k - 3 \geq d_H(v) = 2|\mathrm{Twice}(v)| + |\mathrm{Once}(v)|$, we get $|\mathrm{Once}(v)| \leq 1$. Since $|\mathrm{Twice}(v)| + |\mathrm{Once}(v)| + |\mathrm{Missing}(v)| = k$, we have $|\mathrm{Missing}(v)| \geq 1$ and $|\mathrm{Missing}(v)| + |\mathrm{Once}(v)| = 2$. We are now done by the above claim.   □

By the above lemma, we shall assume from here onwards that $d_G(v) = 2k - 1$. Also, we can assume that $k \geq 2$, as the statement of the theorem can be easily seen to be true for the case $k = 1$. Since v has at most 3 neighbors having degree more than 3, it has at least $2k - 4$ neighbors with degree at most 3. If $k = 2$, then $\Delta(G) \leq 3$, implying that every vertex in $N(v)$ has degree at most 3. If $k \geq 3$, then v has at least $2k - 4 \geq 2$ neighbors having degree at most 3. Thus, in any case, there exists $w \in N(v) \setminus \{u\}$ such that $d_G(w) \leq 3$.

Lemma 2. *If $uw \in E(G)$, then G has a k-linear coloring.*

Proof. Let $H = G - \{u, w\}$. Let x be a neighbor of u other than v, w. Note that x may not exist (i.e. if $d_G(u) = 2$). Similarly, let y be a neighbor of w other than u, v, if such a neighbor exists. We shall assume from here onwards that both x and y exist, since if one of them, say x, does not exist, then we can just add a new vertex x that is adjacent to only u and continue the proof. By the inductive hypothesis, there exists a k-linear coloring c of H. Since $d_H(x) \le 2k - 2$, there exists a color $i \in \mathrm{Missing}(x) \cup \mathrm{Once}(x)$. Color ux with i. Since $d_H(v) = 2k - 3$, we know that either:

(a) $|\mathrm{Once}(v)| = 3$, or
(b) $|\mathrm{Missing}(v)| = 1$ and $|\mathrm{Once}(v)| = 1$.

First let us consider case (a). Color uv, uw with two different colors in $\mathrm{Once}(v) \setminus \{i\}$, and then color vw with the remaining color j in $\mathrm{Once}(v)$. Since $d_H(y) \le 2k - 2$, we have either $|\mathrm{Missing}(y)| \ge 1$ or $|\mathrm{Once}(y)| \ge 2$. In the former case, we color wy with a color in $\mathrm{Missing}(y)$ and in the latter case, we color wy with a color in $\mathrm{Once}(y) \setminus \{j\}$. We now have a k-linear coloring of G.

Now let us consider case (b). We color uv, vw with the color in $\mathrm{Missing}(v)$ and color uw with the color in $\mathrm{Once}(v)$. As $d_H(y) \le 2k - 2$, either $|\mathrm{Missing}(y)| \ge 1$ or $|\mathrm{Once}(y)| \ge 2$. If $|\mathrm{Missing}(y)| \ge 1$, then color wy with a color in $\mathrm{Missing}(y)$. On the other hand, if $|\mathrm{Once}(y)| \ge 2$, then color wy with a color in $\mathrm{Once}(y) \setminus \{i\}$. This gives a k-linear coloring of G. □

Lemma 3. *If u and w have a common neighbor other than v, then G has a k-linear coloring.*

Proof. Let z be a common neighbor of u and w other than v. Note that by Lemma 2, we can assume that $uw \notin E(G)$. As in the proof of Lemma 2, we assume that u has a neighbor x other than v, z and w has a neighbor y other than v, z. Define $H = G - \{u, w\}$. Clearly, $d_H(v) = 2k - 3$ and $d_H(z) \le 2k - 3$. Thus we have either $|\mathrm{Once}(v)| = 3$ or we have both $|\mathrm{Missing}(v)| = 1$ and $|\mathrm{Once}(v)| = 1$. We will use the weaker statement that either $|\mathrm{Once}(v)| \ge 3$ or both $\mathrm{Missing}(v) \ne \emptyset$ and $|\mathrm{Missing}(v) \cup \mathrm{Once}(v)| \ge 2$. Since we also have that either $|\mathrm{Once}(z)| \ge 3$ or both $\mathrm{Missing}(z) \ne \emptyset$ and $|\mathrm{Missing}(z) \cup \mathrm{Once}(z)| \ge 2$, we treat v and z symmetrically, so there are only three cases to consider. Note also that since $d_H(x), d_H(y) \le 2k - 2$, we have $\mathrm{Missing}(x) \ne \emptyset$ or $|\mathrm{Once}(x)| \ge 2$, and we have $\mathrm{Missing}(y) \ne \emptyset$ or $|\mathrm{Once}(y)| \ge 2$.

First, let us consider the case when $\mathrm{Missing}(v) \ne \emptyset$, $|\mathrm{Missing}(v) \cup \mathrm{Once}(v)| \ge 2$, $\mathrm{Missing}(z) \ne \emptyset$, and $|\mathrm{Missing}(z) \cup \mathrm{Once}(z)| \ge 2$. Choose $i \in \mathrm{Missing}(v)$, $j \in (\mathrm{Missing}(v) \cup \mathrm{Once}(v)) \setminus \{i\}$, $p \in \mathrm{Missing}(z)$, and $q \in (\mathrm{Missing}(z) \cup \mathrm{Once}(z)) \setminus \{p\}$. If $i \ne p$, then color uv, vw with i, and wz, uz with p. If $i = p$, then color uv, wz with i, vw with j, uz with q. Let r denote the color so given to vw and ℓ the color so given to uz. Now color wy with a color in $\mathrm{Missing}(y) \cup (\mathrm{Once}(y) \setminus \{r\})$ and ux with a color in $\mathrm{Missing}(x) \cup (\mathrm{Once}(x) \setminus \{\ell\})$. We now have a k-linear coloring of G.

Next, suppose that $\mathrm{Missing}(v) \ne \emptyset$, $|\mathrm{Missing}(v) \cup \mathrm{Once}(v)| \ge 2$, and $|\mathrm{Once}(z)| \ge 3$. Choose $i \in \mathrm{Missing}(v)$ and $j \in (\mathrm{Missing}(v) \cup \mathrm{Once}(v)) \setminus \{i\}$. Color uv with

i, vw with j, wy with a color t in $\mathrm{Missing}(y) \cup (\mathrm{Once}(y) \setminus \{j\})$, ux with a color $s \in \mathrm{Missing}(x) \cup \mathrm{Once}(x)$, wz with a color in $\ell \in \mathrm{Once}(z) \setminus \{t,j\}$, and uz with a color in $\mathrm{Once}(z) \setminus \{s,\ell\}$. We now have a k-linear coloring of G.

Finally, let us suppose that $|\mathrm{Once}(v)|, |\mathrm{Once}(z)| \geq 3$. Choose distinct $i, j, \ell \in \mathrm{Once}(v)$. Color vw with j, wy with a color t in $\mathrm{Missing}(y) \cup (\mathrm{Once}(y) \setminus \{j\})$, uv with a color h in $\{i, \ell\} \setminus \{t\}$, and ux with a color s in $\mathrm{Missing}(x) \cup (\mathrm{Once}(x) \setminus \{h\})$. Then, if $h \in \mathrm{Once}(z)$, let $f = h$, otherwise let f be a color in $\mathrm{Once}(z) \setminus \{t,j\}$. Now color wz with f and uz with a color in $\mathrm{Once}(z) \setminus \{f,s\}$. This gives a k-linear coloring of G.   □

Now we are ready to complete the proof of Theorem 1. By Lemma 3, we can assume that u and w have no common neighbors other than v. assume that w has two neighbors x and y other than v (if not, new vertices of degree one adjacent to w can be added so as to ensure that x and y always exist). By Lemma 2, we can assume that $uw \notin E(G)$, which further implies that x and y are distinct from u.

Let $H = G - \{uv, vw, wx\}$. Let H' be the graph obtained by identifying the vertex w with u in H; i.e. $H' = H/(u,w)$. Figure 1 shows the construction of the graph H' from G. Notice that $d_{H'}(u) \leq 3$ and that the graph $H' - u$ is nothing but $G - \{u, w\}$, which is a 3-degenerate graph as $d_G(u), d_G(w) \leq 3$. Thus, H' is a 3-degenerate graph having $|E(H)| < |E(G)|$. Then by the inductive hypothesis, there exists a k-linear coloring c' of H'. Let c be the coloring of H obtained by "splitting" the vertex u in H' to get back H. Formally, we define for all $e \in E(H)$

$$c(e) = \begin{cases} c'(e) & \text{if } e \neq wy \\ c'(uy) & \text{if } e = wy. \end{cases}$$

It can be seen that c is a k-linear coloring of H, which is a subgraph of G. Note that since $d_H(x) \leq 2k - 2$, either $|\mathrm{Missing}(x)| \geq 1$ or $|\mathrm{Once}(x)| \geq 2$. We first extend c to a coloring of $G - \{uv, vw\}$ by coloring the edge wx by a color in $\mathrm{Missing}(x) \cup (\mathrm{Once}(x) \setminus \{c(wy)\})$. We now describe how the edges uv and vw can be colored so that a k-linear coloring of G can be obtained.

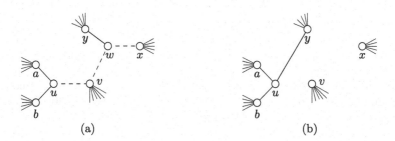

Fig. 1. (a) Deleting edges uv, vw, wx to obtain the graph H, and (b) identifying w with u to get H'.

Since $d_H(v) = 2k - 3$, we have that either both $|\mathrm{Missing}(v)| = 1$ and $|\mathrm{Once}(v)| = 1$, or $|\mathrm{Once}(v)| = 3$. Suppose first that $|\mathrm{Missing}(v)| = 1$ and

$|\text{Once}(v)| = 1$. Let $\text{Missing}(v) = \{i\}$ and $\text{Once}(v) = \{j\}$. First, suppose that i belongs to either $\text{Twice}(u)$ or $\text{Twice}(w)$. We shall assume by the symmetry between u and w that $i \in \text{Twice}(u)$. Color uv with j and vw with i (note that $i \notin \text{Twice}(w)$ as if that were the case, u would have three edges of color i incident with it in c') to obtain a k-linear coloring of G. So let us assume that $i \notin \text{Twice}(u) \cup \text{Twice}(w)$. If $j \in \text{Twice}(u) \cup \text{Twice}(w)$, then we can color uv and vw with i to obtain a k-linear coloring of G. Thus, we can now assume that $i, j \notin \text{Twice}(u) \cup \text{Twice}(v)$. If there is a path of color j having endvertices u and v, then color uv with i and vw with j (we know by Observation 2 that there is no path of color j having endvertices v and w). Otherwise, color uv with j and vw with i. We now have a k-linear coloring of G.

Next, suppose that $|\text{Once}(v)| = 3$. Let L be the set of all the colors for which there is a path of that color having endvertices u and v in c (such colors will all be in $\text{Once}(u) \cap \text{Once}(v)$). The fact that $d_H(u) \le 2$ implies that $0 \le |L| \le 2$ and also that if $\text{Twice}(u) \ne \emptyset$, then $L = \emptyset$. If $|L| = 2$, then color vw with a color in $L \setminus \text{Twice}(w)$ (again, by Observation 2, there is no path having a color in L and having endvertices v and w) and uv with a color in $\text{Once}(v) \setminus L$. Otherwise color vw with a color $r \in \text{Once}(v) \setminus \text{Colors}(w)$ and uv with a color in $\text{Once}(v) \setminus (\{r\} \cup \text{Twice}(u) \cup L)$. We now have a k-linear coloring of G.

This completes the proof of Theorem 1.

4   A Linear Time Algorithm

We now describe how to compute a k-linear coloring of a 3-degenerate graph G having $\Delta(G) \le 2k - 1$. Our algorithm will be a linear-time algorithm; i.e. having a running time of $O(n+m)$, where n and m are the number of vertices and edges in G respectively. Since G is 3-degenerate, we have $m \le 3n - 6$, and therefore our algorithm will also be an $O(n)$-time algorithm. We assume that the input graph G is available in the form of an adjacency list representation.

Our general strategy will be to convert the inductive proof of Theorem 1 into a recursive algorithm, but there are some important differences, the main one being that the algorithm computes a more general kind of edge coloring using k colors. The algorithm follows the proof of Theorem 1 and removes some edges and if needed identifies two vertices to obtain a smaller graph G' for which an edge coloring of the desired kind is found by recursing on it. The graph G' is changed back into G by splitting back any identified vertices and adding the removed edges. The newly added edges are then colored to obtain an edge coloring of the desired kind for G. During this process, we never change the color of an edge that is already colored. We shall first discuss why our algorithm needs to compute a generalized version of k-linear coloring.

If the algorithm were to construct a k-linear coloring of G from a k-linear coloring of G' according to the proof of Theorem 1, and still have overall linear runtime, we would like to be able to decide the right color to be given to an uncolored edge uv of G in $O(1)$ time. This means that we need data structures that allow us to determine in $O(1)$ time a color i for uv such that:

(i) $i \notin \mathrm{Twice}(u) \cup \mathrm{Twice}(v)$, and
(ii) if $i \in \mathrm{Once}(u) \cap \mathrm{Once}(v)$, there is no path colored i having endvertices u, v.

The requirement (i) can be met by storing the sets $\mathrm{Once}(u)$ and $\mathrm{Missing}(u)$ for every vertex u as described in Sect. 4.3 of the full version of this paper [5].

For (ii), we could store a collection of "path objects" representing the monochromatic paths in the current coloring in such a way that by examining these objects, we can determine in $O(1)$ time whether there is a monochromatic path of color i having endvertices u and v. In particular, for a monochromatic path P having endvertices u and v, we could have a path object that stores the pointers to u and v. Further, we store the pointer to this object on the first and last edges of P. In this way, given a vertex u and an edge e colored i incident with u, where $i \in \mathrm{Once}(u)$, we can examine the path object whose pointer is stored on e to determine in $O(1)$ time the other endvertex of the path colored i starting at u. Note that as an edge uv gets colored with color i, a path of color i can get extended (if $i \in \mathrm{Once}(u) \cap \mathrm{Missing}(v)$ or $i \in \mathrm{Missing}(u) \cap \mathrm{Once}(v)$) or two paths of color i can get fused into one path of color i (if $i \in \mathrm{Once}(u) \cap \mathrm{Once}(v)$). When two monochromatic paths get fused, we have to replace the two path objects corresponding to these paths with a single path object representing the new monochromatic path. If we store the pointer to a path object on each edge of the path it represents, then it becomes difficult to fuse a path with another path in $O(1)$ time as we cannot afford to update the pointer stored on every edge of the path so as to point to the new path object. We can get around this difficulty by storing the pointer to a path object only on the first and last edges of the path represented by it. Since we do not need to know what the internal vertices or edges of a path are in order to fuse it with another path, this method could allow us to fuse two paths in $O(1)$ time. As no edge that already has a color is ever recolored, a monochromatic path never gets split into two paths or gets shortened when an edge is colored. But a monochromatic path might need to get split into two monochromatic paths when a vertex is split into two vertices. Since we only split vertices of degree at most 3, at most one monochromatic path gets split during this operation. Suppose that a vertex v that needs to be split into two vertices is an internal vertex of a monochromatic path P. Since the internal vertices or edges of a path do not store the pointer to the corresponding path object, we cannot obtain the pointer to the path object representing P, given just the vertex v. We thus cannot update the collection of path objects so as to replace the monochromatic path P with two new monochromatic paths. We solve this problem by making sure that two paths that meet at a point that will be split later are never fused together into one path. This is explained in more detail below.

We say that a path having an endvertex u and containing the edge uv is "ending at u through uv". Suppose that a vertex w is identified with a vertex u when G' is constructed from G. It is clear from the proof of Theorem 1 that in G', the vertex u has degree at most 3, and there is possibly an edge uy that corresponds to an original edge wy in G. Before recursing to find the coloring for G', we mark the vertex-edge pair (u, uy) as "special" (we call this a "special vertex-edge incidence"; more details given in Sect. 4.1). This mark, which can be stored inside the adjacency list of u, indicates that while computing the coloring for G', a monochromatic path ending at u through uy should not be fused with another monochromatic path ending at u, even if they have the same color. Thus, while splitting the vertex u back into u and w, no path needs to be split. Note that this means that while the coloring for G' is being computed, we might have a path object for a path P colored i ending at u through uy and another path object for a path P' also colored i and ending at u, but through a different edge (as these paths will not be fused). Once this happens, if we denote the other endvertices of P and P' by x and x' respectively, then we can no longer detect that in the coloring constructed so far, there is a monochromatic path colored i starting at x and ending at x', as there is no path object representing the monochromatic path colored i having endvertices x and x'. This means that the edge xx', if it exists, could get colored i, and thus there may be a monochromatic cycle colored i in the coloring of G'. We will allow this to happen, since this monochromatic cycle will anyway get destroyed when the vertex u is split into u and w while recovering G back from G'. Thus, at any stage of the recursion, we compute a coloring for a graph in which certain vertex-edge pairs have been marked as special, and this coloring is not a k-linear coloring any more as it could contain monochromatic cycles. We call this kind of coloring a "pseudo-k-linear coloring". Since the path objects that we store do not correspond to maximal monochromatic paths anymore, we call them "segments" instead of paths. We now define these notions more rigorously.

4.1 Pseudo-k-Linear Colorings and Segments

We define a *vertex-edge incidence* of a graph G to be a pair consisting of a vertex and an edge incident with it; i.e. it is a pair of the form (u, uv) where $u, v \in V(G)$ and $uv \in E(G)$. We say that a cycle contains a vertex-edge incidence (u, uv) if uv is an edge of the cycle.

Given a graph G and a set S of vertex-edge incidences in it, a mapping $c : E(G) \to \{1, 2, \ldots, k\}$ is said to be a *pseudo-k-linear coloring* of (G, S) if each color class is a disjoint union of paths and cycles, in which every cycle contains at least one vertex-edge incidence in S. Given a pair (G, S), we call the set S the *special vertex-edge incidences of G*.

Note that a pseudo-k-linear coloring of $(G, \emptyset)$ is a k-linear coloring of G and also that a k-linear coloring of G is a pseudo-k-linear coloring of (G, S) for any set S of vertex-edge incidences of G. Our algorithm computes a pseudo-k-linear coloring for an input (G, S), where G is a graph with $\Delta(G) \leq 2k - 1$ and S is a set of vertex-edge incidences of G that are marked as special.

Definition 2. *Given a graph G and a set S of vertex-edge incidences in it, a segment of (G, S) is a sequence $\sigma = (u_1, u_2, \ldots, u_s)$ of vertices of G, where $s \geq 2$, such that:*

(i) for each $i \in \{1, 2, \ldots, s-1\}$, $u_i u_{i+1} \in E(G)$,

(ii) the edges $u_1 u_2, u_2 u_3, \ldots, u_{s-1} u_s$ are pairwise distinct and their union gives a path or cycle in G, and

(iii) for each $i \in \{2, 3, \ldots, s-1\}$, $(u_i, u_{i-1} u_i), (u_i, u_i u_{i+1}) \notin S$.

Let $\sigma = (u_1, u_2, \ldots, u_s)$ be a segment of (G, S), We say that the edges $u_1 u_2, u_2 u_3$, $\ldots, u_{s-1} u_s$ are the "edges in the segment σ". If H is the subgraph of G formed by the union of the edges in σ, we say that σ "forms" the subgraph H. Clearly, every segment forms a path or cycle in G. We say that u_1 and u_s are the *terminal vertices* of this segment. Note that the two terminal vertices of a segment can in fact be the same vertex (this happens when the edges in the segment form a cycle in G). Further, we call $(u_1, u_1 u_2)$ and $(u_s, u_{s-1} u_s)$ the *terminal vertex-edge incidences* of this segment. We also sometimes say that this segment is "ending at u_1 through the edge $u_1 u_2$" and "ending at u_s through the edge $u_{s-1} u_s$". Note that every segment has exactly two terminal vertex-edge incidences and they can be in S.

Let c be a pseudo-k-linear coloring of (G, S). A segment of (G, S) is said to be a *monochromatic segment* of color i if every edge in it is colored i. Note that if a monochromatic segment forms a cycle in G, then at least one of its terminal vertex-edge incidences must be in S—otherwise, the cycle formed by the segment will be a monochromatic cycle that does not contain a vertex-edge incidence in S, which contradicts the fact that c is a k-linear coloring. A segment $\sigma = (u_1, u_2, \ldots, u_s)$ is said to be *contained in* a segment $\sigma' = (u_1', u_2', \ldots, u_{s'}')$ if $u_1, u_2, \ldots, u_s$ occur consecutively in that order in σ'. A monochromatic segment σ that is not contained in any other monochromatic segment is called a *maximal monochromatic segment of* (G, S) (the coloring c is assumed to be clear from the context).

Observation 4. *Every edge of G is in exactly one maximal monochromatic segment of* (G, S).

From the above observation (please refer to [5] for a proof), it is clear that the edge set of the graph decomposes into a collection of pairwise edge-disjoint maximal monochromatic segments in a unique way. Observe that a monochromatic path P in the graph decomposes into a collection of $1 + |\{u \in V(P) : d_P(u) \geq 2$ and $\exists e \in E(P)$ such that $(u, e) \in S\}|$ maximal monochromatic segments and a monochromatic cycle C decomposes into a collection of $|\{u \in V(C) : \exists e \in E(C)$ such that $(u, e) \in S\}|$ maximal monochromatic segments.

At a given point of time, we maintain a set of segment objects, one corresponding to each maximal monochromatic segment of (G, S) under the current pseudo-k-linear coloring. The segment object corresponding to a maximal monochromatic segment stores just the terminal vertex-edge incidences of the segment. Given vertices u, v and an edge colored i incident on one of them, these segment objects allow us to determine in $O(1)$ time whether there is a maximal monochromatic segment of color i having terminal vertices u and v. To color an uncolored edge uv, the algorithm determines in $O(1)$ time a color i such that $i \notin \text{Twice}(u) \cup \text{Twice}(v)$ and there is no maximal monochromatic segment of color i having terminal vertices u and v. This ensures that the algorithm always produces a pseudo-k-linear coloring of (G, S).

For a complete description of the algorithm with details of implementation, please refer to the full version of this paper [5].

Acknowledgements. The first author was partially supported by the fixed grant scheme SERB-MATRICS project number MTR/2019/000790.

References

1. Akiyama, J., Exoo, G., Harary, F.: Covering and packing in graphs III: cyclic and acyclic invariants. Math. Slovaca **30**(4), 405–417 (1980)
2. Akiyama, J., Exoo, G., Harary, F.: Covering and packing in graphs IV: linear arboricity. Networks **11**, 69–72 (2006)
3. Alon, N.: The linear arboricity of graphs. Israel J. Math. **62**(3), 311–325 (1988). https://doi.org/10.1007/BF02783300
4. Alon, N., Spencer, J.H.: The Probabilistic Method, 4th edn. Wiley, Hoboken (2016)
5. Basavaraju, M., Bishnu, A., Francis, M., Pattanayak, D.: The linear arboricity conjecture for 3-degenerate graphs. arXiv:2007.06066 (2020)
6. Basavaraju, M., Chandran, L.S.: Acyclic edge coloring of 2-degenerate graphs. J. Graph Theory **69**(1), 1–27 (2012)
7. Cygan, M., Hou, J.-F., Kowalik, Ł., Lužar, B., Jian-Liang, W.: A planar linear arboricity conjecture. J. Graph Theory **69**(4), 403–425 (2012)
8. Diestel, R.: Graph Theory. GTM, vol. 173. Springer, Heidelberg (2017). https://doi.org/10.1007/978-3-662-53622-3
9. Enomoto, H., Peroche, B.: The linear arboricity of some regular graphs. J. Graph Theory **8**(2), 309–324 (1984)
10. Ferber, A., Fox, J., Jain, V.: Towards the linear arboricity conjecture. J. Comb. Theory Ser. B **142**, 56–79 (2019)
11. Gabow, H., Westermann, H.: Forests, frames, and games: algorithms for matroid sums and applications. Algorithmica **7**(5–6), 465–497 (1992). https://doi.org/10.1007/BF01758774
12. Guldan, F.: The linear arboricity of 10-regular graphs. Math. Slovaca **36**(3), 225–228 (1986)
13. Guldan, F.: Some results on linear arboricity. J. Graph Theory **10**(4), 505–509 (1986)
14. Harary, F.: Covering and packing in graphs, I. Ann. New York Acad. Sci. **175**(1), 198–205 (1970)
15. Hsiao, D., Harary, F.: A formal system for information retrieval from files. Commun. ACM **13**(2), 67–73 (1970)
16. Peroche, B.: Complexité de l'arboricité linéaire d'un graphe. RAIRO-Oper. Res. **16**(2), 125–129 (1982)
17. Vizing, V.G.: On an estimate of the chromatic class of a p-graph. Diskret. Analiz **3**, 25–30 (1964)
18. Jian-Liang, W.: On the linear arboricity of planar graphs. J. Graph Theory **31**(2), 129–134 (1999)
19. Jian-Liang, W., Yu-Wen, W.: The linear arboricity of planar graphs of maximum degree seven is four. J. Graph Theory **58**(3), 210–220 (2008)

Node Multiway Cut and Subset Feedback Vertex Set on Graphs of Bounded Mim-width

Bergougnoux Benjamin[1], Charis Papadopoulos[2(⊠)], and Jan Arne Telle[1]

[1] University of Bergen, Bergen, Norway
{benjamin.bergougnoux,jan.arne.telle}@uib.no
[2] University of Ioannina, Ioannina, Greece
charis@uoi.gr

Abstract. The two weighted graph problems NODE MULTIWAY CUT (NMC) and SUBSET FEEDBACK VERTEX SET (SFVS) both ask for a vertex set of minimum total weight, that for NMC disconnects a given set of terminals, and for SFVS intersects all cycles containing a vertex of a given set. We design a meta-algorithm that will allow to solve both problems in time $2^{O(rw^3)} \cdot n^4$, $2^{O(q^2 \log(q))} \cdot n^4$, and $n^{O(k^2)}$ where rw is the rank-width, q the $\mathbb{Q}$-rank-width, and k the mim-width of a given decomposition. This answers in the affirmative an open question raised by Jaffke et al. (Algorithmica, 2019) concerning an XP algorithm for SFVS parameterized by mim-width.

By a unified algorithm, this solves both problems in polynomial-time on the following graph classes: INTERVAL, PERMUTATION, and BI-INTERVAL graphs, CIRCULAR ARC and CIRCULAR PERMUTATION graphs, CONVEX graphs, k-POLYGON, DILWORTH-k and CO-k-DEGENERATE graphs for fixed k; and also on LEAF POWER graphs if a leaf root is given as input, on H-GRAPHS for fixed H if an H-representation is given as input, and on arbitrary powers of graphs in all of the above classes. Prior to our results, only SFVS was known to be tractable restricted only on INTERVAL and PERMUTATION graphs, whereas all other results are new.

Keywords: Mim-width · Rank-width · d-neighbor equivalence · Subset Feedback Vertex Set · Node multiway cut

1 Introduction

Given a vertex-weighted graph G and a set S of its vertices, the SUBSET FEEDBACK VERTEX SET (SFVS) problem asks for a vertex set of minimum weight that intersects all cycles containing a vertex of S. SFVS was introduced by Even et al. [16] who proposed an 8-approximation algorithm. Cygan et al. [14] and

A full version of this paper is available on https://arxiv.org/abs/1910.00887.

I. Adler and H. Müller (Eds.): WG 2020, LNCS 12301, pp. 388–400, 2020.
https://doi.org/10.1007/978-3-030-60440-0_31

Kawarabayashi and Kobayashi [25] independently showed that SFVS is fixed-parameter tractable (FPT) parameterized by the solution size, while Hols and Kratsch [22] provide a randomized polynomial kernel for the problem. As a generalization of the classical NP-complete FEEDBACK VERTEX SET (FVS) problem, for which $S = V(G)$, there has been a considerable amount of work to obtain faster algorithms for SFVS, both for general graphs, where the current best is an $O^*(1.864^n)$ algorithm due to Fomin et al. [18], and restricted to special graph classes [10,17,20,31,32]. Naturally, FVS and SFVS differ in complexity, as exemplified by split graphs where FVS is polynomial-time solvable [11] whereas SFVS remains NP-hard [18]. Moreover note that the vertex-weighted variation of SFVS behaves differently that the unweighted one, as exposed on graphs with bounded independent set sizes: weighted SFVS is NP-complete on graphs with independent set size at most four, whereas unweighted SFVS is in XP parameterized by the independent set size [31].

Closely related to SFVS is the NP-hard NODE MULTIWAY CUT (NMC) problem in which we are given a vertex-weighted graph G and a set T of (terminal) vertices, and asked to find a vertex set of minimum weight that disconnects all the terminals [8,19]. NMC was introduced by Garg et al. in 1994 [19] as a variant in which the terminals were not allowed to be deleted, and both variations are well-studied problems in terms of approximation, as well as parameterized algorithms [8–10,14,15,18]. It is not difficult to see that SFVS for $S = \{v\}$ coincides with NMC in which $T = N(v)$. In fact, NMC reduces to SFVS by adding a single vertex v with a large weight that is adjacent to all terminals, adding a large weight to all non-deletable terminals and setting $S = \{v\}$ [18]. Thus, in order to solve NMC on a given graph one may apply a known algorithm for SFVS on a vertex-extended graph. Observe, however, that through such an approach one needs to clarify that the vertex-extended graph still obeys the necessary properties of the known algorithm for SFVS. Therefore, despite the few positive results for SFVS on graph families [31,32], the complexity of NMC restricted to special graph classes remained unresolved.

In this paper, we investigate the complexity of SFVS and NMC when parameterized by structural graph width parameters. Well-known graph width parameters include tree-width [6], clique-width [13], rank-width [27], and maximum induced matching width (a.k.a. mim-width) [33]. These are of varying strength, with tree-width of modeling power strictly weaker than clique-width, as it is bounded on a strict subset of the graph classes having bounded clique-width, with rank-width and clique-width of the same modeling power, and with mim-width much stronger than clique-width. Belmonte and Vatshelle [1] showed that several graph classes, like interval graphs and permutation graphs, have bounded mim-width and a decomposition witnessing this can be found in polynomial time, whereas it is known that the clique-width of such graphs can be proportional to the square root of the number of vertices [21]. In this way, an XP algorithm parameterized by mim-width has the feature of unifying several algorithms on well-known graph classes.

We give a meta-algorithm that for an input graph G will give parameterized algorithms for several width measures at once, by assuming that we are given a branch decomposition over the vertex set of G. This is a natural hierarchical clustering of G, represented as a subcubic tree T with the vertices of G at its leaves. Any edge of the tree defines a cut of G given by the leaves of the two subtrees that result from removing the edge from T. Judiciously choosing a cut-function to measure the complexity of such cuts, or rather of the bipartite subgraphs of G given by the edges crossing the cuts, this framework then defines a graph width parameter by a minmax relation, minimum over all trees and maximum over all its cuts. Several graph width parameters have been defined this way, like carving-width, maximum matching-width, boolean-width etc. We will in this paper focus on: (i) rank-width [27] whose cut function is the GF[2]-rank of the adjacency matrix, (ii) $\mathbb{Q}$-rank-width [29] a variant of rank-width with interesting algorithmic properties which instead uses the rank over the rational field, and (iii) mim-width [33] whose cut function is the size of a maximum induced matching of the graph crossing the cut. Note that in contrast to e.g. clique-width, for rank-width and $\mathbb{Q}$-rank-width there is a $2^{3k} \cdot n^4$ algorithm that, given a graph and $k \in \mathbb{N}$, either outputs a decomposition of width at most $3k+1$ or confirms that the width of the input graph is more than k [29,30].

Let us mention what is known regarding the complexity of NMC and SFVS parameterized by these width measures. Standard algorithmic techniques give a $k^{O(k)} \cdot n$ time algorithm parameterized by the tree-width of the input graph, so for this reason we do not focus on tree-width. The runtime of our meta-algorithm as a function of a given clique-width expression will be $2^{O(k^2)} \cdot n^{O(1)}$ but we think a faster runtime is achievable through known techniques [4], so we do not focus on clique-width. Since these problems can be expressed in MSO$_1$-logic it follows that they are FPT parameterized by clique-width, rank-width or $\mathbb{Q}$-rank-width [12,28], however the runtime will contain a tower of 2's with more than 4 levels. Moreover, FVS and also SFVS are W[1]-hard when parameterized by the mim-width of a given decomposition [24].

We resolve in the affirmative the question raised by Jaffke et al. [24], also mentioned in [32] and [31], asking whether there is an XP-time algorithm for SFVS parameterized by the mim-width of a given decomposition. For rank-width and $\mathbb{Q}$-rank-width we provide the first explicit FPT-algorithms with low exponential dependency that avoid the MSO$_1$ formulation. Our main results are summarized in the following theorem.

Theorem 1. *We can solve* SUBSET FEEDBACK VERTEX SET *and* NODE MULTIWAY CUT *in time* $2^{O(rw^3)} \cdot n^4$ *and* $2^{O(q^2 \log(q))} \cdot n^4$, *where* rw *and* q *are the rank-width and the* $\mathbb{Q}$-rank-width *of* G, *respectively. Moreover, if a branch decomposition of mim-width* k *for* G *is given as input, we can solve* SUBSET FEEDBACK VERTEX SET *and* NODE MULTIWAY CUT *in time* $n^{O(k^2)}$.

2 Preliminaries

The size of a set V is denoted by $|V|$ and its power set is denoted by 2^V. We write $A \setminus B$ for the set difference of A from B. We let $\min(\varnothing) = +\infty$ and $\max(\varnothing) = -\infty$. The vertex set of a graph G is denoted by $V(G)$ and its edge set by $E(G)$. An edge between two vertices x and y is denoted by xy (or yx). Given $\mathcal{S} \subseteq 2^{V(G)}$, we denote by $V(\mathcal{S})$ the set $\bigcup_{S \in \mathcal{S}} S$. For a vertex set $U \subseteq V(G)$, we denote by $\overline{U}$ the set $V(G) \setminus U$. The set of vertices that are adjacent to x is denoted by $N_G(x)$, and for $U \subseteq V(G)$, we let $N_G(U) = (\bigcup_{v \in U} N_G(v)) \setminus U$.

The subgraph of G induced by a subset X of its vertex set is denoted by $G[X]$. For two disjoint subsets X and Y of $V(G)$, we denote by $G[X, Y]$ the bipartite graph with vertex set $X \cup Y$ and edge set $\{xy \in E(G) \mid x \in X \text{ and } y \in Y\}$. We denote by $M_{X,Y}$ the adjacency matrix between X and Y, i.e., the (X, Y)-matrix such that $M_{X,Y}[x, y] = 1$ if $y \in N(x)$ and 0 otherwise. A *vertex cover* of a graph G is a set of vertices $\mathsf{VC} \subseteq V(G)$ such that, for every edge $uv \in E(G)$, we have $u \in \mathsf{VC}$ or $v \in \mathsf{VC}$. A *matching* is a set of edges having no common endpoint and an *induced matching* is a matching M of edges such that $G[V(M)]$ has no other edges besides M. The *size of an induced matching* M refers to the number of edges in M.

For a graph G, we denote by $\mathsf{cc}_G(X)$ the partition $\{C \subseteq V(G) \mid G[C] \text{ is a connected component of } G[X]\}$. We will omit the subscript G of the neighborhood and components notations whenever there is no ambiguity.

Given a graph G and $S \subseteq V(G)$, we say that a cycle of G is an S-*cycle* if it contains a vertex in S. Moreover, we say that a subgraph F of G is an S-*forest* if F does not contain an S-cycle. Typically, the SUBSET FEEDBACK VERTEX SET problem asks for a vertex set of minimum (weight) size such that its removal results in an S-forest. Here we focus on the equivalent formulation of computing a maximum weighted S-forest, formally defined as follows:

SUBSET FEEDBACK VERTEX SET (SFVS)
Input: A graph G, $S \subseteq V(G)$ and a weight function $\mathsf{w} : V(G) \to \mathbb{Q}$.
Output: The maximum among the weights of the S-forests of G.

Rooted Layout. A *rooted binary tree* is a binary tree with a distinguished vertex called the *root*. Since we manipulate at the same time graphs and trees representing them, the vertices of trees will be called *nodes*. A *rooted layout* (also called a rooted branch decomposition) of G is a pair $\mathcal{L} = (T, \delta)$ of a rooted binary tree T and a bijective function δ between $V(G)$ and the leaves of T. For each node x of T, let L_x be the set of all the leaves l of T such that the path from the root of T to l contains x. We denote by V_x the set of vertices that are in bijection with L_x, i.e., $V_x := \{v \in V(G) \mid \delta(v) \in L_x\}$.

All the width measures dealt with in this paper are special cases of the following one, the difference being in each case the used set function. Given a set function $\mathsf{f} : 2^{V(G)} \to \mathbb{N}$ and a rooted layout $\mathcal{L} = (T, \delta)$, the f-width of a node x of T is $\mathsf{f}(V_x)$ and the f-width of (T, δ), denoted by $\mathsf{f}(T, \delta)$ (or $\mathsf{f}(\mathcal{L})$), is $\max\{\mathsf{f}(V_x) \mid x \in V(T)\}$. Finally, the f-width of G is the minimum f-width over all rooted layouts of G.

For a graph G, we let $\mathsf{mim}, \mathsf{rw}, \mathsf{rw}_\mathbb{Q}$ be functions from $2^{V(G)}$ to $\mathbb{N}$ such that for every $A \subseteq V(G)$, $\mathsf{mim}(A)$ is the size of a maximum induced matching of the graph $G[A, \overline{A}]$ and $\mathsf{rw}(A)$ (resp. $\mathsf{rw}_\mathbb{Q}(A)$) is the rank over $GF(2)$ (resp. $\mathbb{Q}$) of the matrix $M_{A,\overline{A}}$. The *mim-width*, *rank-width* and $\mathbb{Q}$*-rank-width* of G, are respectively, its mim-width, rw-width and $\mathsf{rw}_\mathbb{Q}$-width [33]. The following lemma provides upper bounds between mim-width and the other two parameters.

Lemma 1 ([33]). *Let G be a graph. For every $A \subseteq V(G)$, we have $\mathsf{mim}(A) \leqslant \mathsf{rw}(A)$ and $\mathsf{mim}(A) \leqslant \mathsf{rw}_\mathbb{Q}(A)$.*

d**-neighbor Equivalence.** Let G be a graph and let $A \subseteq V(G)$ and $d \in \mathbb{N}^+$. Two subsets X and Y of A are d-*neighbor equivalent w.r.t.* A, denoted by $X \equiv_A^d Y$, if $\min(d, |X \cap N(u)|) = \min(d, |Y \cap N(u)|)$ for all $u \in \overline{A}$. It is not hard to check that $\equiv_A^d$ is an equivalence relation From the definition of the 2-neighbor equivalence relation we have the following.

Fact 2. *For every $A \subseteq V(G)$ and $W, Z \subseteq A$, if $W \equiv_A^2 Z$, then, for all $v \in \overline{A}$, we have $|N(v) \cap W| \leqslant 1$ if and only if $|N(v) \cap Z| \leqslant 1$.*

For all $d \in \mathbb{N}^+$, we let $\mathsf{nec}_d : 2^{V(G)} \to \mathbb{N}$ where for all $A \subseteq V(G)$, $\mathsf{nec}_d(A)$ is the number of equivalence classes of $\equiv_A^d$. Notice that while nec_1 is a symmetric function [26, Theorem 1.2.3], nec_d is not necessarily symmetric for $d \geq 2$. The following lemma shows how $\mathsf{nec}_d(A)$ is upper bounded by the other parameters.

Lemma 2 ([1,29,33]). *For every $A \subseteq V(G)$ and $d \in \mathbb{N}^+$, we have the following upper bounds on $\mathsf{nec}_d(A)$: (a) $2^{d \cdot \mathsf{rw}(A)^2}$, (b) $(d \cdot \mathsf{rw}_\mathbb{Q}(A)+1)^{\mathsf{rw}_\mathbb{Q}(A)}$, (c) $|A|^{d \cdot \mathsf{mim}(A)}$.*

In order to manipulate the equivalence classes of $\equiv_A^d$, one needs to compute a representative for each equivalence class in polynomial time. This is achieved with the following notion of a representative. Let G be a graph with an arbitrary ordering of $V(G)$ and let $A \subseteq V(G)$. For each $X \subseteq A$, let us denote by $\mathsf{rep}_A^d(X)$ the lexicographically smallest set $R \subseteq A$ such that $|R|$ is minimized and $R \equiv_A^d X$. Moreover, we denote by $\mathcal{R}_A^d$ the set $\{\mathsf{rep}_A^d(X) \mid X \subseteq A\}$. It is worth noticing that the empty set always belongs to $\mathcal{R}_A^d$, for all $A \subseteq V(G)$ and $d \in \mathbb{N}^+$. Moreover, we have $\mathcal{R}_{V(G)}^d = \mathcal{R}_\varnothing^d = \{\varnothing\}$ for all $d \in \mathbb{N}^+$. In order to compute these representatives, we use the following lemma.

Lemma 3 ([7]). *For every $A \subseteq V(G)$ and $d \in \mathbb{N}^+$, one can compute in time $O(\mathsf{nec}_d(A) \cdot n^2 \cdot \log(\mathsf{nec}_d(A)))$, the sets $\mathcal{R}_A^d$ and a data structure that, given a set $X \subseteq A$, computes $\mathsf{rep}_A^d(X)$ in time $O(|A| \cdot n \cdot \log(\mathsf{nec}_d(A)))$.*

Vertex Contractions. In order to deal with SFVS, we will use the ideas of the algorithms for FEEDBACK VERTEX SET from [5,23]. To this end, we will contract subsets of $\overline{S}$ in order to transform S-forests into forests.

To compare two partial solutions associated with $A \subseteq V(G)$, we define an auxiliary graph in which we replace contracted vertices by their representative sets in $\mathcal{R}_A^2$. Since the sets in $\mathcal{R}_A^2$ are not necessarily pairwise disjoint, we will use the following notions of graph "induced" by collections of subsets of vertices.

Given $\mathcal{A} \subseteq 2^{V(G)}$, we define $G[\mathcal{A}]$ as the graph with vertex set $\mathcal{A}$ where $A, B \in \mathcal{A}$ are adjacent if and only if $N(A) \cap B \neq \varnothing$. Observe that if the sets in $\mathcal{A}$ are pairwise disjoint, then $G[\mathcal{A}]$ is obtained from an induced subgraph of G by *vertex contractions* (i.e., by replacing two vertices u and v with a new vertex with neighborhood $N(\{u, v\})$) and, for this reason, we refer to $G[\mathcal{A}]$ as a *contracted graph*. Notice that we will never use the neighborhood notation and connected component notations on contracted graph. Given $\mathcal{A}, \mathcal{B} \subseteq 2^{V(G)}$, we denote by $G[\mathcal{A}, \mathcal{B}]$ the bipartite graph with vertex set $\mathcal{A} \cup \mathcal{B}$ and where $A, B \in \mathcal{A} \cup \mathcal{B}$ are adjacent if and only if $A \in \mathcal{A}$, $B \in \mathcal{B}$, and $N(A) \cap B \neq \varnothing$. Moreover, we denote by $G[\mathcal{A} \mid \mathcal{B}]$ the graph with vertex set $\mathcal{A} \cup \mathcal{B}$ and with edge set $E(G[\mathcal{A}]) \cup E(G[\mathcal{A}, \mathcal{B}])$. Observe that both graphs $G[\mathcal{A}, \mathcal{B}]$ and $G[\mathcal{A} \mid \mathcal{B}]$ are subgraphs of the contracted graph $G[\mathcal{A} \cup \mathcal{B}]$. To avoid confusion with the original graph, we refer to the vertices of the contracted graphs as *blocks*. It is worth noticing that in the contracted graphs used in this paper, whenever two blocks are adjacent, they are disjoint.

The following observation states that we can contract from a partition without increasing the size of a maximum induced matching of a graph. It follows directly from the definition of contractions.

Observation 3. *Let H be a graph. For any partition $\mathcal{P}$ of a subset of $V(H)$, the size of a maximum induced matching of $H[\mathcal{P}]$ is at most the size of a maximum induced matching of H.*

Let (G, S) be an instance of SFVS. The vertex contractions that we use on a partial solution X are defined from a given partition of $X \setminus S$. A partition of the vertices of $X \setminus S$ is called an $\overline{S}$*-contraction* of X. We will use the following notations to handle these contractions.

Given $X \subseteq V(G)$, we denote by $\binom{X}{1}$ the partition of X which contains only singletons, i.e., $\binom{X}{1} = \{\{v\} \mid v \in X\}$. Moreover, for an $\overline{S}$-contraction $\mathcal{P}$ of X, we denote by $X_{\downarrow \mathcal{P}}$ the partition of X where $X_{\downarrow \mathcal{P}} = \mathcal{P} \cup \binom{X \cap S}{1}$. Given a subgraph G' of G such that $V(G') = X$, we denote by $G'_{\downarrow \mathcal{P}}$ the graph $G'[X_{\downarrow \mathcal{P}}]$. It is worth noticing that in our contracted graphs, all the blocks of S-vertices are singletons and we denote them by $\{v\}$.

Given a set $X \subseteq V(G)$, we will intensively use the graph $G[X]_{\downarrow \mathsf{cc}(X \setminus S)}$ which corresponds to the graph obtained from $G[X]$ by contracting the connected components of $G[X \setminus S]$. Observe that, for every subset $X \subseteq V(G)$, if $G[X]$ is an S-forest, then $G[X]_{\downarrow \mathsf{cc}(X \setminus S)}$ is a forest. The converse is not true as we may delete S-cycles with contractions: take a triangle with one vertex v in S and contract the neighbors of v. However, we can prove the following equivalence.

Fact 4. *Let G be a graph and $S \subseteq V(G)$. For every $X \subseteq V(G)$ such that $|N(v) \cap C| \leqslant 1$ for each $v \in X \cap S$ and each $C \in \mathsf{cc}(X \setminus S)$, we have $G[X]$ is an S-forest if and only if $G[X]_{\downarrow \mathsf{cc}(X \setminus S)}$ is a forest.*

3   A Meta-Algorithm for Subset Feedback Vertex Set

In the following, we present a meta-algorithm that, given a rooted layout (T, δ) of G, solves SFVS. We will show that such a meta-algorithm will imply that SFVS can be solved in time $\mathsf{rw}_\mathbb{Q}(G)^{O(\mathsf{rw}_\mathbb{Q}(G)^2))} \cdot n^4$, $2^{O(\mathsf{rw}(G)^3)} \cdot n^4$ and $n^{O(\mathsf{mim}(T,\delta)^2)}$. The main idea of this algorithm is to use $\overline{S}$-contractions in order to employ similar properties of the meta-algorithm for MAXIMUM INDUCED TREE of [5] and the $n^{O(\mathsf{mim}(T,\delta))}$ time algorithm for FEEDBACK VERTEX SET of [23]. In particular, we use the following lemma which is proved implicitly in [3, Lemma 5.5].

Lemma 4. *Let X and Y be two disjoint subsets of $V(G)$. If $G[X \cup Y]$ is a forest, then the number of vertices of X that have at least two neighbors in Y is bounded by $2w$ where w is the size of a maximum induced matching in $G[X, Y]$.*

The following lemma generalizes Fact 4 and presents the equivalence between S-forests and forests that we will use in our algorithm.

Lemma 5. *Let $A \subseteq V(G)$, $X \subseteq A$, and $Y \subseteq \overline{A}$. If the graph $G[X \cup Y]$ is an S-forest, then there exists an $\overline{S}$-contraction $\mathcal{P}_Y$ of Y that satisfies the following conditions:*

(1) $G[X \cup Y]_{\downarrow \mathsf{cc}(X \setminus S) \cup \mathcal{P}_Y}$ is a forest,
(2) for all $P \in \mathsf{cc}(X \setminus S) \cup \mathcal{P}_Y$ and $v \in (X \cup Y) \cap S$, we have $|N(v) \cap P| \leqslant 1$,
(3) the graph $G[X, Y]_{\downarrow \mathsf{cc}(X \setminus S) \cup \mathcal{P}_Y}$ admits a vertex cover VC of size at most $4\mathsf{mim}(A)$ such that the neighborhoods of the vertices in VC are pairwise distinct in $G[X, Y]_{\downarrow \mathsf{cc}(X \setminus S) \cup \mathcal{P}_Y}$.

In the following, we will use Lemma 5 to design some sort of equivalence relation between partial solutions. To this purpose, we use the following notion.

Definition 1 ($\mathbb{I}_x$). *For every $x \in V(T)$, we define the set $\mathbb{I}_x$ of indices as the set of tuples $(X_{\mathsf{vc}}^{\overline{S}}, X_{\mathsf{vc}}^{S}, X_{\overline{\mathsf{vc}}}, Y_{\mathsf{vc}}^{\overline{S}}, Y_{\mathsf{vc}}^{S}) \in 2^{\mathcal{R}_{V_x}^2} \times 2^{\mathcal{R}_{V_x}^1} \times \mathcal{R}_{V_x}^1 \times 2^{\mathcal{R}_{\overline{V_x}}^2} \times 2^{\mathcal{R}_{\overline{V_x}}^1}$ such that $|X_{\mathsf{vc}}^{\overline{S}}| + |X_{\mathsf{vc}}^{S}| + |Y_{\mathsf{vc}}^{\overline{S}}| + |Y_{\mathsf{vc}}^{S}| \leqslant 4\mathsf{mim}(V_x)$.*

In the following, we will define partial solutions associated with an index $i \in \mathbb{I}_x$ (a partial solution may be associated with many indices). In order to prove the correctness of our algorithm (the algorithm will not use this concept), we will also define *complement solutions* (the sets $Y \subseteq \overline{V_x}$ and their $\overline{S}$-contractions $\mathcal{P}_Y$) associated with an index i. We will prove that, for every partial solution X and complement solution $(Y, \mathcal{P}_Y)$ associated with i, if the graph $G[X \cup Y]_{\downarrow \mathsf{cc}(X \setminus S) \cup \mathcal{P}_Y}$ is a forest, then $G[X \cup Y]$ is an S-forest.

Let us give some intuition on these indices by explaining how one index is associated with a solution[1]. Let $x \in V(T)$, $X \subseteq V_x$ and $Y \subseteq \overline{V_x}$ such that $G[X \cup Y]$ is an S-forest. Let $\mathcal{P}_Y$ be the $\overline{S}$-contraction of Y and VC be a vertex cover of $G[X, Y]_{\downarrow \mathsf{cc}(X \setminus S) \cup \mathcal{P}_Y}$ given by Lemma 5. Then, X and Y are associated with $i = (X_{\mathsf{vc}}^{\overline{S}}, X_{\mathsf{vc}}^{S}, X_{\overline{\mathsf{vc}}}, Y_{\mathsf{vc}}^{\overline{S}}, Y_{\mathsf{vc}}^{S}) \in \mathbb{I}_x$ such that:

---

[1] An animation explaining this association can be found here [2].

- X_{vc}^S (resp. Y_{vc}^S) contains the representatives of the blocks $\{v\}$ in VC such that $v \in X \cap S$ (resp. $v \in Y \cap S$) w.r.t. the 1-neighbor equivalence over V_x (resp. $\overline{V_x}$). We will only use the indices where X_{vc}^S contains representatives of singletons, in other words, X_{vc}^S is included in $\{\mathsf{rep}_{V_x}^1(\{v\}) \mid v \in V_x\}$ which can be much smaller than $\mathcal{R}_{V_x}^1$. The same observation holds for Y_{vc}^S. In Definition 1, we state that X_{vc}^S and Y_{vc}^S are, respectively, subsets of $2^{\mathcal{R}_{V_x}^1}$ and $2^{\mathcal{R}_{\overline{V_x}}^1}$, for the sake of simplicity.
- $X_{vc}^{\overline{S}}$ (resp. $Y_{vc}^{\overline{S}}$) contains the representatives of the blocks in $cc(X \setminus S) \cap$ VC (resp. $\mathcal{P}_Y \cap$ VC) w.r.t. the 2-neighbor equivalence relation over V_x (resp. $\overline{V_x}$).
- $X_{\overline{vc}}$ is the representative of $X \setminus V(\mathsf{VC})$ (the set of vertices which do not belong to the vertex cover) w.r.t. the 1-neighbor equivalence over V_x.

Because the neighborhoods of the blocks in VC are pairwise distinct in the graph $G[X,Y]_{\downarrow cc(X \setminus S) \cup \mathcal{P}_Y}$ (Property (3) of Lemma 5), there is a one to one correspondence between the representatives in $X_{vc}^{\overline{S}} \cup X_{vc}^S \cup Y_{vc}^{\overline{S}} \cup Y_{vc}^S$ and the blocks in VC.

While $X_{vc}^{\overline{S}}, X_{vc}^S, Y_{vc}^{\overline{S}}, Y_{vc}^S$ describe VC, the representative set $X_{\overline{vc}}$ describes the neighborhood of the vertices of X which are not in VC. The purpose of $X_{\overline{vc}}$ is to make sure that, for every partial solution X and complement solution $(Y, \mathcal{P}_Y)$ associated with i, the set VC described by $X_{vc}^{\overline{S}}, X_{vc}^S, Y_{vc}^{\overline{S}}, Y_{vc}^S$ is a vertex cover of $G[X,Y]_{\downarrow cc(X \setminus S) \cup \mathcal{P}_Y}$. For doing so, it is sufficient to require that $Y \setminus V(\mathsf{VC})$ has no neighbor in X_{vc} for every complement solution $(Y, \mathcal{P}_Y)$ associated with i.

Observe that the sets $X_{vc}^{\overline{S}}$ and $Y_{vc}^{\overline{S}}$ contain representatives for the 2-neighbor equivalence. We need the 2-neighbor equivalence to control the S-cycles which might disappear after vertex contractions. To prevent this situation, we require, for example, that every vertex in $X \cap S$ has at most one neighbor in $\overline{R}$ for each $\overline{R} \in Y_{vc}^{\overline{S}}$. Thanks to the 2-neighbor equivalence, a vertex v in $X \cap S$ has at most one neighbor in $\overline{R} \in Y_{vc}^{\overline{S}}$ if and only if v has at most one neighbor in the block of $\mathcal{P}_Y$ associated with $\overline{R}$.

In order to define partial solutions associated with i, we need the following notion of auxiliary graph. Given $x \in V(T)$, $X \subseteq V_x$, and $i = (X_{vc}^{\overline{S}}, X_{vc}^S, X_{\overline{vc}}, Y_{vc}^{\overline{S}}, Y_{vc}^S) \in \mathbb{I}_x$, we write $\mathsf{aux}(X,i)$ to denote the graph $G[X_{\downarrow cc(X \setminus S)} \mid Y_{vc}^{\overline{S}} \cup Y_{vc}^S]$. Observe that $\mathsf{aux}(X,i)$ is obtained from the graph induced by $X_{\downarrow cc(X \setminus S)} \cup Y_{vc}^{\overline{S}} \cup Y_{vc}^S$ by removing the edges between the vertices from $Y_{vc}^{\overline{S}} \cup Y_{vc}^S$.

We will ensure that, given a complement solution $(Y, \mathcal{P}_Y)$ associated with i, the graph $\mathsf{aux}(X,i)$ is isomorphic to $G[X_{\downarrow cc(X \setminus S)} \mid Y_{\downarrow \mathcal{P}_Y} \cap \mathsf{VC}]$. We are now ready to define the notion of partial solution associated with an index i.

Definition 2 (Partial Solutions). *Let $x \in V(T)$ and $i = (X_{vc}^{\overline{S}}, X_{vc}^S, X_{\overline{vc}}, Y_{vc}^{\overline{S}}, Y_{vc}^S) \in \mathbb{I}_x$. We say that $X \subseteq V_x$ is a partial solution associated with i if the following conditions are satisfied:*

(a) for every $R \in X_{vc}^S$, there exists a unique $v \in X \cap S$ such that $R \equiv_{V_x}^1 \{v\}$,

(b) for every $R \in X_{vc}^{\overline{S}}$, there exists a unique $C \in \mathsf{cc}(X \setminus S)$ such that $R \equiv_{V_x}^2 C$,

(c) $\mathsf{aux}(X, i)$ is a forest,

(d) for every $C \in \mathsf{cc}(X \setminus S)$ and $\{v\} \in Y_{vc}^S$, we have $|N(v) \cap C| \leqslant 1$,

(e) for every $v \in X \cap S$ and $U \in Y_{vc}^{\overline{S}} \cup \mathsf{cc}(X \setminus S)$, we have $|N(v) \cap U| \leqslant 1$,

(f) $X_{\overline{vc}} \equiv_{V_x}^1 X \setminus V(\mathsf{VC}_X)$, where VC_X contains the blocks $\{v\} \in \binom{X \cap S}{1}$ s.t. $\mathsf{rep}_{V_x}^1(\{v\}) \in X_{vc}^S$ and the blocks $C \in \mathsf{cc}(X \setminus S)$ s.t. $\mathsf{rep}_{V_x}^2(C) \in X_{vc}^{\overline{S}}$.

Similarly to Definition 2, we define the notion of *complement solutions* associated with an index $i \in \mathbb{I}_x$. We use this concept only to prove the correctness of our algorithm.

Definition 3 (Complement Solutions). *Let* $x \in V(T)$ *and* $i = (X_{vc}^{\overline{S}}, X_{vc}^S, X_{\overline{vc}}, Y_{vc}^{\overline{S}}, Y_{vc}^S) \in \mathbb{I}_x$. *We call* complement solutions *associated with* i *all the pairs* $(Y, \mathcal{P}_Y)$ *such that* $Y \subseteq \overline{V_x}$, $\mathcal{P}_Y$ *is an* $\overline{S}$-contraction of Y *and the following conditions are satisfied:*

(a) *for every* $U \in Y_{vc}^S$, *there exists a unique* $v \in Y \cap S$ *such that* $U \equiv_{V_x}^2 \{v\}$,

(b) *for every* $U \in Y_{vc}^{\overline{S}}$, *there exists a unique* $P \in \mathcal{P}_Y$ *such that* $U \equiv_{V_x}^2 P$,

(c) $G[Y]_{\downarrow \mathcal{P}_Y}$ *is a forest,*

(d) *for every* $P \in \mathcal{P}_Y$ *and* $\{v\} \in X_{vc}^S$, *we have* $|N(v) \cap P| \leqslant 1$,

(e) *for every* $R \in X_{vc}^{\overline{S}} \cup \mathcal{P}_Y$ *and* $y \in Y \cap S$, *we have* $|N(y) \cap R| \leqslant 1$,

(f) $N(X_{\overline{vc}}) \cap V(Y_{\overline{vc}}) = \varnothing$, *where* $Y_{\overline{vc}}$ *contains the blocks* $\{v\} \in \binom{Y \cap S}{1}$ *such that* $\mathsf{rep}_{\overline{V_x}}^1(\{v\}) \notin Y_{vc}^S$ *and the blocks* $P \in \mathcal{P}_Y$ *such that* $\mathsf{rep}_{\overline{V_x}}^2(P) \notin Y_{vc}^{\overline{S}}$.

Let us give some explanations on the conditions of Definitions 2 and 3. Let $x \in V(T)$, X be a partial solution associated with i and $(Y, \mathcal{P}_Y)$ be a complement solution associated with i. Conditions (a) and (b) of both definitions guarantee that the set VC described by $X_{vc}^{\overline{S}}, X_{vc}^S, Y_{vc}^{\overline{S}}$ and Y_{vc}^S is included in $X \cup Y$ and that there is an one to one correspondence between the blocks of VC and the representatives in $X_{vc}^{\overline{S}} \cup X_{vc}^S \cup Y_{vc}^{\overline{S}} \cup Y_{vc}^S$.

Condition (c) of Definition 2 guarantees that the connections between sets $X_{\downarrow \mathsf{cc}(X \setminus S)}$ and VC are acyclic. As explained earlier, Conditions (d) and (e) of both definitions control the S-cycles which might disappear with the vertex contractions.

Finally, as explained earlier, the last conditions of both definitions ensure that VC the set described by $X_{vc}^{\overline{S}}, X_{vc}^S, Y_{vc}^{\overline{S}}$ and Y_{vc}^S is a vertex cover of the graph $G[X, Y]_{\downarrow \mathsf{cc}(X \setminus S) \cup \mathcal{P}_Y}$. Notice that $V(Y_{\overline{vc}})$ corresponds to the set of vertices of Y which do not belong to a block of VC. Such observations are used to prove the following two results.

Lemma 6. *For every* $X \subseteq V_x$ *and* $Y \subseteq \overline{V_x}$ *such that* $G[X \cup Y]$ *is an* S-*forest, there exist* $i \in \mathbb{I}_x$ *and an* $\overline{S}$-*contraction* $\mathcal{P}_Y$ *of* Y *such that (1)* $G[X \cup Y]_{\downarrow \mathsf{cc}(X \setminus S) \cup \mathcal{P}_Y}$ *is a forest, (2)* X *is a partial solution associated with* i *and (3)* $(Y, \mathcal{P}_Y)$ *is a complement solution associated with* i.

Lemma 7. *Let* $i = (X_{\mathsf{vc}}^{\overline{S}}, X_{\mathsf{vc}}^{S}, X_{\overline{\mathsf{vc}}}, Y_{\mathsf{vc}}^{\overline{S}}, Y_{\mathsf{vc}}^{S}) \in \mathbb{I}_x$, X *be a partial solution associated with* i *and* $(Y, \mathcal{P}_Y)$ *be a complement solutions associated with* i. *If the graph* $G[X \cup Y]_{\downarrow \mathsf{cc}(X \setminus S) \cup \mathcal{P}_Y}$ *is a forest, then* $G[X \cup Y]$ *is an* S-*forest.*

For each index $i \in \mathbb{I}_x$, we will design an equivalence relation $\sim_i$ between the partial solutions associated with i. We will prove that, for any partial solutions X and W associated with i, if $X \sim_i W$, then, for any complement solution $(Y, \mathcal{P}_Y)$ associated with i, the graph $G[X \cup Y]$ is an S-forest if and only if $G[W \cup Y]$ is an S-forest. Then, given a set of partial solutions $\mathcal{A}$ whose size needs to be reduced, it is sufficient to keep, for each $i \in \mathbb{I}_x$ and each equivalence class $\mathcal{C}$ of $\sim_i$, one partial solution in $\mathcal{C}$ of maximal weight. The resulting set of partial solutions has size bounded by $|\mathbb{I}_x| \cdot (4\mathsf{mim}(V_x))^{4\mathsf{mim}(V_x)}$ because $\sim_i$ generates at most $(4\mathsf{mim}(V_x))^{4\mathsf{mim}(V_x)}$ equivalence classes.

Intuitively, given two partial solutions X and W associated with an index $i = (X_{\mathsf{vc}}^{\overline{S}}, X_{\mathsf{vc}}^{S}, X_{\overline{\mathsf{vc}}}, Y_{\mathsf{vc}}^{\overline{S}}, Y_{\mathsf{vc}}^{S})$, we have $X \sim_i W$ if the blocks of VC (i.e., the vertex cover described by i) are *equivalently connected* in $G[X_{\downarrow \mathsf{cc}(X \setminus S)} \mid Y_{\mathsf{vc}}^{\overline{S}} \cup Y_{\mathsf{vc}}^{S}]$ and $G[W_{\downarrow \mathsf{cc}(W \setminus S)} \mid Y_{\mathsf{vc}}^{\overline{S}} \cup Y_{\mathsf{vc}}^{S}]$. In order to compare these connections, we use the following notion.

Definition 4 ($\mathsf{cc}(X, i)$). *For each connected component* C *of* $\mathsf{aux}(X, i)$, *we denote by* C_{vc} *the following set:*

- *for every* $U \in C$ *such that* $U \subset Y_{\mathsf{vc}}^{\overline{S}} \cup Y_{\mathsf{vc}}^{S}$, *we have* $U \in C_{\mathsf{vc}}$,
- *for every* $\{v\} \in \binom{X \cap S}{1} \cap C$ *such that* $\{v\} \equiv_{V_x}^{1} R$ *for some* $R \in X_{\mathsf{vc}}^{S}$, *we have* $R \in C_{\mathsf{vc}}$,
- *for every* $U \in \mathsf{cc}(X \setminus S)$ *such that* $U \equiv_{V_x}^{2} R$ *for some* $R \in X_{\mathsf{vc}}^{\overline{S}}$, *we have* $R \in C_{\mathsf{vc}}$.

We denote by $\mathsf{cc}(X, i)$ *the set* $\{C_{\mathsf{vc}} \mid C$ *is a connected component of* $\mathsf{aux}(X, i)\}$.

For a connected component C of $\mathsf{aux}(X, i)$, the set C_{vc} contains $C \cap (Y_{\mathsf{vc}}^{\overline{S}} \cup Y_{\mathsf{vc}}^{S})$ and the representatives of the blocks in $C \cap X_{\downarrow \mathsf{cc}(X \setminus S)} \cap$ VC with VC the vertex cover described by i. Consequently, for every $X \subseteq V_x$ and $i \in \mathbb{I}_x$, the collection $\mathsf{cc}(X, i)$ is partition of $X_{\mathsf{vc}}^{\overline{S}} \cup X_{\mathsf{vc}}^{S} \cup Y_{\mathsf{vc}}^{\overline{S}} \cup Y_{\mathsf{vc}}^{S}$, observe that $\mathsf{cc}(X, i)$ is the partition with the blocks $\{R_1, U_1, U_2\}$, $\{R_2, U_3\}$, and $\{U_4\}$.

Now we give the notion of equivalence between partial solutions. We say that two partial solutions X, W associated with i are i-*equivalent*, denoted by $X \sim_i W$, if $\mathsf{cc}(X, i) = \mathsf{cc}(W, i)$. Our next result is the most crucial step. As already explained, our task is to show equivalence between partial solutions under any complement solution with respect to S-forests.

Lemma 8. *Let* $i \in \mathbb{I}_x$. *For every partial solutions* X, W *associated with* i *such that* $X \sim_i W$ *and for every complement solution* $(Y, \mathcal{P}_Y)$ *associated with* i, *the graph* $G[X \cup Y]_{\downarrow \mathsf{cc}(X \setminus S) \cup \mathcal{P}_Y}$ *is a forest if and only if the graph* $G[W \cup Y]_{\downarrow \mathsf{cc}(W \setminus S) \cup \mathcal{P}_Y}$ *is a forest.*

The following theorem proves that, for every set of partial solutions $\mathcal{A} \subseteq 2^{V_x}$, we can compute a small subset $\mathcal{B} \subseteq \mathcal{A}$ such that $\mathcal{B}$ *represents* $\mathcal{A}$, i.e., for every $Y \subseteq \overline{V_x}$, the best solutions we obtain from the union of Y with a set in $\mathcal{A}$ are as good as the ones we obtain from $\mathcal{B}$. Firstly, we formalize this notion of representativity.

Definition 5 (Representativity). *Let* $x \in V(T)$*. For every* $\mathcal{A} \subseteq 2^{V_x}$ *and* $Y \subseteq \overline{V_x}$*, we define* $\mathsf{best}(\mathcal{A}, Y) = \max\{\mathsf{w}(X) \mid X \in \mathcal{A} \, and \, G[X \cup Y] \, is \, an \, S\text{-}forest\}$*. Given* $\mathcal{A}, \mathcal{B} \subseteq 2^{V_x}$*, we say that* $\mathcal{B}$ *represents* $\mathcal{A}$ *if, for every* $Y \subseteq \overline{V_x}$*, we have* $\mathsf{best}(\mathcal{A}, Y) = \mathsf{best}(\mathcal{B}, Y)$*.*

Theorem 5. *Let* $x \in V(T)$*. Then, there exists an algorithm* reduce *that, given a set* $\mathcal{A} \subseteq 2^{V_x}$*, outputs in time* $O(|\mathcal{A}| \cdot |\mathbb{I}_x| \cdot (4\mathsf{mim}(V_x))^{4\mathsf{mim}(V_x)} \cdot \mathsf{s\text{-}nec}_2(V_x) \cdot n^3)$ *a subset* $\mathcal{B} \subseteq \mathcal{A}$ *such that* $\mathcal{B}$ *represents* $\mathcal{A}$ *and* $|\mathcal{B}| \leqslant |\mathbb{I}_x| \cdot (4\mathsf{mim}(V_x))^{4\mathsf{mim}(V_x)}$*.*

We are now ready to prove the main theorem of this paper.

Theorem 6. *There exists an algorithm that, given an* n*-vertex graph* G *and a rooted layout* (T, δ) *of* G*, solves* SUBSET FEEDBACK VERTEX SET *in time* $O\left(\sum_{x \in V(T)} |\mathbb{I}_x|^3 \cdot (4\mathsf{mim}(V_x))^{12\mathsf{mim}(V_x)} \cdot \mathsf{s\text{-}nec}_2(V_x) \cdot n^3\right)$*.*

Algorithmic Consequences. In order to obtain the algorithmic consequences of our meta-algorithm given in Theorem 6, we need the following lemma which bounds the size of each table index with respect to the considered parameters.

Lemma 9. *For every* $x \in V(T)$*, the size of* $\mathbb{I}_x$ *is upper bounded by:* $2^{O(\mathsf{rw}(V_x)^3)}$*,* $\mathsf{rw}_{\mathbb{Q}}(V_x)^{O(\mathsf{rw}_{\mathbb{Q}}(V_x)^2)}$*,* $n^{O(\mathsf{mim}(V_x)^2)}$*.*

Now we are ready to state our algorithms with respect to rank-width, $\mathbb{Q}$-rank-width and mim-width. In particular, with our next result we show that SUBSET FEEDBACK VERTEX SET is in FPT parameterized by $\mathsf{rw}_{\mathbb{Q}}(G)$ or $\mathsf{rw}(G)$.

Theorem 7. *There exist algorithms that solve* SUBSET FEEDBACK VERTEX SET *in time* $2^{O(\mathsf{rw}(G)^3)} \cdot n^4$ *and* $\mathsf{rw}_{\mathbb{Q}}(G)^{O(\mathsf{rw}_{\mathbb{Q}}(G)^2)}) \cdot n^4$*. Moreover, if a rooted layout* $\mathcal{L}$ *of* G *is given as input, we can solve* SUBSET FEEDBACK VERTEX SET *in time* $n^{O(\mathsf{mim}(\mathcal{L})^2)}$*.*

Regarding mim-width, our algorithm given below shows that SUBSET FEEDBACK VERTEX SET is in XP parameterized by the mim-width of a given rooted layout. Note that we cannot solve SFVS in FPT time parameterized by the mim-width of a given rooted layout unless FPT = W[1], since SUBSET FEEDBACK VERTEX SET is known to be W[1]-hard for this parameter even for the special case of $S = V(G)$ [24]. Moreover, contrary to the algorithms given in Theorem 7, here we need to assume that the input graph is given with a rooted layout. However, our next result actually provides a unified polynomial-time algorithm for SUBSET FEEDBACK VERTEX SET on well-known graph classes having bounded mim-width and for which a layout of bounded mim-width can be computed in polynomial time[2] [1].

---

[2] For example: INTERVAL graphs, PERMUTATION graphs, CIRCULAR ARC graphs, CONVEX graphs, k-POLYGON, DILWORTH-k and CO-k-DEGENERATE graphs for fixed k.

Let us relate our results for SUBSET FEEDBACK VERTEX SET to the NODE MULTIWAY CUT. It is known that NODE MULTIWAY CUT reduces to SUBSET FEEDBACK VERTEX SET [18]. In fact, we can solve NODE MULTIWAY CUT by adding a single S-vertex with a large weight that is adjacent to all terminals and, then, run our algorithms for SUBSET FEEDBACK VERTEX SET on the resulting graph. Now observe that any extension of a rooted layout $\mathcal{L}$ of the original graph to the resulting graph has mim-width $\mathsf{mim}(\mathcal{L})+1$. Therefore, all of our algorithms given in Theorem 7 have the same running times for the NODE MULTIWAY CUT problem.

References

1. Belmonte, R., Vatshelle, M.: Graph classes with structured neighborhoods and algorithmic applications. Theoret. Comput. Sci. **511**, 54–65 (2013). https://doi. org/10.1016/j.tcs.2013.01.011
2. Bergougnoux, B.: Animation explaining the relation between an solution and an indice (2019). https://folk.uib.no/bbe089/slides/Explanations.pdf. Accessed 7 Jan 2020
3. Bergougnoux, B., Kanté, M.M.: Rank based approach on graphs with structured neighborhood. CoRR abs/1805.11275 (2018)
4. Bergougnoux, B., Kanté, M.M.: Fast exact algorithms for some connectivity problems parameterized by clique-width. Theor. Comput. Sci. **782**, 30–53 (2019). https://doi.org/10.1016/j.tcs.2019.02.030
5. Bergougnoux, B., Kanté, M.M.: More applications of the d-neighbor equivalence: Connectivity and acyclicity constraints. In: 27th Annual European Symposium on Algorithms, ESA 2019, 9–11 September 2019, Munich, Garching, Germany, pp. 17:1–17:14 (2019). https://doi.org/10.4230/LIPIcs.ESA.2019.17
6. Bodlaender, Hans L.: Treewidth: characterizations, applications, and computations. In: Fomin, Fedor V. (ed.) WG 2006. LNCS, vol. 4271, pp. 1–14. Springer, Heidelberg (2006) https://doi.org/10.1007/11917496_1
7. Bui-Xuan, B.M., Telle, J.A., Vatshelle, M.: Fast dynamic programming for locally checkable vertex subset and vertex partitioning problems. Theoret. Comput. Sci. **511**, 66–76 (2013). https://doi.org/10.1016/j.tcs.2013.01.009
8. Calinescu, G.: Multiway Cut. Springer, Boston (2008). https://doi.org/10.1007/978-0-387-30162-4
9. Chen, J., Liu, Y., Lu, S.: An improved parameterized algorithm for the minimum node multiway cut problem. Algorithmica **55**, 1–13 (2009). https://doi.org/10.1007/s00453-007-9130-6
10. Chitnis, R.H., Fomin, F.V., Lokshtanov, D., Misra, P., Ramanujan, M.S., Saurabh, S.: Faster exact algorithms for some terminal set problems. J. Comput. Sys. Sci. **88**, 195–207 (2017)
11. Corneil, D.G., Fonlupt, J.: The complexity of generalized clique covering. Discrete Appl. Math. **22**(2), 109–118 (1988)
12. Courcelle, B., Makowsky, J.A., Rotics, U.: Linear time solvable optimization problems on graphs of bounded clique-width. Theory Comput. Syst. **33**(2), 125–150 (2000). https://doi.org/10.1007/s002249910009
13. Courcelle, B., Olariu, S.: Upper bounds to the clique width of graphs. Discrete Appl. Math. **101**(1–3), 77–114 (2000)

14. Cygan, M., Pilipczuk, M., Pilipczuk, M., Wojtaszczyk, J.O.: On multiway cut parameterized above lower bounds. TOCT **5**(1), 3:1–3:11 (2013)
15. Dahlhaus, E., Johnson, D.S., Papadimitriou, C.H., Seymour, P.D., Yannakakis, M.: The complexity of multiterminal cuts. SIAM J. Comput. **23**(4), 864–894 (1994)
16. Even, G., Naor, J., Zosin, L.: An 8-approximation algorithm for the subset feedback vertex set problem. SIAM J. Comput. **30**(4), 1231–1252 (2000)
17. Fomin, F.V., Gaspers, S., Lokshtanov, D., Saurabh, S.: Exact algorithms via monotone local search. In: Proceedings of STOC 2016, pp. 764–775 (2016)
18. Fomin, F.V., Heggernes, P., Kratsch, D., Papadopoulos, C., Villanger, Y.: Enumerating minimal subset feedback vertex sets. Algorithmica **69**(1), 216–231 (2014). https://doi.org/10.1007/s00453-012-9731-6
19. Garg, N., Vazirani, V.V., Yannakakis, M.: Multiway cuts in node weighted graphs. J. Algorithms **50**(1), 49–61 (2004)
20. Golovach, P.A., Heggernes, P., Kratsch, D., Saei, R.: Subset feedback vertex sets in chordal graphs. J. Discrete Algorithms **26**, 7–15 (2014)
21. Golumbic, M.C., Rotics, U.: On the clique-width of some perfect graph classes. Int. J. Found. Comput. Sci. **11**(3), 423–443 (2000). https://doi.org/10.1142/S0129054100000260
22. Hols, E.C., Kratsch, S.: A randomized polynomial kernel for subset feedback vertex set. Theory Comput. Syst. **62**, 54–65 (2018). https://doi.org/10.1007/s00224-017-9805-6
23. Jaffke, L., Kwon, O., Telle, J.A.: A unified polynomial-time algorithm for feedback vertex set on graphs of bounded mim-width. In: 35th Symposium on Theoretical Aspects of Computer Science, STACS 2018, 28 February to 3 March 2018, Caen, France, pp. 42:1–42:14 (2018). https://doi.org/10.4230/LIPIcs.STACS.2018.42
24. Jaffke, Lars., Kwon, O-joung, Telle, Jan Arne: Mim-Width II. The Feedback Vertex Set Problem. Algorithmica **82**(1), 118–145 (2019). https://doi.org/10.1007/s00453-019-00607-3
25. Kawarabayashi, K., Kobayashi, Y.: Fixed-parameter tractability for the subset feedback set problem and the S-cycle packing problem. J. Comb. Theory Ser. B **102**(4), 1020–1034 (2012)
26. Kim, K.H.: Boolean Matrix Theory and Applications, vol. 70. Dekker, New York (1982)
27. Oum, S.i.: Rank-width and vertex-minors. J. Combin. Theory Ser. B **95**(1), 79–100 (2005). https://doi.org/10.1016/j.jctb.2005.03.003
28. Oum, S.I.: Approximating rank-width and clique-width quickly. ACM Trans. Algorithms **5**(1), 20 (2009). https://doi.org/10.1007/11604686_5. Art. 10
29. Oum, S.i., Sæther, S.H., Vatshelle, M.: Faster algorithms for vertex partitioning problems parameterized by clique-width. Theoret. Comput. Sci. **535**, 16–24 (2014). https://doi.org/10.1016/j.tcs.2014.03.024
30. Oum, S.i., Seymour, P.: Approximating clique-width and branch-width. J. Combin. Theory Ser. B **96**(4), 514–528 (2006). https://doi.org/10.1016/j.jctb.2005.10.006
31. Papadopoulos, C., Tzimas, S.: Subset feedback vertex set on graphs of bounded independent set size. In: 13th International Symposium on Parameterized and Exact Computation, IPEC 2018, 20–24 August 2018, Helsinki, Finland, pp. 20:1–20:14 (2018)
32. Papadopoulos, C., Tzimas, S.: Polynomial-time algorithms for the subset feedback vertex set problem on interval graphs and permutation graphs. Discrete Appl. Math. **258**, 204–221 (2019)
33. Vatshelle, M.: New width parameters of graphs. Ph.D. thesis, University of Bergen, Bergen, Norway (2012)

Weighted Additive Spanners

Reyan Ahmed[1(✉)], Greg Bodwin[2], Faryad Darabi Sahneh[1],
Stephen Kobourov[1], and Richard Spence[1]

[1] Department of Computer Science, University of Arizona, Tucson, USA
abureyanahmed@email.arizona.edu
[2] Department of Computer Science, Georgia Institute of Technology,
Atlanta, Georgia

Abstract. A *spanner* of a graph G is a subgraph H that approximately preserves shortest path distances in G. Spanners are commonly applied to compress computation on metric spaces corresponding to weighted input graphs. Classic spanner constructions can seamlessly handle edge weights, so long as error is measured *multiplicatively*. In this work, we investigate whether one can similarly extend constructions of spanners with purely *additive* error to weighted graphs. These extensions are not immediate, due to a key lemma about the size of shortest path neighborhoods that fails for weighted graphs. Despite this, we recover a suitable amortized version, which lets us prove direct extensions of classic $+2$ and $+4$ unweighted spanners (both all-pairs and pairwise) to $+2W$ and $+4W$ weighted spanners, where W is the maximum edge weight. Specifically, we show that a weighted graph G contains all-pairs (pairwise) $+2W$ and $+4W$ weighted spanners of size $O(n^{3/2})$ and $O(n^{7/5})$ ($O(np^{1/3})$ and $O(np^{2/7})$) respectively. For a technical reason, the $+6$ unweighted spanner becomes a $+8W$ weighted spanner; closing this error gap is an interesting remaining open problem. That is, we show that G contains all-pairs (pairwise) $+8W$ weighted spanners of size $O(n^{4/3})$ ($O(np^{1/4})$).

Keywords: Additive spanner · Pairwise spanner · Shortest-path neighborhood

1 Introduction

An $f(\cdot)$-*spanner* of an undirected graph $G = (V, E)$ with $|V| = n$ nodes and $|E| = m$ edges is a subgraph H which preserves pairwise distances in G up to some error prescribed by f; that is, $\text{dist}_H(s, t) \leq f(\text{dist}_G(s, t))$ for all nodes $s, t \in V$. Spanners were introduced by Peleg and Schäffer [25] in the setting with multiplicative error of type $f(d) = cd$ for some positive constant c. This setting was quickly resolved, with matching upper and lower bounds [4] on the sparsity of a spanner that can be achieved in general. At the other extreme are (purely) *c-additive spanners* (or $+c$ spanners), with error of type $f(d) = d + c$. More generally, if $f(d) = \alpha d + \beta$, we say that H is an (α, β)-*spanner*. Intuitively, additive error is much stronger than multiplicative error; most applications involve

© Springer Nature Switzerland AG 2020
I. Adler and H. Müller (Eds.): WG 2020, LNCS 12301, pp. 401–413, 2020.
https://doi.org/10.1007/978-3-030-60440-0_32

shrinking enormous input graphs that are too large to analyze directly, and so it is appealing to avoid error that scales with distance.

Additive spanners were thus initially considered perhaps too good to be true, and they were discovered only for particular classes of input graphs [22]. However, in a surprise to the area, a seminal paper of Aingworth, Chekuri, Indyk, and Motwani [3] proved that nontrivial additive spanners actually exist *in general*: every n-node undirected unweighted graph has a 2-additive spanner on $O(n^{3/2})$ edges. Subsequently, more interesting constructions of additive spanners were found: there are 4-additive spanners on $O(n^{7/5})$ edges [7,11] and 6-additive spanners on $O(n^{4/3})$ edges [5,21]. There are also natural generalizations of these results to the *pairwise* setting, where one is given $G = (V, E)$ and a set of demand pairs $P \subseteq V \times V$, where only distances between node pairs $(s, t) \in P$ need to be approximately preserved in the spanner [6,8,10,12,19,20].

Despite the inherent advantages of additive error, multiplicative spanners have remained the more well-known and well-applied concept elsewhere in computer science. There seem to be two reasons for this:

1. Abboud and Bodwin [1] (see also [18]) give examples of graphs that have no c-additive spanner on $O(n^{4/3-\varepsilon})$ edges, for any constants $c, \varepsilon > 0$. Some applications call for a spanner on a near-linear number of edges, say $O(n^{1+\varepsilon})$, and hence these must abandon additive error if they need theoretical guarantees for every possible input graph. However, there is some evidence that many graphs of interest bypass this barrier; e.g.. graphs with good expansion or girth properties [5].
2. Spanners are often used to compress metric spaces that correspond to *weighted* input graphs. This includes popular applications in robotics [9,14,23,28], asynchronous protocol design [26], etc., and it incorporates the extremely well-studied case of Euclidean spaces which have their own suite of applications (see book [24]). Current constructions of multiplicative spanners can handle edge weights without issue, but purely additive spanners are known for unweighted input graphs only.

Addressing both of these points, Elkin et al. [16] (following [15]) recently provided constructions of *near-additive* spanners for weighted graphs. That is, for any fixed $\varepsilon, t > 0$, every n-node graph $G = (V, E, w)$ has a $(1+\varepsilon, O(W))$-spanner on $O(n^{1+1/t})$ edges, where W is the maximum edge weight.[1] This extends a classic unweighted spanner construction of Elkin and Peleg [17] to the weighted setting. Additionally, while not explicitly stated in their paper, their method can be adapted to a $+2W$ purely additive spanner on $O(n^{3/2})$ edges (extending [3]).

The goal of this paper is to investigate whether or not all the other constructions of spanners with purely additive error extend similarly to weighted input graphs. As we will discuss shortly, there is a significant barrier to a direct extension of the method from [16]. However, we prove that this barrier can be overcome with some additional technical effort, thus leading to the following

---

[1] Their result is actually a little stronger: W can be the maximum edge weight on the shortest path between the nodes being considered.

constructions. In these theorem statements, all edges have (not necessarily integer) edge weights in $(0, W]$. Let $p = |P|$ denote the number of demand pairs and $n = |V|$ the number of nodes in G.

Table 1. Table of additive spanner constructions for unweighted and weighted graphs, where W denotes the maximum edge weight.

Unweighted		Weighted	
Stretch	Size	Stretch	Size
+2	$O(n^{3/2})$ [3]	+2W	$O(n^{3/2})$ [this paper], [16]
+4	$O(n^{7/5})$ [7,11]	+4W	$O(n^{7/5})$ [this paper]
+6	$O(n^{4/3})$ [5,21,30]	+6W	?
+c	$\Omega(n^{4/3-\varepsilon})$ [1,18]	+8W	$O(n^{4/3})$ [this paper]

Theorem 1. *For any* $G = (V, E, w)$ *and demand pairs* P, *there is a* $+2W$ *pairwise spanner with* $O(np^{1/3})$ *edges. In the all-pairs setting* $P = V \times V$, *the bound improves to* $O(n^{3/2})$.

Theorem 2. *For any* $G = (V, E, w)$ *and demand pairs* P, *there is a* $+4W$ *pairwise spanner with* $O(np^{2/7})$ *edges. In the all-pairs setting* $P = V \times V$, *the bound improves to* $O(n^{7/5})$.

These two results exactly match previous ones for unweighted graphs [3,11, 19,20], with $+2W$ ($+4W$) in place of $+2$ ($+4$). Theorem 1 is partially tight in the following sense: it implies that $O(n^{3/2})$ edges are needed for a $+2W$ spanner when $p = O(n^{3/2})$, and neither of these values can be unilaterally improved. Relatedly, Theorem 2 implies that $O(n^{7/5})$ edges are needed for a $+4W$ spanner when $p = O(n^{7/5})$; it may be possible to improve this, but it would likely imply an improved $+4W$ all-pairs spanner over [11] which will likely be hard to achieve (see discussion in [7]).

Our next two results are actually a bit weaker than the corresponding unweighted ones [13,19,27]: for a technical reason, we take on slightly more error in the weighted setting (the corresponding unweighted results have $+6$ and $+2$ error respectively).

Theorem 3. *For any* $G = (V, E, w)$ *and demand pairs* P, *there is a* $+8W$ *pairwise spanner with* $O(np^{1/4})$ *edges. In the all-pairs setting* $P = V \times V$, *the bound improves to* $O(n^{4/3})$.

Theorem 4. *For any* $G = (V, E, w)$ *and demand pairs* $P = S \times S$, *there is a* $+4W$ *pairwise spanner with* $O(n|S|^{1/2})$ *edges.*

We summarize our main results in Table 1, contrasted with known results for unweighted graphs.

1.1   Technical Overview: What's Harder with Weights?

There is a key point of failure in the known constructions of unweighted additive spanners when one attempts the natural extension to weighted graphs. To explain, let us give some technical background. Nearly all spanner constructions start with a *clustering* or *initialization* step: taking the latter exposition [21], a *d-initialization* of a graph G is a subgraph H obtained by choosing d arbitrary edges incident to each node, or all incident edges to a node of degree less than d. After this, many additive spanner constructions leverage the following key fact (the one notable exception is the $+2$ all-pairs spanner, which is why one can recover the corresponding weighted version from prior work):

Lemma 1 ([11,13,19,20], **etc.**). *Let G be an undirected unweighted graph, let π be a shortest path, and let H be a d-initialization of G. If π is missing ℓ edges in H, then there are $\Omega(d\ell)$ different nodes adjacent to π in H.*

Proof. For each missing edge $(u,v) \in \pi$, by construction both u and v have degree at least d in H (otherwise, $\deg_H(u) < d$, in which edge (u,v) is added in the d-initialization H). By the triangle inequality, any given node is adjacent to at most three nodes in π. Hence, adding together the $\geq d$ neighbors of each of the ℓ missing edges, we count each node at most three times so the number of nodes adjacent to π is still $\Omega(d\ell)$. □

The difficulty of the weighted setting is largely captured by the fact that Lemma 1 fails when G is edge-weighted. As a counterexample, let π be a shortest path consisting of $\ell + 1$ nodes and ℓ edges of weight ε. Additionally, consider d nodes, each connected to every node along π with an edge of weight $W > \varepsilon\ell$. A candidate d-initialization H consists of selecting every edge of weight W. In this case, all ℓ edges in π are missing in H, but there are still only $d \neq \Omega(d\ell)$ nodes adjacent to π in H (Fig. 1).

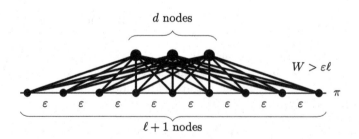

Fig. 1. A counterexample to Lemma 1 for weighted graphs.

The fix, as it turns out, is simple in construction but involved in proof. We simply replace initialization with *light initialization*, where one must specifically add the lightest d edges incident to each node. With this, the proof of Lemma 1 is still not trivial: it remains possible that an external node can be adjacent to arbitrarily many nodes along π, so a direct counting argument fails. However, we show that such occurrences can essentially be amortized against the rising and falling pattern of missing edge weights along π. This leads to a proof that *on average* an external node is adjacent to $O(1)$ nodes in π, which is good enough to push the proof through. We consider this weighted extension of Lemma 1 to be the main technical contribution of this work, and we are hopeful that it may be of independent interest as a structural fact about shortest paths in weighted graphs.

2   Neighborhoods of Weighted Shortest Paths

Here we introduce the extension of Lemma 1. Following the technique in [21], define a *d-light initialization* of a weighted graph $G = (V, E, w)$ to be a subgraph H obtained by including the d lightest edges incident to each node (or all edges incident to a node of degree less than d). Ties between edges of equal weight are broken arbitrarily; for clarity we assume this occurs in the background so that we can unambiguously refer to "the lightest d edges" incident to a node. We prove the weighted analogue of Lemma 1.

Theorem 5. *If H is a d-light initialization of an undirected weighted graph G, and there is a shortest path π in G that is missing ℓ edges in H, then there are $\Omega(d\ell)$ nodes adjacent to π in H.*

We give some definitions and notation which will be useful in the proof of Theorem 5. Let s and t be the endpoints of a shortest path π, and let $M := \pi \setminus E(H)$ be the set of edges in π currently missing in H so that $|M| = \ell$. For convenience we consider these edges to be *oriented* from s to t, so we write $(u, v) \in M$ to mean that $\text{dist}_G(s, u) < \text{dist}_G(s, v)$ and $\text{dist}_G(u, t) > \text{dist}_G(v, t)$. Suppose the edges in M are labeled in order $e_1, e_2, \ldots, e_\ell$ where $e_i = (u_i, v_i)$, and let w_i denote the weight of edge e_i. Given $u \in V$, let $N^*(u)$ denote the *d-neighborhood* of u as follows: $N^*(u) := \{v \in V \mid (u, v) \text{ is one of the lightest } d \text{ edges incident to } u\}$. We will show that the size of the union of the d-neighborhoods of the nodes u_1, $\ldots$, u_ℓ is $\Omega(d\ell)$, that is

$$\left| \bigcup_{(u,v)\in M} N^*(u) \right| = \Omega(d\ell)$$

noting that the above set is a subset of all nodes adjacent to π. In particular, the above set may not contain nodes v' connected to $u \in \pi$ by an edge that is 1) among the d lightest incident to v', 2) *not* among the d lightest incident to u. However, the above set necessarily contains all nodes v' which are connected to some u_i or v_i by an edge among the d lightest incident to u_i or v_i. We remark that if the d-neighborhoods $N^*(u_1)$, $N^*(u_2)$, $\ldots$, $N^*(u_\ell)$ are pairwise disjoint, then $|\bigcup_{(u,v)\in M} N^*(u)| = d\ell$, which immediately implies there are at least $d\ell$ nodes adjacent to π in H. Hence for the remainder of the proof, we assume there exist i and k with $1 \leq i < k \leq \ell$ such that $N^*(u_i) \cap N^*(u_k)$ is nonempty. We use the convention that if a and b are integers with $b < a$, then $\sum_{i=a}^{b} f(i) = 0$. The following lemma holds (see Fig. 2):

Fig. 2. Illustration of Lemma 2. The bold dashed curves represent subpaths in H.

Lemma 2. *Let π be a shortest path, let $x \in V$ be a node such that $x \in N^*(u_i) \cap N^*(u_k)$ for some $1 \leq i < k \leq \ell$, and consider the edges $e_i, \ldots, e_k \in M$ with weights $w_i, \ldots, w_k$. Then $w_k \geq \sum_{i'=i+1}^{k-1} w_{i'}$.*

Proof. Consider the subpath of π from u_i to u_k, denoted $\pi[u_i \rightsquigarrow u_k]$. We have

$$\sum_{i'=i}^{k-1} w_{i'} \leq \text{length}\left(\pi[u_i \rightsquigarrow u_k]\right)$$

$$\leq w(u_i, x) + w(x, u_k) \qquad (\pi[u_i \rightsquigarrow u_k] \text{ is a shortest path})$$

$$\leq w_i + w_k$$

where the last inequality follows from the fact that edges $(u_i, x), (x, u_k)$ are among the d lightest edges incident to u_i and u_k respectively (since $x \in N^*(u_i) \cap N^*(u_k)$), but e_i and e_k are not, since they are omitted from H. Lemma 2 follows by subtracting w_i from both sides of the above inequality. $\square$

For the next part, for edge $e \in M$, say that e is *pre-heavy* if its weight is strictly greater than the preceding edge in M, and/or *post-heavy* if its weight is strictly greater than the following edge in M. For notational convenience, if an edge is not pre-heavy, we say the edge is *pre-light*. Similarly, if an edge is not post-heavy, we say the edge is *post-light*. By convention, the first edge $e_1 \in M$ is pre-light and the last edge $e_\ell \in M$ is post-light. We state the following simple lemma; recall that $|M| = \ell$.

Lemma 3. *Either more than $\dfrac{\ell}{2}$ edges in M are pre-light, or more than $\dfrac{\ell}{2}$ edges in M are post-light.*

Proof. Let S_1 be the set of edges in M which are pre-light, and let S_2 be the set of edges in M which are post-light. Note that $e_1 \in S_1$ and $e_\ell \in S_2$. For each of the $\ell - 1$ pairs of consecutive edges (e_i, e_{i+1}) in M where $i = 1, \ldots, \ell - 1$, it is immediate by definition that either $e_i \in S_2$ or $e_{i+1} \in S_1$ (or both if $w_i = w_{i+1}$). These statements imply $|S_1| + |S_2| \geq \ell + 1$, so at least one of S_1 or S_2 has cardinality at least $\frac{\ell+1}{2} > \frac{\ell}{2}$. $\square$

In the sequel, we assume without loss of generality that more than $\frac{\ell}{2}$ edges in M are pre-light; the other case is symmetric by exchanging the endpoints s and t of π. We can now say the point of the previous two lemmas: together, they imply that *most* edges $(u, v) \in M$ have mostly non-overlapping d-neighborhoods $N^*(u)$. That is:

Lemma 4. *Let π be a shortest path. For any node $x \in V$, there exist at most three nodes u along π such that $x \in N^*(u)$ and edge $(u, v) \in M$ is pre-light.*

Proof. Suppose for sake of contradiction there exist four nodes u_i, u_a, u_b, u_k with $1 \leq i < a < b < k \leq \ell$ such that x belongs to the d-neighborhoods of u_i, u_a, u_b, and u_k, and the edges (u_i, v_i), (u_a, v_a), (u_b, v_b), and (u_k, v_k) are pre-light. In particular, we have $k \geq i + 3$ and $x \in N^*(u_i) \cap N^*(u_k)$. By Lemma 2 and the above observation, we have $w_k \geq \sum_{i'=i+1}^{k-1} w'_i = w_{i+1} + \ldots + w_{k-1} \geq w_{i+1} + w_{k-1}$. By assumption, $e_k = (u_k, v_k)$ is pre-light, so $w_{k-1} \geq w_k$, and the above inequality implies $w_k \geq w_{i+1} + w_{k-1} \geq w_{i+1} + w_k$, or $w_{i+1} = 0$. Since edge weights are strictly positive, we have contradiction, proving Lemma 4. $\qquad\square$

Finally, define set X^* as follows: $X^* := \bigcup_{\substack{(u,v) \in M \\ \text{is pre-light}}} N^*(u)$. By Lemma 3 and the above pre-heavy assumption, there are more than $\frac{\ell}{2}$ pre-light edges (u, v), so the multiset containing all d-neighborhoods $N^*(u)$ contains more than $\frac{d\ell}{2}$ nodes. By Lemma 4, any given node is contained in at most three of these d-neighborhoods, implying $|X^*| > \frac{d\ell}{6}$. Since X^* is a subset of $|\bigcup_{(u,v) \in M} N^*(u)|$, we conclude that there are $\Omega(d\ell)$ nodes adjacent to π in H. proving Theorem 5.

3   Spanner Constructions

We show how Theorem 5 can be used to construct additive spanners on edge-weighted graphs. These constructions are not significant departures from prior work; the main difference is applying Theorem 5 in the right place.

3.1   Subset and Pairwise Spanners

Definition 1 (Pairwise/Subset Additive Spanners). *Given a graph $G = (V, E, w)$ and a set of demand pairs $P \subseteq V \times V$, a subgraph $H = (V, E_H \subseteq E, w)$ is $a+c$ pairwise spanner of G, P if $\text{dist}_H(s, t) \leq \text{dist}_G(s, t) + c$ for all $(s, t) \in P$. When $P = S \times S$ for some $S \subseteq V$, we say that H is $a+c$ subset spanner of G, S.*

In the following results, all graphs G are undirected and connected with (not necessarily integer) edge weights in the interval $(0, W]$, where W is the maximum edge weight. Let $|V| = n$, let $p = |P|$ denote the number of demand pairs (for pairwise spanners), and let $\sigma = |S|$ denote the number of sources (for subset spanners).

Theorem 6. *Any n-node graph $G = (V, E, w)$ with source nodes $S \subseteq V$ has a $+4W$ subset spanner with $O(n\sigma^{1/2})$ edges.*

Proof. The construction of the $+4W$ subset spanner H is as follows, essentially following [21]. Let d be a parameter of the construction, and let H be a d-light initialization of G. Then, while there are nodes $s, t \in S$ such that $\text{dist}_H(s, t) > \text{dist}_G(s, t) + 4W$, choose any $s \rightsquigarrow t$ shortest path $\pi(s, t)$ in G and add all its edges to H. It is immediate that this algorithm terminates with H a $+4W$ subset spanner of G, so we now analyze the number of edges $|E_H|$ in the final subgraph H.

At any point in the algorithm, say that an ordered pair of nodes $(s, v) \in S \times V$ is *near-connected* if there exists v' adjacent to v in H such that $\text{dist}_H(s, v') = \text{dist}_G(s, v')$. We then have the following observation

$$\text{dist}_H(s, v) \leq \text{dist}_H(s, v') + W = \text{dist}_G(s, v') + W. \tag{1}$$

When nodes $s, t \in S$ with shortest path $\pi(s, t)$ are considered in the construction, there are two cases:

1. If there are two nodes v', v'' adjacent in H to a node $v \in \pi(s, t)$, and the pairs (s, v) and (t, v) are near-connected, then we have by triangle inequality and (1):

$$\begin{aligned} \text{dist}_H(s, t) &\leq \text{dist}_H(s, v) + \text{dist}_H(t, v) \\ &\leq (\text{dist}_G(s, v') + W) + (\text{dist}_G(t, v'') + W) \\ &= \text{dist}_G(s, v') + \text{dist}_G(t, v'') + 2W \\ &\leq \text{dist}_G(s, v) + \text{dist}_G(t, v) + 4W \\ &= \text{dist}_G(s, t) + 4W. \end{aligned}$$

   where the last equality follows from the optimal substructure property of shortest paths. In this case, the path $\pi(s, t)$ is not added to H.
2. Otherwise, suppose there is no node v' adjacent in H to a node $v \in \pi(s, t)$ where (s, v) and (t, v) are near-connected. After adding the path $\pi(s, t)$ to H, every such node v' becomes near-connected to both s and t. If there are ℓ edges in $\pi(s, t)$ currently missing in H, then by Theorem 5 we have $\Omega(\ell d)$ nodes adjacent to $\pi(s, t)$, so $\Omega(\ell d)$ node pairs in $S \times V$ go from not near-connected to near-connected. Since there are σn node pairs in $S \times V$, we add a total of $O(\sigma n/d)$ edges to H in this case.

Putting these together, the final size of H is $|E_H| = O\left(nd + \frac{\sigma n}{d}\right)$. Setting $d := \sqrt{\sigma}$ proves Theorem 6. □

We now give our constructions for pairwise spanners. The following lemma will be useful:

Lemma 5 [7]**.** *Let $a, b > 0$ be absolute constants, and suppose there is an algorithm that, on input G, P, produces a subgraph H on $O(n^a |P|^b)$ edges satisfying $\text{dist}_H(s, t) \leq \text{dist}_G(s, t) + c$ for at least a constant fraction of the demand pairs $(s, t) \in P$. Then there is a $+c$ pairwise spanner H' of G, P on $O(n^a |P|^b)$ edges.*

Using the slack to satisfy only a constant fraction of the demand pairs, we have the following proofs.

Theorem 7. *Any graph G with demand pairs P has a $+2W$ pairwise spanner with $O(np^{1/3})$ edges.*

Proof. Let d and ℓ be parameters of the construction, and let H be a d-light initialization of G. For each demand pair $(s,t) \in P$ whose shortest path $\pi(s,t)$ is missing at most ℓ edges in H, add all edges in $\pi(s,t)$ to H. By Theorem 5, any remaining demand pair $(s,t) \in P$ has $\Omega(d\ell)$ nodes adjacent to $\pi(s,t)$. Let R be a random sample of nodes obtained by including each one independently with probability $1/(\ell d)$; thus, with constant probability or higher, there exists $r \in R$ and $v \in \pi(s,t)$ such that nodes r and v are adjacent in H. Add to H a shortest path tree rooted at each $r \in R$. We then compute:

$$\text{dist}_H(s,t) \leq \text{dist}_H(s,r) + \text{dist}_H(r,t)$$
$$= \text{dist}_G(s,r) + \text{dist}_G(r,t)$$
$$\leq \text{dist}_G(s,v) + \text{dist}_G(v,t) + 2W$$
$$= \text{dist}_G(s,t) + 2W.$$

The distance for each pair $(s,t) \in P$ is approximately preserved in H with at least a constant probability, which is sufficient for Lemma 5. The number of edges in the final subgraph H is $|E(H)| = O(nd + \ell p + n^2/(\ell d))$; setting $\ell = n/p^{2/3}$ and $d = p^{1/3}$ proves Theorem 7.   □

Theorem 8. *Any graph G with demand pairs P has a $+4W$ pairwise spanner with $O(np^{2/7})$ edges.*

Proof. Let d and ℓ be parameters of the construction, and let H be a d-light initialization of G. For each demand pair $(s,t) \in P$ whose shortest path $\pi(s,t)$ is missing at most ℓ edges in H, add all edges in $\pi(s,t)$ to H. To handle each $(s,t) \in P$ whose shortest path $\pi(s,t)$ is missing at least n/d^2 edges in H, we let R_1 be a random sample of nodes obtained by including each node independently with probability d^2/n, then add a shortest path tree rooted at each $r \in R_1$ to H. By an identical analysis to Theorem 7, for each such pair, with constant probability or higher we have $\text{dist}_H(s,t) \leq \text{dist}_G(s,t) + 2W$. Finally, we consider the "intermediate" pairs $(s,t) \in P$ whose shortest path $\pi(s,t)$ is missing more than ℓ but fewer than n/d^2 edges in H. We add the first and last ℓ missing edges in $\pi(s,t)$ to the spanner; we will refer to the *prefix* (resp. *suffix*) of $\pi(s,t)$ to mean the shortest prefix (suffix) containing these ℓ missing edges. By Theorem 5, there are $\Omega(\ell d)$ nodes adjacent to the prefix and $\Omega(\ell d)$ nodes adjacent to the suffix. Let R_2 be a random sample of nodes obtained by including each node with probability $1/(\ell d)$, and for each pair $r, r' \in R_2$, add to H all edges in the shortest $r \rightsquigarrow r'$ path in G among the paths that are missing at most n/d^2 edges (ignore any pair r, r' if no such path exists). With constant probability or higher, we sample r, r' adjacent to nodes v, v' in the prefix, suffix respectively. Hence,

$$\text{dist}_H(s,t) \le \text{dist}_H(s,v) + \text{dist}_H(v,v') + \text{dist}_H(v',t)$$
$$= \text{dist}_G(s,v) + \text{dist}_H(v,v') + \text{dist}_G(v',t)$$
$$\le \text{dist}_G(s,v) + \text{dist}_H(r,r') + 2W + \text{dist}_G(v',t).$$

Notice that $\text{dist}_H(r,r') \le 2W + \text{dist}_G(v,v')$, due to the existence of the path $r \circ \pi(s,t)[v,v'] \circ r'$ which is indeed missing $\le n/d^2$ edges. Thus we may continue:

$$\le \text{dist}_G(s,v) + \text{dist}_G(v,v') + 4W + \text{dist}_G(v',t)$$
$$= \text{dist}_G(s,t) + 4W.$$

The distance for each pair $(s,t) \in P$ is approximately preserved in H with at least constant probability, which again suffices by Lemma 5, and the number of edges in H is $|E(H)| = O\left(nd + p\ell + n^3/(\ell^2 d^4)\right)$. Setting $\ell = n/p^{5/7}$ and $d = p^{2/7}$ completes the proof of Theorem 8. $\qquad\square$

Theorem 9. *Any graph G with demand pairs P has a $+8W$ pairwise spanner containing $O(np^{1/4})$ edges.*

Proof. Let ℓ, d be parameters of the construction and let H be a d-light initialization of G. For each $(s,t) \in P$ whose shortest path $\pi(s,t)$ is missing $\le \ell$ edges in H, add all edges in $\pi(s,t)$ to H. Otherwise, like before, we add the first and last ℓ missing edges of $\pi(s,t)$ to H (prefix and suffix). Then, randomly sample a set R by including each node with probability $1/(\ell d)$, and use Theorem 6 to add a $+4W$ subset spanner on the nodes in R. By Theorem 5, the prefix and suffix each have $\Omega(\ell d)$ adjacent nodes. Thus, with constant probability or higher, we sample $r, r' \in R$ adjacent to v, v' in the added prefix and suffix respectively. We then compute:

$$\text{dist}_H(s,t) \le \text{dist}_H(s,v) + \text{dist}_H(v,v') + \text{dist}_H(v',t)$$
$$\le \text{dist}_G(s,v) + \text{dist}_H(v,v') + \text{dist}_G(v',t)$$
$$\le \text{dist}_G(s,v) + \text{dist}_H(r,r') + 2W + \text{dist}_G(v',t)$$
$$\le \text{dist}_G(s,v) + \text{dist}_G(r,r') + 6W + \text{dist}_G(v',t)$$
$$\le \text{dist}_G(s,v) + \text{dist}_G(v,v') + 8W + \text{dist}_G(v',t)$$
$$= \text{dist}_G(s,t) + 8W.$$

Again, the distance for each pair $(s,t) \in P$ is approximately preserved in H with at least constant probability, which suffices by Lemma 5. The number of edges in H is $|E(H)| = O\left(nd + p\ell + n^{3/2}/\sqrt{\ell d}\right)$. Setting $\ell = n/p^{3/4}$ and $d = p^{1/4}$ completes the proof of Theorem 9. $\qquad\square$

4   All-Pairs Additive Spanners

We now turn to the *all-pairs* setting, i.e., demand pairs $P = V \times V$. We use the following lemma from [7]:

Lemma 6 [7]. *Let G be a graph, and suppose one can choose a function π that associates each node pair to a path between them with the following properties:*

- *for all (s,t) the length of the path $\pi(s,t)$ (i.e., the sum of its edge weights) is $\leq \mathrm{dist}_G(s,t) + k$,*
- *π depends only on the input graph G and the number of demand pairs $|P|$ (but not otherwise on the contents of P), and*
- *for some parameter p^* and any $|P| \geq p^*$ demand pairs, we have*

$$\left| \bigcup_{(s,t)\in P} \pi(s,t) \right| < |P|.$$

Then there is an all-pairs k-additive spanner of G containing $\leq p^*$ edges.

Notice that all the above pairwise spanner constructions are *demand-oblivious* – that is, the approximate shortest paths analyzed in order to preserve each demand pair in the spanner depend on the random bits of the construction, which in turn depend on the number of demand pairs $|P|$, but they do not otherwise depend on the contents of P. See [7] for more discussion of this property. Thus we may apply Lemma 6 as follows. For the $+2W$ pairwise bound of $O(np^{1/3})$ provided in Theorem 7, we note that the bound is $< p$ for $p = \Omega(n^{3/2})$ demand pairs (and a sufficiently large constant in the Ω). Hence, taking $p^* = \Theta(n^{3/2})$, Lemma 6 says:

Theorem 10. *Every graph has a $+2W$ spanner on $O(n^{3/2})$ edges.*

Identical logic applied to Theorems 8 and 9 gives:

Theorem 11. *Every n-node graph has a $+4W$ spanner on $O(n^{7/5})$ edges.*

Theorem 12. *Every n-node graph has a $+8W$ additive spanner on $O(n^{4/3})$ edges.*

5   Conclusions and Open Problems

We have shown that most important unweighted additive spanner constructions have natural weighted analogues. At present, the exceptions are the $+4W$ subset spanner on $O(n|S|^{1/2})$ edges (which should probably have only $+2W$ error) and the $+8W$ all-pairs/pairwise spanners (which should probably have only $+6W$ error). Closing these error gaps is an interesting open problem. It would also be interesting to obtain weighted analogues of related concepts, most notably, the Thorup-Zwick emulators [29], which are optimal [2] in essentially the same way that the 6-additive spanner on $O(n^{4/3})$ edges is optimal.

Finally, as mentioned earlier, it would be interesting to find constructions of *purely* additive spanners parametrized by some other statistic besides the maximum edge weight W; a natural parameter is $W(u,v)$, the maximum edge weight along a shortest u-v path.

References

1. Abboud, A., Bodwin, G.: The 4/3 additive spanner exponent is tight. J. ACM (JACM) **64**(4), 1–20 (2017)
2. Abboud, A., Bodwin, G., Pettie, S.: A hierarchy of lower bounds for sublinear additive spanners. In: Proceedings of the 28th Annual ACM-SIAM Symposium on Discrete Algorithms (SODA), pp. 568–576. Society for Industrial and Applied Mathematics (2017)
3. Aingworth, D., Chekuri, C., Indyk, P., Motwani, R.: Fast estimation of diameter and shortest paths (without matrix multiplication). SIAM J. Comput. **28**, 1167–1181 (1999)
4. Althöfer, I., Das, G., Dobkin, D., Joseph, D.: Generating sparse spanners for weighted graphs. In: Gilbert, J.R., Karlsson, R. (eds.) SWAT 1990. LNCS, vol. 447, pp. 26–37. Springer, Heidelberg (1990). https://doi.org/10.1007/3-540-52846-6_75
5. Baswana, S., Kavitha, T., Mehlhorn, K., Pettie, S.: Additive spanners and (α, β)-spanners. ACM Trans. Algorithms (TALG) **7**(1), 5 (2010)
6. Bodwin, G.: Linear size distance preservers. In: Proceedings of the Twenty-Eighth Annual ACM-SIAM Symposium on Discrete Algorithms (SODA), pp. 600–615. Society for Industrial and Applied Mathematics (2017)
7. Bodwin, G.: A note on distance-preserving graph sparsification. arXiv preprint arXiv:2001.07741, 2020
8. Bodwin, G., Williams, V.V.: Better distance preservers and additive spanners. In: Proceedings of the Twenty-Seventh Annual ACM-SIAM Symposium on Discrete Algorithms (SODA), pp. 855–872. Society for Industrial and Applied Mathematics (2016)
9. Cai, L., Keil, J.M.: Computing visibility information in an inaccurate simple polygon. Int. J. Comput. Geometr. Appl. **7**(6), 515–538 (1997)
10. Chang, H.-C., Gawrychowski, P., Mozes, S., Weimann, O.: Near-optimal distance emulator for planar graphs. In: Proceedings of 26th Annual European Symposium on Algorithms (ESA 2018), vol. 112, pp. 16:1–16:17 (2018)
11. Chechik, S.: New additive spanners. In: Proceedings of the Twenty-Fourth Annual ACM-SIAM Symposium on Discrete Algorithms (SODA), pp. 498–512. Society for Industrial and Applied Mathematics (2013)
12. Coppersmith, D., Elkin, M.: Sparse sourcewise and pairwise distance preservers. SIAM J. Discrete Math. **20**(2), 463–501 (2006)
13. Cygan, M., Grandoni, F., Kavitha, T.: On pairwise spanners. In: Proceedings of 30th International Symposium on Theoretical Aspects of Computer Science (STACS 2013), vol. 20, pp. 209–220 (2013)
14. Dobson, A., Bekris, K.E.: Sparse roadmap spanners for asymptotically near-optimal motion planning. Int. J. Robot. Res. **33**(1), 18–47 (2014)
15. Elkin, M.: Computing almost shortest paths. ACM Trans. Algorithms (TALG) **1**(2), 283–323 (2005)
16. Elkin, M., Gitlitz, Y., Neiman, O.: Almost shortest paths and PRAM distance oracles in weighted graphs. arXiv preprint arXiv:1907.11422 (2019)
17. Elkin, M., Peleg, D.: $(1 + \epsilon, \beta)$-spanner constructions for general graphs. SIAM J. Comput. **33**(3), 608–631 (2004)
18. Huang, S.-E., Pettie, S.: Lower bounds on sparse spanners, emulators, and diameter-reducing shortcuts. In: Proceedings of 16th Scandinavian Symposium and Workshops on Algorithm Theory (SWAT), pp. 26:1–26:12 (2018)

19. Kavitha, T.: New pairwise spanners. Theory Comput. Syst. **61**(4), 1011–1036 (2017)
20. Kavitha, T., Varma, N.M.: Small stretch pairwise spanners and approximate d-preservers. SIAM J. Discrete Math. **29**(4), 2239–2254 (2015)
21. Knudsen, M.B.T.: Additive spanners: a simple construction. In: Ravi, R., Gørtz, I.L. (eds.) SWAT 2014. LNCS, vol. 8503, pp. 277–281. Springer, Cham (2014). https://doi.org/10.1007/978-3-319-08404-6_24
22. Liestman, A., Shermer, T.: Additive graph spanners. Networks **23**, 343–363 (1993)
23. Marble, J.D., Bekris, K.E.: Asymptotically near-optimal planning with probabilistic roadmap spanners. IEEE Trans. Robot. **29**(2), 432–444 (2013)
24. Narasimhan, G., Smid, M.: Geometric Spanner Networks. Cambridge University Press, New York (2007)
25. Peleg, D., Schäffer, A.A.: Graph spanners. J. Graph Theory **13**(1), 99–116 (1989)
26. Peleg, D., Upfal, E.: A trade-off between space and efficiency for routing tables. J. ACM (JACM) **36**(3), 510–530 (1989)
27. Pettie, S.: Low distortion spanners. ACM Trans. Algorithms (TALG) **6**(1), 7 (2009)
28. Salzman, O., Shaharabani, D., Agarwal, P.K., Halperin, D.: Sparsification of motion-planning roadmaps by edge contraction. Int. J. Robot. Res. **33**(14), 1711–1725 (2014)
29. Thorup, M., Zwick, U.: Spanners and emulators with sublinear distance errors. In: Proceedings of the 17th Annual ACM-SIAM Symposium on Discrete Algorithms (SODA), pp. 802–809. Society for Industrial and Applied Mathematics (2006)
30. Woodruff, D.P.: Additive spanners in nearly quadratic time. In: Abramsky, S., Gavoille, C., Kirchner, C., Meyer auf der Heide, F., Spirakis, P.G. (eds.) ICALP 2010. LNCS, vol. 6198, pp. 463–474. Springer, Heidelberg (2010). https://doi.org/10.1007/978-3-642-14165-2_40

Author Index

Printed in the United States
By Bookmasters